Mechanisms of Elimination Reactions

MECHANISMS
OF ELIMINATION
REACTIONS

WILLIAM H. SAUNDERS, JR. *Department of Chemistry*
The University of Rochester
Rochester, New York

ANTHONY F. COCKERILL *Lilly Research Center, Ltd.*
Erl Wood Manor, Windlesham
Surrey, England

A Wiley-Interscience Publication

JOHN WILEY & SONS
New York · London · Sydney · Toronto

Library of Congress Cataloging in Publication Data:

Saunders, William Hundley, 1926-
    Mechanisms of elimination reactions.

    "A Wiley-Interscience publication."
    1. Elimination reactions.   2. Chemistry, Physical
organic.   I. Cockerill, Anthony F., joint author.
II. Title.

QD281.E4S28   547'.1'393   73-6809
ISBN 0-471-75516-8

Printed in the United States of America

10-9 8 7 6 5 4 3 2 1

PREFACE

Along with substitution, addition, and rearrangement
reactions, elimination reactions are one of the major
classes of reactions of organic compounds and rank
among the more thoroughly documented types of reactions
in organic chemistry. The most well-known eliminations
are those giving alkenes, but other important categories include alkyne, carbonyl, and imine-forming eliminations. All of these unsaturated products possess
widely varying reactivities, and they are extremely
useful not only as end products but also as intermediates in organic synthesis. To optimize the yield
and to control the stereochemical and orientational
fate of the elimination product, an understanding of
the reaction mechanism is essential. Often several
methods can be used to prepare a desired elimination
product. For example, amines can be converted directly
into alkenes in variable and poor yield by deamination
with nitrous acid. Initial conversion to the dimethylamine oxide, followed by a thermal reaction, gives a
high yield of alkene in a stereospecific syn elimination. Hofmann decomposition of the quaternary ammonium
hydroxide also gives good yields of the alkene, the
elimination ranging from anti to syn stereoselective,
depending on the nature of the reactant, solvent, and
base. Clearly, the choice of elimination reaction
determines the yield and the structure of the product.
    In an elimination reaction, two bonds must be
broken and one or more new bonds (including the double
bond) must be formed. These multiple changes afford
a variety of mechanistic possibilities ranging from
multistep reactions to concerted processes which exhibit transition states with varying electron distributions and geometries. Nearly all of the fundamental
mechanistic problems of organic chemistry are

encountered in elimination reactions, and the principles deduced from studies of the mechanisms of these
reactions often have wide application to the general
theory of organic reactions.

Alkene-forming eliminations have attracted the
most attention, and we devote our first chapter to the
major reaction mechanisms. In the next three chapters
we discuss, in detail, the most interesting of these
mechanisms, namely the E2 reaction. The topics covered
include the influence of structure, solvent, and base
on the electron distribution and geometry of the transition state, on the stereochemistry of elimination,
and on orientation. The mechanisms and their finer
points, which we discuss in detail in these chapters,
are applicable to many of the reactions considered in
later chapters. Chapter V concerns the unimolecular
eliminations of alkyl halides and esters, which occur
under solvolytic conditions. Chapter VI is devoted to
the mechanisms of dehydration reactions. Chapter VII
completes the discussion on eliminations involving
carbonium-ion intermediates, two of its sections being
devoted to deamination and deoxidation reactions.
There are a number of significant differences in the
reactivity of carbonium ions formed in these reactions
compared with those involved in the mechanisms discussed in the two preceding chapters. Our discussion
on the mechanisms of alkene-forming 1,2-eliminations is
completed with an outline of dehalogenations and
thermal eliminations.

Chapter IX is concerned with 1,2-eliminations that
do not give alkenes. The reaction products include
carbonyl, imine, nitrile, thiocarbonyl, and alkyne
derivatives. To limit the size of this chapter, we
exclude the large number of 1,2-eliminations in which
hydrogen is removed from an oxygen or a nitrogen atom,
and a leaving group is removed from an adjacent carbon
atom. These eliminations constitute only part of the
reactions of compounds with carbon-oxygen and carbon-
nitrogen multiple bonds, and the mechanisms of the
overall reactions are beyond the scope of a treatise
on elimination reactions.

Chapter X contains details of the mechanisms of
elimination reactions other than 1,2-processes.
Finally Chapter XI deals with the mechanism of photoeliminations, an expanding area of interest in recent
times.

Although our purpose is to cover all the important
aspects of the elimination reactions, we make no
effort to be exhaustive. Where there are numerous
studies in a particular area, we often choose one or a

few illustrative cases. No judgment on the quality or importance of the other studies is intended in this process. Doubtless we overlook some important contributions to the subject. We apologize for omissions of this kind, and hope that they are not too numerous.

We are indebted to the many individuals who contributed in a variety of ways to this book. WHS is especially grateful to Professor Lars Melander of the Institutionen för Organisk Kemi, Göteborgs Universitet och Chalmers Tekniska Högskola, Göteborg, Sweden, for his hospitality and for many stimulating discussions during the 1970 to 1971 academic year. Professor Melander also read, and made many helpful suggestions on, Chapters I to IV. Suggestions on the other chapters and help in checking the final text from Professor Irving Feit, Dr. David M. Rackham, Dr. John R. Jones, Dr. William J. Ross, Dr. Eric Wildsmith, and Professor J. Christopher Dalton were also appreciated. Information in advance of publication was furnished by many, including F. G. Bordwell, A. J. Parker, J. Závada, A. N. Bourns, A. C. Frosst, P. J. Smith, J. P. Lowe, R. A. Bartsch, R. L. Klimisch, J. A. Kampmeier, and Z. Rappoport. The typing of the final manuscript was done mainly by Mrs. Evelyn Scott, Mrs. Hazel V. Pearson, and Miss Janette Hayward. Finally we thank our wives Nina and Sheila for their continued patience during our commitment to this book and for their considerable help in the checking of the manuscript.

William H. Saunders, Jr.
Anthony F. Cockerill

Rochester, New York
Windlesham, Surrey, England
December 1972

CONTENTS

Mechanisms of Elimination Reactions

I MECHANISMS OF BASE- OR SOLVENT-PROMOTED 1,2-ELIMINATIONS GIVING ALKENES

A FUNDAMENTAL MECHANISMS

1 The E2 Mechanism

By far the most common type of elimination reaction is the base-promoted loss of HX from adjacent carbon atoms of an organic compound (Eq. 1). It has been known for well over a century,[1] and the

$$B \ + \ H\text{-}\underset{|}{\overset{|}{C}}\text{-}\underset{|}{\overset{|}{C}}\text{-}X \longrightarrow BH^+ \ + \ >C=C< \ + \ X^- \qquad (1)$$

presently accepted mechanism was first proposed in 1927 by Hanhart and Ingold.[2] The reaction follows second-order kinetics, first order in base and first order in substrate.[3] This dependency was proposed to result from the attack of base on the β-hydrogen accompanied by the loss of the leaving group, X, with its pair of electrons (1). The mechanism was

1

given the name E2,[2] signifying "elimination, bimolecular." While other mechanisms are consistent with the observed kinetics, subsequent investigations have shown the concerted process to be of very wide applicability. The present chapter deals with the scope of the E1 and E2 reactions and how they can be distinguished from other mechanisms for elimination. More subtle points of mechanism, such as stereochemistry, relative reactivities and the details of the bond-making and bond-breaking processes, will be treated in later chapters.

Ingold's original paper on the E2 reaction[2] was followed up by a long series of investigations from his research group, many of them in collaboration with E. D. Hughes. The behavior of quaternary ammonium salts was explored,[2-5] as well as dialkylsulfones,[6] sulfonium hydroxides,[7] and phosphonium ethoxides.[8] An accompanying, and often major, side reaction with the phosphonium salts led to a saturated hydrocarbon and a phosphine oxide rather than the expected olefin. While this interesting reaction is beyond the scope of this book, it is worth mentioning that its mechanism has been clarified by later work from McEwen and his group.[9]

2   The E1 Mechanism

The Hughes-Ingold group also recognized that elimination reactions in solution could occur in the absence of added bases under certain conditions. A series of papers was published in which the structural and environmental factors favoring such eliminations in alkyl halides were explored.[10-15] The name E1, for "elimination, unimolecular," was proposed for these

reactions, and it was suggested that they involved two
steps: a slow ionization (Eq. 2), followed by fast
loss of a proton from the resulting ion (Eq. 3).[10]

$$H-\overset{|}{\underset{|}{C}}-\overset{|}{\underset{|}{C}}-X \rightleftharpoons H-\overset{|}{\underset{|}{C}}-\overset{|}{\underset{|}{C}}{}^{+} + X^{-} \qquad (2)$$

$$H-\overset{|}{\underset{|}{C}}-\overset{|}{\underset{|}{C}}{}^{+} \longrightarrow H^{+} + >C=C< \qquad (3)$$

The early investigations on both the E1 and E2
mechanisms were reviewed in 1941.[16] In a subsequent
series of papers in 1948, further studies of reaction
conditions, rates, and product proportions in E1 and
E2 reactions of sulfonium salts[17-19,23-26] and alkyl
halides,[18-22,26] were made. The findings and conclusions of the Hughes-Ingold group are summarized in the
final article of the series,[27] and also, in a wider
context, in Ingold's book.[28] Details of these investigations will be given at the appropriate points later
in the volume.

3 Scope of E1 and E2 Reactions

A very wide variety of compounds will undergo E1 or E2
reactions (or both) under appropriate conditions. The
only structural requirements are that the leaving
group (X in Eqs. 1 and 2) not be so strongly bound as
to make reaction impossible except under conditions
leading to nonspecific decomposition, and that the
molecules possess at least one hydrogen β to the leaving group. Even this latter requirement need not be
fulfilled in an E1 reaction if the carbonium ion produced in the first step (Eq. 2) can rearrange to an
isomeric carbonium ion possessing a hydrogen β to the
positive charge. An example would be the production
of rearranged olefins from neopentyl[29] and neophyl[30]
derivatives (Eqs. 4 to 6, R = Me and Ph, respectively).
The substituents attached to the α- and β-carbon atoms
in E1 and E2 reactions can be hydrogen, alkyl, aryl, or
almost any other group that does not itself react under
the experimental conditions.
The most common leaving groups have been trialkylammonium,[2-5] dialkylsulfonium,[7,17-19,23-26] and
halide.[18-22,26] By far the most common halide leaving
group has been bromide, which offers a good balance between availability and reactivity. Fluorides were
studied considerably later.[31-33] Other leaving groups
that have been used include trialkylphosphonium,[8]
alkylsulfonyl,[6] sulfonate ester,[34-36] hindered

$$\underset{\underset{Me}{|}}{\overset{\overset{Me}{|}}{R-C-}}CH_2X \longrightarrow \underset{\underset{Me}{|}}{\overset{\overset{Me}{|}}{R-C-}}CH_2{}^+ + X^- \qquad (4)$$

$$\underset{\underset{Me}{|}}{\overset{\overset{Me}{|}}{R-C-}}CH_2{}^+ \longrightarrow \underset{\underset{Me}{|}}{\overset{\overset{+}{}}{Me-C-}}CH_2R \qquad (5)$$

$$\underset{\underset{Me}{|}}{\overset{\overset{+}{}}{Me-C-}}CH_2R \longrightarrow \underset{\underset{Me}{|}}{\overset{}{Me-C}}=CHR + CH_2=\underset{\underset{Me}{|}}{\overset{}{C-}}CH_2R \qquad (6)$$

carboxylate,[37] acetoxyl,[38] alkoxyl,[39],[40] phenoxyl,[41],[42] phenylthio,[42] phenylseleno,[42] and hydroxyl.[43] The last six or seven leaving groups in this list are relatively unreactive, and in many instances probably react under basic conditions not by the E2 mechanism, but by a stepwise mechanism (E1cB) that is described shortly. Ethers, alcohols, and carboxylate esters generally do not undergo E1 reactions unless acid is present, in which case the leaving group is doubtless protonated. Nitrous acid converts amines to diazonium ions which can decompose rapidly to give, among other products, olefins,[44] at least in part by way of carbonium ions. Similar products are obtained when alcohols are treated with a haloform and a base.[45]

The solvents most commonly used for E2 reactions have been water and alcohols. For a long time, the standard recipe for preparative use of the E2 reaction was alcoholic potassium hydroxide. This mixture is not desirable for mechanistic studies, however, because of uncertainty over whether the reactive base is hydroxide or ethoxide, although it appears that ethoxide predominates in the equilibrium.[46]

The usual way of bringing about an E2 reaction on a quaternary ammonium salt involves conversion of the salt to the hydroxide, either by treatment of the halide with a silver oxide suspension,[47] or by passing the solution through a strong-base ion exchange resin.[48] The hydroxide solution is then concentrated to a syrup by distillation or by freeze-drying.[49] Heating the syrup decomposes the quaternary hydroxide to olefin and a tertiary amine, which usually distill at the reaction temperature. It should be emphasized that this syrup has quite different properties from a dilute aqueous solution of hydroxide. The decomposition occurs under mild conditions (often less than

100°,[49] compared to 180 to 200° for dilute solutions[50]),
and the product proportions usually differ from
those in dilute solutions. It is generally assumed
that the hydroxide ion is a considerably stronger
base in the syrup than in dilute solution because fewer
water molecules are available to solvate it by hydrogen
bonding.

Alcohols other than ethanol have been widely used.
Almost always the base is the corresponding sodium or
potassium alkoxide. Tertiary alcohol-alkoxide combinations[51] appear to possess definite advantages. There
is less tendency toward an accompanying E1 reaction
with compounds prone to solvolysis,[52] and the product
proportions are often quite different from those with
n-alkoxides in n-alcohols. More is said later about
the nature of and reasons for these changes.

Within the past decade, increasing attention has
been given to dipolar aprotic solvents as media for
elimination reactions. Even earlier, it was known that
metal halides in dipolar aprotic solvents were unusually effective as bases in E2 reactions. The early
literature is summarized by Parker.[53] Winstein and
Parker have made thorough studies of the mechanisms of
elimination reactions promoted by weakly basic anions
in dipolar aprotic solvents. These studies will be
described in detail later (Sections II.B.5 and II.
C.2.c). From his observations of the effectiveness
of dimethyl sulfoxide in enhancing the basicity of
alkoxides, Cram[54] suggested this solvent for eliminations and other base-promoted reactions. This suggestion was followed up by Schriesheim and his co-workers,[55,56] who studied elimination reactions of halides, sulfones, sulfoxides, sulfides, and other compounds in dimethyl sulfoxide with alkali metal t-
butoxides, and showed that they possessed the kinetic
characteristics of E2 reactions. Even the addition of
modest amounts of dimethyl sulfoxide to potassium t-
butoxide in t-butyl alcohol markedly accelerates elimination reactions of 2-arylethyl bromides.[57] A dipolar
aprotic solvent may also be used when one wishes to
vary the base without varying the solvent.[58]

When a protic solvent is used, the base is most
often the conjugate base of the solvent. When any
other base is used, an equilibrium is set up between
the added base and the conjugate base of the solvent
(recall the example given above of potassium hydroxide
in ethanol), and one cannot be sure which is the
effective base. If a system is chosen such that this
equilibrium (Eq. 7) will lie far on one side or the
other, this danger can be minimized. The best way of

$$B \ + \ ROH \ \rightleftharpoons \ BH^+ \ + \ RO^- \qquad\qquad (7)$$

accomplishing this is to use a much weaker base than the conjugate base of the solvent, preferably along with some of its conjugate acid to suppress further the tendency of the solvent to ionize. In this manner, phenoxides[59] and thiophenoxides[60] have been used to promote elimination. The introduction of ring substituents permits systematic variation of basicity without concomitant change in the steric requirements of the base. For the phenoxides this is the only important advantage, since they are generally much more sluggish than alkoxides as reagents for promoting eliminations. Thiolates, however, can be even more reactive than alkoxides toward some classes of substrates,[61] and can consequently be useful in preparative work.

For El reactions, protic solvents with good ionizing ability are the best reaction media. Water, alcohol, or mixtures of the two are widely used,[12-14] as well as carboxylic acids.[11] Strongly ionizing aprotic solvents, such as sulfur dioxide,[11] can be used, but have no advantage over the protic solvents. Hydroxylic solvents are usually preferred because the strong acid (such as hydrogen bromide, toluenesulfonic acid, etc.) produced in the El reaction forms ROH_2^+ and has less tendency to complicate the reaction by adding to or isomerizing the olefins produced. In fact, it is advisable to add sufficient weak base to the reaction mixture to cause it to react with all of the strong acid. Pyridines, particularly sterically hindered ones such as 2,6-lutidine, can either be added to alcoholic solvents or can serve as solvents themselves. In carboxylic acids, the conjugate base of the solvent (e.g., sodium acetate in acetic acid) is used. One must, of course, make sure that bases are not so reactive that they change the mechanism of the process by direct attack on the substrate to give E2 products. Failure of changes in the base concentration to affect the rate or products significantly is good evidence that this problem has not arisen.

4 Distinguishing Between El and E2 Mechanisms

Distinguishing between El and E2 mechanisms is generally easy from the kinetics. Both will be first order in substrate, but E2 rates will show a dependence on the base concentration while El rates will not.

One possibly ambiguous situation is that in which only substrate and solvent are present. The reaction could be E1, but it could also be E2 with solvent functioning as base. A distinction can usually be made by adding a little of a stronger base. For example, ethanol is unlikely to be functioning as a base if added sodium ethoxide does not affect the rate, and acetic acid is unlikely to be functioning as a base if added sodium acetate has no effect.

While the E2 reaction will always be first order in base and first order in substrate, the E1 reaction may show somewhat more complex kinetics than a simple first-order dependence on substrate. If the ionization of Eq. 2 is reversible, the steady-state approximation can be applied to the carbonium-ion concentration (eq. 8). From this the carbonium-ion

$$\frac{d(R^+)}{dt} = 0 = k_1(RX) - k_{-1}(R^+)(X^-) - k_2(R^+) \tag{8}$$

concentration and thence the rate of the reaction can be determined (Eqs. 9 and 10). If the ionization is irreversible,

$$(R^+) = \frac{k_1(RX)}{k_{-1}(X^-) + k_2} \tag{9}$$

$$rate = k_2(R^+) = \frac{k_1 k_2(RX)}{k_{-1}(X^-) + k_2} \tag{10}$$

$k_2 \gg k_{-1}(X^-)$, and the equation reduces to the usual one for an E1 reaction (Eq. 11). If it is strongly reversible, $k_{-1}(X^-) \gg k_2$, and the rate becomes dependent on the inverse

$$rate = k_1(RX) \tag{11}$$

of the anion concentration (mass-law effect, Eq. 12).

$$rate = \frac{k_1 k_2(RX)}{k_{-1}(X^-)} \tag{12}$$

If ion-pair intermediates as well as free carbonium ions are involved,[62] the kinetic situation can become even more complex,[62] and the product proportions can also be influenced by the nature of the counter ion.[63] Ion pairs as intermediates in eliminations is discussed later (Sections I.B.2 and V.A.2). Although there is no difficulty in distinguishing E1 from E2 reactions kinetically, the demonstration of first-order kinetics in substrate and first-order kinetics in base does not prove an E2 reaction, because a number of other mechanisms can show the same kinetics. These mechanisms and possible ways of distinguishing them from the E2 mechanism will now be described.

B   OTHER MECHANISMS AND THEIR EXPERIMENTAL VERIFICATION

1   The ElcB Mechanism

(a)   Rate-Determining Steps and the Exchange Criterion. Perhaps the most frequently discussed alternative to the E2 reaction is the ElcB,[5] for "elimination, unimolecular, from the conjugate base." Instead of simultaneous removal of the β-proton and loss of the leaving group, the β-proton is removed first to give a carbanion intermediate (eq. 13), and the leaving group is lost in a subsequent step (eq. 14).

$$B + H-\overset{|}{\underset{|}{C}}-\overset{|}{\underset{|}{C}}-X \underset{k_{-1}}{\overset{k_1}{\rightleftharpoons}} BH^+ + {}^-\overset{|}{\underset{|}{C}}-\overset{|}{\underset{|}{C}}-X \qquad (13)$$

$$ {}^-\overset{|}{\underset{|}{C}}-\overset{|}{\underset{|}{C}}-X \overset{k_2}{\longrightarrow} {\scriptstyle >}C{=}C{\scriptstyle <} + X^- \qquad (14)$$

The kinetic possibilities can be discussed with the aid of a steady-state treatment. Assuming a steady-state concentration for the carbanion gives Eqs. 15 and 16, from which the overall rate of reaction is given by Eq. 17. In these equations SHX is the substrate and SX^- the carbanion.

$$\frac{d(SX^-)}{dt} = 0 = k_1(SHX)(B) - k_{-1}(SX^-)(BH^+) - k_2(SX^-) \qquad (15)$$

$$(SX^-) = \frac{k_1(SHX)(B)}{k_{-1}(BH^+) + k_2} \qquad (16)$$

$$rate = k_2(SX^-) = \frac{k_1 k_2(SHX)(B)}{k_{-1}(BH^+) + k_2} \qquad (17)$$

If the carbanion goes on to give olefin much more rapidly than it picks up a proton to revert to starting material, $k_2 \gg k_{-1}$ (BH^+), and Eq. 18 results. This rate law is kinetically indistinguishable from

$$\text{rate} \quad = \quad k_1 (SHX)(B) \qquad\qquad (18)$$

that for the E2 reaction, the only difference being in the significance of the rate constant. In the E2 reaction, the constant refers to concerted cleavage of the β-carbon-hydrogen and α-carbon-X bonds, while in the ElcB, only the first of these two processes is occurring. This difference does provide the basis for some nonkinetic approaches to distinguishing between the two mechanisms. These approaches are discussed shortly (Section I.B.1.c).

A second limiting case of the ElcB reaction occurs when return of the carbanion to starting material is faster than its decomposition to products. In this case, k_{-1} $(BH^+) \gg k_2$, and Eq. 19 results. The rate is still first order in substrate and first order in base,

$$\text{rate} \quad = \quad \frac{k_1 k_2 (SHX)(B)}{k_{-1} (BH^+)} \qquad\qquad (19)$$

but the inverse dependence on the concentration of the conjugate acid of the base affords the opportunity of a kinetic distinction.

Most discussions of ElcB reactions cite only these two possibilities, but Bordwell[64] and Rappoport[143] have emphasized that there is a third, kinetically distinguishable case. If the substrate is very acidic but the leaving group not very labile, an excess of strong base will rapidly convert the substrate completely or almost completely to the carbanion, which will then slowly undergo a first-order decomposition to products. Further addition of base will not change the concentration of carbanion and thus will not affect the rate of reaction. Under these conditions, the reaction will be first order in stoichiometric substrate but zero order in base. Energy profiles for the three cases are given in Figure 1.

While this third category can easily be distinguished kinetically from the E2 reaction, the first cannot be distinguished at all and the second can be distinguished only under some circumstances. Consequently, numerous more or less successful nonkinetic methods have been developed in an effort to make the distinction. The classic method is isotope exchange. If the solvent contains deuterium (or tritium),

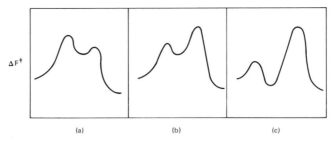

$\Delta F^{\ddagger}$

(a)   (b)   (c)

Reaction coordinate

Figure 1   Energy profiles for three limiting cases of the ElcB mechanism: (a) proton removal rate determining; (b) decomposition of carbanion rate determining; and (c) reactant essentially completely converted to carbanion, decomposition of carbanion rate determining.

reversal of Eq. 13 can be replaced by Eq. 20.   If there is appreciable reversal of the first step, then the reactant should become isotopically labeled.   Similarly,

$$BD^+ \; + \; -\overset{|}{\underset{|}{C}}-\overset{|}{\underset{|}{C}}-X \; \underset{k_1^{\,D}}{\overset{k_{-1}^{\,D}}{\rightleftharpoons}} \; D-\overset{|}{\underset{|}{C}}-\overset{|}{\underset{|}{C}}-X \; + \; B \quad (20)$$

a labeled reactant would lose label under the same circumstances.   If there is no exchange, however, an ElcB reaction of the first category could still be occurring.   A negative result does not exclude the ElcB mechanism.

Breslow[65] has argued that a positive result is not conclusive, either, because exchange could occur parallel to a normal E2 process without the intermediate in the exchange playing a role in the elimination.   Formally, his argument is quite correct.   Detection of a species in a reaction mixture does not demonstrate that it is an intermediate in the main reaction rather than in an irrelevant side reaction.   Nevertheless, the situation described by Breslow should be extremely rare or nonexistent.   An ElcB reaction might possibly coexist with an E2 reaction that gained very little driving force from cleavage of the α-carbon-X bond.   If the ElcB reaction were of category b (Figure 1), then small but detectable amounts of exchange might result.   Clearly, a very specific and unlikely set of relative values of the various rate constants is required for this situation.   As was pointed out by

Hine,[66] a fully formed carbanion should always be more effective in promoting departure of the leaving group than the partially free electron pair in a concerted process. The coexistence of E2 and ElcB paths, then, implies an unstable carbanion that is more likely to decompose to product than to exchange. It is virtually impossible to conceive of an E2 mechanism accompanied by independent and rapid exchange, unless the exchange does not involve a carbanion at all. To our knowledge, there is no precedent for such an exchange mechanism under conditions normally used for base-promoted eliminations.

Early efforts to observe exchange in elimination reactions showed that it was rather uncommon. Elimination from 2-phenylethyl bromide promoted by ethoxide in ethanol[67] or methoxide in methanol[68] was not accompanied by uptake of deuterium from the solvent. Even the more strongly basic medium, potassium amide in liquid ammonia, did not exchange with 2-ethylbutyl-2-d bromide.[68] Extensive deuterium exchange, at least, can be excluded for eliminations of a number of 2-phenylethyl derivatives (bromide, tosylate, dimethyl-sulfonium ion, and trimethylammonium ion) through the fact that the β-deuterated compounds gave good second-order rate constants, with no sign of the upward drift during reaction that would be expected if there were appreciable contamination with undeuterated material produced by exchange.[69] An examination of starting material confirms the absence of exchange in the 2-phenylethyl system for the trimethylammonium ion[70] and the fluoride.[71]

The significance of one early positive result in an exchange study, the dehydrochlorination of 1,1,2,2,-tetrachloroethane-d_2, is beclouded by the fact that the product also undergoes exchange under the reaction conditions.[72] Another early positive result, the 0.08% uptake of deuterium from solvent during elimination from β-benzene hexachloride,[73] has been much discussed because the deuterium content is close to the limit of experimental detectability. This interesting case is considered later.

It should be emphasized that positive results in an exchange experiment may be misleading if care is not taken to be sure that the exchange is occurring at the β-position. Sulfonium salts, for example, exchange readily at the α-position.[74] 2-Phenylethyl-1,1-d_2-dimethylsulfonium ion has been shown to undergo extensive α-exchange during elimination.[75]

From the early results, it is evident that the reversible ElcB mechanism is not the normal mode of elimination. The conditions necessary to achieve this mechanism might be expected to be[76] a poor leaving group, a rather stable carbanion intermediate (so that it can survive long enough to exchange), and a product olefin that is not markedly more stable than the reactant (so that the carbanion will not go irreversibly to product). Most of the well-established examples of exchange possess all of these characteristics.

One of the most thoroughly studied examples is the dehydrofluorination and exchange of 2,2-dihalo-1,1,1-trifluoroethanes[66] (Scheme 1). Exchange is a much faster process than elimination.[76]

Scheme 1

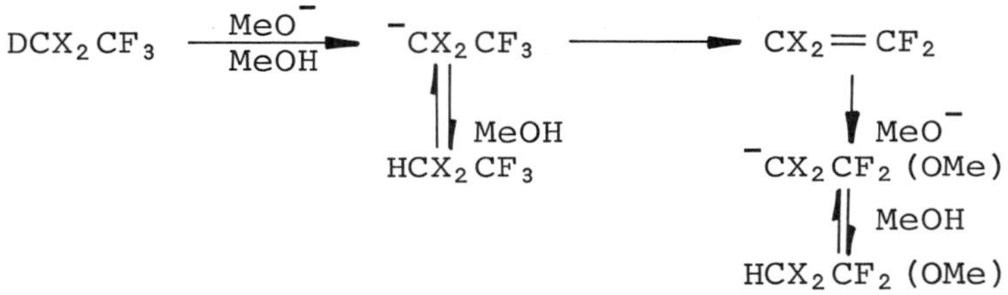

$$DCX_2CF_3 \xrightarrow[\text{MeOH}]{\text{MeO}^-} {}^-CX_2CF_3 \longrightarrow CX_2{=}CF_2$$

$$\Big\updownarrow \text{MeOH}$$
$$HCX_2CF_3$$

$$\Big\downarrow \text{MeO}^-$$
$${}^-CX_2CF_2(\text{OMe})$$

$$\Big\updownarrow \text{MeOH}$$
$$HCX_2CF_2(\text{OMe})$$

Here one has a poor leaving group, halogen atoms to stabilize the carbanion, and a relatively unstable product, as shown by the fact that it rapidly and irreversibly adds methanol in a nucleophilic addition.[77] This observation constitutes a possible ambiguity, since failure to isolate olefin could mean that olefin is not a primary product of the reaction, and that the observed product results from a direct S_N2 displacement rather than elimination-addition. To meet this objection, Hine[66] pointed out that the reactions were much faster than known S_N2 displacements of primary monofluorides,[31] and that the two additional fluorine atoms should make the displacement reactions <u>slower</u> than those of primary monofluorides. Another possible mechanism, α-elimination (loss of H and X from the same carbon atom, Section I.B.3), cannot be more than a minor side reaction, for only 4 to 11% of halide ions other than fluoride could be detected in the reaction mixture. Another fluoride which appears to follow the E1cB mechanism is 9-trifluoromethylfluorene.[78]

(b) <u>Kinetic and Isotope-Effect Criteria</u>. As noted above, a kinetic demonstration of the E1cB mechanism is possible when the carbanion returns to starting material faster than it goes on to olefin, since the rate then shows an inverse dependence on the concentration

of the conjugate acid of the base (Eq. 19). If a suitable buffer can be found, the rate will be dependent
on the buffer ratio, (B)/(BH$^+$), rather than the base
concentration alone. Increasing the buffer concentration without changing the buffer ratio thus should not
affect the rate of a reversible ElcB reaction, but
should cause a linear increase in rate of either an E2
or an irreversible ElcB reaction (Eq. 19 vs. Eq. 18).
The former behavior is analogous to a specific lyate-
ion catalyzed reaction, and the latter is analogous to
a general-base catalyzed reaction[79-80] (we say "analogous to" simply because elimination reactions consume
base and so are not really base catalyzed).

Whether this criterion can be applied depends on
specific circumstances. If the conjugate acid of the
base is the solvent, as is so often the case, its concentration cannot be varied without changing the medium. Only when the reaction can be effected at a reasonable rate by usable concentrations of good buffers
is the criterion really reliable. Specific lyate-ion
promoted reactions cannot be demonstrated simply by
showing that the lyate ion is the only effective
agent; it may be that rates with other bases are too
small to be observed under the reaction conditions. If
the reaction goes at an appreciable rate in neutral or
slightly acidic solution (i.e., the solvent is the
effective base), added strong acid will depress the
rate of a reversible ElcB reaction by increasing the
proportion of carbanions that revert to reactant.

Another possible situation is that a change from
one type of dependence on base concentration to another may occur as the nature of the base, the base concentration, or the buffer concentration is changed.
Referring to Eq. 17, for example, one can expect
$k_2 > k_{-1}$(BH$^+$) at very low buffer concentrations, but at
higher concentrations one may find k_{-1}(BH$^+$)$> k_2$ (unless
k_2 is so much larger than k_{-1} that no practically
attainable concentration of BH$^+$ can produce this result), and the reaction will become independent of the
buffer concentration as the concentration increases.

A considerable number of efforts have been made to
apply the base-catalysis criterion of mechanism to elimination reactions. Some reverse Michael additions
have been studied and appear to be ElcB processes, including the decomposition of 1,1,1,3-tetranitro-2-
phenylpropane (Eq. 21)[81] and 4,4-dicyano-3-p-nitro-
phenyl-1-phenylbutanone-1 (Eq. 22).[82] Both reactions
were studied in neutral, acidic, and buffered methanol.
Rates in buffered solution depended in both cases on
buffer concentration, showing either an E2 or

$$C_6H_5CHCH_2NO_2 \rightleftharpoons C_6H_5CH=CHNO_2 + HC(NO_2)_3 \quad (21)$$
$$|$$
$$C(NO_2)_3$$

$$p\text{-}NO_2C_6H_4CHCH_2COC_6H_5 \rightleftharpoons p\text{-}NO_2C_6H_4CH=CHCOC_6H_5$$
$$|$$
$$CH(CN)_2$$

$$+ \quad CH_2(CN)_2 \qquad (22)$$

irreversible ElcB mechanism. Strong acid, however, depressed the rates of both reactions, pointing to the
reversible ElcB mechanism with solvent as base in acidic solution. The reasonable conclusion that the reaction was irreversible ElcB in the buffered solutions
was then drawn.

In other cases, however, similar observations have
proved to be misleading. The conversion of 2-(p-nitrophenyl)ethyltrimethylammonium ion to p-nitrostyrene in
aqueous solution (Eq. 23) is accelerated by base but
retarded by acid.[3] A carbanion mechanism was suggested

$$p\text{-}NO_2C_6H_4CH_2CH_2NMe_3{}^+ \longrightarrow p\text{-}NO_2C_6H_4CH=CH_2 + NMe_3$$

$$(23)$$

to account for the effect of acid, but much later there
was found to be no exchange of tritiated reactant with
solvent in neutral solution.[83] This excludes reversible carbanion formation in neutral solution, but the
acidic region was unfortunately not studied in the exchange experiments.

A still more puzzling situation is found in the
elimination reaction shown in Eq. 24.[84] A thorough
investigation of the kinetics over the entire

$$p\text{-}ClC_6H_4COCH-CH(p\text{-}ClC_6H_4) \longrightarrow$$
$$\qquad\qquad | \quad\; |$$
$$\qquad\qquad Cl \;\; Cl$$

$$p\text{-}ClC_6H_4COC=CH(p\text{-}ClC_6H_4) \qquad + HCl \quad (24)$$
$$\qquad\qquad |$$
$$\qquad\qquad Cl$$

accessible pH range was carried out. The pH-rate profile is shown in Figure 2. The reaction was studied in

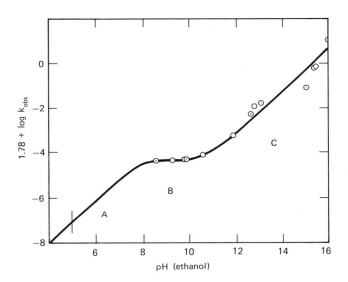

Figure 2 pH - rate profile for the elimination reaction of 4,4'-Dichlorochalcone dichloride (reproduced by permission of the authors[84] and the Journal of the American Chemical Society).

absolute ethanol with acetate buffers for the middle range of pH, and with added strong acid and strong base, respectively, for the low and high values of pH. The results could be correlated with Eq. 25 (a modified form[79] of Eq. 17).   Here K_{SHX} is the ionization constant of the substrate, and the sums are over all the bases present in solution, with $k_1{}^B$ being the

$$k = \frac{k_2 K_{SHX} \sum k_1{}^B (B)}{k_2 K_{SHX} + (H^+) \sum k_1{}^B (B)}$$
(25)

specific rate constant for proton removal by a particular base, B.   When $k_1 EtOH = 8 \times 10^{-7}$ sec^{-1}, $k_1 EtO^-$ $= 1.8 \times 10^2$ M^{-1} sec^{-1}, $k_1 OAc^- = 1.3 \times 10^{-3}$ M^{-1} sec^{-1}, and $k_2 K_{SHX} = 1.5 \times 10^{-14}$ sec^{-1}, the experimental curve of Figure 2 is reproduced by Eq. 25.

The three regions of Figure 2 can be accounted for qualitatively as follows.   For pH 4-8 (Case A), the second term in the denominator is large and the second step (elimination of Cl$^-$ from the carbanion) is rate controlling.   For pH 8-10 (Case B), the second term

becomes small, and formation of the carbanion becomes
rate controlling with ethanol as the most important
base, hence the independence of pH. Above pH 10 (Case
C), ethoxide becomes the most important base, and the
rate rises as its concentration increases. A fourth
region, in which the carbanion is formed rapidly and
in high concentration, and k_2 becomes rate controlling
again, is in principle expected. The leveling off of
the pH rate profile (Eq. 25 no longer applies at high
carbanion concentration) predicted for this situation
was not observed, however.

The kinetic results given thus far are quite consistent with a carbanion mechanism, but there is a major difficulty. At low pH (Case A), extensive return
of the carbanion to reactant is required by the observation that acid depresses the rate. Since the carbanion should not hold its configuration, interconversion of threo and erythro isomers of the reactant
should occur. None was observed. In addition, deuterium incorporation by reactant is expected in a deuterated solvent, and was not observed. The suggestion,[84] that the proton abstracted remains close to the
carbanion and is returned to it in the reprotonation,
is not very convincing. Although such phenomena have
been postulated to explain observed results in certain
studies of racemization and exchange,[85] the conditions
were quite different from those of the present investigation. It is difficult to believe that an ethoxonium-
carbanion ion pair would have an appreciable lifetime
in ethanol, and the observed rate depression in acid
should be found only if protonation is by an external
proton source. For these reasons, the mechanism of
the elimination reaction of dichlorochalcone dichloride, in acid solution at least, must be regarded as
unsolved in spite of the large amount of careful work
that has gone into it.

In some systems, a simple change of leaving group
will produce a change from general-base to lyate-ion
promoted reaction. Such a situation helps to strengthen the presumption that the lyate-ion dependence does
indicate a reversible ElcB process, and not just a
general-base promoted reaction in which lyate ion happens to be the only effective base. For example, elimination reactions of β-acetoxy ketones are general-
base promoted by water, hydroxide ion, and by a variety of amines,[38] behavior consistent with either an E2
or an irreversible carbanion mechanism. When the
leaving group is changed from acetoxy to methoxy, however, the elimination becomes specific lyate-ion promoted.[39] Furthermore, exchange in deuterium oxide is

much faster than elimination. The exchange is general-
base catalyzed, strongly suggesting that it occurs via
rate-determining proton abstraction to give a carbanion.

The deuterium solvent isotope effect, k_r^D/k_r^H,
runs 1.15 to 1.30. For an ElcB mechanism, the isotope
effect is given by Eq. 26, where K_W is the autoproto-
lysis constant of water, K_S is the ionization constant
of the ketone, and k_2 is the rate at which carbanion
gives product. It is argued[39] that the observed

$$\frac{k_r^D}{k_r^H} = \frac{k_2^D \cdot K_S^D \cdot K_W^H}{k_2^H \cdot K_S^H \cdot K_W^D} \tag{26}$$

effects are consistent with this equation, since there
should be little or no isotope effect on k_2, and the
isotope effect on K_S should be a little smaller than
that on K_W. If the elimination process were E2 (which
assumes that the exchange is an irrelevant side reaction[65]), an isotope effect in the opposite direction
($k_r^D/k_r^H<1$) would be expected, on the reasonable assumption that the isotope effect favoring proton over
deuteron transfer would more than counterbalance the
solvent effect favoring deuteroxide in deuterium oxide
over hydroxide in water.[86]

A similar approach was taken by O'Ferrall and
Slae[43] in their study of the elimination of water from
9-fluorenylmethanol. They determined rates of elimination of the β-hydrogen and β-deuterium compounds in
water and deuterium oxide, and rates of exchange of the
β-tritium compound in both solvents. Exchange was
faster than elimination. Kinetic analysis, using initial and limiting slopes of the rate plots, allowed
evaluation of the specific rates of elimination of both
hydrogen and deuterium, and the specific rate of exchange of tritium, in water and deuterium oxide. These
quantities were used in an effort to decide whether exchange was part of the main reaction path or an irrelevant side reaction. Scheme 2 incorporates both possibilities. If $k_2 = 0$, the reaction is entirely ElcB,
and if $k = 0$, the elimination is E2 and the exchange is
a side reaction. From this scheme one obtains Eqs. 27
to 30, where k_E refers to elimination and k_X to exchange. If one assumes that the reaction is E2, then
$k_2 \gg \alpha k_1$, and Eqs. 31 and 32 result (the numbers are
the experimentally determined values). The only
difference between the two is that Eq. 32 incorporates,
in addition to the primary isotope effect of Eq. 31,

Scheme 2

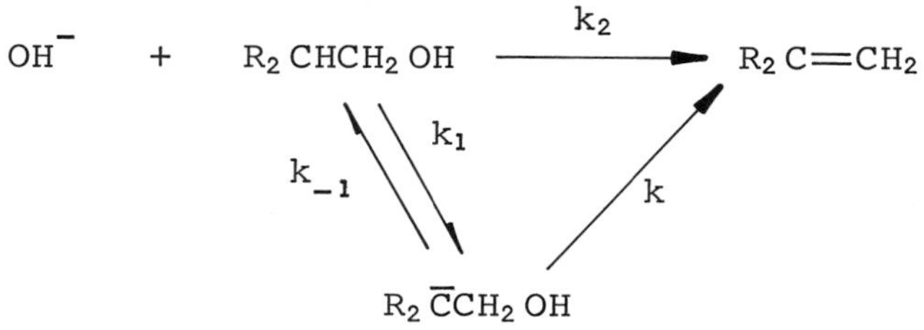

$$\text{OH}^- \;+\; R_2\,\text{CHCH}_2\,\text{OH} \xrightarrow{\; k_2 \;} R_2\,\text{C}{=}\text{CH}_2$$

with k_1, k_{-1} interconverting to $R_2\,\overline{\text{C}}\text{CH}_2\,\text{OH}$ which proceeds by k.

$$k_E{}^H \;=\; k_2{}^H \;+\; \alpha_H k_1{}^H \tag{27}$$

$$k_E{}^D \;=\; k_2{}^D \;+\; \alpha_D k_1{}^D \tag{28}$$

$$k_X{}^D \;=\; k_2{}^D \;+\; k_1{}^D \tag{29}$$

$$\alpha \;=\; k/(k + k_{-1}) \tag{30}$$

$$\frac{k_E{}^H(\text{D}_2\text{O})}{k_E{}^D(\text{D}_2\text{O})} \;=\; \frac{k_2{}^H(\text{D}_2\text{O})}{k_2{}^D(\text{D}_2\text{O})} \;=\; 7.2 \tag{31}$$

$$\frac{k_E{}^H(\text{H}_2\text{O})}{k_E{}^D(\text{D}_2\text{O})} \;=\; \frac{k_2{}^H(\text{H}_2\text{O})}{k_2{}^D(\text{D}_2\text{O})} \;=\; 0.92 \tag{32}$$

solvent effects arising from the difference between hydroxide and deuteroxide as base and leaving group.
These effects would have to be unreasonably large for
Eq. 32 to incorporate a primary effect of 7.2.
     On the other hand, $\alpha k_1 \gg k_2$ if the reaction is
ElcB, and Eqs. 33 and 34 result. The former is simply
the isotope effect on carbanion formation, and is entirely consistent with a primary effect of 7.2. Discussion of Eq. 34 is facilitated by a simplification.

$$\frac{k_E^H(D_2O)}{k_E^D(D_2O)} = \frac{k_1^H(D_2O)}{k_1^D(D_2O)} \tag{33}$$

$$\frac{k_E^H(H_2O)}{k_E^D(D_2O)} = \frac{\alpha_H}{\alpha_D} \frac{k_1^H(H_2O)}{k_1^D(D_2O)} \tag{34}$$

Because exchange is much faster than elimination, $k_{-1} >> k$, and neglecting k in the numerator of α gives Eq. 35. The isotope effect on k should be small and

$$\frac{k_E^H(H_2O)}{k_E^D(D_2O)} = \frac{k_1^H}{k_1^D} \frac{k_{-1}^D}{k_{-1}^H} \frac{k(H_2O)}{k(D_2O)} \tag{35}$$

secondary, while the primary effects on k_1 and k_{-1} should approximately cancel. Thus, an ElcB mechanism is consistent with a value of 0.92 for the product of the three effects.

The mechanism of the elimination could be changed by varying the conditions.[87] In methanol with methoxide, the situation was qualitatively the same as in water, but with t-butyl alcohol and t-butoxide, the exchange was completely suppressed, indicating either an E2 or, more probably, an irreversible ElcB reaction.

Isotope effects, as noted above, can be very useful in elucidating the course of elimination reactions, but one must take care to analyze the results carefully. It has been argued, in an otherwise excellent review of ElcB reactions,[88] that small primary deuterium isotope effects are to be expected in ElcB reactions. O'Ferrall's results[87] show that this need not be true, and a more thorough examination of the theoretical background shows why. It is now generally accepted that deuterium isotope effects on proton transfers are largest when the proton is half transferred in the transition state, and are small when it is either very little or almost completely transferred.[89] The argument and the evidence for it will be discussed later (Section II.C.3a). McLennan[88] argues that the carbanion intermediate in an ElcB reaction will be of very high energy and, hence, the transition state leading to it should not differ much from it in energy or structure.[90]   But the ElcB reaction is most probable under those conditions in which a relatively stable carbanion can be formed.   Such a carbanion may differ markedly in structure from the transition state leading to it, and it is not at all necessary to assume that the proton transfer is nearly complete in the transition state.   In fact, a whole range of isotope effects, depending on the stability of the carbanion and the strength of the attacking base, is to be expected.   There is no need to take a moderate-to-large isotope effect as evidence against an ElcB process.

A more reliable conclusion from the magnitude of a single isotope effect can be drawn if the isotopic substitution is in the leaving group.   If there is a sizable leaving-group isotope effect, weakening of the bond to the leaving group in the transition state, and hence an E2 process, is indicated.   Since the extent of weakening of this bond may vary markedly from one E2 reaction to another, however, one cannot safely draw the converse conclusion that a small isotope effect indicates an ElcB process (see Section II.C.3 for a discussion of isotope effects and structure of the E2 transition state).   Because determination of heavy-atom isotope effects is considerably more difficult than the determination of deuterium isotope effects, this tool has rather seldom been applied to the differentiation of E2 and ElcB processes.   It has been found, for example, that the syn elimination (to give 1-phenyl-cyclohexene) of trans-2-phenylcyclohexyltrimethyl-ammonium ion occurs with a nitrogen isotope effect of 0.4%, and is thus probably concerted E2.[91]   Even the earlier value of 0.2%[92] appears high for an ElcB reaction.   Although the contrary has sometimes been assumed, a true ElcB process should not involve significant weakening of the bond to the leaving group in the proton-removal step.   If formation of the carbanion gained energetic advantage from such weakening, it seems very likely that a concerted (E2) process would be still more advantageous.

(c)   The Leaving-Group Effect.   This discussion leads us to another type of argument that has been used to distinguish between the E2 and irreversible ElcB processes.   If the reaction is ElcB, the only influence of the leaving group on the rate should be through an inductive or field effect.   If the reaction is E2, the weakening of the bond to the leaving group should lower the energy of the transition state below that for a stepwise process.   The reaction should thus be faster

than expected from a simple electrostatic effect of the
leaving group. The key to the successful use of this
method is a way of estimating what the electrostatic
effect of the leaving group on rate should be.
     Weinstock, Pearson, and Bordwell[93] studied the
base-promoted eliminations of cis- and trans-2-(p-
tolylsulfonyl) cyclohexyl tosylates (2a and 2b), and
the corresponding cyclopentyl derivatives (3a and 3b).
Both cis and trans reactants yielded the conjugated

2a (cis)   3a (cis)

2b (trans)   3b (trans)

4   5

olefins (4 and 5), so that the trans reactants must
have eliminated by a syn mechanism (see Chapter III for
a discussion of stereochemistry of elimination).   At
that time, syn eliminations in E2 reactions were unusual and unexpected, and an ElcB mechanism was suggested.   Later, however, a study of the rate of deuterium
exchange of cyclohexyl-1-d 1-p-tolyl sulfone (6) was
made.[94]   The rate was much slower than that of the elimination -- so much slower that no reasonable correction for the electrostatic effect of the tosylate group
appeared capable of closing the gap.   Hence, the elimination was considered to be concerted.
     This conclusion was disputed by Hine and Ramsay,[95]
who studied deuterium exchange in cis-2-methoxycyclo-
hexyl-1-d 1-p-tolyl sulfone (7), a closer model than
(6) for 2a and 2b.   The methoxyl group was found to
accelerate the exchange 500-fold, a considerably
larger figure than expected from the qualitative arguments used earlier.[94]   A plot of the Taft substituent constants (σ^*) for hydrogen, methoxyl, and
tosylate versus the logarithms of the rates of exchange

6   7

for 6 and 7, and the rate of elimination of 2b, gave a
good straight line, with 2b actually a little slower
than predicted by the line. On the other hand, 2a
reacted more than 100-fold faster than predicted, so
Hine concluded that 2b followed the ElcB, but 2a the
E2, mechanism.

It should be apparent from the differing conclusions reached by the two groups that this is not a
method which is capable of fine distinctions. The uncertainties in the estimation of the rate of proton
removal are such that a reaction should be at least
one, and preferably two, orders of magnitude faster
than the estimated rate before one concludes it is
concerted. Conversely, a rate close to the estimated
rate merely permits, but does not require, the conclusion that the reaction is ElcB. Small accelerations
due to weakening of the bond to the leaving group cannot be reliably detected.

Hine also applied his method of estimating rates
of proton exchange to the elimination reaction of β-
benzone hexachloride.[96] In earlier studies by
Cristol[73] (see above), very small amounts of exchange
had been detected. Hine concluded that the observed
rate of elimination was consistent with rate-determining carbanion formation, but cautioned that the
uncertainties were large enough that the exchange could
easily be a minor side reaction accompanying a concerted syn elimination.

Another prediction one can make for a reaction involving no weakening of the bond to the leaving group
in the rate determining step is that changing the
leaving group should have a relatively minor effect on
rate. For example, carbon-halogen bond strengths
increase markedly along the series I<Br<Cl<F, and a
concerted reaction would be expected to show steeply
decreasing relative rates in the order RI>RBr>RCl>RF.
Differences in inductive effect should be rather small,
however, so that changing the leaving group from one
halogen to another should have only a minor effect on

the rate of an irreversible ElcB process.

Changing the leaving halogen does, indeed, have a marked effect on rates of eliminations that can be presumed to be concerted. The 2-phenylethyl halides react with ethoxide ion in ethanol at rates that span a range of more than 25,000 between fluoride and iodide.[35,97,98] The major differences, however, are between fluoride and chloride (70-fold) and between chloride and bromide (60-fold). The difference between bromide and iodide is only a little over sixfold.

Among other possible comparisons would be the tosylate/halide ratio,[99,100] or the effect on rate of ring substituents in substituted benzenesulfonate esters. In the latter case, the substituent is sufficiently remote from the site of proton removal that its effect should be very small if the elimination is not concerted. Hammett rho values for substituents in a leaving benzenesulfonate group are, in fact, moderately large for known E2 reactions such as those of the 2-pentyl[101] and 2,2-diphenylethyl[102] benzenesulfonates (+1.35 and +1.1, respectively). Bordwell[103] finds considerably smaller rho values for eliminations of the 2-(p-tolysulfonyl)cyclohexyl arenesulfonates (+0.3 to +0.6), and argues, on the basis of this and other evidence, that both the cis and trans compounds may react by the carbanion mechanism.

There are two sources of difficulty in applying this criterion of mechanism. One is that a change of leaving group may change the mechanism (recall the change between β-acetoxy and β-methoxy ketones described earlier[38-39]), so that rate comparisons would be invalid. The other is that it is not easy to decide when a leaving group effect is small enough to indicate no bond weakening to the leaving group. Bordwell's[103] examples nicely illustrate this dilemma. His effects are certainly smaller than those in known concerted reactions, but it is very hard to say whether they are small enough for an ElcB reaction.

(d) Dependence on the H_ Acidity Function. The acidity function H_ has also been suggested as a kinetic criterion for distinguishing between reversible ElcB reactions, on the one hand, and irreversible ElcB or E2 reactions, on the other.[104,105] This function describes the tendency of a basic medium to accept a proton from a neutral acid BH and is defined by Eq. 36.

$$H_- = -\log \frac{a_{H^+} f_{B^-}}{f_{BH}} = pK_{BH} + \log \frac{(B^-)}{(BH)} \quad (36)$$

The relationships between H_- and rate to be expected for the two cases have been derived.[106]

For an ElcB reaction with the second-step rate determining (eqs. 37 to 38), the rate is given by Eq. 39, where the f's are activity coefficients of SX^- and

$$SHX \quad + \quad RO^- \quad \xrightarrow{\text{fast}} \quad SX^- \quad + \quad ROH \qquad (37)$$

$$SX^- \quad \xrightarrow[\text{slow}]{k'} \quad \text{products} \qquad (38)$$

$$\text{rate} \quad = \quad k'(SX^-) \quad \frac{f_{SX^-}}{f^{\ddagger}} \qquad (39)$$

the transition state. The ionization constant for SHX is given by Eq. 40.

$$K_{SHX} \quad = \quad \frac{a_{H^+} \, a_{SX^-}}{a_{SHX}} \quad = \quad \frac{(SX^-) f_{SX^-} \, a_{H^+}}{(SHX) f_{SHX}} \qquad (40)$$

If $f_{SX^-}/f_{SHX} = f_{B^-}/f_{BH}$, then Eq. 40 becomes Eq. 41.

$$K_{SHX} \quad = \quad \frac{(SX^-) h_-}{(SHX)} \qquad (41)$$

(where h_- is defined by $H_- = -\log h_-$). Substitution of Eq. 41 into Eq. 39 gives Eq. 42.

$$\text{rate} \quad = \quad k'(SHX) \, K_{SHX} \cdot \frac{1}{h_-} \cdot \frac{f_{SX^-}}{f^{\ddagger}} \qquad (42)$$

If $f_{SX^-}/f^{\ddagger} = 1$ (because SX^- and the transition state have the same composition and the same charges), the observed rate constant is given by Eq. 43, and its logarithm by Eq. 44. A plot of $\log k_{obs}$ versus H_- should thus be linear with unit slope.

$$k_{obs} \quad = \quad k' \, K_{SH} \, \frac{1}{h_-} \qquad (43)$$

$$\log k_{obs} \quad = \quad \log k' \, K_{SH} \quad + \quad H_- \qquad (44)$$

For rate-determining proton transfer, Eq. 45, the observed rate constant at any particular base concentration is given by Eq. 46.

$$SH \quad + \quad OH^- \quad \xrightarrow{\quad k_2 \quad} \quad products \qquad (45)$$

$$k_{obs} \quad = \quad k_2 \cdot \frac{f_{OH^-} \, f_{SH}}{f^{\ddagger}} \cdot (OH^-) \qquad (46)$$

Multiplication by $(H_2O)f_{H_2O}$ and division by a_{H_2O} gives Eq. 47. One then assumes that the activity coefficient

$$k_{obs} \quad = \quad k_2 \cdot \frac{f_{H_2O} \, f_{SH}}{f^{\ddagger}} \cdot \frac{a_{OH^-}}{a_{H_2O}} \cdot (H_2O) \qquad (47)$$

ratio equals f_{BH}/f_{B^-}, which permits one to relate its behavior to h_- (Eq. 48). Substitution of Eq. 48 into

$$\frac{f_{H_2O} \, f_{SH}}{f^{\ddagger}} \quad = \quad \frac{f_{BH}}{f_{B^-}} \quad = \quad \frac{a_{H^+}}{h_-} \quad = \quad \frac{K_W}{h_-} \cdot \frac{a_{H_2O}}{a_{OH^-}} \qquad (48)$$

Eq. 47 yields Eq. 49 for k_{obs}, and Eq. 50 for $\log k_{obs}$.

$$k_{obs} \quad = \quad k_2 \, K_W \cdot \frac{1}{h_-} \cdot (H_2O) \qquad (49)$$

$$\log k_{obs} \quad = \quad \log k_2 \, K_W \quad + \quad \log (H_2O) \quad + \quad H_- \qquad (50)$$

Correlations with Eq. 50 have been found for the E2 reactions of DL-serine phosphate with hydroxide ion in water,[106] and of 2-phenylethyldimethylsulfonium ion with hydroxide ion in aqueous dimethyl sulfoxide.[107] Correlation with Eq. 44, albeit with slight downward curvature, was found for the reaction of chloroform with methoxide ion.[108] Although this is not an E1cB reaction, it was known to involve reversible formation of the trichloromethyl anion, followed by rate determining reaction with solvent.[109-111] A known E2 reaction, the action of methoxide ion on 2-phenylethyl chloride, increased considerably less rapidly in rate as methoxide concentration was increased, which is qualitatively what would be expected from Eq. 50. On the other hand, the rate of another known E2 reaction, of 2-phenylethyl bromide with t-butoxide ion in mixtures of t-butyl alcohol and dimethyl sulfoxide, increases much _faster_ than the H_ value of the medium.[57] Particularly in such mixed solvents, it seems unwise to assume a relation between H_ and mechanism as simple as that embodied in Eqs. 44 and 50.

(e) _Entropy Effects_. Qualitatively, one might expect a more negative entropy of activation for an E2 or irreversible E1cB than for a reversible E1cB process because of the loss of translational entropy that occurs when the base and substrate come together to form a single species in the activated complex. Indeed, E2 dehydrohalogenations with ethoxide in ethanol of the 2-phenylethyl halides have entropies of activation some 20-30 eu more negative than those of the E1cB reactions of 1,1-dihalo-2,2,2-trifluoroethanes with methoxide in methanol.[66,67] There have not been any comprehensive studies to test the reliability of this criterion. Clearly, care must be taken that reactants and conditions do not differ too radically, and particularly that reactants of different charge types are not compared.

(f) _Stereochemistry_. At one time it was considered that eliminations by the E2 mechanism should occur stereospecifically with anti stereochemistry, and that lack of stereospecificity strongly suggested an E1cB process. Since the stereochemistry of eliminations is discussed later (Chapter III), only brief comments will be made here. First, enough exceptions to the anti rule for E2 reactions have been discovered in recent years that lack of stereospecificity need not implicate the E1cB mechanism. On the other hand, high stereospecificity still carries a rather strong suggestion that the process is E2. It should not be taken as proof of an E2 process, however, since there are circumstances under which stereospecificity can be entirely compatible with an E1cB process.
    Bordwell[103] has compared rates of elimination from the cis and trans isomers of 2-phenylsulfonylcyclohexyl and 2-phenylsulfonylcyclopentyl derivatives. The preference for anti elimination from the cis isomers (2a and 3a) over syn elimination from the trans isomers (2b and 3b) ran much higher for the cyclohexyl than for the cyclopentyl system. The anti eliminations in the two systems were of comparable rate (actually,

cyclopentyl was a little faster); the difference in
stereospecificity was mainly the result of a much slower syn elimination in the cyclohexyl than in the cyclopentyl series. Bordwell argues from this that the
stereospecificity arises not from a rate advantage for
the anti elimination (as is commonly supposed), but
from some factor that makes the syn elimination abnormally slow. He argues further that 1,2-diequatorial
substituents (as in the trans isomer) can relieve steric compressions by bending away from each other, and
that the resulting ring deformation hinders abstraction of an axial proton. For instance, trans-2-phenyl-
1-nitrocyclohexane (8) reacts with methoxide ion to
lose a proton 350 times slower than the cis isomer
(9).[112],[113] The preference for anti over syn elimination in 1,2-disubstituted cyclohexanes could thus

8   9

persist in an ElcB mechanism. One need not accept
Bordwell's conclusion, that many reactions formerly
regarded as E2 are really ElcB, to admit that stereospecificity can no longer be regarded as definitive
proof for a concerted process.

A recent study has shown that the ElcB reaction
can also prefer a syn pathway under some conditions.
The stereochemistry of elimination from 1-methoxyace-
naphthene, and the accompanying exchange, depend on the
cation associated with the alkoxide.[113b] It was suggested that a strongly ion-paired alkoxide favors syn
elimination because there is association of the cation
with the methoxyl leaving group of the substrate.

Before leaving the ElcB mechanism, some comment on
relative reactivities is in order. We noted above[76]
the conditions favorable to the ElcB process: a poor
leaving group, a stable carbanion intermediate, and a
relatively unstable product. There has, however, been
very little systematic exploration of structural
effects in ElcB reactions. A nice demonstration of the
effect of carbanion stability is provided by the recent
work of Crosby and Stirling[41] on elimination of

phenoxide from 2-substituted ethyl phenyl ethers. A
number of criteria were applied to demonstrate that the
reaction was ElcB.[114] Changes in the β-substituent had
a dramatic effect on rate: a total span of 10^{11} between the fastest and slowest substituents was observed. Furthermore, there was a very good correlation
with the resonance, but not the inductive, effect of
the substituent. For example, dimethylsulfonium was
one of the best substituents and trimethylammonium one
of the worst, though they should have nearly identical
inductive effects. There was also a good quantitative
correlation with Taft's $\underline{\sigma}_R$- constants, where
available.[115]

2   The Ion-Pair Mechanism

A mechanism which lies at the opposite extreme from the
ElcB mechanism is the ion-pair mechanism. In this
mechanism, the reactant RX first ionizes to R^+X^-, which
then reacts with base. Very little work has gone into
exploring the possible existence and characteristics of
such a mechanism. Sneen and Robbins[116] have examined
the reaction of 1-phenylethyl bromide with ethoxide in
ethanol, which had been considered by earlier workers
to involve concurrent unimolecular (S_N1 and E1) and bimolecular (S_N2 and E2) paths. He showed that this assumption, by which the observed pseudo-first-order rate
constant for product formation should follow Eq. 51,
led to values of k_2 that decreased sharply as the

$$k_{obs} = k_1 + k_2(OEt^-) \qquad (51)$$

ethoxide concentration increased (the rate was halved
between 0.114 and 0.533 M sodium ethoxide). In con-
contrast, k_2 for ethyl bromide decreased only 17%
over the same range. Sneen and Robbins concluded that
this discrepancy was far too great to be accounted for
by salt effects, and proposed an alternate mechanism
(Scheme 3).
     The mechanism leads to a rather complex set of ki-
netic equations which will not be reproduced here.
These equations give a rather good, though not highly
precise, fit to the observed data, certainly far better
than the assumption of concurrent unimolecular and bimolecular processes. They do require numbers for four
ratios of rate constants. These ratios are not adjus-
table parameters, however, but are evaluated from product ratios. Given the accumulated errors in the rate
and product determinations, the fit must be regarded
as good evidence for the ion-pair mechanism.

Scheme 3

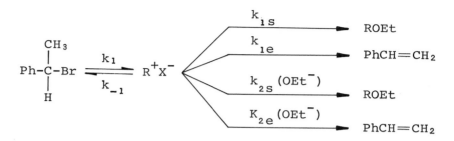

Judgment on the importance of the ion-pair mechanism should await further work. Thus far, Sneen has considered only the two extremes of concurrent unimolecular and bimolecular mechanism, on the one hand, and a mechanism in which all products arise from an ion pair, on the other. It is conceivable that some products may arise from an ion pair and others directly from unionized substrate. Another possible weakness is the neglect of salt effects. Ethyl bromide does not seem to be a very good model for 1-phenylethyl bromide, which has a much stronger tendency to ionize. In fact, McLennan has shown that the effect of base concentration on k_{2e} for 1-phenylethyl bromide can be accounted for by the variation in the fractions of ethoxide ion in the ion paired ($Na^+ OEt^-$) and free (OEt^-) states as the total base concentration is varied.[116b]

Finally, Sneen points out that the ion-pair mechanism becomes kinetically indistinguishable from the conventional picture if the ion pair returns to reactant much faster than it goes on to products. He then suggests that even reactions following normal E2 kinetics may involve ion pairs. While this interesting speculation does not seem very likely in the light of the accumulated evidence on the E2 mechanism, it should certainly be kept in mind as further information on the occurrence of ion pairs in eliminations is developed.

3   α-Elimination

An apparent 1,2-elimination can result if H and X are lost from the same carbon atom, followed by migration of a hydrogen atom (or an alkyl or aryl group) from the adjacent carbon (Eqs. 52 and 53). The intermediate carbene of this mechanism has been implicated in a variety of organic reactions, and there is no doubt at all that carbenes exist.[117] However, they seem to be important as intermediates in elimination reactions

$$RCH_2CH_2X \quad + \quad B \longrightarrow RCH_2\overset{\cdot\cdot}{C}H \qquad (52)$$

$$RCH_2\overset{\cdot\cdot}{C}H \longrightarrow RCH=CH_2 \qquad (53)$$

only under rather special circumstances. With reactants possessing β-hydrogens, and in protic solvents with the usual bases, α-elimination occurs at most to a small extent, and normally not at all.

α-Elimination cannot be distinguished kinetically from β-elimination, but two other criteria permit a reliable distinction. First, deuterium labeling gives characteristically different results with the two mechanisms. When the α-hydrogens of the substrate are replaced by deuterium, the product from a normal β-elimination will not have lost any deuterium unless there was exchange in the reactants or products. On the other hand, one of these deuterium atoms must be lost in an α-elimination. Since an independent check can be made for exchange, this technique distinguishes unambiguously between the two mechanisms. Replacement of the β-hydrogens by deuterium leads to the loss of one deuterium atom in a normal β-elimination, but no deuterium is lost in an α-elimination, the only change in labeling being the migration of a deuterium from the β to the α-position of the intermediate carbene in the formation of the final product (Eq. 54).

$$RCD_2\overset{\cdot\cdot}{C}H \longrightarrow RCD=CHD \qquad (54)$$

The other method of detecting α-elimination is to look for products characteristic of carbene reactions. In particular, carbenes undergo a very characteristic intramolecular insertion reaction to give cyclopropanes, hence, appearance of cyclopropanes in base-promoted elimination reactions is excellent evidence that, at least, part of the reaction involves a carbene intermediate. This mode of reaction is illustrated in Eq. 55. The product criterion was applied by Friedman and Berger[118] in a survey of the reactions of butyl halides with phenylsodium in decane and with sodium

$$RCH_2CH_2\overset{\cdot\cdot}{C}H \longrightarrow RCH\overset{-CH_2}{\underset{-CH_2}{\mid}} \qquad (55)$$

amide or sodium methoxide in diethyl carbitol. n-Butyl
and isobutyl chlorides labeled with deuterium in the α
-position were used to calibrate the study, the yields
of cyclopropanes obtained with the other compounds
being compared with those obtained from these compounds
with phenylsodium. Under these conditions, the labeling studies gave 94% α-elimination from n-butyl chloride and 89% α-elimination from isobutyl chloride. The
isobutyl chloride also gave substantially more methyl-
cyclopropane (32%) than the n-butyl chloride (3.7%).
The bromides and the iodides gave less α-elimination
than the chlorides. Sodium amide gave considerably
less α-elimination than did phenylsodium, and sodium
methoxide gave detectable α-elimination only with n-
butyl chloride.

Sodium amide in liquid ammonia is a still less
effective agent for promoting α-elimination. n-Octyl
chlorides and bromides deuterated in the α- and β-posi-
tions were found to yield at most 23% α-elimination
(from the α-deuterated bromide), and no α-elimination
at all was detected with 2-ethylbutyl-1,1-d$_2$ bromide.[119]

The only compounds other than alkyl halides that
have been examined for possible occurrence of α-elimi-
nation are sulfonium salts. Treatment of 1-butyl-1,1-
d$_2$-diphenylsulfonium fluoborate with tritylsodium in
ether led to butenes (mainly 1-butene) and to triphenylmethane which was 50% deuterated.[120] It was concluded that 50% of the elimination involved a carbene,
but the deuterium content of the butenes was not determined to check this conclusion. The possibility remains that the attack by tritylsodium on the α-deu-
terium atoms did not lead to olefin via a carbene.
Some 1,1,1-triphenylpentane was isolated, and it could
have resulted from attack of the carbene on triphenylmethane or tritylsodium. That a carbene can indeed be
formed under the conditions used by Franzen[120] is
shown by the occurrence of both isobutylene and methyl-
cyclopropane among the products from isobutyldiphenyl-
sulfonium fluoborate and tritylsodium.

4    $\alpha'-\beta$ Elimination

Sulfonium or ammonium salts possessing α-hydrogens on
more than one of the alkyl groups attached to the central atom can undergo another type of elimination reaction in which an α-hydrogen is first removed, and the
resulting anionic center then functions as a base to
remove a β-proton from another alkyl group and, thus,
effect a β-elimination. The mechanism is illustrated
for quaternary ammonium salts in Eqs. 56 and 57.

Kinetically this mechanism should show the same

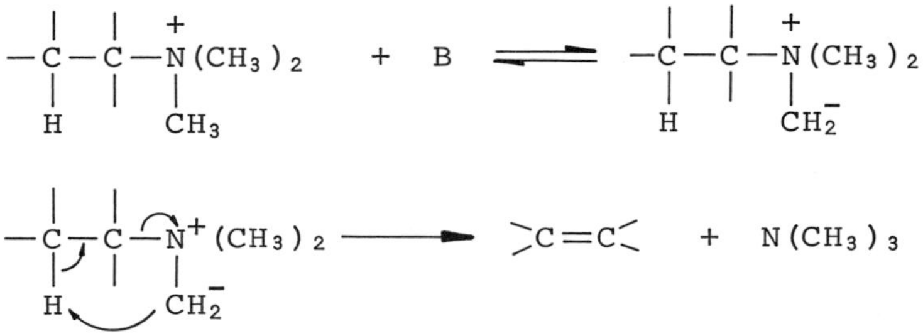

$$-\overset{\underset{|}{H}}{\underset{|}{C}}-\overset{\underset{|}{CH_3}}{\underset{|}{C}}-\overset{+}{N}(CH_3)_2 \quad + \quad B \quad \rightleftharpoons \quad -\overset{\underset{|}{H}}{\underset{|}{C}}-\overset{\underset{|}{CH_2^-}}{\underset{|}{C}}-\overset{+}{N}(CH_3)_2 \quad (56)$$

$$-\overset{\underset{|}{H}}{\underset{|}{C}}-\overset{\underset{|}{CH_2^-}}{\underset{|}{C}}-\overset{+}{N}(CH_3)_2 \quad \longrightarrow \quad \rangle C=C\langle \quad + \quad N(CH_3)_3 \quad (57)$$

behavior as either the E2 or ElcB mechanisms, depending on whether the intermediate ylid (a term used to des- cribe species possessing positive and negative charges on adjacent atoms) is formed irreversibly or reversibly. Recognition of the mechanism thus requires nonkinetic methods. The most reliable method is isotopic label- ing, but conclusions have also been drawn from the na- ture of the products.
    The mechanism was first suggested to explain an elimination observed when a quaternary ammonium salt containing an iodomethyl group was treated with phenyl- lithium (Eqs. 58 and 59).

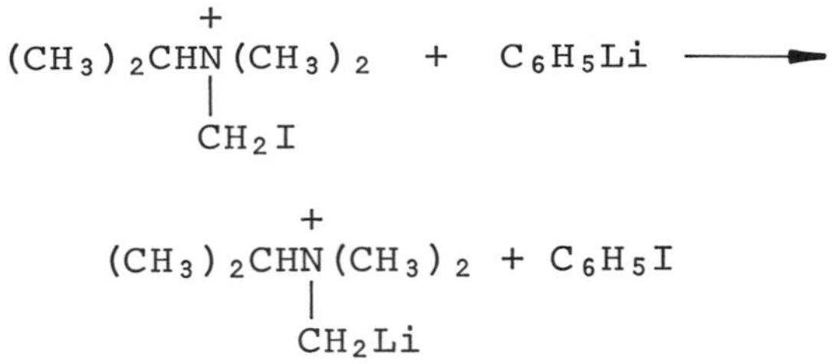

$$(CH_3)_2CH\overset{+}{\underset{\underset{CH_2I}{|}}{N}}(CH_3)_2 \quad + \quad C_6H_5Li \quad \longrightarrow$$

$$(CH_3)_2CH\overset{+}{\underset{\underset{CH_2Li}{|}}{N}}(CH_3)_2 \quad + \quad C_6H_5I \quad (58)$$

$$(CH_3)_2CH\overset{+}{\underset{\underset{CH_2Li}{|}}{N}}(CH_3)_2 \quad \longrightarrow$$

$$CH_2{=}CHCH_3 \quad + \quad N(CH_3)_3 \quad + \quad Li^+ \quad (59)$$

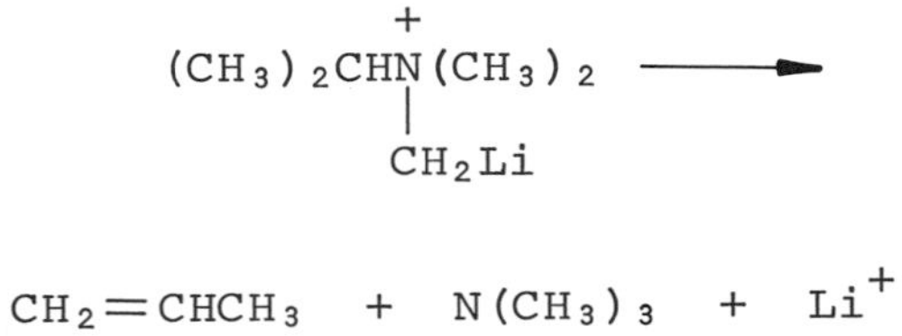

If the elimination had been a simple E2 process with phenyllithium functioning as base, benzene rather than iodobenzene would have been obtained.[121] Since iso-propyltrimethylammonium iodide also gives elimination on treatment with phenyllithium, the reaction was sug-gested by analogy to involve initial removal of an α-proton from one of the methyl groups by the phenylli-thium. The reaction of Eq. 59 could then follow.[121]

Tracer studies have shown clearly that the α'-β mechanism is not the usual route for elimination re-actions of 'onium salts conducted under the normal con-ditions of a protic solvent and hydroxide or an alkox-ide as base. The technique involves the preparation of a β-deuterated reactant. In a normal E2 reaction this deuterium will be lost to the solvent. In an α'-β reaction of an ammonium salt it will be found in the trimethylamine, and in an α'-β reaction of a sul-fonium salt, in the dimethyl sulfide. The only pre-caution needed is to demonstrate whether any deuterium found in the amine or sulfide comes from exchange of the 'onium salt with deuterium from the solvent rather than from a genuine α'-β mechanism. Since the solvent will be deuterium-free at the beginning of the reac-tion, a deuterium content that increases with extent of reaction and extrapolates to a very small value at the beginning of the reaction can be taken as indi-cating exchange. In dilute solution, deuterium uptake from this source will be very small, but it can be quite marked in the concentrated syrup of the quater-nary hydroxide used in the Hofmann elimination.[122]

By the tracer technique, 10 and 11 have been shown to undergo the Hofmann elimination by the E2 mechanism.[122,123] The sulfonium salt 12 also reacts with hydroxide ion in water entirely by an E2

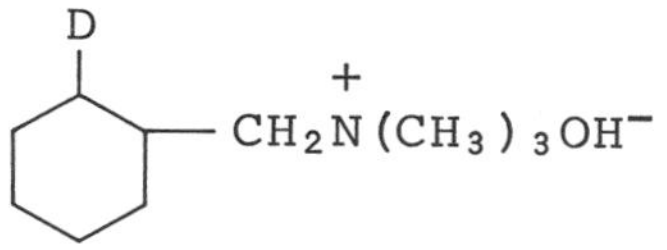

$$\text{(cyclohexyl-D)}-CH_2N(CH_3)_3OH^- \qquad\qquad CD_3CH_2N(CH_3)_3OH^-$$

10   11

$$PhCD_2CH_2S(CH_3)_2Br^-$$

12

mechanism.[124]   On the other hand, <u>13</u> reacts with tritylsodium in ether to yield triphenylmethane which is

$$\overset{+}{PhCH_2 CD_2 S(CD_3)_2}\ \ Br^-$$

<u>13</u>

79% deuterated.[125]   If all of the ylid led to styrene
(the styrene yield was not determined), nearly 80% of
the reaction followed the α'-β pathway. That sulfonium
salts can react via the α'-β mechanism in hydroxylic
solvents is shown by recent work in which 3-pentyl-2,2,
4,4-d$_4$-dimethylsulfonium ion gives methyl sulfide which
is 65% deuterated with potassium t-butoxide in t-butyl
alcohol, but only 1 to 2% deuterated with potassium
n-butoxide in n-butyl alcohol.[144]   The more strongly
basic medium probably favors the α'-β mechanism by increasing both the rate of formation and the equilibrium
concentration of the sulfonium ylid. The quaternary
ammonium hydroxide <u>14</u> undergoes a Hofmann elimination
to give 75% deuterated trimethylamine.[126]   Apparently
the t-butyl groups offer sufficient hindrance to the

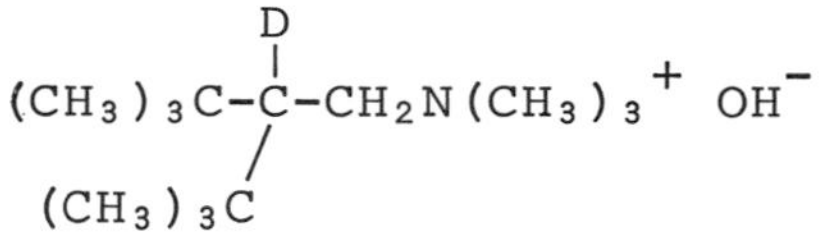

$$\begin{array}{c} D \\ | \\ (CH_3)_3C-\overset{}{\underset{}{C}}-CH_2N(CH_3)_3{}^+\ OH^- \\ / \\ (CH_3)_3C \end{array}$$

<u>14</u>

approach to the β-deuterium that the ylid pathway becomes preferred. It is often noted in organic reactions that intramolecular mechanisms are less sensitive
to steric congestion than their intermolecular counterparts. In contrast to <u>14</u>, the bicyclic compounds <u>15</u>
and <u>16</u> yield trimethylamine containing only 6% and 13%,

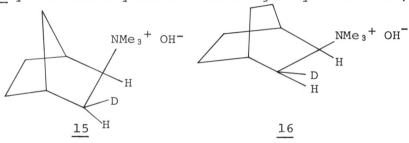

<u>15</u>   <u>16</u>

respectively, of deuterium.[127]   The olefins contain no
deuterium, so the reaction must have been entirely syn,
but the major part of it was apparently E2.   Thus
steric hindrance to the E2 mechanism seems to be more
effective than steric preference for the ylid mechanism.
    Most of the early conclusions concerning the
applicability of the α'-β mechanism were drawn from
examination of products and product ratios.   For example, the compounds 17 were subjected to normal Hofmann

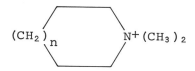

$(CH_2)_n$ $N^+(CH_3)_2$

17 (n = 0,1,2)

reactions and gave good yields of the dimethylamino-
olefins expected by the E2 mechanism.   On the other
hand, treatment with butyllithium gave ylids that decomposed only at 150 to 160°, except for that from
17 (n = 2), which decomposed at room temperature, and
gave much lower yields of the dimethylamino-olefins.
The major products for 17 (n = 0,1) were the N-methyl
and N-ethyl cyclic amines, resulting from demethyla-
tion and Stevens rearrangement, respectively.   From
these pronounced differences, it was concluded that
the Hofmann elimination did not proceed through a
ylid.[128]   A possible weakness in such arguments is that
the solvation of the intermediates and transition
states is very different between ether and hydroxylic
solvents, and solvent is totally absent in the pyrolysis of the ylid at elevated temperatures.   Under such
different conditions, even the same mechanism might
give quite different product mixtures.   One cannot be
sure, for example, that the ylid decomposes unimolecularly.
    The cycloalkyltrimethylammonium bromides 18 and
the cycloalkyldimethyl(bromomethyl)ammonium bromides
19 gave similar product mixtures on treatment with
methyllithium or phenyllithium.[128-130]   In contrast,
treatment of 18 under Hofmann conditions or with
potassium amide in liquid ammonia gave different product mixtures.   Particularly revealing were the results
with the cyclooctyl compounds (n = 4), which yielded
mainly cis-cyclooctene with the organolithium reagent,
but mainly trans-cyclooctene under Hofmann conditions
or with amide ion.   A predominance of cis-cyclooctene

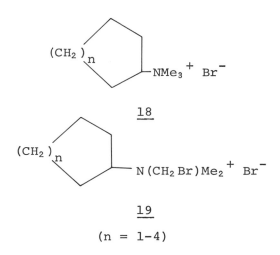

$$18$$

$$19$$

$$(n = 1-4)$$

was also found when cyclooctyldimethylsulfonium ion was treated with tritylsodium.[125] The difference was again taken as indicating α'-β elimination with the organo-metallic reagent, but a normal E2 reaction with the other bases.[125,128] It was further pointed out that conformational considerations favored the cis olefin in an α'-β elimination, but that an anti elimination should give trans olefin.[125]

While it seems likely that 18 and 19 react with organolithium compounds by the α'-β mechanism, it is interesting to note that the difference between these reactions and the ones in protic solvents is not a simple matter of difference in stereochemistry of eli-mination. Contrary to former belief, 18(n = 1) and 18(n = 3) both undergo Hofmann elimination to a marked extent with syn stereochemistry.[49] With 18(n = 4), trans-cyclooctene by 51% syn and 49% anti elimination under Hofmann conditions.[131] It should be emphasized that syn elimination need not occur by the α'-β mechanism, and that a high ratio of cis to trans cyclo-octent does not of itself indicate the α'-β mechanism. When 18(n = 4) is treated with t-butoxide, methoxide, or phenoxide in various solvents, a high proportion of trans-cyclooctene is obtained with t-butoxide, a high proportion of cis-cyclooctene is obtained with pheno-xide, and intermediate proportions are obtained with methoxide.[132] These reactions are probably mainly or entirely E2, but differ in the amount of syn elimina-tion. In fact, both analogy and theory (Section III. F) suggest that the weakest base, phenoxide, gives the highest proportion of "normal" anti E2 reaction.

5   Intramolecular Mechanisms

A final class of mechanisms should be mentioned, pyrolytic eliminations, although they are covered later in
Chapter VIII.   These are intramolecular, single-step
reactions that could in principle be confused with E1
reactions, although the reaction conditions are sufficiently different that there is usually no danger in
practice.   The E1 mechanism usually operates at relatively low temperatures in ionizing solvents and with
substrates that readily give carbonium ions.   Pyrolytic
eliminations, on the other hand, are usually carried
out in the absence of solvent, often in the gas phase.
Some substrates, such as alkyl halides, can follow
either the E1 or intramolecular mechanisms, depending
on reaction conditions.
    Pyrolytic eliminations are believed to occur via
cyclic transition states, as illustrated in Eqs. 60 to
63 for the major classes of compounds undergoing such
reactions, carboxylic esters,[133] xanthate esters,[134]
amine oxides[47] and alkyl halides.[135]   This provides
another way of distinguishing them from E1 reactions,
since the cyclic process will enforce a syn stereochemistry, in contrast to the less stereospecific E1
reaction (Section V.B).

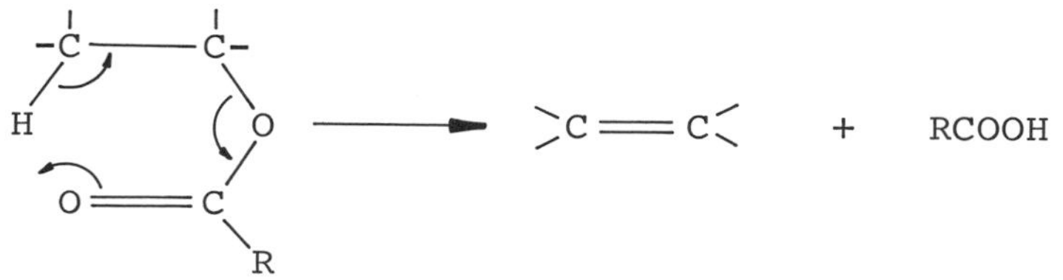

$$>C = C< \quad + \quad RCOOH \qquad (60)$$

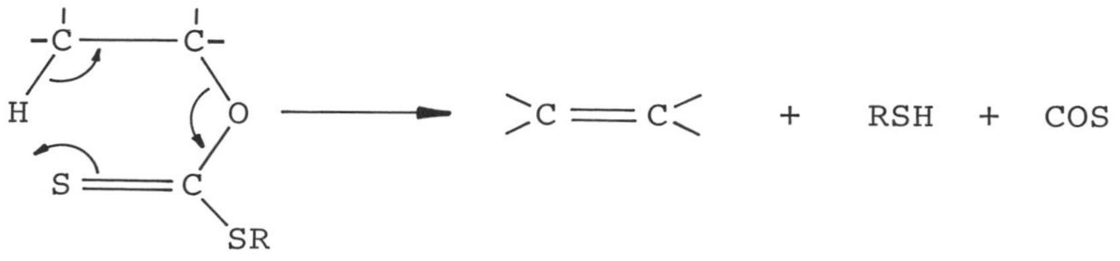

$$>C = C< \quad + \quad RSH \quad + \quad COS \quad (61)$$

$$>C = C< \quad + \quad (CH_3)_2NOH$$

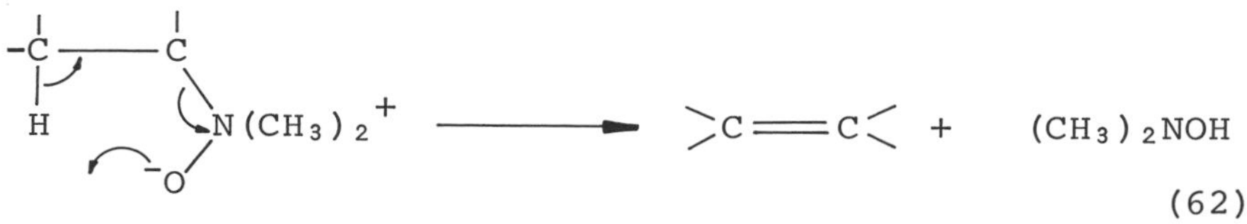

$$(62)$$

$$-\underset{\underset{H}{|}}{\overset{|}{C}}\text{---}\underset{\underset{X}{|}}{\overset{|}{C}}\text{-} \quad\longrightarrow\quad \underset{}{\overset{}{>}}C = C\underset{}{\overset{}{<}} \quad + \quad HX \qquad (63)$$

C   COMPETING REACTIONS

1   Substitution

The competition between the E2 reaction and unimolecular (S_N1 - E1) solvolysis is discussed elsewhere (Section V.C.1). Apart from solvolysis, the most common competitor to the E2 reaction is substitution by the S_N2 mechanism. So many factors can affect this competition that it is difficult to make any precise predictions. A number of trends are apparent, however, and it is not too difficult to decide whether a given reaction will yield sufficient olefin to be practicable for synthetic use of mechanistic study.

Olefin yields increase in the order primary < secondary < tertiary, the order being the result of both increasing rate of elimination and decreasing rate of substitution. For example, n-alkyl bromides with ethoxide ion in ethanol yield less than 10% olefin,[20] s-alkyl bromides yield around 60%,[24] and t-alkyl bromides yield more than 90%.[22] A 2-aryl group strongly accelerates elimination. Ethyl bromide gives only 1% olefin, and 2-arylethyl bromides around 99%.[20,136] A less marked increase results from branching at the β-position. i-Butyl bromide gives 60% olefin.[20] Similar trends are found with alkyldimethylsulfonium salts, except that β-alkyl groups lower rather than raise olefin proportions.[23-25] Pyrolysis of quaternary ammonium hydroxides can give satisfactory olefin yields even from primary reactants. n-Alkyltrimethylammonium hydroxides give 75 to 100% olefin.[137]

Sulfonium and ammonium salts can yield two types of substitution product as illustrated in Eq. 64. The proportions of these two products have seldom been determined. One would expect on steric grounds mainly substitution on methyl for simple alkyl derivatives, but comparable amounts of the two types have been observed with 1-arylethyldimethylsulfonium ions.[138]

Factors other than alkyl structure that influence olefin yield are summarized briefly. Increasing temperature[19] and decreasing solvent polarity[18] increase olefin yield. No systematic exploration of the role of the leaving group has been made, but bromides give considerably better yields of olefin than do

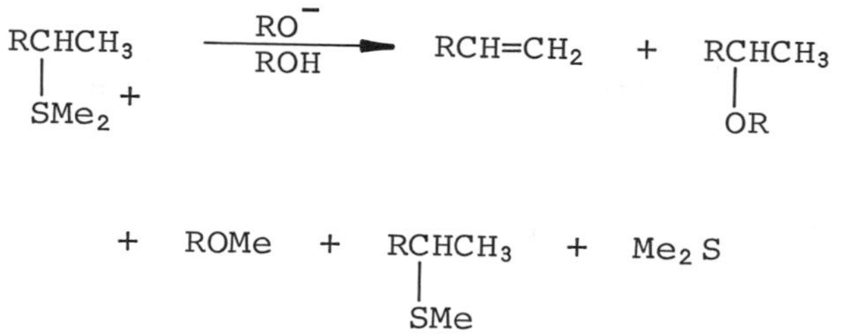

$$RCHCH_3 \xrightarrow[ROH]{RO^-} RCH=CH_2 \quad + \quad RCHCH_3$$
$$\underset{SMe_2^+}{|} \qquad\qquad\qquad\qquad\qquad \underset{OR}{|}$$

$$+ \quad ROMe \quad + \quad RCHCH_3 \quad + \quad Me_2S \qquad\qquad (64)$$
$$\underset{SMe}{|}$$

tosylates.[139] As noted above, quaternary ammonium
salts tend to give higher olefin yields than halides.
One can generally say that the stronger the base the
better the olefin yield, but good nucleophiles can give
high olefin yields with tertiary and, to some extent,
with secondary halides (the "E2C" elimination; see
Sections II.A.5, II.C.2.c, and IV.C.2).

2   Other Reactions

Competing base-promoted reactions other than substitution are relatively uncommon under the conditions used
for E2 reactions. Eliminations proceeding by pathways
other than the E2 mechanism offer more opportunities
for side reaction. The α-elimination mechanism
(Section I.B.3), for example, can lead to cyclopropanes
via an intramolecular insertion reaction of a carbene.
The E1cB mechanism (Section I.B.1) can, of course, be
accompanied by exchange, and could also be accompanied
by carbanion oxidation or rearrangement. There is,
however, no good evidence on the role of the last two
processes in elimination reactions.
     The α'-β elimination (Section I.B.4) occurs via
an ylid intermediate, and might be expected to be
accompanied by ylid rearrangements. A few examples of
these rearrangements have been found. For example,
1-arylethyldimethylsulfonium ions and base give o-
ethylbenzyl methyl sulfides, presumably via a Sommelet
rearrangement (Eq. 65).[138] Other ylid rearrangements,
such as the Stevens rearrangement (Eq. 66),[140,141]

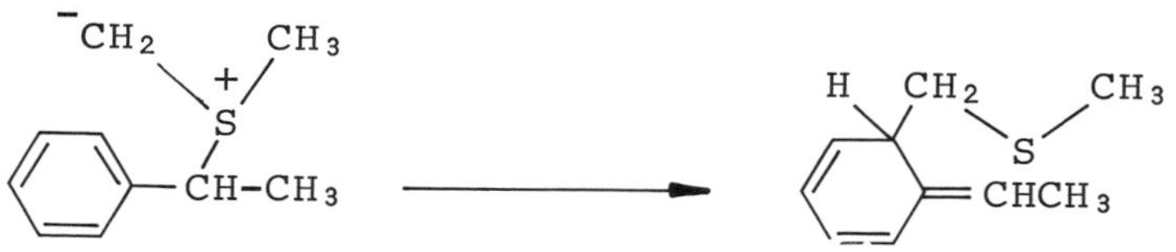

$$(65)$$

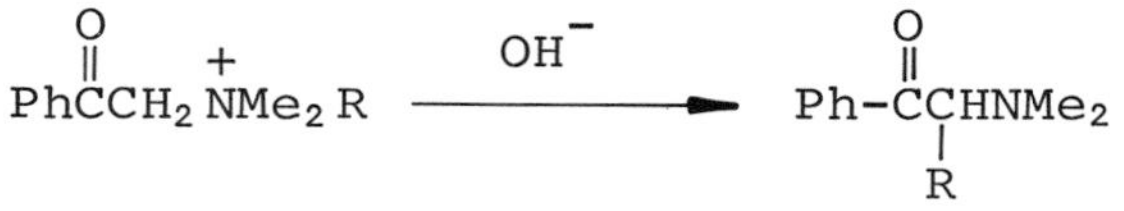

$$\underset{\substack{\| \\ PhCCH_2NMe_2R}}{O} + \quad OH^- \quad \longrightarrow \quad \underset{\substack{\| \\ Ph-CCHNMe_2 \\ | \\ R}}{O} \tag{66}$$

and electrocyclic rearrangements[142] have been observed
with ammonium and sulfonium salts but seldom with those
that are also capable of undergoing 1,2-eliminations
(Section I.B.4).[128]

REFERENCES

1. A. W. Hofmann, Ann. Chem., **78**, 253 (1851); **79**, 11 (1851).
2. W. Hanhart and C. K. Ingold, J. Chem. Soc., 997 (1927).
3. E. D. Hughes and C. K. Ingold, J. Chem. Soc., 523 (1933).
4. C. K. Ingold and C. C. N. Vass, J. Chem. Soc., 3125 (1928).
5. E. D. Hughes, C. K. Ingold, and C. S. Patel, J. Chem. Soc., 526 (1933).
6. G. W. Fenton and C. K. Ingold, J. Chem. Soc., 3127 (1928).
7. C. K. Ingold, J. A. Jessop, K. I. Kuriyan, and A. M. M. Mandour, J. Chem. Soc., 533 (1933).
8. L. Hey and C. K. Ingold, J. Chem. Soc., 531 (1933).
9. K. F. Kumli, W. E. McEwen, and C. A. VanderWerf, J. Am. Chem. Soc., **81**, 3805 (1959); M. Zanger, C. A. VanderWerf, and W. E. McEwen, J. Am. Chem. Soc., **81**, 3806 (1959); and subsequent papers.
10. E. D. Hughes, J. Am. Chem. Soc., **57**, 708 (1935).
11. E. D. Hughes, C. K. Ingold, and A. D. Scott, J. Chem. Soc., 1271 (1937).
12. E. D. Hughes, C. K. Ingold, and U. G. Shapiro, J. Chem. Soc., 1277 (1937).
13. K. A. Cooper, E. D. Hughes, and C. K. Ingold, J. Chem. Soc., 1280 (1937).
14. E. D. Hughes and B. J. McNulty, J. Chem. Soc., 1283 (1937).
15. E. D. Hughes, C. K. Ingold, S. Masterman, and B. J. McNulty, J. Chem. Soc., 899 (1940).
16. E. D. Hughes and C. K. Ingold, Trans. Far. Soc., **37**, 657 (1941).
17. K. A. Cooper, E. D. Hughes, C. K. Ingold, and B. J. McNulty, J. Chem. Soc., 2038 (1948).
18. K. A. Cooper, M. L. Dhar, E. D. Hughes, C. K. Ingold, B. J. McNulty, and L. I. Woolf,

J. Chem. Soc., 2043 (1948).

19. K. A. Cooper, E. D. Hughes, C. K. Ingold, G. A. Maw, and B. J. McNulty, J. Chem. Soc., 2049 (1948).

20. M. L. Dhar, E. D. Hughes, C. K. Ingold, and S. Masterman, J. Chem. Soc., 2055 (1948).

21. M. L. Dhar, E. D. Hughes, and C. K. Ingold, J. Chem. Soc., 2058 (1948).

22. M. L. Dhar, E. D. Hughes, and C. K. Ingold, J. Chem. Soc., 2065 (1948).

23. E. D. Hughes, C. K. Ingold, and G. A. Maw, J. Chem. Soc., 2072 (1948).

24. E. D. Hughes, C. K. Ingold, G. A. Maw, and L. I. Woolf, J. Chem. Soc., 2077 (1948).

25. E. D. Hughes, C. K. Ingold, and L. I. Woolf, J. Chem. Soc., 2084 (1948).

26. E. D. Hughes, C. K. Ingold, and A. M. M. Mandour, J. Chem. Soc., 2090 (1948).

27. M. L. Dhar, E. D. Hughes, C. K. Ingold, A. M. M. Mandour, G. A. Maw and L. I. Woolf, J. Chem. Soc., 2093 (1948).

28. C. K. Ingold, "Structure and Mechanism in Organic Chemistry," Cornell University Press, Ithaca, Chap. 8, First ed., 1953; second ed., 1969.

29. I. Dostrovsky and E. D. Hughes, J. Chem. Soc., 166 (1946).

30. W. H. Saunders, J. and R. H. Paine, J. Am. Chem. Soc., $\underline{83}$, 882 (1961).

31. N. B. Chapman and J. L. Levy, J. Chem. Soc., 1673 (1952).

32. W. H. Saunders, Jr., S. R. Fahrenholtz, E. A. Caress, J. P. Lowe, and M. Schreiber, J. Am. Chem. Soc., $\underline{87}$, 3401 (1965).

33. R. A. Bartsch and J. F. Bunnett, J. Am. Chem. Soc., $\underline{90}$, 408 (1968).

34. D. J. Cram, J. Am. Chem. Soc., $\underline{74}$, 2149 (1952).

35. C. H. DePuy and C. A. Bishop, J. Am. Chem. Soc., $\underline{82}$, 2532 (1960).

36. I. N. Feit and W. H. Saunders, Jr., J. Am. Chem. Soc., $\underline{92}$, 1630 (1970).

37. D. Y. Curtin and D. B. Kellom, J. Am. Chem. Soc., $\underline{75}$, 6011 (1953).

38. L. R. Fedor, J. Am. Chem. Soc., $\underline{89}$, 4479 (1967).

39. L. R. Fedor, J. Am. Chem. Soc., $\underline{91}$, 908 (1969).

40. D. H. Hunter and D. J. Cram, J. Am. Chem. Soc., $\underline{88}$, 5765 (1966).

41. J. Crosby and C. J. M. Stirling, J. Chem. Soc., $\underline{B}$, 671 (1970).

42. R. A. Bartsch and J. F. Bunnett, J. Am. Chem. Soc., $\underline{91}$, 1376 (1969).

43. R. A. More O'Ferrall and S. Slae, J. Chem. Soc., $\underline{B}$, 260 (1970).

44. A. Streitwieser, Jr., J. Org. Chem., 22, 861 (1957).
45. P. A. Skell and I. Starer, J. Am. Chem. Soc., 81, 4117 (1959).
46. E. F. Caldin and G. Long, J. Chem. Soc., 3737 (1954).
47. A. C. Cope and E. R. Trumbull in "Organic Reactions," Vol. 11, Wiley, New York, 1960, Chap. 5.
48. J. Weinstock and V. Boekelheide, J. Am. Chem. Soc., 75, 2546 (1953).
49. M. P. Cooke, Jr., and J. L. Coke, J. Am. Chem. Soc., 90, 5556 (1968).
50. W. H. Saunders, Jr. and T. A. Ashe, J. Am. Chem. Soc., 91, 4473 (1969).
51. H. C. Brown, I. Moritani, and Y. Okamoto, J. Am. Chem. Soc., 78, 2193 (1956).
52. H. C. Brown and I. Moritani, J. Am. Chem. Soc., 76, 455 (1954).
53. A. J. Parker, Q. Rev., 16, 163 (1962).
54. D. J. Cram., B. Rickborn, C. A. Kingsbury, and P. Haberfield, J. Am. Chem. Soc., 83, 3678 (1961).
55. T. J. Wallace, J. E. Hofmann, and A. Schriesheim, J. Am. Chem. Soc., 85, 2739 (1963).
56. J. E. Hofmann, T. J. Wallace, and A. Schriesheim, J. Am. Chem. Soc., 86, 1561 (1964).
57. A. F. Cockerill, S. R. Rottschaefer, and W. H. Saunders, Jr., J. Am. Chem. Soc., 89, 901 (1967).
58. D. H. Froemsdorf and D. M. Robbins, J. Am. Chem. Soc., 89, 1737 (1967).
59. R. F. Hudson and G. Klopman, J. Chem. Soc., 5 (1964).
60. B. D. England and D. J. McLennan, J. Chem. Soc., B, 696 (1966).
61. P. B. D. de la Mare and C. A. Vernon, J. Chem. Soc., 41 (1956).
62. S. Winstein, P. E. Klinedinst, Jr., and E. Clippinger, J. Am. Chem. Soc., 83, 4986 (1961), and preceding papers.
63. M. Cocivera and S. Winstein, J. Am. Chem. Soc., 85, 702 (1963).
64. F. G. Bordwell, A. C. Knipe, and K. C. Yee, J. Am. Chem. Soc., 92, 5945 (1970).
65. R. S. Breslow, Tetrahedron Lett., 399 (1964).
66. J. Hine, R. Wiesboeck, and O. B. Ramsay, J. Am. Chem. Soc., 83, 1222 (1961).
67. P. S. Skell and C. R. Hauser, J. Am. Chem. Soc., 67, 1661 (1945).
68. D. G. Hill, B. Stewart, S. W. Kantor, W. A. Judge, and C. R. Hauser, J. Am. Chem. Soc., 76, 5129 (1954).

69. W. H. Saunders, J. and D. H. Edison, J. Am. Chem. Soc., $\underline{82}$, 138 (1960).

70. D. V. Banthorpe and J. H. Ridd, Proc. Chem. Soc., 365 (1964).

71. W. H. Saunders, Jr. and M. R. Schreiber, Chem. Commun., 145 (1966).

72. L. C. Leitch and H. J. Bernstein, Can. J. Res., $\underline{28B}$, 35 (1950).

73. S. J. Cristol and D. D. Fix, J. Am. Chem. Soc., $\underline{75}$, 2647 (1953).

74. W. von E. Doering and A. K. Hoffman, J. Am. Chem. Soc., $\underline{77}$, 521 (1955).

75. S. Asperger, N. Ilakovac, and D. Pavlović, J. Am. Chem. Soc., $\underline{83}$, 5032 (1961); S. Asperger, L. Klasinc, and D. Pavlović, Croat. Chim. Acta, $\underline{36}$, 159 (1964).

76. J. Hine, R. Wiesboeck and R. G. Ghirardelli, J. Am. Chem. Soc., $\underline{83}$, 1219 (1961).

77. W. T. Miller, Jr., E. W. Fager, and P. H. Griswald, J. Am. Chem. Soc., $\underline{70}$, 431 (1948).

78. A. Streitwieser, Jr., A. P. Marchand, and A. H. Pudjaatmaka, J. Am. Chem. Soc., $\underline{89}$, 693 (1967).

79. R. P. Bell, "The Proton in Chemistry," Cornell University Press, Ithaca, 1959, p. 134.

80. A. A. Frost and R. G. Pearson, "Kinetics and Mechanism," 2nd Ed., Wiley, New York, 1961, p. 213.

81. J. Hine and L. A. Kaplan, J. Am. Chem. Soc., $\underline{82}$, 2915 (1960).

82. S. Patai, S. Weinstein, and Z. Rappoport, J. Chem. Soc., 1741 (1962).

83. E. M. Hodnett and J. Flynn, J. Am. Chem. Soc., $\underline{79}$, 2300 (1957).

84. T. I. Crowell, R. Kemp, R. E. Lutz, and A. A. Wall, J. Am. Chem. Soc., $\underline{90}$, 4638 (1968).

85. D. J. Cram, "Fundamentals of Carbanion Chemistry," Academic Press, New York, 1965, Chap. III, p. 28.

86. L. J. Steffa and E. R. Thornton, J. Am. Chem. Soc., $\underline{89}$, 6149 (1967).

87. R. A. More O'Ferrall, J. Chem. Soc., $\underline{B}$, 268 (1970).

88. D. J. McLennan, Q. Rev. (London), $\underline{21}$, 490 (1967).

89. F. H. Westheimer, Chem. Rev., $\underline{61}$, 265 (1961); L. Melander, "Isotope Effects on Reaction Rates," Ronald Press, New York, 1960, pp. 24-32.

90. G. S. Hammond, J. Am. Chem. Soc., $\underline{77}$, 334 (1955).

91. A. N. Bourns and A. Frost, Private Communication.

92. G. Ayrey, E. Buncel, and A. N. Bourns, Proc. Chem. Soc., 458 (1961).

93. J. Weinstock, R. G. Pearson, and F. G. Bordwell,

J. Am. Chem. Soc., <u>78</u>, 3468, 3473 (1956).

94.   J. Weinstock, J. L. Bernardi, and R. G. Pearson, J. Am. Chem. Soc., <u>80</u>, 4961 (1958).

95.   J. Hine and O. B. Ramsay, J. Am. Chem. Soc., <u>84</u>, 973 (1962).

96.   J. Hine, R. D. Weimar, Jr., P. B. Langford, and O. B. Ramsay, J. Am. Chem. Soc., <u>88</u>, 5522 (1966).

97.   C. H. DePuy and C. A. Bishop, J. Am. Chem. Soc., <u>82</u>, 2535 (1960).

98.   C. H. DePuy and D. H. Froemsdorf, J. Am. Chem. Soc., <u>79</u>, 3710 (1957).

99.   H. M. R. Hoffmann, J. Chem. Soc., 6753, 6762 (1965).

100. G. M. Fraser and H. M. R. Hoffmann, J. Chem. Soc., <u>B</u>, 265 (1967).

101. A. K. Colter and R. D. Johnson, J. Am. Chem. Soc., <u>84</u>, 3289 (1962).

102. A. V. Willi, Helv. Chim. Acta, <u>49</u>, 1735 (1966).

103. F. G. Bordwell, J. Weinstock, and T. F. Sullivan, J. Am. Chem. Soc., <u>93</u>, 4728 (1971).

104. K. Bowden, Chem. Rev., <u>66</u>, 119 (1966).

105. C. H. Rochester, Q. Rev. (London), <u>20</u>, 511 (1966).

106. M. Anbar, M. Bobtelsky, D. Samuel, B. Silver, and G. Yagil, J. Am. Chem. Soc., <u>85</u>, 2380 (1963).

107. A. F. Cockerill, J. Chem. Soc., B, 964 (1967).

108. R. A. More O'Ferrall and J. H. Ridd, J. Chem. Soc., 5035 (1963).

109. J. Hine, J. Am. Chem. Soc., <u>72</u>, 2438 (1950).

110. J. Hine, R. C. Peek, Jr., and B. D. Oakes, J. Am. Chem. Soc., <u>76</u>, 827 (1954).

111. J. Hine, A. D. Ketley, and K. Tanabe, J. Am. Chem. Soc., <u>82</u>, 1398 (1960).

112. F. G. Bordwell and M. M. Vestling, J. Am. Chem. Soc., <u>89</u>, 3906 (1967).

113. F. G. Bordwell and K. C. Yee, J. Am. Chem. Soc., <u>92</u>, 5933 (1970).

113b. D. H. Hunter and D. J. Shearing, J. Am. Chem. Soc., <u>93</u>, 2348 (1971).

114. J. Crosby and C. J. M. Stirling, J. Chem. Soc., B, 679 (1970).

115. R. W. Taft, Jr., N. C. Deno, and P. S. Skell, Ann. Rev. Phys. Chem., <u>9</u>, 292 (1958).

116. R. A. Sneen and H. M. Robbins, J. Am. Chem. Soc., <u>91</u>, 3100 (1969).

116b. D. J. McLennan, Tetrahedron Lett., 2317 (1971).

117. J. Hine, "Physical Organic Chemistry," 2nd ed., McGraw-Hill, New York, 1962, Chap. 8.

118. L. Friedman and J. G. Berger, J. Am. Chem. Soc., <u>83</u>, 492 (1961).

119. D. G. Hill, W. A. Judge, P. S. Skell, S. W. Kantor, and C. R. Hauser, J. Am. Chem. Soc., 74, 5599 (1952).

120. V. Franzen, H. J. Schmidt, and C. Mertz, Chem. Ber., 94, 2942 (1961).

121. G. Wittig and R. Polster, Ann., 599, 13 (1956).

122. A. C. Cope, N. A. LeBel, P. T. Moore, and W. R. Moore, J. Am. Chem. Soc., 83, 3861 (1961).

123. V. J. Shiner, Jr. and M. L. Smith, J. Am. Chem. Soc., 80, 4095 (1958).

124. W. H. Saunders, Jr. and D. Pavlović, Chem. Ind. (London), 180 (1962).

125. V. Franzen and H. J. Schmidt, Chem. Ber., 94, 2937 (1961).

126. A. C. Cope and A. S. Mehta, J. Am. Chem. Soc., 85, 1949 (1963).

127. J. L. Coke and M. P. Cooke, Jr., J. Am. Chem. Soc., 89, 6701 (1967).

128. G. Wittig and T. F. Burger, Ann., 632, 85 (1960).

129. G. Wittig and R. Polster, Ann., 612, 102 (1958).

130. J. Rabiant and G. Wittig, Bull. Soc. Chim. Fr., 798 (1957).

131. J. L. Coke and M. C. Mourning, J. Am. Chem. Soc., 90, 5561 (1968).

132. J. Sicher and J. Závada, Coll. Czech. Chem. Commun., 33, 1278 (1968).

133. C. H. DePuy and R. W. King, Chem. Rev., 60, 431 (1960).

134. H. R. Nace, in "Organic Reactions," Vol. 12, Wiley, New York, 1962, Chap. 2.

135. A. Maccoll, in S. Patai, Ed., "The Chemistry of Alkenes,: Wiley, New York, 1964, Chap. 3.

136. W. H. Saunders, Jr. and R. A. Williams, J. Am. Chem. Soc., 79, 3712 (1957).

137. P. A. S. Smith and S. Frank, J. Am. Chem. Soc., 74, 509 (1952).

138. F. L. Roe, Jr., Ph.D. thesis, University of Rochester, 1971.

139. P. Veeravagu, R. T. Arnold, and E. W. Eigenmann, J. Am. Chem. Soc., 86, 3072 (1964).

140. T. S. Stevens, J. Chem. Soc., 2107 (1930).

141. H. Hellerman and D. Eberle, Ann., 662, 188 (1963).

142. J. E. Baldwin, R. E. Hackler and D. P. Kelley, Chem. Commun., 538 (1968).

143. Z. Rappoport and E. Shohamy, J. Chem. Soc., B, 2060 (1971).

144. J. K. Borchardt, R. Hargreaves, and W. H. Saunders, Jr., Tetrahedron Lett., 2307 (1972).

II RATE EFFECTS IN E2 REACTIONS -- THE VARIABLE

TRANSITION STATE

A   INTERPRETATION OF EFFECTS OF STRUCTURE AND
CONDITIONS ON RATE

1   General

Before any reliable predictions of the effects of reactant structure and reaction conditions on rate can be
made, it is necessary to understand the factors affecting rates. The achievement of this understanding is
particularly difficult in a complex process such as bimolecular elimination, where effects on rate can differ
widely between reactions which appear to proceed by the
same gross mechanism. One need only consider orientation effects to appreciate the magnitude of the problem (Section IV.A.1-5). Where there is a choice between elimination into two different branches of the
same molecule, some E2 reactions show a preference for
the less-substituted olefin (Hofmann rule), while others show a preference for the more-substituted olefin
(Saytzev rule). A mechanistic theory capable of
accounting for such differences must obviously possess
a considerable degree of complexity and flexibility.
    In physical organic chemistry, one is usually con-
cerned not with the prediction of absolute rates but
with the considerably easier task of predicting changes
in rate when some alteration is made in reactant structure or in reaction conditions. Even so, one must
still be able to predict effects on the energies of
both reactant and transition state in order to predict
activation energy and hence rate. Where one is comparing two different paths of the same reactant (as in
orientation studies), the problem reduces to that of
predicting only transition-state energies. In other
cases it is often assumed that only changes in transition-state energies are of importance, but this assumption may not be justified, particularly if there are
differences in solvation or in internal steric strains
between the two reactants. On the other hand, the
looser structure of the transition state, with its partially formed and partially broken bonds, is likely to
respond more to perturbations than is the more tightly
bonded reactant molecule. For this reason, transition-
state effects will often be controlling when changes in
reactant structure are not great.

2   The Variable Transition State Theory

The unifying concept in our present theoretical picture
of elimination reactions is the idea that the balance
and timing of the bond-making and bond-breaking pro-

cesses involved may vary with reactant structure and reaction conditions even though the mechanism remains a single-stage E2 process with no detectable intermediates. The reaction involves breaking a carbon-leaving group bond, forming a carbon-carbon multiple bond, breaking a carbon-hydrogen bond, and forming a base-hydrogen bond. According to the variable transition state theory, one or more of these processes may be farther advanced than the others in the transition state. The less advanced process(es) must, of course, "catch up" on the subsequent downhill path to products if the mechanism is a single-stage process, but what happens after the transition state has no effect on transition-state structure or energy. The result of the theory in general terms is that transition states for different E2 reactions may have quite different electron distributions and atom positions, and consequently may respond differently to electronic and steric effects.

Assigning credit for such a simple but general idea as the variable transition state theory is not easy, for many persons have contributed in different ways to its development and application. The basic idea was suggested many years ago by Hanhart and Ingold,[1] who pointed out that the bond changes occurring in an elimination reaction need not be precisely simultaneous. They did not, however, explore the consequences of the idea, nor did anyone else for a long time. Letsinger, Schnizer, and Bobko[2] clearly saw the relationship between the orientation rules and the structure of the transition state in their discussion of the cleavage of 2-methoxyoctane by organometallic reagents. The first really detailed statement of the theory was by Cram, Greene, and DePuy,[3] who explained variations in the relative rates of elimination from the threo and erythro isomers of 1,2-diphenyl-1-propyl derivatives as resulting from variations in eclipsing effects in the transition state as the nature of the leaving group, the base, and the solvent were changed. Evidence that the postulated changes in electron distribution and bond strength in the transition state really were occurring came from groups led by Saunders[4-6] and DePuy,[7] who used substituent effects[5,7] hydrogen isotope effects,[6] and sulfur isotope effects[4] to demonstrate and measure the changes. The early development of the theory culminated in Bunnett's 1962 review.[8] Although the basic theory and many of its applications had already been discussed by other workers, Bunnett was the first to follow up systematically a wide range of implications of the theory and

to predict the influence on transition-state structure
and rate of numerous changes in reactant structure and
reaction conditions.

3   The Variation from E1-Like to E1cB-Like Transition State

We discuss a number of different, though related, for-
mulations of the variable transition state theory in
the next few pages, starting with the original, and
still most common, version of it. In this approach, it
is assumed that the major changes in transition state
structure result from changes in the relative extents
of cleavage of the carbon-hydrogen and carbon-leaving
group bonds. One can envision a "spectrum" of transition states, ranging from the "E1-like" (1), in which
only the bond to the leaving group is appreciably
stretched, through the synchronous (2), in which the
carbon-hydrogen and carbon-leaving group bonds are
stretched to about the same extent, and finally to the
"E1cB-like" (3), in which only the carbon-hydrogen bond
is appreciably stretched. It should be emphasized that
1 and 3 do not represent the E1 and E1cB mechanisms,
which are two-stage processes, but rather concerted
mechanisms whose transition states bear some

$$
\begin{array}{ccc}
\underset{\overset{|}{H}}{\overset{}{>}C} \!-\!\!-\!\! C\!< \quad \overset{X}{\underset{}{\vdots}} & \qquad >C\!=\!=\!=\!=\!=\! C\!< \quad \overset{X}{\underset{}{\vdots}} & \qquad >C\!-\!\!-\!\! C\!< \quad \overset{X}{\underset{}{\vdots}} \\
\end{array}
$$

B (1)   B (2)   B (3)

resemblance to that of the E1 and that of the E1cB
mechanism, respectively. In actual fact there must be
some alteration of the carbon-hydrogen bond in 1 and of
the carbon-leaving group bond in 3, for otherwise these
transition states would possess no energetic advantages
over the two-stage processes whose rate-determining
transition states they resemble. A somewhat different
nomenclature has been advocated by Bunnett,[9] who calls

1 the paenecarbonium extreme, 2 the central transition
state, and 3 the paenecarbanion extreme. The prefix
comes from the Latin word for almost. While this nomenclature does have the advantage of providing a precise description in one word of each of the transition
states, it has the disadvantage of being somewhat puzzling on first encounter to the reader who possesses a
chemical, but lacks a classical, education. Consequently, we use the more immediately obvious terms
"E1-like" and "E1cB-like" in our subsequent discussion.
    Although we do not at this point discuss in any
detail the conclusions to be drawn from this version of
the theory, it is evident that it can accommodate a
wide variety of structural and environmental effects.
Transition state 2, for example, possesses double-bond
character and should be stabilized by the same factors
that stabilize the final product. On the other hand,
1 possesses carbonium-ion character at the α-carbon
and should be stabilized by factors that stabilize carbonium ions. Similarly, 3 possesses carbanion character at the β-carbon and should be stabilized by electron-withdrawing and should be destabilized by electron-repelling substituents. In addition, 2 is expected to resemble the products geometrically,[3] while 1
and 3 are not.

4   The Variation from Reactantlike to Productlike Transition State

Another possible variation in transition-state structure is that the extents of carbon-hydrogen and carbon-
leaving group bond rupture can vary synchronously,
thereby giving different degrees of double-bond character even for transition states of the central type.
For example, if each bond is 25% ruptured the double
bond can be only 25% formed, while if each is 75%
ruptured the double bond can be 75% formed. For un-
equal degrees of bond rupture, the extent of formation
of the double bond will essentially be governed by the
extent of rupture of the least broken of the two. This
possibility was recognized by Bunnett[8] in his first
review, but he regarded it as a factor of secondary
importance, and continued to hold to this position in
his more recent review.[9] On the other hand, there is
no experimental evidence demonstrating that such variation does not occur or is of little practical consequence, so it should be included in any complete theory
of the E2 mechanism.
    It is, in fact, rather easy to combine the views
of the transition state presented in this and the

preceding sections. One can draw a two-dimensional
diagram like that shown in Figure 1, where the extent
of rupture of the carbon-hydrogen bond is represented

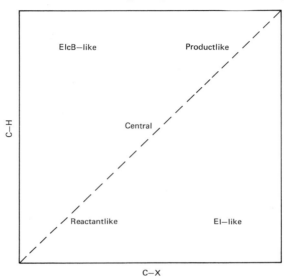

Figure 1 Schematic representation of the structure of
the E2 transition state.

along one axis and the extent of rupture of the carbon-
leaving group bond along the other. This type of diagram has been used by Závada, Krupička, and Sicher[10]
and by More O'Ferrall.[11] It incorporates all possible
variations in transition-state structure, apart from
stereochemical variations (anti vs. syn elimination)
and variations in the position of the base with respect
to the substrate molecule (linear vs. nonlinear B-H-C).
Even in these latter two cases, it can still describe
the electron distribution and geometry about the α-
and β-carbons. It can also describe the first steps
of E1 and ElcB reactions by means of points on the
abscissa and ordinate, respectively.
     We should not conclude that E2 reactions will
necessarily utilize the entire variety of transition
states theoretically available in Figure 1, any more
than we would conclude that all areas depicted on a
map were necessarily populated. Some types of transition states will be more common than others, and some
may not be found at all. We see later, for example,
that good evidence for an E1-like transition state is
lacking. We attempt in the course of this chapter to
summarize present knowledge about which types of

transition states are utilized in E2 reactions, as well as when and why they are utilized.

5   E2H and E2C Transition States

Although it is generally assumed that strong bases should be more effective than weak bases in promoting elimination reactions, there are important exceptions to this generalization.   It was noted some time ago that bases normally regarded as weak, such as halide ions in dipolar aprotic solvents[12] or thiolate ions in alcoholic solvents,[13] were unexpectedly effective in bringing about eliminations from certain substrates. Because these bases had in common a high degree of effectiveness as nucleophiles, it was suggested that the S_N2 and E2 paths might proceed from a common intermediate in which the nucleophile was bound both to the α-carbon and β-hydrogen atoms.[12]   This "merged" mechanism was later abandoned for a number of reasons, mainly because the stereochemistry of elimination remained strictly anti even when the base and the leaving group were identical and where a symmetrical intermediate would be expected.[14]   The idea that the elimination might involve interaction of the base with the α-carbon atom was, however, retained and developed further.

The picture favored by Winstein and Parker and their co-workers is that strong bases attack mainly at the β-hydrogen atom to give an E2H transition state (4), while good nucleophiles attack mainly at the α-carbon atom to give an E2C transition state (6).[15] The actual transition state in most elimination reactions is viewed as falling somewhere between these two extremes, and is represented by a structure such as 5.

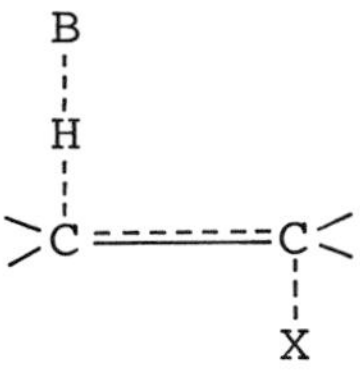

| 4 | 5 | 6 |

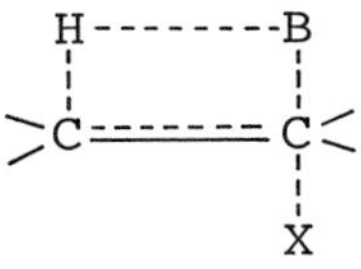

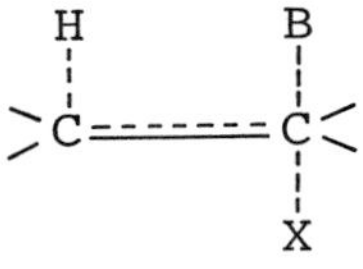

The characteristics of elimination reactions promoted by weak bases but good nucleophiles have been explored in detail in a series of papers that we discuss later (Section II.C.2.c).   For the present, we only consider

the theoretical implications of the E2H-E2C spectrum.
The main support for the proposed interaction of
the base with the α-carbon atom in the transition state
is the fact that the reactivity order for different
bases correlates well with the order of their reactivity in nucleophilic displacement, but not at all with
their equilibrium basicity.[15] Eck and Bunnett have
argued that the interaction cannot be significant on
the grounds that 7 reacts somewhat faster than 8 with
halide ions in acetone or dioxane, whereas steric

hindrance with the neopentyl-like α-carbon atom of 7
should dramatically reduce the rate of nucleophilic
attack at that atom.[16] Cook and Parker replied that
the neopentyl effect would be expected to be small for
the loose transition state involved in a nucleophilic
attack at a tertiary carbon atom, and that the evidence
of Bunnett and Eck consequently did not disprove the
E2C mechanism.[17] The present situation, then, is that
any nucleophilic interaction of the base with the α-
carbon atom in the E2C transition state must be rather
loose, but there is no really decisive evidence either
for or against such interaction. The correlation of
reactivity with nucleophilicity[15] is certainly suggestive, but it only proves that similar properties of the
bases are called into play in both cases, not that
attack at the same atom of the substrate is necessarily
involved in both.
    If the interaction of the base with the α-carbon
atom is negligible or nonexistent, the E2C mechanism
constitutes an unnecessary elaboration of the variable
transition state theory. Even if it is significant,
the method of depicting the transition state shown in
Figure 1 is still applicable in many ways. For example, a discussion of structural changes within the
substrate requires only a consideration of the geometry and electron distribution of that part of the
transition state, and the position of the base need not
be specified. This is not to say that a knowledge of
the position of the base is in all respects unimportant
in a discussion of mechanism. There are many cases
where it is important, and it is to be hoped that more

definitive evidence will become available in the future.

B   APPROACHES TO THE PREDICTION OF TRANSITION-STATE STRUCTURE

1   The Hammond Postulate

In 1955, Hammond discussed the application of some ideas from transition-state theory to the prediction of the structure of the transition state in organic reactions.[18] The basic idea of the theory is that two states which occur consecutively along a reaction path will differ little in molecular structure if they differ little in energy. The principle behind this statement is simply that the energy is a function of the positions of the nuclei and electrons of a species, and that small changes in energy are likely to correspond to small changes in positions, and vice versa. The predictions of the postulate for single-stage reactions, or for single steps of more complex reactions, run as follows. A highly endothermic reaction is expected to have a transition state that resembles products, because it will be much closer to products than to reactants in energy. A highly exothermic reaction is similarly expected to have a transition state that resembles reactants, at least, on the condition that it have a relatively low activation energy also. More nearly thermoneutral reactions are expected to have transition states that are between reactants and products in structure, and that resemble neither closely.

The application to simple processes is straight-forward. Thus an ionization of an alkyl halide to produce a high-energy carbonium ion is expected to have a transition state resembling the carbonium ion. In a bimolecular reaction of a reagent with a substrate, the more reactive the reagent the lower the activation energy and, consequently, the more reactant-like the transition state. When a base abstracts a proton from a substrate, for example, one predicts that the extent of transfer of the proton to the base in the transition state will be less for strong than for weak bases.

For a complex process such as a bimolecular elimination, where several bonds are being ruptured and several are being formed in a single step, the application of Hammond postulate reasoning is much less straightforward. Since most E2 reactions require a substantial activation energy, one can say in a

general sense that the transition state will not closely resemble reactants. On the other hand, this does
not exclude the possibility that some of the reacting
bonds will have changed very little between reactant
and transition state. For an E1cB-like transition
state, for instance, the proton transfer will be well
advanced, but the bond to the leaving group will be
little weakened. Bunnett suggests that the adjustment
between the degrees of C-H and C-X rupture in the
transition state will be such that there will be more
extensive rupture of the more easily broken bond.[9]
There is a problem, however, in deciding which is
really the more easily broken bond. This will depend
on local conditions such as strength of the attacking
base and possibilities for resonance stabilization of
developing charges or unsaturation, and cannot be deduced simply from bond energies. Under these circumstances, a priori predictions become very difficult.
The best one can do is to see whether in practice certain bond changes follow the predictions of the postulate (the extent of proton transfer as a function of
base strength seems to be in this category), but to
keep in mind that other bond changes may not.

2   The Swain-Thornton Rules

A theory first advanced by Swain and Thornton,[19] and
later revised by Thornton,[20] considers the effect of
electron donation or electron withdrawal at various
atoms of the transition state. A detailed presenta-
tion of the underlying theory would be too space-con-
suming and, in any event, numerous approximations and
more or less subjective judgments are required to make
the theory useful for qualitative predictions. For
these reasons, we treat the rules as extensions of the
Hammond postulate, and refer those interested in more
detail to the original papers.[19,20]
    An increase in base strength of a base B would be
expected to strengthen and shorten the B-H bond in the
stable conjugate acid of B. Note that the effect is
opposite for the proton-transfer transition state: increasing the base strength so as to strengthen the B-H
bond lengthens it. Thornton argues that, where effects on progress along the reaction coordinate are
under discussion, a perturbation producing a change in
bond strength will generally affect the length of a
reacting bond in a sense opposite to that expected for
stable molecules.[20] Reacting bonds other than the one
which is the site of the change will then follow along
so that overall progress along the reaction coordinate

is increased or decreased, depending on whether a
breaking or a forming bond is strengthened (and vice
versa when the bond is weakened). For example, an increase in base strength in the E2 transition state (9)
is predicted to lengthen the B-H and C-C, and shorten

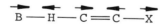

$$\underline{9}$$

the C-H and C-X bonds. The net effect is a more reactantlike transition state, just as in the simple three-
center proton transfer.

It is not always a straightforward matter to decide what effect a given change will have upon a bond.
Thornton suggests that a polar bond should be strengthened by an increase and weakened by a decrease in polarity, based on Pauling's arguments that ionic resonance forms increase bond strength.[69] Thus an electron-repelling substituent in B increases the polarity
of the B-H bond and strengthens it. Thornton proposes
that the basicity of B (or whatever other atom or group
is under consideration) be taken as a measure of its
electron-repelling character, while Swain, Kuhn, and
Schowen[21] suggest that electron-repelling character
should vary inversely with ability as a leaving group.
There is no good way of deciding which, if either, is
correct.

Thornton also makes predictions of the effects of
perturbations at other positions in the reacting system. He concludes that increasing the basicity of the
leaving group X should strengthen, and thereby length-
en, the C-X bond in the transition state. This conclusion assumes that X will always be the more electron-rich end of the C-X bond, but this may not be
true for positively-charged leaving groups. Electron
release at the α-carbon should, by Thornton's reasoning, shorten the C-X bond, but this prediction also
assumes invariance in the direction of polarity of the
C-X bond. The β-carbon, on the other hand, should always be more electron-rich than the β-hydrogen, so
electron repelling groups at the β-carbon should
strengthen, and lengthen, the C-H bond.

It should be kept in mind that Thornton's predictions tacitly assume a definite relationship between
bond lengths, bond strengths, and force constants in
the transition state. Such relationships are often

observed in stable molecules, but are by no means required by basic theory. The larger the structural
change, the more likelihood the relationship will
break down. A change from an oxygen to a sulfur base,
for example, makes predictions more uncertain than does
a change from one oxygen base to another.

In view of the approximations and assumptions that
go into Thornton's treatment, it is best to regard it
as a framework for discussion of the experimental
facts, not as a set of inflexible predictions that must
stand or fall as a unit. Consequently, we review the
experimental evidence before any further discussion of
the predictions.

C   EXPERIMENTAL EVIDENCE ON RATE EFFECTS

1   Design and Interpretation of Experiments

A typical change in reactant structure or reaction conditions may affect the rate of reaction in several
different ways, thereby complicating considerably the
interpretation of the results. The introduction of an
alkyl group in the β-position of the substrate of an
E2 reaction can increase electron density at that position, can increase eclipsing effects, can introduce
steric hindrance to proton removal, can increase nonbonding interactions with the leaving group, and can
stabilize the developing double bond of the transition
state. To sort out the relative importance of each of
these effects, one would like to be able to design reactants in which each was varied in turn while the
others remained constant. Such an ideal situation is
seldom possible, but one can achieve some success in
this direction. A common approach is to choose reactants with the site of substitution sufficiently remote
from the site of reaction that the substituent can exert an electronic, but not a steric effect on the rate
of the reaction. Having formed an impression of the
magnitude of the electronic effect in this fashion, one
is in a better position to decide the relative contributions of steric and electronic effects in similar reactants where both are varied at the same time.
Another very useful technique, isotopic substitution,
enables one to draw conclusions about the extent to
which a bond is weakened (or strengthened) in the transition state. An appreciable kinetic isotope effect
will be observed only if there are changes in force
constants at or near the isotopic atom between the reactants and the transition state. The magnitude
of the effect measures how large the changes have been.

2 Variations in the Structure of the Substrate

(a) <u>Electronic Effects in E2 Reactions</u>. The simplest and most commonly used method of isolating electronic effects in elimination reactions is to use reactants containing m- or p-substituted phenyl groups. The substituent in the phenyl group is sufficiently far removed from the site of reaction that it exerts only an electronic effect without changing the steric requirements of the group. A great many organic reactions of compounds bearing m- and p-substituted phenyl groups have been shown to follow the Hammett equation (Eq. 1)[22],[23] where k is the rate (or equilibrium) constant

$$\log \frac{k}{k_O} = \rho\sigma \tag{1}$$

for the reaction of the compound having a substituted phenyl group, and k_O is the rate (or equilibrium) constant for the reaction of the compound having an unsubstituted phenyl group. ρ is a constant characteristic of the reaction and reaction conditions, while σ is a constant characteristic of the substituent and its position (m or p). σ is defined by taking as a standard the ionization equilibrium for benzoic acids in water, for which ρ is taken as unity. The definition of σ in terms of the ionization constants is then given by Eq. 2. The reaction constant ρ measures the sensitivity of the reaction to changes in electron density at

$$\log \frac{K}{K_O} = \sigma \tag{2}$$

the reaction site. Since the ionization of benzoic acids is favored by electron-withdrawing substituents, a positive value of ρ for the kinetic case signifies a reaction accelerated by electron withdrawal. This, in turn, is interpreted as a reaction in which electron density at the site of the phenyl substituent increases from reactant to transition state. The magnitude of ρ is considered to measure the magnitude of this increase. Similarly, a negative value of ρ signifies a decrease in electron density in the transition state at the site of substitution.

As we see shortly, most of the available Hammett correlations of rates of elimination reactions involve reactants with the phenyl substitution on the β-carbon. Here ρ is taken as a measure of carbanion character at the β-carbon in the transition state.

While one can use the ρ values to arrange a series of
reactions studied under comparable conditions in the
order of carbanion character, it is much more difficult
to interpret any single value of ρ. Hybridization at
the α- and β-carbon atoms changes during an elimination
reaction from sp³ to sp², with the result that the
electron affinity of these atoms increases.[24] The
change in hybridization should be aided by electron-
repelling groups and thereby should contribute a negative factor to ρ, but there is no way of estimating
this factor. The best one can say is that there is no
evidence that it affects conclusions about relative
degrees of carbanion character. The failure thus far
to observe any negative ρ values for β-substitution in
elimination reactions suggests that it is never a controlling factor. We can do somewhat better at estimating the value of ρ to be expected in a process that
forms a carbanion without any concomitant weakening of
the bond to the leaving group. The rates of methyl-
hydrogen exchange of substituted toluenes with lithium
cyclohexylamide in cyclohexylamine at 50° give a ρ
of +4.0, while anionic polymerization of substituted
styrenes in tetrahydrofuran at 25° gives +5.0.[25,26]
The ionization constants of substituted fluorenes at
25° lead to somewhat higher ρ values, +6.3 with and
+7.5 without the unsubstituted compound.[27] The values
derived from rates are probably better indications of
what to expect in elimination reactions, since it is
unlikely that the transition state would closely resemble a fully formed carbanion even in an E1cB process. One should keep in mind in considering all ρ
values, of course, that they may vary with solvent
polarity and with reaction temperature, as well as with
the nature of the reaction.[22,23]
    The 2-phenylethyl system is by far the most inten-
sively studied type of substrate for Hammett correlations of elimination rates. Table 1 is a collection
of data from a number of sources. Different temperatures and different sets of substituents were used, so
that care in comparison is needed. Where possible,
values extrapolated to 30° are quoted if measured
values at that temperature are not available. Only
values calculated from data on at least three different substituents (including unsubstituted phenyl) are
quoted.
    A number of regularities are evident in the data.
The ρ values for the halogens increase in the order
I<Br<Cl<F. This is just what one would expect, since
the difficulty of breaking the carbon-halogen bond increases in the same order, and progressively greater

TABLE 1 Hammett Reaction Constants for E2 Reactions of 2-Phenylethyl Derivatives

$$YC_6H_4CH_2CH_2X \longrightarrow YC_6H_4CH=CH_2$$

X	ρ[a]	T, °C	Y[b]	Reference
Ethanol/Ethoxide:				
I	2.07 ± 0.09	30	c	28
Br	2.14 ± 0.15	30	c	28
Br	2.15 ± 0.24	30	d	5
Cl	2.58 ± 0.33	60	e	30
Cl	2.61	30	e	30
F	3.10 ± 0.07	60	e	30
F	3.12[f]	30	e	30
OTs	2.27 ± 0.08	30	g	31
SMe$_2^+$	2.64 ± 0.16	30	d	5
SMe$_2^+$	2.75 ± 0.21	30	c	28
NMe$_3^+$	3.77 ± 0.21[f]	30	d	32
NMe$_3^+$	3.09 ± 0.15	60	d	32
t-Butyl Alcohol/t-Butoxide:				
I	1.88 ± 0.06	30	c	31
Br	2.08 ± 0.02	30	c	31
OTs	3.39 ± 0.29	30	c	31
OTs	2.49 ± 0.03	40	h	33
OSO$_2$Ph	2.50 ± 0.02	40	h	33
OBs^i	2.36 ± 0.02	40	h	33
ONs^j	2.03 ± 0.04	40	h	33
NMe$_3^+$	3.04 ± 0.03	30	k	32
SOMe	4.4	52	d	123
Water/Hydroxide:				
SMe$_2^+$	2.21 ± 0.08[f]	30	d	34
SMe$_2^+$	2.03 ± 0.04	60	d	34
SMe$_2^+$	2.11[f]	40	d	35
SMe$_2^+$	2.53 ± 0.05[l]	40	d	35
SMe$_2^+$	2.55 ± 0.04[l]	40	d	35
SMe$_2^+$	2.59 ± 0.05[l]	40	d	35

[a] Value determined by least-squares method with standard deviation.
[b] Each set of Y substituents is listed in a separate footnote.
[c] p-Cl, m-Br, p-MeO, H.

[d] p-Cl, p-MeO, p-Me, H.
[e] p-Cl, m-Br, H.
[f] Extrapolated from data at higher temperatures.
[g] p-MeO, p-Cl, H.
[h] p-MeO, p-Me, H, m-MeO, p-Cl, m-Cl.
[i] p-bromobenzenesulfonate.
[j] p-nitrobenzenesulfonate.
[k] p-MeO, p-Me, H.

[l] Last three reactions carried out in mixtures of water and dimethyl sulfoxide containing, respectively, 19.1, 38.0, and 60.3 mole % of dimethyl sulfoxide.

buildup of negative charge on the β-carbon atom in the transition state should be required to "displace" the leaving group. An analogous effect is noted on orientation in elimination from alkyl halides (Section IV. B.2). By the same reasoning, tosylate seems to be a somewhat poorer leaving group than bromide or iodide. The charged groups, dimethylsulfonium and, especially, trimethylammonium, react via transition states of relatively high carbanion character. Such a conclusion is in agreement with the fact that β-alkyl groups retard elimination reactions of these classes or compounds (Section IV. A.4). Another leaving-group effect is seen in the data on the various substituted benzenesulfonates with t-butoxide in t-butyl alcohol. The more electron withdrawing the substituent in the leaving group, the smaller the value of ρ. Again, less carbanion character is found with the better (more reactive) leaving groups.

There are three instances where data from two different laboratories, using different sets of substituents, are reported for the same leaving group. In two of these cases, the bromide and the sulfonium salt in ethanol, the agreement is excellent. In the third, the tosylate in t-butyl alcohol, the disagreement is far beyond experimental error, and comparison at a common temperature would be expected to reduce the difference only slightly. It is likely that the 2.49 at 40° is more reliable than the 3.39 at 30°, for the former is derived from six substituents compared to four for the latter, and the relationship shows less scatter as well.

Changes in ρ with changes in the solvent-base system are less regular than changes with the nature of the leaving group. The ρ values for iodide and bromide decrease slightly on going from ethoxide in ethanol to t-butoxide in t-butyl alcohol, while that for tosylate increases (either markedly or slightly, depending on which of the two discordant values in the latter

solvent is chosen -- see the preceding paragraph).
There is a marked decrease in ρ for the trimethylammonio group the same solvent change. The sulfonium
salt gives a smaller ρ with hydroxide in water than
with ethoxide in ethanol. Addition of dimethyl sulfoxide to the aqueous medium produces at first an increase in ρ for the sulfonium salt, but further addition produces no change.

In interpreting these mixed results, one must
remember that both the solvent and the base are changing. It has been suggested that ρ should vary inversely with dielectric constant,[36] and there does seem to
be a general tendency in this direction.[23] The changes
in electron distribution in the transition state that
doubtless accompany many solvent changes make any precise evaluation of the solvent effect on the transmission of electronic effects very difficult. Among
the present data, such a solvent effect may be responsible for at least part of the changes in ρ for the
sulfonium and ammonium salts. The change from water
to ethanol for the sulfonium salt is in the right direction, but not the change from ethyl to t-butyl
alcohol for the ammonium salt. On the other hand,
there is good evidence that the decrease in ρ for the
ammonium salt on changing from ethanol-ethoxide to t-
butyl alcohol - t-butoxide is caused by a decrease in
the extent of proton transfer in the transition
state.[32] We noted previously (Section II.B.1,2), that
theory predicts a decrease in the extent of proton
transfer with increasing base strength. The relatively small changes in ρ for the uncharged substrates between ethanol and t-butyl alcohol can be accommodated
by the assumption (Section II.B.2) that a decrease in
extent of proton transfer is accompanied by an approximately compensating decrease in stretching of the
carbon-leaving group bond.

Table 2 lists ρ values for miscellaneous E2 reactions. Here there are wider variations both of substrate structure and reaction conditions than in Table
1, but some interesting conclusions are still possible.
Substitution of alkyl or phenyl on the α-carbon atom
lower ρ. The decrease in carbanion character is expected, since the substituents should stabilize a
developing double bond. It is interesting to note that
the ρ value for 2-phenyl-1-arylethyl chloride is positive. This reaction is close to the E1-E2 borderline
(correction for a competing E1 process is necessary),
yet the α-carbon does not seem to be more positive in
the transition state than in the ground state. This
reaction may be an example of the E2C mechanism, which

TABLE 2 Reaction Constants for Miscellaneous E2 Reactions

Substrate	ρ	T, °C	Solvent/Base	Reference
$YC_6H_4CH_2CMe_2Cl$	1.00	66.5	$MeOH/MeO^-$	37
trans-2-Arylcyclopentyl OTs	2.77	50	$t-BuOH/t-BuO^-$	38
cis-2-Arylcyclopentyl OTs	1.48	50	$t-BuOH/t-BuO^-$	38
cis-2-Arylcyclopentyl OTs	0.99	50	$EtOH/EtO^-$	38
$YC_6H_4CH_2CHPhCl$	1.98	50	$EtOH/EtO^-$	39
$PhCH_2CH(C_6H_4Y)Cl$	0.77	50	$EtOH/EtO^-$	39
$(YC_6H_4)_2CHCCl_3$	2.43	30	$92.6\% \; EtOH/OH^-$	47
$(YC_6H_4)_2CHCHCl_2$	2.26	30	$92.6\% \; EtOH/OH^-$	47
$YC_6H_4SO_2CH_2CH_2Cl$	1.81	50	$MeCN/Et_3N$	43
$YC_6H_4SCH_2CH_2Cl$	1.98	60	$t-BuOH/t-BuO^-$	44
$YC_6H_4SCH_2CH_2Cl$	2.14	60	$EtOH/EtO^-$	44
$YC_6H_4OCH_2CH_2Cl$	1.33	60	$t-BuOH/t-BuO^-$	44
$YC_6H_4OCH_2CH_2Cl$	1.50	60	$EtOH/EtO^-$	44
$YC_6H_4OCH_2CH_2CH_2Br$	0.0	50	$t-BuOH/t-BuO^-$	45
$YC_6H_4SCH_2CH_2CH_2Br$	0.37	50	$t-BuOH/t-BuO^-$	45
$p-MeOC_6H_4CH_2CH_2OSO_2C_6H_4Y$	1.24	40	$t-BuOH/t-BuO^-$	33
$p-MeC_6H_4CH_2CH_2OSO_2C_6H_4Y$	1.24	40	$t-BuOH/t-BuO^-$	33
$PhCH_2CH_2OSO_2C_6H_4Y$	1.08	40	$t-BuOH/t-BuO^-$	33
$m-MeOC_6H_4CH_2CH_2OSO_2C_6H_4Y$	1.06	40	$t-BuOH/t-BuO^-$	33
$p-ClC_6H_4CH_2CH_2OSO_2C_6H_4Y$	1.01	40	$t-BuOH/t-BuO^-$	33
$m-ClC_6H_4CH_2CH_2OSO_2C_6H_4Y$	0.94	40	$t-BuOH/t-BuO^-$	33
2-pentyl $OSO_2C_6H_4Y$	1.35	50	$EtOH/EtO^-$	40
2-methyl-3-pentyl $OSO_2C_6H_4Y$	1.51	50	$50\% \; t-BuOH-$ dioxane$/t-BuO^-$	41
2-methyl-3-pentyl $OSO_2C_6H_4Y$	2.40	25	$25\% \; t-BuOH-$ DMSO$/t-BuO^-$	41

TABLE 2 Continued

Substrate	ρ	T, °C	Solvent/Base	Reference
$Ph_2CHCH_2OSO_2C_6H_4Y$	1.11	50	Me cellosolve/$\underline{MeO}^-$	46
cis-2-(p-tolylsulfonyl)cyclo-hexyl $OSO_2C_6H_4Y$	0.42	0	50% dioxane/OH^-	42
cis-2-(p-tolylsulfonyl)cyclo-hexyl $OSO_2C_6H_4Y$	0.50	25	50% dioxane/Me_3N	42
trans-2-(p-tolylsulfonyl)cyclo-hexyl $OSO_2C_6H_4Y$	0.56	0	50% dioxane/OH^-	42
trans-2-(p-tolylsulfonyl)cyclo-hexyl $OSO_2C_6H_4Y$	0.59	25	50% dioxane/OH^-	42
trans-2-(p-tolylsulfonyl)cyclo-hexyl $OSO_2C_6H_4Y$	0.33	25	50% dioxane/Me_3N	42

is discussed in more detail subsequently (Section II. C.2.c). An attempt to determine ρ values for α-aryl-ethyl bromides failed because the Hammett plots were curved, but the ρ values would clearly be positive and increase in magnitude in the order: ethoxide in ethanol< t-butoxide in t-butyl alcohol≤ t-butoxide in 10% t-butyl alcohol-dimethyl sulfoxide.[29]

The structural effects in the 2-arylcyclopentyl tosylates are more or less as expected. The ρ values for the anti eliminations run somewhat lower than those for the 2-arylethyl tosylates, again probably due to the greater degree of substitution. It is interesting that the ρ value for the syn elimination is not larger. This was taken as evidence that the syn elimination was a concerted and not an E1cB process (see Section III.E.2).[38]

The effect of an aryl substituent still seems to be fairly large when an oxygen, and especially a sulfur, is interposed between it and the β-carbon atom. The effect predictably drops off markedly when a methylene group is interposed as well, but the advantage of sulfur over oxygen is retained. It was suggested that there may be some direct interaction of the d-orbitals of sulfur with the developing p-orbital on the β-carbon atom in the transition state.[45]

The ρ value for substitution in the leaving group is shown for a number of benzenesulfonates. The effect of changing the substituent in the leaving group would be expected to be larger, the greater the extent of bond breaking to the leaving group in the transition state. On this basis, the results form a reasonable pattern. The reactants with the strongly acidifying p-tolylsulfonyl group on the β-carbon would be expected to react via E1cB-like transition states with relatively little cleavage of the bond to the leaving group, and the ρ values are, in fact, small. The question as to whether these should be regarded as genuine E1cB reactions is discussed earlier (Section I.B.1.c). An electron-repelling substituent in the β-phenyl group of the 2-arylethyl benzenesulfonates increases the leaving-group ρ value, while an electron-withdrawing substituent decreases it. Thus electron-repelling groups at the β-carbon seem to increase the extent of cleavage of the bond between the α-carbon and the leaving group in the transition state. The Swain-Thornton rules predict that an electron-repelling group at the β-carbon should lengthen the carbon-hydrogen bond (Section II.B.2), and a parallel change in the carbon-leaving group bond is reasonable. The very large increase in ρ value on change of solvent from t-butyl alcohol - dioxane to t-butyl alcohol - dimethyl sulfoxide for the 2-methyl-3-pentyl benzenesulfonates is intriguing, but there is no obvious explanation for it.

(b) <u>Relative Reactivities of Bromides and Tosylates.</u> If changes in relative reactivities of different ben- zenesulfonates can afford information about transition- state structure, one would expect the same to hold for wider variations of the leaving group. It was noted some time ago that the relative reactivities of alkyl bromides and alkyl tosylates varied widely for diffe- rent types of elimination and substitution reactions, but no explanation was offered for this phenomenon.[48], [49] The first systematic treatment was by Hoffmann,[50] who noted that the ratio k_{OTs}/k_{Br} varied between 0.36 and 5000 for various S_N1, S_N2, and E2 reactions. He pointed out that the tosylate ion, with its three oxy- gen atoms, could delocalize negative charge much more effectively than could bromide ion, and that the high ratios were, in fact, observed in those reactions where one would expect a high degree of cleavage of the bond to the leaving group in the transition state. For ex- ample, the ratio increases very sharply along the series primary<secondary<tertiary for S_N1 reactions.[51] Values for E2 reactions of 2-arylethyl derivatives are recorded in Table 3.[52] The ratio is increased by electron-withdrawing groups on the aryl group, and is

TABLE 3   Tosylate-Bromide Rate Ratios for E2 Reactions of 2-Arylethyl Derivatives

$$YC_6H_4CH_2CH_2X \longrightarrow YC_6H_4CH=CH_2$$

Y	Solvent/Base	k_{OTs}/k_{Br} (E2)
p-MeO	t-BuOH/t-BuO$^-$	0.15 ± 0.01
H	t-BuOH/t-BuO$^-$	0.22 ± 0.01
p-Cl	t-BuOH/t-BuO$^-$	0.44 ± 0.03
m-Br	t-BuOH/t-BuO$^-$	1.19 ± 0.07
p-NO$_2$	t-BuOH/t-BuO$^-$	1.57 ± 0.06
p-MeO	EtOH/EtO$^-$	0.065 ± 0.002
H	EtOH/EtO$^-$	0.10 ± 0.01

also larger with t-butoxide in t-butyl alcohol than with ethoxide in ethanol. Hoffmann points out that

both electron-withdrawing β-substituents and stronger
bases are considered to increase the ElcB-like character of the transition state. According to his interpretation of the k_{OTs}/k_{Br} ratio, however, these changes
also seem to produce increased cleavage of the bond between the α-carbon atom and the leaving group, which
makes it very difficult to see how a true ElcB mechanism could be attained.[52] To make matters worse, this
trend in extent of carbon-leaving group cleavage with
the nature of the 2-aryl group is precisely the opposite of that we drew from the data in Table 2 on the
ρ values for substitution in the benzenesulfonate ring
of 2-arylethyl benzenesulfonates (Section II.C.2.a).[33]

A plausible resolution of this discrepancy was
suggested by Cockerill,[53] who reported α-deuterium isotope effects for the reactions of several arylethyl tosylates with t-butoxide in t-butyl alcohol. These ran
1.047 for p-methoxy, 1.043 for unsubstituted, and 1.017
for p-chloro, suggesting the same trend in extent of
cleavage of the bond to the leaving group as was deduced
from the ρ values. He suggested that the k_{OTs}/k_{Br}
ratios had simply been incorrectly interpreted because
it was wrong to assume that they would be a monotonic
function of the extent of cleavage of the bond to the
leaving group. In the limiting case where there is no
weakening in the transition state of the bond to the
leaving group, the ratio should be somewhat greater
than unity because tosylate is more electron withdrawing than bromide. With slight bond weakening, the
greater polarizability of the carbon-bromine bond makes
the ratio decrease. Finally, the charge on the leaving
group becomes sufficient that the greater ability of
tosylate to delocalize charge takes control, and the
ratio rises. If the elimination reactions under discussion are on the descending portion of the curve
(a-b in Figure 2), then the k_{OTs}/k_{Br} ratios are consistent with the conclusion that electron-repelling
β-substituents increase the extent of cleavage of the
bond between the α-carbon atom and the leaving group
in the transition state.

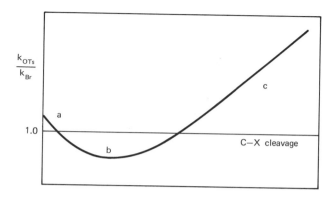

<u>Figure 2</u> Relationship between the tosylate-bromide rate ratio and the extent of C-X cleavage (reproduced by permission of the author[53] and Tetrahedron Letters).

(c) <u>Substituent Effects in E2C Reactions</u>. We mentioned earlier (Section II.A.5) that Winstein and Parker and their co-workers gave the name E2C to the mechanism for elimination reactions promoted by halide ions in dipolar aprotic solvents and by certain other weak bases. The basis for this name is the conclusion that the base may be interacting with the α-carbon as well as with the β-hydrogen in the transition state. Although evidence on this point cannot be regarded as conclusive, these reactions show distinctive features in their response to substituent effects and deserve separate discussion.

A very thorough survey of substituent effects in both the E2C and the E2H (or normal E2) mechanisms has been published.[54] Product distributions and overall rates were determined to arrive at partial rates for the elimination and substitution processes in each reaction. It is difficult to abstract from the large mass of data representative numbers to illustrate the generalizations, but an effort to do this is made in Table 4. The figures are corrected for statistical factors where appropriate. The rates with halide ion in acetone are quoted as representative of results under E2C conditions, and the rates with t-butoxide ion in t-butyl alcohol are quoted as representative of results under E2H conditions. Methyl groups at both the α- and β-positions strongly accelerate the E2C reactions, but either hinder or have little effect on the E2H reactions (lines 1-2,5-7). A phenyl group at the β-position has about the same effect as methyl on an E2C reaction, but strongly accelerates an E2H reaction (lines 3, 9). An α-phenyl group, on the other

TABLE 4   Effects of Substituents on Elimination Rates
under E2C and E2H Conditions

R^β	R^α	R^α	Log k_E for	
			$\overline{Bu_4NBr/}$ Me_2CO, 75°	t-BuOK/ t-BuOH, 40°
Me	H	H	−5.94	−4.82
Me	Me	H	−4.80	−5.6
Me	Ph	H	−3.38	−4.5
Me	p-NO$_2$Ph	H	−3.59	--
Me	Me	Me	−2.0	−5.4
H	Me	Me	−4.00[a]	−6.2
Me	Me	Me	−2.33[a]	−6.3
Ph	Me	Me	−2.6 [a]	--
Ph	Ph	H	−3.16	−1.9

[a]Reagent was tetrabutylammonium chloride under same
conditions.

hand, accelerates an E2C reaction but has little effect
on an E2H reaction (lines 1, 3). Furthermore, substitution on the α-phenyl group has little effect on the
E2C reaction; p-nitrophenyl and phenyl increase rate by
about the same amount (lines 1, 3, 4).
    The main conclusion to be drawn from these rate
effects, and others quoted in the original paper,[54]
is that E2C reactions possess a transition state having
little charge development at either the α- or β-carbon
atoms but a great deal of double-bond character. The
latter point is also forcefully demonstrated by orientation data (Section IV. C.2), in which the proportion
of the more stable (Saytzev-rule) olefin approaches and
sometimes even exceeds the proportion in the equilibrium mixture of product olefins. In the general formulation of the variable transition-state theory described
earlier (Section II.A.4), the E2C transition state
would lie near the upper right-hand corner of Figure 1.
    All of the rate effects do not fit equally well
into this picture. It is somewhat suprising, for example, that a phenyl group is not much superior to an

alkyl group in the β-position, for it should be considerably more effective in stabilizing a highly developed double bond. It is more effective in the α-position, but not by a wide margin. The authors point out
this problem, and they suggest that the phenyl groups
may be prevented by steric hindrance from attaining
the coplanarity with the double bond that would be
necessary for most effective overlap.[54] In such a case
one might expect the phenyl group at the more congested
α-carbon to be less effective than that at the β-carbon, contrary to fact. A basic problem in interpreting
the broad pattern of results is the diversity of substituents, which makes it difficult to separate steric
and electronic effects on rate. Studies aimed at doing
this, such as the careful Hammett correlations that
have been done on alkoxide-promoted eliminations, are
clearly needed.

3   Kinetic Isotope Effects

(a) _Theory_. The basic idea behind the application of
isotope effects to the study of reaction mechanisms is
very simple. Two reactant molecules that differ only
in the substitution of a heavy isotope at some particular position in the second molecule will differ somewhat in energy, the main contribution to this difference arising from the effect of isotopic substitution
on the zero-point vibrational energy. If the bonding
to the isotopic atom remains the same in the transition
state for a reaction as it is in the reactant, the
zero-point-energy difference between the two isotopic
transition states will be the same as between the two
isotopic reactants. Under these circumstances, isotopic substitution will have a negligible effect on reaction rate. If the bonding to the isotopic atom
changes in the transition state, however, the zero-
point-energy difference in the transition state will no
longer cancel that in the reactants, and an isotope
effect on rate will result. The isotope effect thus
affords a very powerful tool for exploring chemical
bonding at particular positions in the transition state
of a reaction.

Quantitative expressions for the effect of isotopic substitution on rate can be derived from absolute
rate theory.[55-58] This theory regards the reactants as
being in equilibrium with an activated complex (or
transition state) which is treated as a normal molecule
except that one vibrational degree of freedom becomes
the "motion along the reaction coordinate." This is
the motion that converts the transition state to

products, and it lacks the restoring force of a normal vibration. If two reactants A_1 and A_2, which differ only isotopically, react with one or more other reagents according to Eqs. 3 and 4, the rate constants are

$$A_1 + B + \cdots \underset{}{\overset{K_1^{\ddagger}}{\rightleftharpoons}} \left[C_1\right]^{\ddagger} \longrightarrow \text{products} \quad (3)$$

$$A_2 + B + \cdots \underset{}{\overset{K_2^{\ddagger}}{\rightleftharpoons}} \left[C_2\right]^{\ddagger} \longrightarrow \text{products} \quad (4)$$

given by Eqs. 5 and 6. The ratio of these Eq. 7, then gives the ratio of rates of reaction of the two

$$k = \kappa_1 \left(\frac{kT}{h}\right) K_1^{\ddagger} \quad (5)$$

$$k = \kappa_2 \left(\frac{kT}{h}\right) K_2^{\ddagger} \quad (6)$$

$$\frac{k}{k} = \left(\frac{\kappa_1}{\kappa_2}\right) \left(\frac{K_1^{\ddagger}}{K_2^{\ddagger}}\right) \quad (7)$$

isotopes (strictly, one should call this the "isotopic rate ratio" and reserve the term "isotope effect" for the departure of the ratio from unity, but the terms have been used interchangeably for so long that there seems little point in trying to reverse the generally-accepted usage). Since $K_1^{\ddagger}$ and $K_2^{\ddagger}$ are treated as normal equilibrium constants, they can be expressed in terms of either concentrations or partition functions of the species involved. The transmission coefficient κ is usually assumed to be independent of isotopic substitution. (Symmetry numbers are ignored in this derivation, so the resulting isotope effects will not be corrected for differences in symmetry between the isotopic species.) The rate ratio can then be expressed simply as a ratio of ratios involving the partition functions of the isotopic reactant (those for non-isotopic reactants cancel) and the isotopic transition states, Eq. 8. The electronic partition function is

$$\frac{k_1}{k_2} = \frac{(Q_{A_2}/Q_{A_1})}{(Q_2^{\ddagger}/Q_1^{\ddagger})} \quad (8)$$

insensitive to isotopic substitution, so that the partition function ratios can be expressed in terms of translational, rotational, and vibrational partition functions. Many terms cancel in taking the isotopic ratios, and one is left with the relatively simple expression of Eq. 9.

$$\frac{Q_2}{Q_1} = \left(\frac{I_{X_2} I_{Y_2} I_{Z_2}}{I_{X_1} I_{Y_1} I_{Z_1}}\right)^{1/2} \cdot \left(\frac{M_2}{M_1}\right)^{3/2} \cdot \prod_i^{3n-6}$$

$$\left[\exp(\Delta u_i/2) \cdot \cdot \frac{1 - \exp(-u_{1i})}{1 - \exp(-u_{2i})}\right] \tag{9}$$

The first term contains the principal moments of inertia of each species and comes from the rotational partition function. Only the atomic masses, bond lengths, and bond angles are needed to compute these quantities. The second term is simply the ratio of the molecular masses of the two species and comes from the translational partition function. These two terms are often referred to together as the mass-and-moment-of-inertia term (MMI). In the product term, $u_i = h\nu_i/kT$, where ν_i is one of the $3n - 6$ normal vibrational frequencies of the reactant, or $3n - 7$ of the transition state, where n is the number of atoms. The first term in the product is usually referred to as the zero-point-energy (ZPE) term, and the second as the vibrational partition function or excitation (EXC) term. Various forms of the complete expression for the isotopic rate ratio are to be found in the literature.[55-58] Our main concern here is to emphasize the type of information needed to compute isotope effects. The MMI term is quite easy to compute but is not often the major contributor to the observed effect. The single most important term, at least with hydrogen isotope effects, is the ZPE term, and a complete knowledge of all the fundamental frequencies of both the isotopic reactants and isotopic transition states are needed for rigorous computation of this and the EXC terms. While these frequencies are in principle obtainable from analysis of infrared and Raman spectra for stable molecules, they must be calculated from assumed force fields and molecular geometries for the transition states. Calculation of the fundamental frequencies of all but the simplest molecules requires a high-speed computer. As a result, early applications of isotope-effect theory utilized highly simplified models, and it is only in recent years that more complete calculations have been undertaken.

The bimolecular elimination reaction constitutes a particular challenge to the theory because of the numbers of reacting bonds, and the possibilities for isotopic substitution in different parts of the reacting system. To have tractable models for the simplified approach, it has been customary to consider the parts of the reacting system in isolation from each other. Thus, the isotope effect on replacement of hydrogen by deuterium in the β-position has been treated as a simple proton-transfer isotope effect. Isotopic substitution in the leaving group has been treated as a simple cleavage of the bond between the leaving group and the α-carbon. These models assume that what is happening at the other end of the reacting system has a negligible influence on the isotope effect under consideration. Before we inquire into the legitimacy of this assumption, it is advisable to look briefly at the models themselves.

We consider as an example of leaving-group isotope effects the $^{32}S/^{34}S$ effect in the decomposition of sulfonium salts in the presence of base. The hypothetical molecule C-S is taken as the reactant. It has only a single vibrational frequency which can be calculated for each isotopic species from the expression for a simple harmonic oscillator (Eq. 10). In the equation, F is the force constant, a measure of the stiffness of the bond, and μ is the reduced mass of C-S, $m_C m_S / (m_C + m_S)$. The ZPE term for the reactants can now readily be calculated. The one vibrational mode

$$\nu = (\tfrac{1}{2}\Pi)(\tfrac{F}{\mu})^{1/2} \tag{10}$$

becomes the "motion along the reaction coordinate" in the transition state and does not contribute any zero-point-energy difference. From this model one calculates that the ^{32}S species will react about 1.5% faster than the ^{34}S species, the exact value depending on the assumed value of the force constant.[4] As we note below, the experimental values bear a reasonable relationship to the calculated one.

This very simple model probably works reasonable well because there are no new covalent bonds being formed to the leaving group in the transition state. Early efforts to apply an analogous model to the hydrogen/deuterium isotope effect in the cleavage of the β-carbon-hydrogen bond led to erroneous predictions because they ignored the base that is forming a covalent bond to hydrogen at the same time the carbon-hydrogen bond is breaking. Here the simplest permissible model of the transition state is the triatomic system B---H---C, as was first pointed out by Melander[59] and Westheimer.[60] One can arrive at the stretching frequencies of the isotopic reactants, H-C and D-C, in the same manner as above. The ratio of frequencies will be approximately $2^{1/2}$, because carbon is much heavier than hydrogen or deuterium, and the reduced mass of each species will be close to that of the lighter atom. The transition state is treated as a linear triatomic molecule with four vibrational modes (3n - 5 for linear species). These are shown in 10 - 13. The two bending modes are generally ignored on the assumption

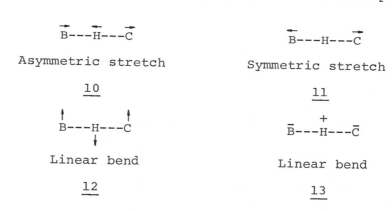

Asymmetric stretch

10

Symmetric stretch

11

Linear bend

12

Linear bend

13

that they will approximately cancel bending modes in the reactant (which is usually not the diatomic species it is assumed to be above). The asymmetric stretch converts the transition state into products if allowed to proceed without restoring force, and is thus the "motion along the reaction coordinate." The symmetric stretch is a real vibration, however, and contributes to the zero-point energy of the transition state. When the proton is equally bonded to B and C in the transition state, it will move little or none at all in the symmetric stretching motion if B and C are of similar or equal masses. In that case, the frequency of the symmetric stretch will be unaffected by isotopic substitution, and there will be no difference in zero-point energy from this source for the two isotopic transition states. The isotope effect will then reflect the full difference in stretching zero-point energies of the two isotopic reactants. If the proton if more strongly bonded to B than to C, or vice versa, in the transition state, it will move in the symmetric stretching motion. There will then be a difference in zero-point energy between the isotopic transition states, and this difference will partially cancel the difference between the isotopic reactants. The isotope effect will reflect only part of the difference in zero-point energy between the isotopic reactants and will thus be smaller than when the proton is half transferred in the transition state. The overall picture from this model is that the isotope effect will be small for small extents of proton transfer, will attain a maximum when the proton is half transferred, and will become small again for large extents of proton transfer. There results an ambiguity in the interpretation of small isotope effects, the resolution of which requires the use of other information, such as substituent effects.

Although these models are in qualitative accord with the experimental results that we discuss shortly, it is by no means certain that the leaving-group isotope effect will be independent of the extent of proton transfer, or that the hydrogen isotope effect will be independent of the extent of cleavage of the bond to the leaving group. In general, there is no assurance that reactions with complex reaction coordinates, such as the elimination reaction, will give isotope effects that are similar in form or magnitude to those from simpler reactions. For these reasons, calculations on more complex models have been performed. Katz and Saunders[61] calculated hydrogen and sulfur isotope effects on the 12-atom transition state 14. The hydrogen isotope effect was found to go through a maximum

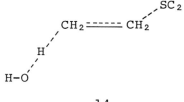

<u>14</u>

at half transfer just as in the simple model, though the magnitude of the effect depended on the extent to which the proton transfer was coupled to the C-C and C-S motions. When the extent of proton transfer was kept constant and the extent of carbon-sulfur cleavage varied, there was little change in the hydrogen isotope effect. Replacement of SC_2 by other leaving groups (Cl, Br, and NC_3) also left the hydrogen isotope effect essentially unchanged.[62] The sulfur isotope effect was originally reported to be inverse for small extents of carbon-sulfur cleavage,[61] but this was subsequently found to arise from an unrealistic model for the reactant.[62] According to the more recent work, the sulfur isotope effect varies almost linearly with extent of carbon-sulfur cleavage, up to a maximum of about 1.3%, and is essentially unaffected by changes in geometry of the model (from sp^3 to sp^2 carbon atoms).[62] There was a variation of a few tenths of a percent with the extent of proton transfer at a constant extent of carbon-sulfur cleavage, but not enough to complicate seriously the interpretation of sulfur isotope effects. The general conclusion to be drawn at this point is that the earlier qualitative interpretations of hydrogen and sulfur isotope effects on the basis of simplified models were probably rather close to the mark.

Calculations were also made for isotopic substitution in other positions of the molecule.[62] In most cases no experimental results were available for comparison, and the aim was to decide which isotope effects would be most useful for studying transition-state structure. The β-carbon isotope effect was particularly interesting, decreasing at first and then rising with extent of proton transfer. As a result, the effect remains below 1 to 2% until the proton is more than half transferred, and should be a good way of deciding whether small hydrogen isotope effects indicate small or large extents of proton transfer. The other effects calculated (α-carbon, α- and β-secondary deuterium) increase steadily with extent of bond cleavage at the position in question, and are little affected by bond changes elsewhere in the molecule. Unlike the hydrogen isotope effect, the carbon and secondary hydrogen isotope effects can vary considerably with the nature of the leaving group.

Although the hydrogen and sulfur isotope effects were little affected by a change from sp^3 to sp^2 geometry at the carbons, the other isotope effects ran somewhat higher for the sp^2 than for the sp^3 model. There was no significant difference in any of the isotope effects between syn and anti transition states. This is not to say that such differences will not be observed experimentally, but that they will be due to changes in force constants, not just changes in geometry. Hydrogen isotope effects were distinctly lower for transition states in which the O---H---C system is nonlinear, such as those for the E2C (Section II. A.5) or α'-β (Section I.B.4) mechanisms.[62,64]

None of these calculations took into account the possibility of a contribution to the isotope effect

from tunneling, which is more probable for the light
than the heavy isotope. This phenomenon is expected to
be significant only for hydrogen isotope effects, and
is generally considered to manifest itself in the form
of unusually large isotope effects or an unusual temperature dependence of the isotope effect.[63] Thus far,
no clear-cut examples of such abnormalities in elimination reactions have been reported. Even if tunneling
does contribute to some extent to observed hydrogen
isotope effects in elimination reactions, there is no
reason to believe that it affects the form of the dependence of the effect on the extent of proton transfer.

Isotope effects can be observed even when the bond
to the isotopic atom is not breaking in the rate-determining step of the reaction, provided it is at least
undergoing some change. Such effects are called secondary isotope effects. Secondary deuterium isotope
effects tend to be rather small (generally from a few
percent up to about 30% per deuterium), and the exact
causes of their origin are not always immediately
apparent. Among the suggested causes have been hyperconjugation, change in hybridization, inductive effects, and steric effects. A detailed discussion of
these causes is beyond the scope of this book, and the
reader is referred to reviews of the topic for further
information.[65,67] The most likely causes of secondary
deuterium isotope effects in eliminations are changes
in hybridization at either the α- or β-carbon atoms,
and possibly the inductive effect of deuterium on partial charges at the α- or β-positions in the transition
state. In the former case, a change from sp^3 to sp^2
hybridization results in some loosening of the bending
vibrations of the carbon-hydrogen bonds in the transition state, leading to a normal secondary isotope
effect (hydrogen favored over deuterium). In the
latter case, the anharmonicity of the carbon-hydrogen
stretch (i.e., the bond is a little easier to stretch
than it is to compress) means that the carbon-deuterium
bond will, on the average, be a little shorter than the
carbon-hydrogen bond. Consequently, the electrons of
the carbon-deuterium bond will be a little closer to
the carbon atom than the electrons of the carbon-hydrogen bond, and deuterium will be electron releasing relative to hydrogen. The direction of an isotope effect
from this source will depend on the charge density at
the carbon atom in the transition state relative to
the reactant. If there is an increase in electron density from the reactant to the transition state, the
hydrogen compound will react faster, but if there is

a decrease, the deuterium compound will react faster.

(b) Experimental Results on Deuterium and Tritium Isotope Effects. The largest body of experimental results that can be fairly readily compared with each other has been obtained on the 2-phenylethyl system, which also permits comparisons of isotope effects and Hammett ρ values (Section II.c.2.a). Data on primary deuterium isotope effects in this system are recorded in Table 5. The first four entries in the Table show that the magnitude of the isotope effect with respect to the leaving group runs in the order $Br > OTs > SMe_2 > NMe_3$. This could mean either that the proton is not

TABLE 5 Primary Deuterium Isotope Effects for E2 Reactions of 2-Phenylethyl Derivatives[a]

X	Solvent/Base	T $^{\circ}$C	k_H/k_D	Reference
Br	EtOH/EtO$^-$	30	7.11 ± 0.17	66
OTs$_+$	EtOH/EtO$^-$	30	5.66 ± 0.52	66
SMe$_2$$_+$	EtOH/EtO$^-$	30	5.07 ± 0.22	66
NMe$_3$	EtOH/EtO$^-$	50	2.98 ± 0.08	66
Br	t-BuOH/t-BuO$^-$	30	7.89 ± 0.46	66
OTs$_+$	t-BuOH/t-BuO$^-$	30	8.01 ± 0.32	66
NMe$_3$	t-BuOH/t-BuO$^-$	30	6.96 ± 0.13	32
SOMe	t-BuOH/t-BuO$^-$	52	2.7	123
Br$_+$	t-BuOH-DMSO/t-BuO$^-$	30	8.19 ± 0.05	68
SMe$_2$$_+$	H$_2$O/OH$^-$	30	5.93[b]	66
SMe$_2$$_+$	H$_2$O/OH$^-$	50	5.03[c]	66
SMe$_2$$_+$	19% DMSO-H$_2$O/OH$^-$	50	5.29 ± 0.09	35
SMe$_2$$_+$	26% DMSO-H$_2$O/OH$^-$	50	5.55 ± 0.12	35
SMe$_2$$_+$	38% DMSO-H$_2$O/OH$^-$	50	5.73 ± 0.03	35
SMe$_2$$_+$	50% DMSO-H$_2$O/OH$^-$	50	5.61 ± 0.02	35
SMe$_2$$_+$	60% DMSO-H$_2$O/OH$^-$	50	5.63 ± 0.03	35
SMe$_2$$_+$	72% DMSO-H$_2$O/OH$^-$	50	4.79 ± 0.09	35
NMe$_2$$_+$	17% DMSO-H$_2$O/OH$^-$	60	3.80 ± 0.15	70
NMe$_3$$_+$	23% DMSO-H$_2$O/OH$^-$	60	3.86 ± 0.20	70
NMe$_3$$_+$	30% DMSO-H$_2$O/OH$^-$	60	4.77 ± 0.18	70
NMe$_3$$_+$	34% DMSO-H$_2$O/OH$^-$	60	5.22 ± 0.11	70
NMe$_3$$_+$	41% DMSO-H$_2$O/OH$^-$	60	4.94 ± 0.08	70
NMe$_3$$_+$	44% DMSO-H$_2$O/OH$^-$	60	4.83 ± 0.04	70
NMe$_3$$_+$	51% DMSO-H$_2$O/OH$^-$	60	4.43 ± 0.04	70
NMe$_3$$_+$	57% DMSO-H$_2$O/OH$^-$	60	3.83 ± 0.01	70
NMe$_3$$_+$	i-PrOH/i-PrO$^-$	30	3.80	71
NMe$_3$$_+$d	EtOH/EtO$^-$	40	3.23 ± 0.04	92
NMe$_3$$_+$e	EtOH/EtO$^-$	40	2.64 ± 0.04	92
NMe$_3$	EtOH/EtO$^-$	40	3.48 ± 0.06	92

TABLE 5   Continued

X	Solvent/Base	T $^\circ$C	k_H/k_D	Reference
NMe$_3^{+f}$	EtOH/EtO$^-$	40	4.15 ± 0.05	92
Cl	MeCN/F$^-$	25	3.99	72
OTs^g	C$_6$H$_6$/t-BuO$^-$	80	2.3 (syn) 3.9 (anti)	73
Cl^g	C$_6$H$_6$/t-BuO$^-$	80	2.9 (syn) 4.8 (anti)	73

[a]The comparison is between PhCH$_2$CH$_2$X and PhCD$_2$CH$_2$X unless otherwise specified.
[b]Extrapolated from data at higher temperatures.
[c]Interpolated from data in Ref. 66.
[d]Reactant was the 2-(p-methoxyphenyl)ethyl derivative.
[e]Reactant was the 2-(p-chlorophenyl)ethyl derivative.
[f]Reactant was the 2-(p-trifluoromethylphenyl)ethyl derivative.
[g]Reactant was the threo-2-phenylethyl-1,2-d$_2$ derivative. Appreciable syn elimination is observed under those reaction conditions, so isotope effects for both the syn and anti components are reported.

yet half transferred but is most completely transferred with the bromide, or that it is more than half transferred and most completely transferred with the quaternary ammonium salt. If we refer to the Hammett ρ values in Table 1, we find that they run in the order NMe$_3^+$>SMe$_2^+$>OTs>Br. A large ρ value means a high degree of carbanion character, which is most easily achieved by a high degree of proton transfer. Consequently, it is very probable that the second explanation of the isotope effects if the correct one, and that the extent of proton transfer in the transition state runs in the order Br<OTs<SMe$_2^+$<NMe$_3^+$. Such an order is quite reasonable, since it is also the order of increasing difficulty of displacement of the leaving group, and one would expect an increasing buildup of electron density on the β-carbon to be necessary to effect the displacement.

Results on the quaternary ammonium salts in ethanol/ethoxide show that electron-repelling substituents in the 2-phenyl group decrease and electron-withdrawing substituents increase the isotope effect. In view of our conclusion above that the proton transfer is well past the midpoint with the quaternary ammonium salt, we can conclude that electron repulsion increases and electron withdrawal decreases the extent of proton transfer in the transition state. This conclusion is in accord with both the Hammond postulate and the Swain-Thornton rules (Section II.B.1,2), for electron-repelling substituents increase the electron density and "basicity" of the β-carbon atom.

Most of the remaining data illustrate the effect of base and solvent on transition-state structure. The change from ethanol/ethoxide to more branched alcohol/alkoxide media is seen to lead to an increase in the isotope effect. The change probably results from a decrease in the extent of proton transfer as a result of the increased basicity of the reagent, though it could also reflect a looser transition state caused by the greater steric requirements of the base, or the incursion of some proton tunneling.[63] Evidence that the base strength effect is decisive is afforded by studies in which increasing concentrations of dimethyl sulfoxide are added to hydroxide in water. With both the sulfonium and ammonium salts, the isotope effect increases initially and then decreases, precisely the behavior expected from the three-center model of the proton-transfer transition state (Section II.C.3a). It is known that dimethyl sulfoxide increases the basicity of the medium,[98] hence, we seem to have a clear case of decreasing extent of proton transfer with increasing basicity of the medium.

The data on the two threo-2-phenylethyl-1,2-d_2 derivatives were obtained under conditions where both syn and anti elimination (see Section III.A for definitions) contribute to the overall reaction, and separate isotope effects for the two processes can be calculated. The larger isotope effect for anti elimination in each case very probably indicates that proton transfer has progressed farther in the syn than in the anti transition state. The data in Table 6 on deuterium isotope effects in eliminations from three cyclic quaternary ammonium salts support this view. The isotope effect for syn is smaller than for anti elimination in all three cases, and the evidence is strong that the proton transfer is well past the midpoint in the anti eliminations from the quaternary ammonium salts. Although it is possible that a factor other than extent of proton transfer is responsible for the differences between the syn and anti isotope effects,[73] theoretical calculations show that the geometric difference per se cannot be responsible,[62] unless one assumes a nonlinear proton transfer for the syn process.

It is generally more difficult to interpret the data in Table 6 than those in Table 5, for reactant

TABLE 6 Primary Deuterium and Tritium Isotope Effects for Miscellaneous E2 Reactions

Substrate	Solvent/Base, T °C	k_H/k_D (T)	Reference
(p-ClPh)$_2$CDCCl$_3$	EtOH/EtO$^-$, 25	3.8	74
(p-ClPh)$_2$CDDCl$_3$	EtOH/EtO$^-$, 45	3.4	74
(p-ClPh)$_2$CDCCl$_3$	EtOH/PhS$^-$, 45	3.1	74
PhCD(CH$_3$)CH$_2$Br	EtOH/EtO$^-$, 25	7.82	75,76
PhCD(CH$_3$)CH$_2$Br	EtOH/EtO$^-$, 35	6.81	75,76
trans-2-d-Cyclohexyl OTs	EtOH/EtO$^-$, 50	4.47	77
trans-2-d-Cyclohexyl OTs	t-BuOH/t-BuO$^-$, 50	7.53	77
Ph$_2$CDCH$_2$OSO$_2$Ph-p-Me	Me cellosolve/MeO$^-$, 50	5.27	78
Ph$_2$CDCH$_2$OSO$_2$Ph	Me cellosolve/MeO$^-$, 50	5.42	78
Ph$_2$CDCH$_2$OSO$_2$Ph-p-NO$_2$	Me cellosolve/MeO$^-$, 50	6.70	78
PhCD$_2$CMe$_2$Cl	MeOH/MeO$^-$, 76	2.6	79
PhCD$_2$CMe$_2$Cl	MeOH/EtS$^-$, 76	2.4	79
9-Br-9,9'-bifluorenyl-9'-d	t-BuOH/t-BuO$^-$, 30	8.0	80
9-Br-9,9'-bifluorenyl-9'-d	t-BuOH/OH$^-$, 30	7.0	80
9-Br-9,9'-bifluorenyl-9'-d	t-BuOH/MeO$^-$, 30	6.2	80
9-Br-9,9'-bifluorenyl-9'-d	t-BuOH/piperidine, 30	6.0	80
CD$_3$CH$_2$NMe$_3$$^+$	(HOCH$_2$CH$_2$)$_2$O, RO$^-$, 137	3.9	81
MeCHTCH$_2$NMe$_2$$^+$	pyrol. RNMe$_3$$^+OH^-$, 60	2.9[d]	82,83
CH$_2$TCH$_2$NMe$_3$$^+$	pyrol. RNMe$_3$$^+OH^-$, 60	3.0[d]	82,83
p-NO$_2$PhCHTCH$_2$NMe$_3$$^+$	MeOH/MeO$^-$, 60	8.0[d]	82,83
p-NO$_2$PhCHTCH$_2$NMe$_3$$^+$	pH 6.0 buffer, 98	1.8[d]	84
p-NO$_2$PhCHTCH$_2$NMe$_3$$^+$	pH 6.6 buffer, 98	2.0[d]	84
p-NO$_2$PhCHTCH$_2$NMe$_3$$^+$	EtOH-6% HCONH$_2$, 98	2.4[d]	84
p-NO$_2$PhCHTCH$_2$NMe$_3$$^+$	EtOH-5% HCONMe$_2$, 98	2.3[d]	84
trans-cyclopentyl-2-d-NMe$_3$$^+$(anti elim.)[a]	t-BuOH/t-BuO$^-$, 70	4.7	85

TABLE 6 Continued

Substrate	Solvent/Base, T °C	k_H/k_D (T)	Reference
cis-3,3-Dimethylcyclo-pentyl-2-d-NMe₃⁺(syn elim.)b	t-BuOH/t-BuO⁻, 70	1.8	85
cis-2-Phenylcyclohexyl-2-d-NMe₃⁺(anti elim.)c	EtOH/EtO⁻, 60	5.40	94
trans-2-Phenylcyclohexyl-2-d-NMe₃⁺(syn elim.)c	EtOH/EtO⁻, 60	2.63	94
cis-2-Phenylcyclopentyl-2-d-NMe₃⁺(anti elim.)c	EtOH/EtO⁻, 60	4.44	94
trans-2-Phenylcyclopentyl-2-d-NMe₃⁺(syn elim.)c	EtOH/EtO⁻, 60	3.08	94
Me₂CDCH(OTs)CH₃	EtOH/EtO⁻, 50	2.6	99
Me₂CHCH(OTs)CD₃	EtOH/EtO⁻, 50	6.3	99
Me₂CDCH(OTs)CH₃	Me₂CO/Cl⁻	2.3	99
trans-2-Methylcyclo-hexyl-2,6,6-d₃	Me₂CO/Cl⁻	2.7	99
cis-2-Methylcyclo-hexyl-2,6,6-d₃	Me₂CO/Cl⁻	3.0	99
trans-4-t-Butylcyclo-hexyl-2,2,6,6-d₄	Me₂CO/Cl⁻	3.2	99

aBoth anti and syn elimination are observed. The figure given is corrected for the syn component.
bCorrected for the anti component.
cThe elimination product is the 1-phenylcycloalkene.
dThe tritium isotope effects are intramolecular; that is, the comparison is between loss of H or T from the same molecule.

83

structure, leaving group and reaction conditions are
all being changed. The increased isotope effect with
increasing base strength is illustrated in a number of
entries in Table 6. There are no discernible trends
among the tritium isotope effects. They are mostly
rather small, particularly when one remembers that
tritium isotope effects should be larger than their
deuterium counterparts, and can be regarded as consistent with a high degree of proton transfer in the transition state. The one large value involves a stronger
base than is used in the other reactions of 2-(p-nitro-
phenylethyl)trimethylammonium ion, but so large a
difference is suprising.

Table 6 contains a number of examples of reactions
that may belong to the E2C category (Section II.A.5),
particularly the reactions with thiolate ions in ethanol and chloride ion in acetone. The isotope effects
are generally small, a fact consistent with either a
nonlinear transition state[62,64] or with a linear transition state having an unsymmetrically located proton.
The most interesting entries are those for elimination
into the two branches of 3-methyl-2-butyl tosylate
with ethoxide in ethanol, where a much larger isotope
effect is found for formation of 3-methyl-1-butene
than 3-methyl-2-butene. It has been suggested that
the Saytzev-rule product is formed via a more E2C-like
transition state than the Hofmann-rule product.[99]

Some secondary and solvent isotope effects are recorded in Table 7. The α-deuterium isotope effects
would be expected to be related to the extent of cleavage of the bond to the leaving group in the transition state. The pattern is consistent with this expectation, since the better leaving groups, such as bromide and tosylate, give larger effects than the poorer
leaving groups such as trimethylammonio and dimethylsulfonio. An electron-withdrawing substituent in the
phenyl group of 2-phenylethyl tosylate is seen to lead
to decreased cleavage of the carbon-tosylate bond in
the transition state. This result is as expected if
the extent of cleavage of this bond parallels the extent of cleavage of the carbon-hydrogen bond (Section
II.B.2). It also provides striking suppprt for
Cockerill's theory of the tosylate/bromide rate ratio
(Section II.C.2.b).[87] There does not seem to be any
clear dependence of the α-deuterium isotope effect on
the nature of the base and solvent.

The secondary β-deuterium and tritium isotope
effects are somewhat larger and more variable than
the α-effects. There is a rough trend toward larger
effects with the stronger bases, but the data are too

TABLE 7 Secondary Deuterium and Tritium Isotope Effects in Some E2 Reactions

Substrate	Solvent/Base, T °C	k_H/k_D (T)	Reference
$PhCH_2CD_2Br$	$EtOH/EtO^-$, 60	1.09	86
$p-MeOPhCH_2CD_2OTs$	$t-BuOH/t-BuO^-$, 30	1.047	87
$PhCH_2CD_2OTs$	$t-BuOH/t-BuO^-$, 30	1.043	87
$p-ClPhCH_2CD_2OTs$	$t-BuOH/t-BuO^-$, 30	1.017	87
Cyclohexyl-1-d OTs	$EtOH/EtO^-$, 50	1.14	77
Cyclohexyl-1-d OTs	$t-BuOH/t-BuO^-$, 50	1.15	77
Cyclohexyl-cis-2-d OTs	$EtOH/EtO^-$, 50	1.36	77
Cyclohexyl-cis-2-d OTs	$t-BuOH/t-BuO^-$, 50	1.51	77
$PhCH_2CD_2SMe_2^+$	H_2O/OH^-, 80	1.00a	86
$PhCH_2CD_2NMe_3^+$	H_2O/OH^-, 97	1.02	88
$PhCH_2CD_2NMe_3^+$	$EtOH/EtO^-$, 40	1.03	88
$CH_3CH_2CHTNMe_3^+$	Pyrol. $RNMe_3^+$, OH^-, 50	1.10	83
$CH_3CHTCH_2NMe_3^+$	Pyrol. $RNMe_3^+$, OH^-, 60	1.33	83
$TCH_2CH_2NMe_3^+$	Pyrol. $RNMe_3^+$, OH^-, 60	1.26	83
$TCH_2CH_2CH_2NMe_3^+$	Pyrol. $RNMe_3^+$, OH^-, 60	1.11	83
$p-NO_2PhCHTCH_2NMe_3^+$	$MeOH/MeO^-$, 22	1.18	83
$p-NO_2PhCHTCH_2NMe_3^+$	pH 6.0 buffer, 98	1.20	84
$p-NO_2PhCHTCH_2NMe_3^+$	pH 6.6 buffer, 98	1.20	84
$p-NO_2PhCHTCH_2NMe_3^+$	$EtOH$, 6% $HCONH_2$, 98	0.97	84
$p-NO_2PhCHTCH_2NMe_3^+$	$EtOH$, 5% $HCONMe_2$, 98	0.97	84
$PhCH_2CH_2SMe_2^+$	H_2O versus D_2O/ OH^- versus OD^-, 80	0.64b	97
$PhCH_2CH_2NMe_2Ph^+$	H_2O versus D_2O/ OH^- versus OD^-, 80	0.62b	97
$p-ClPhCH_2CH_2NMe_3^+$	H_2O versus D_2O/ OH^- versus OD^-, 80	0.58b	97

TABLE 7 Continued

Substrate	Solvent/Base, T °C	k_H/k_D (T)	Reference
$PhCH_2CH_2NMe_3$	H_2O versus D_2O/ OH⁻ versus OD⁻, 80	0.56^b	97

[a]Extensive exchange of deuterium with solvent during reaction, so actual figure for isotope effect may be somewhat higher.
[b]These are solvent isotope effects, k_{OH^-}/k_{OD^-} in the respective solvents.

miscellaneous to be sure. One would again expect a
larger effect the greater the extent of carbon-hydrogen
cleavage in the transition state, but both theory and
the other evidence we have discussed point to decreasing carbon-hydrogen cleavage with increasing base
strength. The figures on cyclohexyl tosylate should
perhaps be viewed with caution because a small contribution from syn elimination could introduce some primary isotope effect, and the precision of the tritium
effects is difficult to judge. A more systematic and
comprehensive set of experimental results on β-secondary effects would be desirable.

The solvent isotope effects appear to be due mainly to the isotope effect on the strength of the attacking base. Steffa and Thornton[97] point out that the
ratio of the base strength of hydroxide ion to that of
deuteroxide ion under the conditions of their experiments is expected to be 0.53. This difference should
be fully reflected in the rate ratio only when the proton is essentially completely transferred to the base
in the transition state. The actual values suggest
that proton transfer is extensive but not quite complete, a conclusion in harmony with primary isotope
effects quoted in Table 5 for similar reactants and
conditions. The order, both with respect to the leaving group and the p-substituent, is as expected from
the primary isotope effects and the Swain-Thornton
rules.[19,20,97]

(c) Experimental Results on Heavy-Atom Isotope
Effects. Measurement of heavy-atom isotope effects in
elimination reactions is feasible with respect to the
α- and β-carbon atoms and the leaving group. Most published studies have concentrated on the leaving group,
although there are a few carbon isotope effects in the
literature. Theoretical calculations indicate (Section
II.C.3.a) that the β-carbon isotope effect should at
first decrease and then increase with extent of proton
transfer, while the α-carbon isotope effect should increase almost linearly with the extent of cleavage of
the bond to the leaving group. The numbers reported in
Table 8 are reasonable but do not permit any very detailed set of conclusions. Miscellaneous conditions
and miscellaneous temperatures are used, and conditions
for pyrolysis of the quaternary ammonium hydroxide are
not easy to reproduce precisely from one experiment to
another. Nonetheless, the results are reasonable.
Most of the α-carbon effects seem to correspond to
about 50% cleavage (i.e., about 50% reduction of the
stretching force constant) of the carbon-nitrogen bond

TABLE 8 Carbon-14 Isotope Effects in Some E2 Reactions[a]

Substrate	Solvent/Base, T °C	$(k_{12}/k_{14} - 1)100$
$CH_3C^*H_2NMe_3^+$	$(HOCH_2CH_2)_2O/RO^-$, 139	3.3 ± 0.4 (5)[b]
$CH_3C^*H_2NMe_3^+$	Pyrol. $RNMe_3OH$, 40	6.5 (6)
$CH_3CH_2C^*H_2NMe_3^+$	Pyrol. $RNMe_3OH$, 50	7.5 ± 0.6 (8)
$(CH_3)_3C^*NMe_3^+$	Triethylene glycol/ RO^-, 91	5.2 (6)
$p\text{-}NO_2PhCH_2C^*H_2NMe_3^+$	pH 7 buffer, 100	2.6 (3)
$CH_3C^*H_2CH_2NMe_3^+$	pyrol. $RNMe_3OH$, 51	3.6 (4)

[a]From Refs. 83 to 84.
[b]Figures in parentheses are estimated effects for a common temperature of 40°, using the temperature dependence derived for a simple bond-cleavage model.

in the transition state, which is somewhat more than one would judge from most of the nitrogen isotope effects that have been measured (see below and Table 9). The one β-carbon isotope effect appears smaller than the α-effects, but the form of the dependence on extent of carbon-hydrogen bond cleavage is such that it probably represents, at least, 50% cleavage.[62]

Table 9 records results on leaving-group isotope effects. Thus far, the data cover only sulfonium and ammonium salts, but there is no reason to believe that the qualitative conclusions drawn for them will not apply to other leaving groups. The effects appear to run distinctly below those expected for complete rupture of the bond to the leaving group. With sulfur, for example, cleavage of the hypothetical diatomic molecule C—S leads to an isotope effect of about 1.5% at 25°, while more sophisticated calculations on the ethyldimethylsulfonium ion give effects ranging from 1.1 to 1.3% in most cases.[61,62,101] Thus the largest sulfur isotope effects in Table 9 seem to correspond to a carbon-sulfur bond that is approximately half broken in the transition state. With nitrogen, the simple C—N and the more sophisticated ethyltrimethylammonium ion models both give values between 4.0 and 4.5%.[62,101] A comparison of these figures with the experimental values suggests that the carbon-nitrogen bond is somewhat less completely broken in the transition state than the carbon-sulfur bond. In most of the experiments, about 20 to 40% cleavage would fit the results.

Increasing the strength of the attacking base lowers the leaving-group isotope effect, as can be seen for the addition of dimethyl sulfoxide to aqueous hydroxide in eliminations from the 2-phenylethyldimethylsulfonium ion, and in the change from ethoxide in ethanol to t-butoxide in t-butyl alcohol in eliminations from ethyltrimethylammonium ion. Electron-withdrawing substituents on the β-phenyl group decrease the isotope effect in eliminations from 2-arylethyltrimethylammonium ions. These data, in conjunction with hydrogen isotope effects in the same systems (Section II.C.2.b), allow us to conclude that the extent of proton transfer and the extent of cleavage of the bond between the α-carbon and the leaving group are decreasing in both cases. This conclusion is consistent with predictions from the Swain-Thornton rules (Section II.B.2).

The last four entries in Table 9 compare isotope effects for syn and anti eliminations. From them, the extent of carbon-nitrogen cleavage seem to be distinctly less in the syn elimination, while the hydrogen isotope effects for the same reactions indicate that the

TABLE 9 Leaving-Group Isotope Effects for Some E2 Reactions

Substrate	Solvent/Base, T °C	$(k_L/k_H - 1)100$[a]	Reference
PhCH$_2$CH$_2$SMe$_2$$^+$	H$_2$O/OH$^-$, 59	0.64 ± 0.012	89
PhCH$_2$CH$_2$SMe$_2$$^+$	H$_2$O/OH$^-$, 40	0.74	90
PhCH$_2$CH$_2$SMe$_2$$^+$	H$_2$O-1.4M DMSO/OH$^-$, 40	0.65 ± 0.025	90
PhCH$_2$CH$_2$SMe$_2$$^+$	H$_2$O-2.8M DMSO/OH$^-$, 40	0.64 ± 0.023	90
PhCH$_2$CH$_2$SMe$_2$$^+$	H$_2$O-4.2M DMSO/OH$^-$, 40	0.38 ± 0.021	90
PhCH$_2$CH$_2$SMe$_2$$^+$	H$_2$O-7.0M DMSO/OH$^-$, 40	0.11 ± 0.056	90
Me$_3$CSMe$_2$$^+$	97% EtOH/EtO$^-$ -OH$^-$, 24	0.72 ± 0.04	100
PhCH$_2$CH$_2$NMe$_3$$^+$	EtOH/EtO$^-$, 40	1.17 ± 0.07	91
PhCH$_2$CH$_2$NMe$_3$	EtOH/EtO$^-$, 40	1.42 ± 0.04	92,93
p-MeOPhCH$_2$CH$_2$NMe$_3$$^+$	EtOH/EtO$^-$, 40	1.37 ± 0.09	92,93
p-ClPhCH$_2$CH$_2$NMe$_3$$^+$	EtOH/EtO$^-$, 40	1.14 ± 0.09	92,93
p-CF$_3$PhCH$_2$CH$_2$NMe$_3$$^+$	EtOH/EtO$^-$, 40	0.88 ± 0.06	92
p-NO$_2$PhCH$_2$CH$_2$NMe$_3$$^+$	pH 6 buffer, 98	2.4	84
PhCH$_2$CH$_2$NMe$_3$$^+$	H$_2$O/OH$^-$, 99	0.9	93
CH$_3$CH$_2$NMe$_3$$^+$	EtOH/EtO$^-$, 60	1.86 ± 0.04	95
CH$_3$CH$_2$NMe$_3$$^+$	t-BuOH/t-BuO$^-$, 60	1.41 ± 0.03[b]	95
5-nonyl-NMe$_3$$^+$	t-BuOH/t-BuO$^-$, 55	1.00 ± 0.03[b]	95
cyclodecyl-NMe$_3$$^+$	t-BuOH/t-BuO$^-$, 55	1.31 ± 0.08[c]	95
cis-2-Phenylcyclo-hexyl-NMe$_3$$^+$ (anti elimination)[d]	95% EtOH/EtO$^-$-OH$^-$, 60	1.23 ± 0.07	94,96
trans-2-Phenylcyclo-hexyl-NMe$_3$$^+$ (syn elimination)[d]	95% EtOH/EtO$^-$-OH$^-$, 60	0.39 ± 0.03	94,96

cis-2-Phenylcyclo-pentyl-NMe$_3$+(anti elimination)[d]	95% EtOH/EtO$^-$--OH$^-$, 60	1.08 ± 0.02	94
trans-2-Phenylcyclo-pentyl-NMe$_3$+(syn elimination)[d]	95% EtOH/EtO$^-$--OH$^-$, 60	0.64 ± 0.06	94

[a] The light and heavy (L and H) isotopes are S^{32} and S^{34} for the sulfonium salts, and N^{14} and N^{15} for the ammonium salts.
[b] Reacts partly by syn elimination (Section III.E.6).
[c] Reacts mainly by syn elimination (Section III.E.5).
[d] The product in each case is the 1-phenylcycloalkene.

extent of proton transfer is greater in the syn elimination (see Table 6 and Section II.C.3.b). Thus the
syn transition state appears to have considerably more
carbanion character than the anti.[85] One discordant
note is the isotope effect for cyclodecyltrimethyl-
ammonium ion, which is rather large even though there
is good evidence that it reacts mainly or entirely via
a syn transition state (Section III.E.5).

4 Variations in the Base and the Solvent

(a) Base, Solvent, and Ion-Pairing Effects. We have
laredy noted (Section II.C.2,3) changes in transition-
state structure that were brought about by changes in
solvent and base. In this section we consider the extent to which base and solvent effects can be separated, and to comment briefly on the potential importance
of ion-pairing effects.
     Most elimination reactions are conducted in protic
solvents, where there is always the possibility that
added base will interact with the solvent to produce
significant concentrations of the conjugate base of the
solvent (Eq. 11). In such cases, the concentration of

$$ROH \quad + \quad B \rightleftharpoons RO^- \quad + \quad BH^+ \tag{11}$$

the conjugate base of the solvent will depend on the
ratio of the concentrations of B and BH^+. Whether the
rate of base-promoted reaction depends on the concentration of B or on the ratio $(B)/(BH^+)$, can be used
to decide whether the effective base is B or RO^-. Of
course, there will also be cases where both are involved. Only when B is very much weaker than RO^- is it
safe to assume that reaction via RO^- is insignificant.
Eliminations promoted by phenoxides in ethanol, for
example, are accompanied by ethoxide-promoted reaction,
an interference that is absent in eliminations promoted by the more weakly basic thiophenoxides.[102]
     When the added base is the conjugate base of the
solvent, these complications are avoided, but new ones
are introduced. A change in base must then be accompanied by a change in solvent, and one is left with the
problem of deciding whether resulting changes in the

reaction under study were caused by the change of base, the change of solvent, or both. Furthermore, a series of relative base strengths can be rigorously establish- ed only with reference to a common solvent. The con- cept of the basicity of the medium offers a good prac- tical solution to the latter problem. We have already discussed (Section I.B.1.d) the acidity function, H_-, which is a measure of the tendency of a basic medium to accept a proton from a neutral acid.[106] While there is not always a simple quantitative relationship between H_- and the rates of elimination reactions, it is quali- tatively clear that an increase in H_- results in an in- crease in the rate of elimination. In addition, a num- ber of deuterium isotope-effect studies (Section II. C.3.b) have shown that increasing H_- leads to less com- plete proton transfer in the transition state. The H_- value is probably the best objective criterion of bas- icity available, and values have been determined for a variety of base/solvent systems, including cases where the solvent is a mixture of a hydroxylic and dipolar aprotic solvent.[98,106]

From the above, it is clear that the major effect of a solvent change on an E2 reaction lies in the change of the basicity of the medium. It is very pro- bable that the solvent also exerts an effect on the ease of departure of the leaving group via hydrogen bonding or other types of solvation. Thus far there has been no work aimed at quantitative separation of these effects, and it is often not even easy to distin- guish them qualitatively. A possible case is the reac- tion of 2-phenylethyl bromide with potassium t-butoxide in mixtures of t-butyl alcohol and dimethyl sulfox- ide.[68] The rate of reaction rises much faster than the H_- of the medium as dimethyl sulfoxide is added, suggesting the operation of some effect other than base strength alone. H_- is generally measured with nitrogen acids as indicators while elimination reactions involve proton removal from carbon, but this does not seem capable of accounting for major deviations. An H_- scale based on substituted fluorenes as indicators was very close in form to one based on nitrogen acids.[107, 108]

Because t-butoxide in t-butyl alcohol behaves as a much stronger base than ethoxide in ethanol, it has sometimes been assumed that adding a little ethanol to the former medium will convert all of the t-butoxide to ethoxide.[31,109] This is, in effect, the same situation as that encountered with an added base in protic sol- vent, and the possibilities noted earlier for reaction via either or both of the possible basic species exist.

Recent work has shown that there is little difference
in acidity of different alcohols measured in t-butyl
alcohol as solvent, so that attempts to prepare solutions of alkoxides of these alcohols in t-butyl alcohol
lead to mixtures of the alkoxide with t-butoxide.[110]
Earlier measurements of the acidities of common alcohols in isopropyl alcohol as solvent revealed a similarly narrow range.[111]

An almost unexplored aspect of elimination reactions is the effect of ion pairing involving the anionic base or cationic substrates. In, at least, some
of the media used for elimination reactions, it is
clear that extensive ion pairing is possible. The
basicity function of potassium t-butoxide, for example,
is parallel to but lies significantly above that of
sodium t-butoxide in t-butyl alcohol.[106] The rates of
elimination reactions with 2-arylethyl bromides and 2-
arylethyltrimethylammonium salts are higher by up to
2.6-fold with potassium than with sodium t-butoxide in
t-butyl alcohol.[32,68] Further evidence for extensive
ion pairing is afforded by the fact that a 0.1-M solution of sodium t-butoxide in t-butyl alcohol has a conductance only 6% higher than that of pure t-butyl alcohol.[32] In contrast, quaternary ammonium salts markedly increase the conductance. A plot of conductance
versus concentration for benzyltrimethylammonium chloride in t-butyl alcohol is linear from 0.02 to 0.1 M
with unit slope.[32] The reaction of 2-phenylethyltri-
methylammonium bromide with potassium t-butoxide in t-
butyl alcohol does not follow simple second-order kinetics. Its behavior is qualitatively consistent with
the equilibrium of Eq. 13, with the free t-butoxide ion
being considerably more effective as a base than the

$$t\text{-BuO}^-M^+ \; + \; PhCH_2CH_2NMe_3^+ \; + \; Br^- \; \rightleftharpoons$$

$$M^+Br^- \; + \; PhCH_2CH_2NMe_3^+ \; + \; t\text{-BuO}^- \quad (13)$$

t-butoxide ion paired with the metal ion.[32] A major
question raised by these observations is whether ion
pairing phenomena affect differently the transition
states of different E2 processes and, thereby, relative rates of different processes as well as the overall rate. Recent work has shown that the proportion
of syn elimination from 3-hexyl-4-d-trimethylammonium
ion depends markedly on the nature of the positive ion
associated with the base.[112] Further exploration may
well reveal other interesting and useful effects of
ion pairing on stereochemistry and orientation in elimination reactions. Some further aspects of the effect

of ion pairing on stereochemistry are mentioned later in this volume (Section III.E.5,6 and Section III.F).

(b) The Brønsted Correlation. Nearly fifty years ago, it was suggested by Brønsted and Pedersen[113],[114] that the effectiveness of the catalyst in an acid- or base-catalyzed reaction should be related to its strength as an acid or a base. For a base-promoted process such as the E2 reaction, the relation takes the form of Eq. 14, where k is the rate constant for the reaction, K_b

$$\log k = \beta \log K_b + \log G \qquad (14)$$

is the ionization constant for the base, G is a constant, and β is a proportionality factor. The interesting aspect of the correlation from our point of view is that β can be taken as a measure of the extent of transfer of the proton from substrate to base in the transition state. The argument for this assertion can be presented in the form of molecular potential-energy curves,[115] but a simpler qualitative formulation suffices to demonstrate its reasonableness. If the proton is completely transferred to the base in the transition state, the difference in activation energies for reaction with two different bases should reflect fully the difference in their base strength, and β is expected to be unity. If the proton is not completely transferred in the transition state, the effect of a change in base strength on the activation energy will be less. In the limit of an insignificant degree of proton transfer in the transition state, it will be negligible. Thus β is expected to vary between zero and unity, and should depend on the degree of bond formation between the attacking base and the proton in the transition state. This is not quite the same as the extent of proton transfer, but it should be close to it provided that the total bonding to the proton does not change too much with the extent of proton transfer.

More recently, doubt has been cast on the utility of Brønsted coefficients by the discovery that they can be less than zero or greater than unity for reactions of nitroalkanes with bases where the structural variation is in the nitroalkane rather than in the base.[116] Additional data on nitroalkanes and ketones have been presented to support the contention that β is not a reliable guide to transition-state structure.[117] In a theoretical treatment of the Brønsted correlation, Marcus points out that two factors determine β values: the standard free energy of the reaction in question, and the intrinsic barrier to reaction when the free

energy change is zero.[118]   Only when the first of these
terms is dominant is β a reliable measure of the extent
of proton transfer. The second term is expected to remain nearly constant when the structural variation is
in oxygen or nitrogen acids or bases, which undergo
proton transfers at rates that are diffusion controlled
or nearly so, but not when the variation is in pseudo
acids such as nitroalkanes, which undergo slow proton
transfers. Such a conclusion is in line with a number
of recent studies which show that β can be a good measure of the extent of proton transfer and can correlate
well with isotope effects.[119-121]   One would conclude
that the results of Bordwell et al. point up the need
for caution in interpreting Brønsted coefficients, but
do not require that they be abandoned as measure of
transition-state structure.

     Table 10 records β values for some E2 reactions.
The range of base types is necessarily rather small because of the requirement that the base be much weaker
than the conjugate base of the solvent to avoid complications arising from reactions promoted by the latter.
In general, the results are reasonably in accord with
expectations and with other measures of transition-
state structure. The poorer the leaving group, the
greater the extent of proton transfer, as is also demonstrated by substituent effects and isotope effects
(Section II.C.2,3). The extent of proton transfer is
greatest for the primary and least for the tertiary
substrates. One conflict with other data is the larger β for 2-(p-nitrophenyl)ethyl than for phenylethyl
bromide. Isotope effects indicate decreased proton
transfer with electron-withdrawing substituents in the
2-arylethyltrimethylammonium ion, in agreement with expectations from the Swain-Thornton rules (Section II.
C.3.b and Table 5). The results with the bromides may
indicate a real breakdown in the relation between β
and transition-state structure, but it should be kept
in mind that the difference in β values is not large,
and the two values are not derived from data on precisely the same set of bases.[102]   It is interesting
that the β values for the thiophenoxide bases cover
such a wide range. These bases are considered to favor
E2C-like transition states (Section II.A.5) possessing
a high degree of double-bond character but little or no
carbanion character. The reaction with DDT (line 4 of
Table 10) seems unlikely to belong in this category,
indicating that thiolates can be effective in normal
E2 reactions if their propensity for substitution is
blocked. The low β values for the tertiary halides,
coupled with evidence that E2C-like reactions show high

TABLE 10 Brønsted β Values for Some E2 Reactions

Substrate	Base/Solvent,	T °C	β	Reference
PhCH$_2$CH$_2$Br	ArO$^-$/EtOH,	60	0.56(0.54)[a]	102
p-NO$_2$PhCH$_2$CH$_2$Br	ArO$^-$/EtOH,	60	0.72(0.67)[a]	102
(p-ClPh)$_2$CHCCl$_3$	ArO$^-$/EtOH,	45	0.88 ± 0.05	74
(p-ClPh)$_2$CHCCl$_3$	ArS$^-$/EtOH,	75	0.77 ± 0.05	74
Cyclo-C$_6$H$_{11}$OTs	ArS$^-$/EtOH,	35	0.27	103
Cyclo-C$_6$H$_{11}$Br	ArS$^-$/EtOH,	55	0.36	103
Cyclo-C$_6$H$_{11}$Cl	ArS$^-$/EtOH,	55	0.39	103
1,1-cyclo-C$_6$H$_{10}$Br$_2$	ArS$^-$/EtOH,	55	0.51	103
1,1-cyclo-C$_6$H$_{10}$Cl$_2$	ArS$^-$/EtOH,	55	0.58	103
t-BuCl	ArS$^-$/EtOH,	45	0.17	104
t-BuSMe$_2$$^+$	ArS$^-$/EtOH,	25	0.46	104
EtMe$_2$CCl	ArS$^-$/EtOH,	55	0.19[b]	105
EtMe$_2$CCl	ArS$^-$/i-PrOH,	55	0.16[b]	105
EtMe$_2$CCl	ArS$^-$/t-BuOH,	55	0.13[b]	105

[a] Values in parentheses corrected for reaction via ethoxide in equilibrium with phenoxide.
[b] Based on pK$_a$ values for ArSH measured in ethanol.

double-bond character (Section II.C.2.c) suggests a
loose transition state in which the sum of the bonds to
the proton may be considerably less than a full single
bond.

D   CONCLUSION

All in all, there is a remarkable degree of consistency
between different experimental measures of the structure of transition states in elimination reactions, as
well as between the experimental measures and theoretical predictions from the Swain-Thornton rules and the
Hammond postulate. Even though quantitative predictions still elude us, qualitative ones can be made with
considerable confidence, especially for small and regular changes in reactant structure or reaction conditions.
     Dealing with large changes, particularly those
that promote changes in mechanism, is more difficult.
We still cannot predict reliably when an elimination
will choose a syn over an anti pathway, and we know
much less about electronic effects on syn than on anti
eliminations. E2C-like mechanisms are still puzzling
in many respects. Even though the basic structural
and environmental effects have been thoroughly explored, we are still unsure of the position of the attacking base and whether it is interacting significantly
with the α-carbon atom of the substrate. We cannot
explain why thiolate bases seem to give transition
states with a lesser degree of proton transfer than do
alkoxide bases, even though the thiolates are much
weaker, and both theoretical predictions and other experimental data indicate that the degree of proton
transfer should be greater with the weaker base. Another difficult-to-explain point is that the E1-like extreme of the spectrum of E2 transition states does not
seem ever to be utilized in real reactions.
     The basic factor behind many of the problems is
doubtless the fact that it is much more difficult to
predict or correlate the effects of large changes in
structure or conditions. It should be emphasized that
one of the theoretical frameworks we have used in our
discussions, the Swain-Thornton rules, arises from a
perturbation treatment that is intended to be applicable only to small changes. We are all familiar with
the fact that quantitative correlations of rates and
equilibria, such as the Hammett equation and the
Brønsted relationship, become poorer, the wider the
range of the data to which they are applied. We have

seen earlier in this chapter (Section II.C.3.b) numerous examples of the correlation of hydrogen isotope effects with the extent of proton transfer, including two cases in which the maximum effect predicted for half transfer of the proton is observed. Yet Bordwell and Boyle[122] have recently pointed out that a plot of observed isotope effects against the difference in pK values of the donor and the conjugate acid of the base gives bad scatter in which a maximum can be discerned, but not located with any precision. On this basis, they conclude that the isotope effect is of little if any value in determining transition-state structure. The results certainly emphasize that it is risky to draw conclusions about the extent of proton transfer when a large change in the structure of the base or the substrate produces a small change in isotope effect. We still believe, however, that isotope effects are our best single probe for transition-state structure when used with due caution.

Finally, there is still much that is uncertain about efforts to apply our conclusions concerning the electronic structure of the transition state to reactions where both steric and electronic factors are varying. This is true, as noted above, with respect to prediction of stereochemistry, and it is equally true with respect to prediction of orientation. These problems are discussed in more detail in Chapters III and IV.

REFERENCES

1. W. Hanhart and C. K. Ingold, J. Chem. Soc., 997 (1927).
2. R. L. Letsinger, A. W. Schnizer, and E. Bobko, J. Am. Chem. Soc., 73, 5708 (1951).
3. D. J. Cram, F. D. Greene, and C. H. DePuy, J. Am. Chem. Soc., 78, 790 (1956).
4. W. H. Saunders, Jr. and S. Ašperger, J. Am. Chem. Soc., 79, 1612 (1957).
5. W. H. Saunders, Jr. and R. A. Williams, J. Am. Chem. Soc., 79, 3712 (1957).
6. W. H. Saunders, Jr. and D. H. Edison, J. Am. Chem. Soc., 82, 138 (1960).
7. C. H. DePuy and D. H. Froemsdorf, J. Am. Chem. Soc., 79, 3705 (1957)
8. J. F. Bunnett, Angew. Chem., 74, 731 (1962); Angew. Chem. Int. Ed. Engl., 1, 225 (1962).
9. J. F. Bunnett, "Olefin-Forming Elimination Reactions," Vol. 5, in A. F. Scott, Ed., "Survey

of Progress in Chemistry," Academic Press, New York, 1969, p. 53.

10. J. Závada, J. Krupička, and J. Sicher, Coll. Czech. Chem. Commun., <u>33</u>, 1393 (1968).

11. R. A. More O'Ferrall, J. Chem. Soc., <u>B</u>, 274 (1970).

12. S. Winstein, D. Darwish, and N. J. Holness, J. Am. Chem. Soc., <u>78</u>, 2915 (1956).

13. P. B. D. de la Mare and C. A. Vernon, J. Chem. Soc., 41(1956).

14. G. Biale, A. J. Parker, S. G. Smith, I. D. R. Stevens, and S. Winstein, J. Am. Chem. Soc., <u>92</u>, 115 (1970).

15. A. J. Parker, M. Ruane, G. Biale, and S. Winstein, Tetrahedron Lett., 2113 (1968).

16. D. Eck and J. F. Bunnett, J. Am. Chem. Soc., <u>91</u>, 3099 (1969).

17. D. Cook and A. J. Parker, Tetrahedron Lett., 4901 (1969).

18. G. S. Hammond, J. Am. Chem. Soc., <u>77</u>, 334 (1955).

19. C. G. Swain and E. R. Thornton, J. Am. Chem. Soc., <u>84</u>, 817 (1962).

20. E. R. Thornton, J. Am. Chem. Soc., <u>89</u>, 2915 (1967).

21. C. G. Swain, D. A. Kuhn, and R. L. Schowen, J. Am. Chem. Soc., <u>87</u>, 1553 (1967).

22. L. P. Hammett, "Physical Organic Chemistry," McGraw-Hill, New York, 1940, Chap. 7. 2nd ed., New York, 1970, Chap. 11.

23. H. H. Jaffé, Chem. Rev., <u>53</u>, 191 (1953).

24. C. A. Coulson, "Valence," Oxford University Press, Oxford, England, 1952, p. 206.

25. A. Streitwieser, Jr. and H. F. Koch, J. Am. Chem. Soc., <u>86</u>, 404 (1964).

26. M. Shima, D. N. Bhattacharyya, J. Smid, and M. Szwarc, J. Am. Chem. Soc., <u>85</u>, 1306 (1963).

27. K. Bowden, A. F. Cockerill, and J. R. Gilbert, J. Chem. Soc., <u>B</u>, 179 (1970).

28. C. H. DePuy and D. H. Froemsdorf, J. Am. Chem. Soc., <u>79</u>, 3710 (1957).

29. T. Yoshida, Y. Yano, and S. Oae, Tetrahedron, <u>27</u>, 5343 (1971).

30. C. H. DePuy and C. A. Bishop, J. Am. Chem. Soc., <u>82</u>, 2535 (1960).

31. C. H. DePuy and C. A. Bishop, J. Am. Chem. Soc., <u>82</u>, 2532 (1960).

32. W. H. Saunders, Jr., D. G. Bushman, and A. F. Cockerill, J. Am. Chem. Soc., <u>90</u>, 1775 (1968).

33. J. Banger, A. F. Cockerill, and G. L. O. Davies, J. Chem. Soc., <u>B</u>, 498 (1971).

34. W. H. Saunders, Jr., C. B. Gibbons, and R. A. Williams, J. Am. Chem. Soc., $\underline{80}$, 4099 (1958).

35. A. F. Cockerill, J. Chem. Soc., $\underline{B}$, 964 (1967).

36. L. P. Hammett, J. Am. Chem. Soc., $\underline{59}$, 96 (1937).

37. L. F. Blackwell, A. Fischer, and J. Vaughan, J. Chem. Soc., $\underline{B}$, 1084 (1967).

38. C. H. DePuy, G. F. Morris, J. S. Smith, and R. J. Smat, J. Am. Chem. Soc., $\underline{87}$, 2421 (1965).

39. J. G. Griepenburg, Ph.D. thesis, University of Rochester, 1970.

40. A. K. Colter and R. D. Johnson, J. Am. Chem. Soc., $\underline{84}$, 3289 (1962).

41. A. K. Colter and D. R. McKelvey, Can. J. Chem., $\underline{43}$, 1282 (1965).

42. F. G. Bordwell, J. Weinstock, and T. F. Sullivan, J. Am. Chem. Soc., $\underline{93}$, 4728 (1971).

43. Y. Yano and S. Oae, Tetrahedron, $\underline{26}$, 27 (1970).

44. S. Oae and Y. Yano, Tetrahedron, $\underline{24}$, 5721 (1968).

45. Y. Yano and S. Oae, Tetrahedron, $\underline{26}$, 67 (1970).

46. A. V. Willi, Helv. Chim. Acta, $\underline{49}$, 1725 (1966).

47. S. J. Cristol, N. L. Hause, A. J. Quant, H. W. Miller, K. R. Eilar, and J. S. Meek, J. Am. Chem. Soc., $\underline{72}$, 3333 (1952); See also Ref. 5.

48. C. A. Bishop and C. H. DePuy, Chem. Ind. (London), 297 (1959).

49. P. Veeravagu, R. T. Arnold, and E. W. Eigenmann, J. Am. Chem. Soc., $\underline{86}$, 3072 (1964).

50. H. M. R. Hoffmann, J. Chem. Soc., 6753 (1965).

51. H. M. R. Hoffmann, J. Chem. Soc., 6762 (1965).

52. G. M. Fraser and H. M. R. Hoffmann, J. Chem. Soc., $\underline{B}$, 265 (1967).

53. A. F. Cockerill, Tetrahedron Lett., 4913 (1969).

54. G. Biale, D. Cook, D. J. Lloyd, A. J. Parker, I. D. R. Stevens, J. Takahashi, and S. Winstein, J. Am. Chem. Soc., $\underline{93}$, 4735 (1971).

55. J. Bigeleisen, J. Chem. Phys., $\underline{17}$, 675 (1949).

56. J. Bigeleisen and M. Goeppert-Mayer, J. Chem. Phys., $\underline{15}$, 261 (1947).

57. J. Bigeleisen and M. Wolfsberg, "Theoretical and Experimental Aspects of Isotope Effects in Chemical Kinetics," in "Advance in Chemical Physics," Interscience, New York, 1958.

58. L. Melander, "Isotope Effects on Reaction Rates," Ronald Press, New York, 1960.

59. Ref. 58, pp. 24-32.

60. F. H. Westheimer, Chem. Rev., $\underline{61}$, 265 (1961).

61. A. M. Katz and W. H. Saunders, Jr., J. Am. Chem. Soc., $\underline{91}$, 4469 (1969).

62. W. H. Saunders, Jr., unpublished results, and Abstracts of Papers, p. 232, 14th Nordic Chemical

Meeting, Umeå, Sweden, June 18-22, 1971.

63. E. F. Caldin, Chem. Rev., 69, 135 (1969).
64. R. A. More O'Ferrall, J. Chem. Soc., B, 785 (1970).
65. E. A. Halevi, in "Progress in Physical Organic Chemistry," Vol. I, S. G. Cohen, A. Streitwieser, Jr., and R. W. Taft, eds., Interscience, New York, 1963.
66. W. H. Saunders, Jr. and D. H. Edison, J. Am. Chem. Soc., 82, 138 (1960).
67. E. R. Thornton, Ann. Rev. Phys. Chem., 17, 349 (1966).
68. A. F. Cockerill, S. Rottschaefer, and W. H. Saunders, Jr., J. Am. Chem. Soc., 89, 901 (1967).
69. L. Pauling, "The Nature of the Chemical Bond," Third ed., Cornell University Press, Ithaca, 1960, Chap. 3.
70. K. C. Brown and W. H. Saunders, Jr., unpublished results.
71. A. F. Cockerill and J. Kendall, unpublished results.
72. J. I. Hayami, N. Ono, and A. Kaji, Bull. Soc. Chem. Jap., 44, 1628 (1971).
73. W. F. Bayne and E. I. Snyder, Tetrahedron Lett., 571 (1971).
74. B. D. England and D. J. McLennan, J. Chem. Soc., B, 696 (1966).
75. V. J. Shiner, Jr. and M. L. Smith, J. Am. Chem. Soc., 83, 593 (1961).
76. V. J. Shiner, Jr. and B. Martin, Pure Appl. Chem., 8, 371 (1964).
77. K. T. Finley and W. H. Saunders, Jr., J. Am. Chem. Soc., 89, 898 (1967).
78. A. V. Willi, Helv. Chim. Acta, 49, 1725 (1966).
79. J. F. Bunnett, G. T. Davis, and H. Tanida, J. Am. Chem. Soc., 84, 1606 (1962).
80. D. Bethell and A. F. Cockerill, J. Chem. Soc., B, 917 (1966).
81. V. J. Shiner, Jr. and M. L. Smith, J. Am. Chem. Soc., 80, 4095 (1958).
82. H. Simon and G. Mullhöfer, Chem. Ber., 97, 2202 (1964).
83. H. Simon and G. Mullhöfer, Pure Appl. Chem., 8, 379 (1964).
84. E. M. Hodnett and J. J. Sparapany, Pure Appl. Chem., 8, 385 (1964).
85. K. C. Brown and W. H. Saunders, Jr., J. Am. Chem. Soc., 92, 4292 (1970).
86. S. Ašperger, N. Ilakovac, and D. Pavlović, J. Am. Chem. Soc., 83, 5032 (1961).

87. A. F. Cockerill, Tetrahedron Lett., 4913 (1969).

88. S. Ašperger, L. Klasinc, and D. Pavlović, Croat. Chem. Acta, 36, 159 (1964).

89. W. H. Saunders, Jr., A. F. Cockerill, S. Ašperger, L. Klasinc, and D. Stefanović, J. Am. Chem. Soc., 88, 848 (1966).

90. A. F. Cockerill and W. H. Saunders, Jr., J. Am. Chem. Soc., 89, 4985 (1967).

91. G. Ayrey, A. N. Bourns, and V. A. Vyas, Can. J. Chem., 41, 1759 (1963).

92. P. J. Smith, Ph.D. thesis, McMaster University, 1965.

93. A. N. Bourns, and P. J. Smith, Proc. Chem. Soc., 366 (1964).

94. A. C. Frosst, Ph.D. thesis, McMaster University, 1968.

95. P. J. Smith, private communication.

96. G. Ayrey, E. Buncel, and A. N. Bourns, Proc. Chem. Soc., 458 (1961).

97. L. J. Steffa and E. R. Thornton, J. Am. Chem. Soc., 89, 6149 (1967).

98. K. Bowden, Chem. Rev., 66, 119 (1966).

99. G. Biale, A. J. Parker, I. D. R. Stevens, J. Takahashi, and S. Winstein, J. Am. Chem. Soc., 94, 2235 (1972).

100. W. H. Saunders, Jr. and S. E. Zimmerman, J. Am. Chem. Soc., 86, 3789 (1964).

101. W. H. Saunders, Jr., Chem. Ind. (London), 1661 (1963).

102. R. F. Hudson and G. Klopman, J. Chem. Soc., 5 (1964).

103. D. J. McLennan, J. Chem. Soc., B, 705 (1966).

104. D. J. McLennan, J. Chem. Soc., B, 709 (1966).

105. D. S. Bailey, Ph.D. thesis, University of Rochester, 1969.

106. D. Bethell and A. F. Cockerill, J. Chem. Soc., B, 913 (1966).

107. A. F. Cockerill and D. Bethell, J. Chem. Soc., B, 920 (1966).

108. K. Bowden and A. F. Cockerill, J. Chem. Soc., B, 173 (1970).

109. D. H. Froemsdorf and M. D. Robbins, J. Am. Chem. Soc., 89, 1737 (1967).

110. G. K. Zwolinski and D. L. Griffith, unpublished results.

111. J. Hine and M. Hine, J. Am. Chem. Soc., 74, 5266 (1952).

112. J. Borchardt and W. H. Saunders, Jr., Tetrahedron Lett., 3439 (1972).

113. J. N. Brønsted and K. J. Pedersen, Z. Phys.

Chem., 108, 185 (1924).

114. J. N. Brønsted, Chem. Rev., 5, 322 (1928).
115. R. P. Bell, "The Proton in Chemistry," Cornell University Press, Ithaca, 1959, Chap. 10.
116. F. G. Bordwell, W. J. Boyle, Jr., J. A. Hautala, and K. C. Yee, J. Am. Chem. Soc., 91, 4002 (1969).
117. F. G. Bordwell and W. J. Boyle, Jr., J. Am. Chem. Soc., 93, 511 (1971).
118. R. A. Marcus, J. Am. Chem. Soc., 91, 7224 (1969).
119. F. Hibbert, F. A. Long, and E. A. Walters, J. Am. Chem. Soc., 93, 2829 (1971).
120. R. P. Bell and B. G. Cox, J. Chem. Soc., B, 194 (1970).
121. L. Melander and N. Å. Bergman, Acta Chem. Scand., 25, 2264 (1971).
122. F. G. Bordwell and W. J. Boyle, Jr., J. Am. Chem. Soc., 93, 512 (1971).
123. R. Baker and M. J. Spillett, J. Chem. Soc., B, 481 (1969).

III STEREOCHEMISTRY OF E2 REACTIONS

A DEFINITIONS

The orbitals involved in almost all chemical reactions
have directional properties. As a consequence, the
atoms in and around the reaction site usually prefer
to occupy certain definite positions with respect to
each other in the transition state of a reaction. The
E2 reaction is particularly interesting in this respect
because the reaction site encompasses four atoms of the
substrate and one of the attacking base.
        There are two extreme cases for the location of
the leaving group and the β-hydrogen with respect to
each other in the transition state: on opposite sides
of the molecule (1) or on the same side of the molecule
(2). The term anti-periplanar means that the dihedral
angle between the carbon-X and the carbon-hydrogen
bonds is approximately 180°; the term syn-periplanar

1

Anti-periplanar

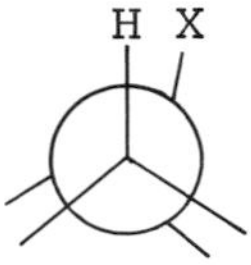

2

Syn-periplanar

means that the angle is approximately 0°. The nomenclature is due to Klyne and Prelog.[1] These two arrangements are referred to in much of the literature as
trans and cis, respectively. This terminology has the
disadvantage of making no distinction between the names
for the stereochemical course of the reaction and the
stereochemistry of the products (and sometimes of the
reactants, as well).

Two further conformations are defined in the
Klyne-Prelog[1] nomenclature: anti-clinal (3), where the
dihedral angle is approximately 120°, and syn-clinal
(4), where it is approximately 60°. In principle, of
course, any angle between 0 and 180° is possible. This

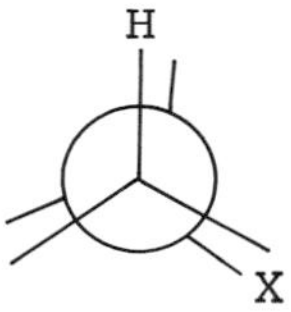

H

X

3

Anti-clinal

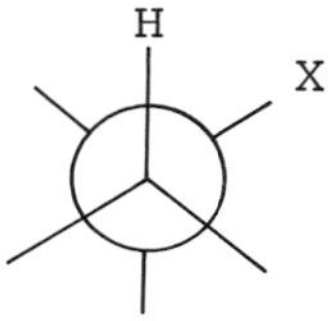

H

X

4

Syn-clinal

nomenclature encompasses all the possibilities, however, for each definition includes all angles within
±30° of the stated one. In fact, the nomenclature
makes more distinctions than are generally useful in
discussing the stereochemistry of elimination. Usually
it is not possible to devise experiments that will distinguish between anti-periplanar and anti-clinal on
the one hand, or syn-periplanar and syn-clinal on the
other. In addition, there are good theoretical reasons
for believing that angles near 180° or 0° will be preferred in most E2 reactions. Consequently, we normally

use simply the terms anti and syn.

B   THEORETICAL EXPECTATIONS

As we see later, the early experimental evidence on E2 reactions pointed to a strong preference for anti elimination, and the first theoretical discussions were consequently attempts to explain this stereochemistry. Hückel[2] suggested that electrostatic repulsion between the negatively charged base and the leaving group would make a syn arrangement less favorable than an anti. Calculations of this effect based on simple electrostatic theory were made by Cristol[3] for reaction of an alkoxide ion with an alkyl chloride. He concluded that the energy difference was too small to account for his observed anti/syn preference, which was in the range of 7000 to 24,000.

Although subsequent research has shown that the observed preference for anti elimination is by no means always this large (indeed, there are now numerous instances where syn elimination is preferred), it still seems unlikely that the electrostatic factor is the sole or even a major effect on stereochemistry in most elimination reactions. Perhaps the most compelling argument on this point is that a preference for anti elimination persists in certain cases where the electrostatic effect should favor syn elimination, such as reactions involving a negative base and a positive leaving group, or a neutral base and a neutral leaving group.

Quantum-chemical reasoning has been frequently invoked in discussions of the stereochemistry of elimination. It is easy to demonstrate that dihedral angles of either 0° or 180° between the bond to the leaving group and the bond to the β-hydrogen should be preferred to other angles. The α- and β-carbon atoms rehybridize from sp^3 to sp^2, and the carbon-leaving-group and carbon-hydrogen σ-bonds become p-orbitals, in the course of the reaction. These p-orbitals will overlap most effectively to form a π-orbital if they are parallel to each other. By this reasoning, one can conclude that either anti-periplanar or syn-periplanar elimination should be preferred over other modes, but one cannot choose between them.

Essentially the same argument has been advanced by Eliel, Allinger, Angyal, and Morrison[4] in somewhat more sophisticated terms. They consider the energetics of addition of a diatomic species, X_2, to an olefin and the elimination of X_2, which is the reverse of

this process. They express the wave function of the
transition state by Eq. 1, where Ψ_r is the wave

$$\Psi = \lambda(t)\ \Psi_p\ +\ \Psi_r \tag{1}$$

function of the reactant, Ψ_p the wave function of the
products (olefin plus X_2), and $\lambda(t)$ a parameter that
increases with progress along the reaction coordinate.
It should be noted that this is not a simple matter of
assuming the transition state to be an electronic resonance hybrid of reactants and products, for reactants
and products are not identical in geometry as is required for the contributors to a true resonance hybrid.
The wave functions concerned must be complete, including both electronic and nuclear parts. Since an olefin with the p-orbitals parallel is 60 kcal/mole more
stable than one with them orthogonal, a transition
state with any degree of double-bond character will
prefer to have the bonds to the leaving group and to
the β-hydrogen parallel. According to this formulation, the strength of the stereochemical preference
will depend upon the amount of double-bond character
possessed by the transition state, a point we return to
later.

Although periplanar elimination is obviously preferred on quantum-chemical grounds, how to deal with
the anti versus syn question is much less clear. The
same authors[4] present a simple analogy in which they
liken the anti arrangement of X-C-C-X to a linear conjugated system such as butadiene, and the syn arrangement to a cyclic conjugated system such as cyclobuta-
diene. They point out that the former is of lower
energy than the latter, and argue that this preference
should persist regardless of the interatomic distances
in the real transition state. The model is so far from
reality, however, that it is very difficult to judge
whether the effect would be large enough to explain the
sometimes imposing predominance of anti over syn elimination (for example, Cristol's[3] work quoted above).

Another approach has been made by Fukui,[5] who
used the extended Hückel method to calculate frontier
electron densities (the electron density of the lowest
vacant orbital) of the hydrogen atoms in the staggered
conformations of ethyl chloride, 1,2-dichloroethane
and 2-chlorobutane. In each case, the hydrogen atoms
anti to the chlorine atom had the highest values, and
the bonds to them should thus have the least bonding
character. On the contrary, the hydrogen with the
highest value in exo-chloronorbornane was the

exo-β-hydrogen, which is syn-periplanar to the chlorine. As we see later, this exception is in accord with experimental evidence on bridged-ring systems.

Perhaps the most common electronic argument for anti elimination has been that the bonding electrons of the carbon-hydrogen bond are performing a sort of displacement of the leaving group, and that, by analogy to the S_N2 reaction, a backside displacement is preferred.[3] The argument has been developed in more detail by Ingold[6] and Sicher.[7] The normal course of the reaction is depicted in 5, where a nucleophilic displacement with inversion occurs at the α-carbon and an electrophilic displacement with retention at the β-carbon, the net result being anti elimination. When the carbon-hydrogen bond is very weak in the transition state, however, the electrophilic displacement may proceed with inversion also.[6] The process may be depicted as in 6, where formation of the p-orbital on the β-carbon has proceeded so far that there is a well-formed lobe of electron density opposite the carbon-hydrogen bond. These electrons are now envisioned as displacing the leaving group.[7]

5   6

This type of argument could also be couched in terms of electron repulsion effects. The electrons of the bond to the β-hydrogen and of the bond to the leaving group will prefer to stay as far apart as possible while the system progresses along the reaction coordinate, and this is usually best accomplished in an anti arrangement, except possibly when the carbon-hydrogen bond is so ionic that there is greater electron density on the opposite side.

Dixon[8] suggested from simple molecular orbital theory that a correlation should exist between spin-spin coupling in nmr spectra and the rates and stereochemistries of elimination reactions because the ease of delocalization of spin density and of charge should run parallel. This suggestion was followed up by Lowe,[9] who performed CNDO/2 calculations on eclipsed and staggered ethane and ethyl fluoride, and on the perturbed systems resulting from approach of a hydride ion to ethane and a fluoride ion to ethyl fluoride. He concluded that a site having strong positive spin coupling to a given proton is a site to which negative charge is preferentially transferred upon attack of base at that proton. Charge transfer to an anti is more effective than to a syn leaving group, according to these calculations, except when both base and leaving group are neutral. The syn transition state is expected to have more carbanion character than the anti.

The basic conclusion from quantum-chemical arguments, then, is that there will usually be a preference for anti-periplanar elimination, with syn-periplanar elimination being next most favorable, and all intermediate conformations of the transition state being still less favorable. These preferences can be relaxed, however, when there is little or no π-overlap between the α- and β-carbon atoms in the transition state. It is by no means safe to conclude that electronic factors always play a major role in determining stereochemistry of elimination.

A nonelectronic factor that leads to similar conclusions is the principle of least motion.[10,11] According to the principle, those elementary reactions will be favored that involve the least change in atomic positions and electron configuration.[10] When applied to elimination reactions, the principle predicts an order of preference that runs anti-periplanar>syn-periplanar>nonplanar.[12,13] One would also expect that the least-motion preference would be small for a reactant-like transition state. At present there seems to be no good way of distinguishing experimentally between the quantum-chemical and least-motion explanations of anti elimination.

Steric effects can also have an important influence on stereochemistry of elimination. The details are best discussed later in connection with specific examples, but a few of the more general types of steric effects will be mentioned here. Perhaps most important is a simple eclipsing effect. The anti-periplanar conformation is staggered, but the syn-periplanar conformation is eclipsed. Although the conformation with the leaving group and the β-hydrogen anti-periplanar may not be the most stable of the possible staggered conformations, it will certainly be more stable than an eclipsed conformation unless the eclipsed conformation is enforced by a rigid ring system. McLennan[14] suggests that the rotational barrier of about 3 kcal/mole in ethane can be taken as a minimum eclipsing effect for syn eliminations, and points out that this factor alone is sufficient to make the anti/syn rate ratio about 160 at the usual temperatures for elimination reactions. This figure assumes that the transition state is geometrically similar to reactants. If it is geometrically similar to products, the difference will be narrowed because nonbonded repulsions between the groups attached to the developing double bond will grow for the anti process as a result of increased eclipsing, and will diminish for the syn process as a result of widened bond angles.

Conformational effects can be particularly important in cyclic systems. An anti-periplanar arrangement can be achieved in a cyclohexane ring only when the leaving group and the β-hydrogen are trans-diaxial (7). A syn-periplanar arrangement is enforced in rigid ring compounds such as exo-norbornyl chloride (8). We discuss later cases in which conformational effects seem

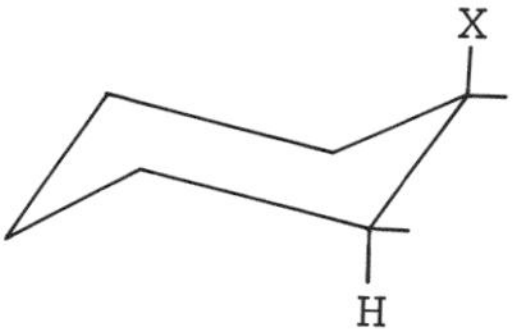

7 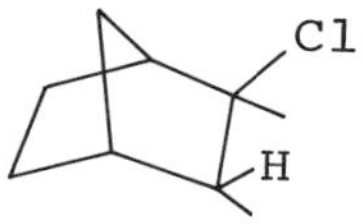   8

to act so as to hinder approach of base to anti or syn β-hydrogens in cyclic and acyclic systems.

C   METHODS OF STUDYING STEREOCHEMISTRY

1   Cyclic Compounds

To deduce the stereochemistry of an elimination reaction, one must be able to determine the relative positions of the leaving group and the β-hydrogen which is lost. A simple way of doing this is to incorporate the carbons bearing them in a ring system so that rotation is restricted. For example, the trans-2-substituent in 9 means that an anti elimination can give 12, while the cis-2-substituent in 11 would permit it to give either 10 or 12 by an anti elimination but only 12 by a syn elimination. This method has been very widely used, but has two drawbacks affecting the generality of the results. One is that the substituent affects the rate of elimination into the substituted

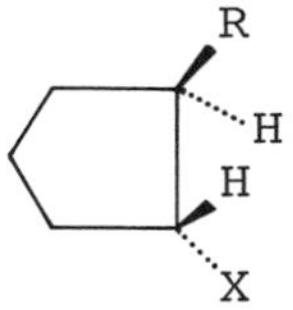

9

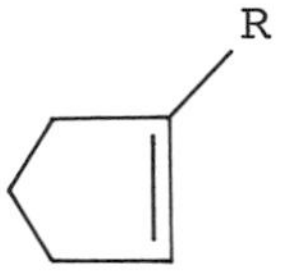

10

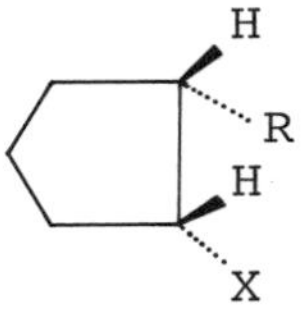

11

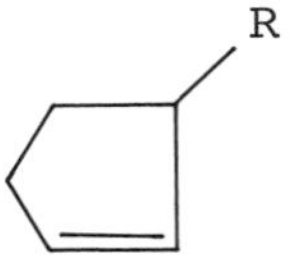

12

relative to the unsubstituted branch (i.e., the orientation), and this effect may be difficult to disentan-
gle from the stereochemistry. Thus conclusions cannot
be drawn with any confidence about the intrinsic stereochemistry in the absence of substituents. This problem can be surmounted if the 2-substituent is a deuterium atom. In this case either protium or deuterium
can be lost, depending on the stereochemistry of the
reaction and the isotope effect. If an independent
measure of the latter can be obtained, the stereochemistry of elimination from the undeuterated compound
can be calculated from the isotope effect plus the
deuterium content of the product.[15] When the stereochemistry is either pure syn or anti, the isotope
effect is not needed. For example, a pure anti elimination from 11 (R = D) would give a product containing
all of the original deuterium.

Even when the perturbation due to the substituent
can be dealt with, however, it is unwise to make sweep-
ing generalizations from the results. Changing the
ring size can change drastically the stereochemistry
of elimination.[15] Indeed, the fact that cyclohexane
rings were used in much of the early work on stereochemistry created the impression that anti elimination
was much more general than it really is, because the
tendency toward anti elimination is stronger in the
cyclohexane system than in any other common ring.[15]
It should also be mentioned that different stereochemical results may be obtained with different leaving
groups.

With the cyclohexane and larger rings, one can
sometimes draw conclusions about the conformation of
the transition state by using appropriately designed
reactants. The compound 13, for example, can in principle react through either conformation 13a or 13b.
The conformation of a cyclohexane ring can be fixed

13a   13b

either by incorporating it in a rigid ring system (such
as two trans-fused cyclohexane rings) or by placing a
bulky substituent such as t-butyl, which strongly prefers an equatorial position, on the ring at a position
far enough from the reaction site to minimize direct
steric effects on the reaction. One can then observe
the relative reactivities of the various fixed conformations. We describe later specific examples of this
approach. One should keep in mind that this is not a
foolproof method. Conformations are never absolutely
fixed, so that one can only attain a situation where
the desired conformation is of much lower energy than
the other possibilities. If the system is very unreactive in this favored conformation, it may react
through an energetically unfavorable conformation such
as the twist-boat.

2   Open-Chain Diastereomers

Another very common approach is to use an open-chain
reactant in which both the α- and β-carbon atoms are
asymmetric. The two diastereomeric forms of the reactant will then give different products in a stereospecific elimination reaction. For example, an anti
elimination from 14 will give 15, in which the two R
groups are cis to each other, while an anti elimination
from 16 will give 17, in which the two R groups are
trans to each other. The experimental problem then
simply amounts to preparing and identifying the two
diastereomeric reactants, distinguishing between the
two isomeric products if the stereochemistry is clean,
and analyzing mixtures of them if it is not.

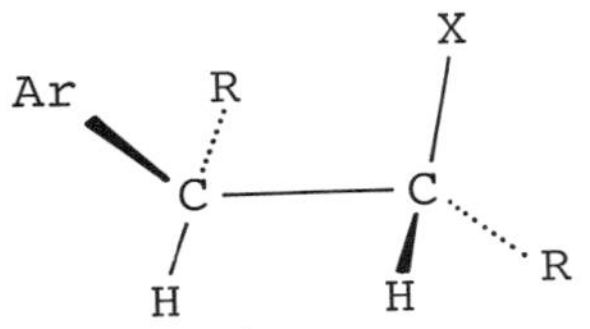

14

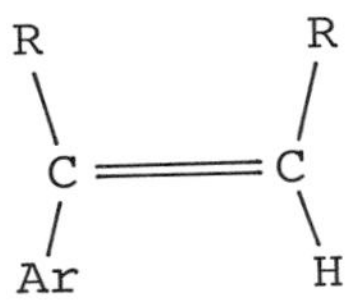

15

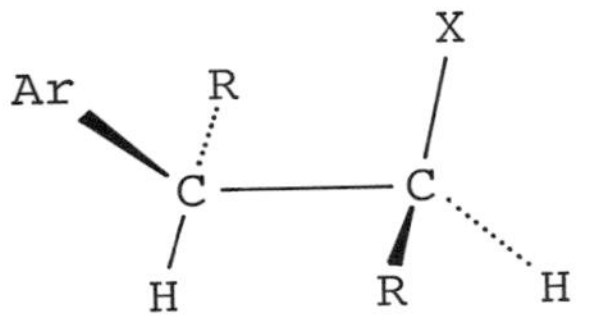

16

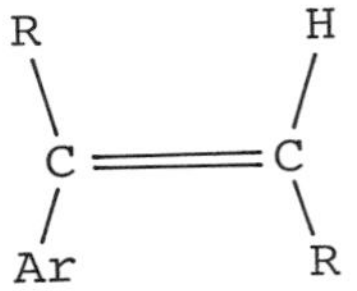

17

An interesting and useful variant on this approach
is to use reactants in which the asymmetry at the β-
carbon is due to the presence of a deuterium atom in
place of one of the protium atoms. Such compounds can
readily be obtained by stereospecific addition of DX to
an olefin. A number of direct and indirect methods of
accomplishing this process are available. Perhaps the
simplest and most versatile is deuteroboration with
B_2D_6, a reaction that takes advantage of the stereospecific cis addition which occurs in hydroboration.[16]

When such a stereospecifically deuterated compound
is subject to elimination, the cis and trans olefins
resulting from elimination into the deuterated branch
will be formed, one with loss and one with retention
of deuterium. For example, 18 will undergo an anti
elimination to give 19 containing deuterium and 20 with
loss of deuterium. This is best seen by viewing 18 in
the two conformations 18a and 18b. Separation of the
products, followed by analysis for deuterium enables
one to conclude whether the elimination has been syn or
anti in each case. If an estimate or an independent
measure of the isotope effect is available, the figures
for the stereochemistry can be converted to what they
would be in the absence of isotope effects.[15]

Semi-quantitative information on stereochemistry
can be obtained by an ingenious variant of the above
procedure that does not require deuterium analyses.[17,18]
The basic assumption is that elimination into an
undeuterated branch of a reactant will not be affected

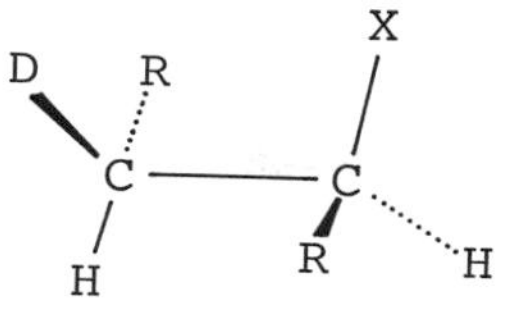

18a

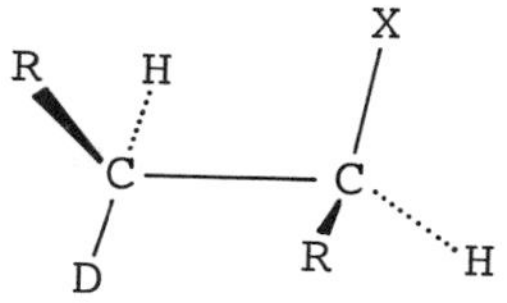

18b

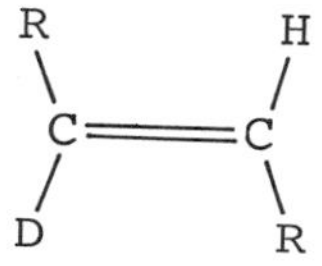

19

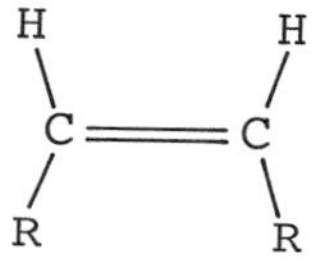

20

by deuterium in the other branch, an assumption that is certainly safe to within a few percent. Thus the rate of elimination into the left-hand branch of 21 will be the same as for the undeuterated compound, but the rate of elimination into the right-hand branch may be much

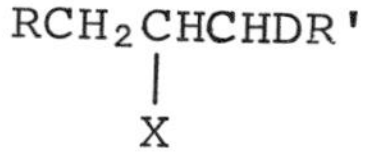

$$RCH_2CHCHDR'$$
$$|$$
$$X$$

21

slower if deuterium is lost, and perhaps slightly slower (from a secondary isotope effect) if it is not. Thus the percentage of a particular product of elimination into the right-hand branch (say the trans olefin) divided by the percentage of elimination into the left-hand branch will be proportional to the rate of formation of that particular product for both the deuterated and undeuterated compounds. The ratio of these relative rates for the undeuterated and deuterated compounds then is k_H/k_D for the formation of that product. If k_H/k_D is near unity for a given product, it is assumed that the product was formed with loss of hydrogen; if it is substantial, it is assumed that the product was formed with loss of deuterium. The only experimental data needed, then, are the compositions of the olefin mixtures from the deuterated and undeuterated compounds. In principle the method could be

indecisive if the isotope effect for loss of deuterium
were small, or if the secondary isotope effect for e-
limination of hydrogen from the deuterated branch were
unusually large, but such complications have not been
encountered thus far in application of the method, and
are fairly unlikely.

3 Rate Profiles

This approach is based on the old idea that reactions
of similar mechanism should show similar dependence of
reactivity on structure. The application of this idea
to medium-ring systems was pointed out some time ago
by Brown,[19] and its use in elucidating stereochemistry
of reactions has been brilliantly explored by Sicher
and his co-workers.[18,20] The details of the reasoning
are best left to specific examples which are discussed
later (Section III.E.4) under stereochemistry of eliminations from medium-ring compounds. In simple terms,
one merely determines the effect of ring size on rate
(the rate profile) for a reaction of unknown stereochemistry, and compares this rate profile with those
for reactions of known stereochemistry. Where the
profiles are qualitatively similar, a similarity in
stereochemistry is assumed. While one cannot obtain
quantitative information on stereochemistry in this
fashion, the method is certainly useful in providing
qualitative indications that can later be checked by
more rigorous methods. It was such studies that
prompted reexamination of the widely accepted view that
anti elimination predominated except under unusual circumstances.

D THE ANTI RULE

1 Early Evidence

The first results on stereochemistry of elimination
appeared about the turn of the century, and consisted
of product studies on eliminations from isomeric reactants. For example, meso-stilbene dibromide with eth-
anolic potassium hydroxide gives cis-bromostilbene,
while the dl-dibromide gives trans-bromostilbene (Eqs.
2 and 3, respectively).[21] Other early workers found
that chlorofumaric acid reacted with base fifty times
faster than chloromaleic acid to give acetylenedicar-
boxylic acid,[22] and that cis-1,2-dichloroethylene reacted with base twenty times faster than trans-1,2-
dichloroethylene to give chloroacetylene.[23] In each

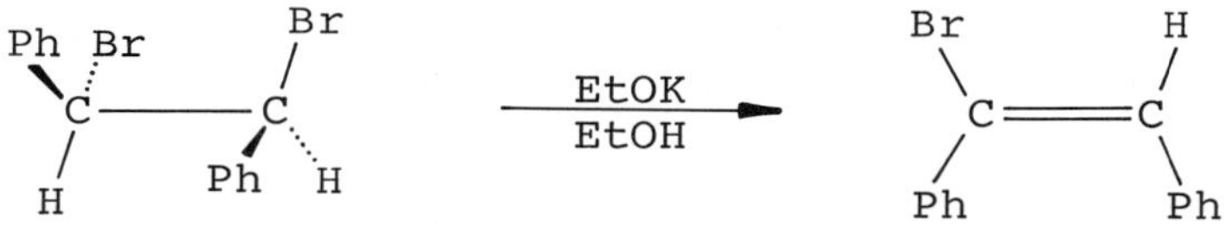

$$\text{(2)}$$

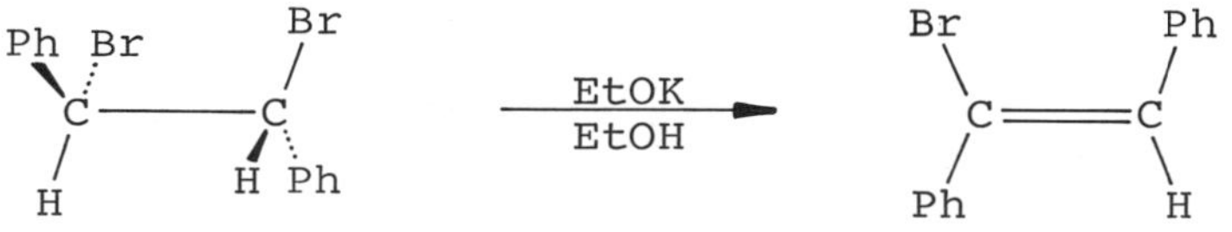

$$\text{(3)}$$

case, the anti elimination is favored. It should be kept in mind, however, that these two studies used unsaturated halides, which may not react by the same mechanisms as their saturated counterparts. Some other early results are summarized by Frankland.[24] Another point to keep in mind is that methods of product isolation were not well developed at that time, so that failure to isolate a given product did not always mean that it was not present in appreciable quantity.

The next important study of stereochemistry of elimination was by Hückel[2] in 1940, who studied solvolyses and reactions with base of menthyl (22) and neomenthyl (23) derivatives, and who determined the proportions of 2-menthene (24) and 3-menthene (25) produced. Although the reactions were not studied kinetically to establish whether they were E2 or E1, this

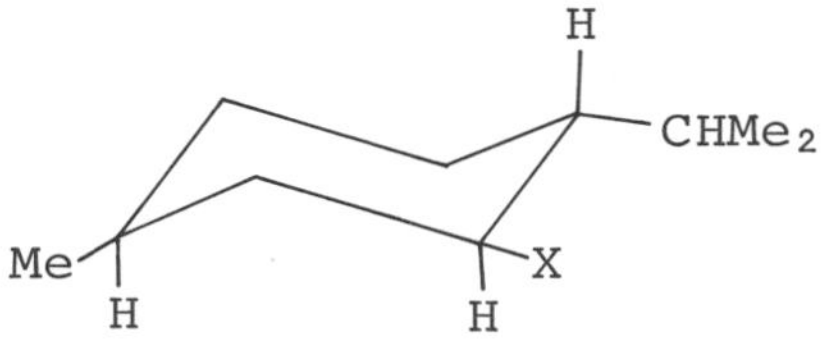

22

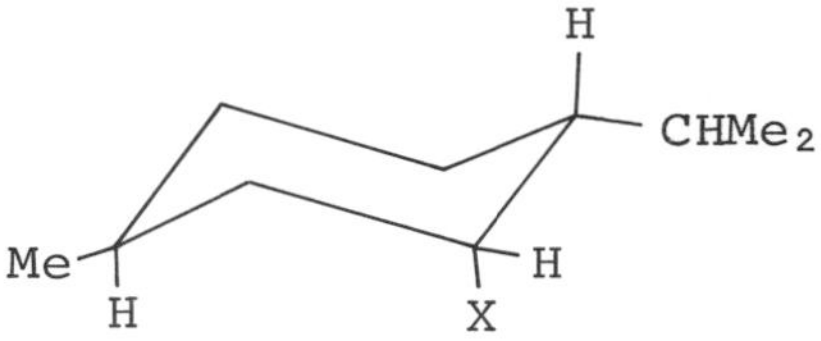

23

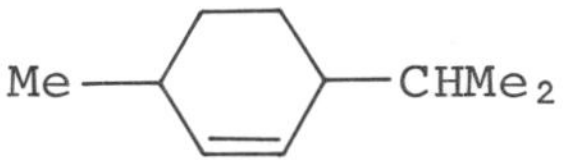

24

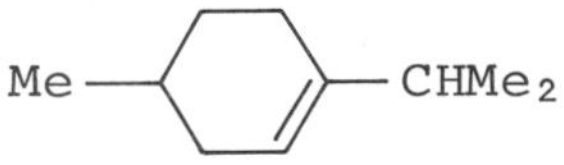

25

has subsequently been done for a number of them.[25,26]
The results are summarized in Table 1. The neomenthyl
derivatives can undergo an anti elimination in either

TABLE 1   E2 Reactions of Menthyl and Neomenthyl
Derivatives

Reactant	% 2-Menthene	% 3-Menthene	Reference
Menthyl Cl	100	0	25
Neomenthyl Cl	22	78	25
Menthyl NMe_3^+	100	0	26
Neomenthyl NMe_3^+	12	88	26
Menthyl OTs[a]	100	0	26

[a]Probably E2, but not kinetically established.

direction, and 3-menthene is the preferred product in
both of the examples in the table. The menthyl derivatives must adopt a conformation with both alkyl groups
axial before an anti-periplanar elimination can occur,
and even then only 2-menthene can be formed. Within
the experimental precision of the measurements (probably a few percent of 3-menthene could be detected)
the menthyl derivatives give all anti elimination.

   2   More Recent Studies

Although the early work clearly indicated a preference
of 50:1 or more for anti elimination in the cases examined, a more precise figure for the preference could
not be obtained because of limitations on detection of
minor products. Where the preference is very large, in
fact, even modern methods such as gas chromatography
cannot give really good answers. The best way to establish limits in these cases is to compare rates of
reaction for a reactant that is unable to undergo anti
elimination with another that can. This was the approach that was adopted in the classic studies of the
benzene hexachloride isomers.[27-31] One of them, β-benzene hexachloride (26) has all of the chlorine atoms on
adjacent carbon atoms trans to each other, so that it
is impossible to bring adjacent hydrogen and chlorine
into an anti relationship without causing very severe
ring strain. In agreement with the expectations for
the anti mechanism, 26 was by far the slowest of the
benzene hexachloride isomers. The others underwent
elimination 7000 to 24,000 times faster,[28] and the

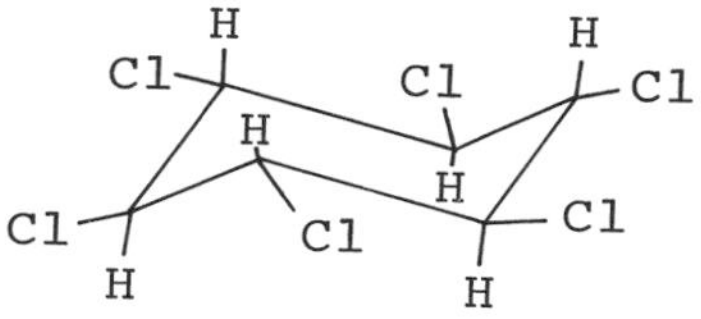

26

energies of activation were 11 to 12 kcal/mole lower.[31]
Furthermore, 26 was found to pick up 0.08% deuterium
when the elimination was carried out in a deuterated
solvent.[30] This result was taken to indicate that an
ElcB mechanism, with loss of chloride from the carbanion to give product much faster than return to reactant,[30] was preferred to a syn E2 mechanism. If this
interpretation is correct (some have argued that the
deuterium uptake is too small to be relied upon[31]), the
preference for anti over syn elimination by an E2 mechanism is even greater than the figures quoted above.
    A very high degree of stereospecificity is also
observed in the dehydrohalogenation of the 9,10-dichlo-
ro-9,10-dihydrophenanthrenes.[32] The trans isomer
(27), which cannot undergo an anti elimination, reacts

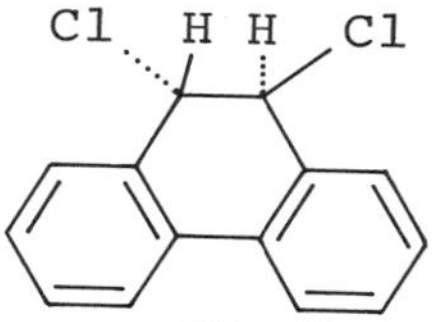

27

even slower than β-benzene hexachloride with sodium hydroxide in ethanol. In contrast, the cis isomer reacts
faster than any of the benzene hexachloride isomers.
While we see later that rigid ring systems behave differently, there is obviously a dramatic preference for
anti elimination from cyclohexyl chlorides in the absence of special features. This preference colored
thinking on stereochemistry of eliminations for a long
time. Its generality was considerably overestimated,
probably because it was assumed that no common structural change could produce an effect strong enough to
counter it.
    Most of the remaining support for the anti rule
is based on product proportions, so that the limit on
syn elimination depends on the lowest detectable amount

of the minor product. This is generally of the order
of 1 to 2%, though a careful analysis by gas chromatography is capable of setting considerably lower
limits, say 0.01 to 0.1%. Studies on the menthyl, 2-
methylcyclohexyl, 2-methylcycloheptyl, and decalyl tosylates up to 1967 are summarized by Hückel and
Hanack,[33] who also mention some results on the corresponding quaternary ammonium salts. Usually, the isomers with a 2-alkyl group trans to the leaving group
eliminate entirely in the direction away from the alkyl
group, but in a few cases small amounts of elimination
toward the alkyl group are noted. The trans-2-methyl-
cyclohexyltrimethylammonium ion, for example, yields
a few percent of 1-methylcyclohexene under most conditions, and as much as 8% with t-butoxide ion in t-butyl
alcohol.[34] This result was originally interpreted as
indicating the incursion of a small amount of E1 reaction,[26,34] but it could also mean that the E2 reaction
goes to a minor extent by a syn mechanism. Similar results were obtained in a more recent study of the reactions of menthyltrimethylammonium ion (22 X = NMe₃⁺)
with a variety of basic reagents.[35] Although the product was mainly 24, substantial amounts of 25 resulted
under some conditions. Solid potassium t-butoxide gave
27%, and phenyllithium in ether 50%, of 25. Tracer
experiments showed that the latter was mainly an α'-β
elimination (Section I.B.4). Adherence to the anti
rule in the cyclohexyl series, then, is considerably
less one-sided with the quaternary ammonium salts than
with the chlorides.
      There are two possible conformations of a cyclo-
hexyl derivative in which the leaving group and β-hydrogen are trans-related: diaxial and diequatorial.
Which of these is preferred can be determined by substituting the ring so that one conformation will be
strongly favored (Section III.C.1). The 4-t-butylcyc-
lohexyl system has been used in a number of such studies. The group X will be constrained almost entirely
to an equatorial position in 28 and to an axial position in 29 (unless its effective bulk is comparable to
or greater than that of the t-butyl group). When the
tosylates are treated with sodium ethoxide in ethanol,

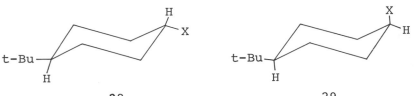

28   29

<u>29</u> readily undergoes an E2 reaction, while <u>28</u> reacts by
S_N1 and E1 pathways.[36]   The quaternary ammonium salts
(X = $NMe_3{}^+$) behave similarly.   Treatment of <u>28</u> with
base results only in an S_N2 displacement on one of the
methyl groups of the NMe_3 group, while <u>29</u> gives 90% of
the E2 product, 4-t-butylcyclohexene, and only 10%
displacement.[37]   It should be noted that this indicates
a definite preference for elimination in the absence of
conformational effects, since the trimethylammonio
group is comparable in size to the t-butyl and would by
no means be locked in an axial position in <u>29</u>.   It
should be kept in mind that there is no necessary relation between the population of ground state conformations and the proportion of reaction occurring via the
corresponding transition state conformations; the latter is determined solely by the relative energies of
the transition states (the Curtin - Hammett principle[38]).   More detailed analysis of the results shows
that both the elimination and substitution reactions
occur faster when the trimethylammonio group is in an
axial conformation -- the former by a wide margin and
the latter by about 20:1.

Even earlier evidence for the axial preference
had been found in studies on the quaternary salts derived from some steroidal tertiary amines.   In the
chloestene and allopregnane systems (the two differ only in the nature of the alkyl group attached to ring D),
the 3-α isomers, with axial $NMe_3{}^+$, gave mainly elimination, while the 3-β isomers, with equatorial $NMe_3{}^+$,
gave mainly displacement on the methyl of the trimeth-
ulammonio group.[39]   Similar results were observed in
the coprostane system.[40]   The preference was somewhat
less marked, presumably because the cis fusion of the
A and B rings in coprostane is more flexible than the
trans fusion in cholestane and alopregnane.   Rate
studies were performed on the reaction with potassium
hydroxide in ethanol of a number of cholestanyltri-
methylammonium salts.[41]   These included the 3α, 3β,
6α, 6β, and 7α isomers (a partial formula giving the
significance of this notation is shown in 30).   The
axial isomers always underwent elimination faster, and
the equatorial isomers underwent substitution faster.
The 6β isomer eliminated much faster than any of the
other isomers, presumably because of the additional
strain of a 1,3-diaxial interaction between the trimethylammonio group and the methyl group at the ring
juncture.

Eliminations in a variety of open-chain systems
have been found to follow the anti rule more or less
firmly.   The treatment of erythro-3-phenyl-2-butyl

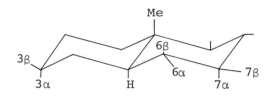

30

tosylate (31) with sodium ethoxide in ethanol yields
cis-2-phenyl-2-butene, while the threo tosylate (32)
yields the trans olefin.[42]   While product determination

31   32

was not quantitative, none of the other isomer could be
found in either case. The first application of deute-
rium-induced asymmetry was in a study of eliminations
from derivatives of 1,2-diphenylethanol-2-d.[43] The
main results of the study dealt with pyrolytic eliminations, which is treated later in this volume (Chapter
VIII), but a few E2 reactions were also run. The 2,4,
6-triethylbenzoate with potassium t-butoxide in t-
butyl alcohol and the bromide with ethoxide in ethanol
gave mainly anti elimination, though the acetate and
triethylbenzoate with potassium amide in liquid ammonia
gave mixed stereochemistry. In no case was the reaction entirely stereospecific, however, and epimerization or a cyclic mechanism was suspected in the reactions with potassium amide.
     One of the most thorough studies of stereochemis-
try of elimination utilized the 1,2-diphenyl-1-propyl
system.[44] The bromide, chloride, and trimethylammonium
salt were treated with primary, secondary, and tertiary
alkoxide-alcohol media, and sodium 2-octyloxide in benzene was used in one experiment with the chloride.
With only one exception, the threo isomer (33) gave
trans olefin, and the erythro isomer (34) gave cis olefin, the result expected for anti elimination. The exception was the trimethylammonium salt with t-butoxide
in t-butyl alcohol, where both isomers gave trans olefin. The elimination from the erythro isomer, then,

was formally a syn process. It was established that
the cis was not isomerized to the trans olefin under
the reaction conditions. The presumption in favor of
anti elimination was sufficiently strong at that time,
however, that it was assumed without proof that the
erythro reactant must have undergone epimerization at
either the α- or β-carbon atom prior to or during reaction. Subsequent studies have revealed that there is
no exchange at either position under the reaction conditions, so a carbanion that can return to reactant after inversion cannot be involved.[44b] Whether the reaction is a genuine concerted syn elimination remains uncertain.

One could argue that simple alkyl derivatives need
not show the same stereochemical behavior as these phen-
yl-substituted systems, but there are also a number of
reports in the literature on one of the simplest systems that allows stereochemical distinctions, the
threo- and erythro-2-butyl-3-d derivatives. The earliest study was by Skell and Allen,[45] who found that
2-butyl bromide reacted with potassium ethoxide in ethanol to yield 2-butene by an exclusively anti route.
Further studies on 2-butyl bromide have recently been
reported by Bartsch.[46] He examined potassium ethoxide,
s-butoxide and t-butoxide in the corresponding alcohols, as well as potassium t-butoxide in dimethyl sulfoxide and in tetrahydrofuran, and tetra-n-butylammo-
nium fluoride in dimethylformamide. In all cases the
elimination was anti within experimentally-detectable
limits.

A similar result is obtained when 2-butyl-3-d tosylate is treated with potassium ethoxide or potassium
t-butoxide in dimethyl sulfoxide.[47] The eliminations
were at least 98 to 99% anti. The longer-chain 5-
decyl-6-d tosylate reacts with potassium t-butoxide in
dimethylformamide, t-butyl alcohol, or benzene to give
predominantly, but not exclusively, anti elimination.
The overall proportions of syn elimination ranged from
5% in dimethylformamide to 19% in benzene. The general
picture, then, is that open-chain tosylates obey the

anti rule, though not so firmly as the bromides.
     The reaction of threo-2-butyl-3-d-trimethylammo-
nium tosylate with potassium ethoxide in dimethyl sulfoxide goes with anti stereochemistry,[49] and there is
indirect evidence that 2-butyltrimethylammonium ion also reacts with a number of alkoxide-alcohol media with
anti stereochemistry.[50] We see later in this chapter,
however, that there are numerous exceptions to the anti
rule among open-chain quaternary ammonium salts. The
point to keep in mind for the present is that simple
systems do show a predilection for anti elimination
that may vary in strength, but is still a very common
stereochemical result.

     E   EXCEPTIONS TO THE ANTI RULE

     1   Bridged-Ring Reactants

Among the earliest recognized exceptions to the anti
rule were eliminations from bridged-ring compounds.
These compounds have the great advantage from the theoretical point of view of being conformationally rigid.
It follows that elimination must either proceed through
a conformation very much like that of the ground state,
or else must incur considerable extra strain energy in
attaining a conformation that is more favorable for
elimination. In either case, the relative rates of
different isomers tell a great deal about conformational requirements for elimination. In bridged bicyclic systems, the usual situation is that cis substituents on adjacent carbon atoms (except when one carbon
atom is at the bridgehead) are constrained to be coplanar, while trans substituents cannot be coplanar. E-
limination of cis-related groups thus tends to be syn-
periplanar, and elimination of trans-related groups
tends to be anti-clinal.
     A typical example of the result of such con-
straints is found in the reactions of the 9,10-ethano-
anthracene derivatives 35 and 36. The reaction with
sodium hydroxide in 50% dioxane-ethanol at 110° C
proceeds 7.8 times faster with 36 (syn elimination)
than with 35 (anti elimination).[51] The anti elimination is favored by the energy of activation, but the
entropy of activation more than compensates for the
energy. In both instances the reaction is rather sluggish compared to similar nonrigid systems. That a high
degree of conformational rigidity is needed to make
syn elimination favored with dichlorides is shown by
the behavior of the cis- and trans-dichloroacenaphthenes

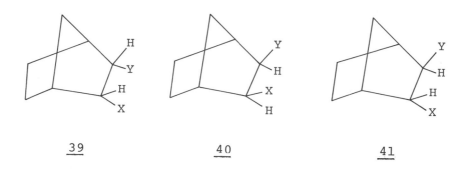

35   36

(37 and 38, respectively); anti is favored over syn elimination by a factor of 740.

37   38

    The bicyclo(2.2.1)heptane (norbornane) system has been used in a considerable number of studies of the effect of conformational rigidity on the stereochemistry of elimination. Suitable derivatives are readily obtained, and the molecules cannot be distorted appreciably from their equilibrium geometry without introducing serious strain. The 2,3-dihalonorbornanes

39   40   41

(39-41) have been especially thoroughly investigated. The trans-dichloride 41(X = Y = Cl), which must eliminate syn, reacts about eighty times faster than the endo-cis dichloride 39(X = Y = Cl) with sodium pentoxide in n-pentanol.[53] The preference for syn elimination is considerably more marked than that observed with 35 and 36. It is also possible to distinguish between elimination of exo and endo hydrogen in this system. The order of preference is syn-exo > anti-exo > anti-endo.[55,56] The position of syn-endo cannot be so quantitatively fixed, but two lines of evidence show that it is much less favorable than syn-exo. First, 41(X = Br, Y = Cl) reacts mainly with loss of hydrogen chloride (syn-exo) in spite of the fact that loss of bromine is preferred by a factor of at least twenty over loss of chlorine in the cases where both are in the same stereochemical situation. Second, deuterium-labeled dibromides 41(X = Y = Br) react with a substantial isotope effect when the deuterium is exo, but with little or none when it is endo. The isotope effect, k_H/k_D, for the 2,3-dideuterated derivative of 41(X = Y = Br) is 3.6, while for the endo-3-deuterated derivative it is only 1.02. The temperature for the former was 126.7°, so the effect is a substantial one that indicates a transition state with the proton transfer not far from the halfway point. For the 2,3-dideuterated derivative of 39(X = Y = Br), k_H/k_D is 3.4, so the extent of proton transfer is similar for both the syn-exo and anti-exo eliminations.

The authors[54] argue against the suggestion that the elimination takes an ElcB path when an anti-periplanar arrangement cannot be attained.[53] They found no deuterium exchange or isomerization of reactant molecules, and loss of bromine was easier than loss of chlorine, which would not be expected if the bond to the leaving group was not weakened in the transition state. They also believe that the isotope effects are too large for an ElcB mechanism. As is pointed out elsewhere in this volume (Section I.B.1.b), small deuterium isotope effects are by no means necessary in an ElcB reaction. On the other hand, the similarity of the two isotope effects certainly suggests that the syn and anti eliminations occur by the same mechanism, and it seems likely that this mechanism is E2. The original suggestion of an ElcB mechanism[53] was based in part on the assumption that E2 reactions would always show a strong preference for anti elimination, and we now know that there are numerous exceptions to this assumption.

A preference for syn-exo elimination is also found

with simple monosubstituted norbornane derivatives (42). Kwart, Takeshita, and Nyce[56] examined the reactions of deuterium-labeled 42 (X = OTs) and 42 (X = Br) with potassium t-hexoxide in p-cymene (the unusual choice of solvent was made because solvolysis of the tosylate could not be suppressed in t-hexyl alcohol). Using an estimated isotope effect, they concluded that syn-exo was preferred over anti-exo elimination by about 3.5:1 for the tosylate and 49:1 for the bromide. The tosylate was also studied by Brown and Liu[57] with sodium 2-cyclohexylcyclohexoxide in triglyme, and was found to give at least 98% syn-exo elimination. Even the substitution of methyl groups at the 7-position (43), which would be expected to increase hindrance to

42   43

abstraction of an exo hydrogen, only lowered the preference for syn-exo elimination to 95%. The action of potassium t-hexoxide in t-hexyl alcohol on 42 (X = Cl) also give 98% syn-exo elimination.[58] With 44, the same reagents favor anti-exo over syn-endo elimination by about 6:1.[58] This result, and Brown's[57] on 43, show that a specific preference for attack at an exo hydrogen exists in the norbornane system. The preference is probably due to the greater steric hindrance to attack at an endo hydrogen.[57] On the other hand, it cannot be the sole reason for the syn elimination that is observed in these systems. Both 42 (X = NMe$_3$+) and 45 (X = NMe$_3$+) give exclusively syn elimination in a Hofmann-type reaction, in spite of the fact that the syn hydrogen in 45 is, if anything, more hindered than the anti.[59,60] With 45 (X = OTs), treatment with potassium t-butoxide gives 90% syn elimination in benzene, 65% in dimethylformamide, and 45% in dimethyl sulfoxide.[61] The greater preference for syn elimination shown by 45 (X = NMe$_3$+) is probably due to electrostatic attraction between the leaving group and the base.

44   45

2 Activation at the β-Position

If a molecule, which would normally prefer anti elimination, is substituted in the β-position by a strong
activating group, syn elimination will often be faster
than unactivated anti elimination. In extreme cases,
the preference for anti elimination may disappear alto-
gether. The activating group can be any substituent
that increases the acidity of the β-hydrogen, either by
its electron-withdrawing ability, its ability to stabilize the resulting anion by resonance, or both.

A particularly well-documented example of this effect is found with the 2-phenylcyclohexyltrimethylammo-
nium ions. Both the cis and trans isomers give 1-phen-
ylcyclohexene when the corresponding quaternary ammonium hydroxides are heated.[61] It has been shown that
3-phenylcyclohexene is stable under the conditions of
the reaction, so that the trans reactant could not have
given 3-phenylcyclohexene which then isomerized to
1-phenylcyclohexene.[62,63] Furthermore, epimerization
of the trans reactant via formation of an α-carbanion
(or ylid) is excluded by the fact that the 1-deutero
compound gives 1-phenylcyclohexene containing all the
deuterium of the reactant.[64] The activated anti elimination from the cis-2-phenyl isomer is 133 times faster than the activated syn elimination from the trans-
2-phenyl isomer, but both are faster than the unactivated anti elimination to give 3-phenylcyclohexene,
which is sufficiently slow to be unobservable.[65] We
noted previously (Section I.B.1.b) that an E1cB mechanism for the syn elimination can be excluded by the observation of an appreciable nitrogen isotope effect.
The α'-β mechanism has been excluded by tracer experiments.[66]

When the leaving group is tosylate, there is less
tendency toward syn elimination. The 2-phenylcyclo-
hexyl tosylates cannot be studied with ethoxide ion in
ethanol because of interference from concurrent solvolysis, but with t-butoxide ion in t-butyl alcohol the
cis isomer reacts at a convenient rate under conditions

where the trans isomer undergoes no detectable reaction over a period of 22 days.[67] From this result, the preference for anti over syn elimination can be estimated to be at least 10^4.

The 2-phenylcyclopentyl tosylates present a different picture. Both isomers yield 1-phenylcyclopentene when treated with t-butoxide in t-butyl alcohol, and the cis isomer (anti elimination) reacts at only 9.1 times the rate of the trans isomer (syn elimination).[67] The difference between the 2-phenylcyclohexyl and 2-phenylcyclopentyl systems is probably the result of steric factors. On the one hand, the cis-2-phenylcyclopentyl system (46) cannot readily attain a conformation in which the leaving group and the β-hydrogen are anti-periplanar, so that anti elimination in this system is relatively slow for the same reasons as in the bridged-ring systems discussed above (Section III. E.2). On the other hand, syn elimination in the trans-2-phenylcyclohexyl system (47) may be slow because of difficulty in attaining a syn-periplanar conformation, or because of difficulty in abstracting an axial proton from a 1,2-diequatorial substituted cyclohexane (Section I.B.1.f).[68,69]

46   47

The Hammett ρ values at 50° for reaction of the 2-phenylcyclopentyl tosylates with t-butoxide in t-butyl alcohol are 1.48 for the anti and 2.76 for the syn elimination.[67] While the syn elimination seems to proceed via a transition state with greater carbanion character than that for the anti, the difference is not dramatic, and an E1cB mechanism for the syn elimination seems unlikely. Although the ρ value for the 2-phenylethyl system cited to bolster this argument (3.39 at 30°)[70] has been reported on reinvestigation to be considerably smaller (2.49 at 40°),[71] all of the ρ values are still well within the range expected for concerted E2 reactions.

A sulfonyl group in the β-position provides considerably more activation than an aryl group, but

qualitatively its effect is the same: activated syn is
faster than unactivated anti elimination, and the preference for anti over syn elimination is reduced substantially in the cyclopentane and less in the cyclohexane series. These points are nicely illustrated by
the 2-arylsulfonyl derivatives 48-49. They all eliminate toward the sulfonyl group when treated with base,
which means syn elimination for 48b and 49b.[72] The
evidence on whether these reactions, particularly the
syn eliminations, are E2 or ElcB has already been discussed (Section I.B.1.f) and will not be repeated here.

48a (cis)   49a (cis)

48b (trans)   49b (trans)

The rates of reaction of 48, 49 and 50 (where Ar =
p-tolyl in all cases) with hydroxide ion and with tertiary amines in dioxane-water have been determined and
are recorded in Table 2.[73] With hydroxide ion, 48a
reacts at about 400 times the rate of 48b, but 49a reacts at only twenty times the rate of 49b. The preferences are considerably reduced when trimethylamine is
the base, to about 20 and 1.2, respectively. A probable explanation is electrostatic interaction between
the base and the leaving group. The leaving group
bears a partial negative charge in the transition state
(even if the bond to the leaving group is little broken, the oxygens of the sulfonate will bear partial
charges) and, consequently, repels the hydroxide ion
but attracts the amino nitrogen because it has begun
to acquire a positive charge. Triethylamine shows a
somewhat higher preference for anti elimination than
trimethylamine (a factor of 6.5), probably because of
its greater steric requirements. All of the cyclic
compounds react slower than the acyclic 50.
      Very similar results were obtained in the reaction
of trimethylamine with the corresponding brosylates.[74]
At the same time, the threo and erythro diastereomers
of the open-chain compound 51(R = Me, Ar = p-tolyl,
X = OBs) were found to give exclusively anti elimination, though at a rate advantage for the erythro over
the threo of only 2.3 (the erythro isomer has only a

TABLE 2 Rates of Reaction of 2-p-Tolylsulfonyl Tosylates with Bases in Dioxane–Water at 25°

Compound	Stereochemistry of Elimination	Relative Rates with			
		Hydroxide	Trimethylamine	Triethylamine
48a	anti	81.	21.7	15.7
48b	syn	0.19	0.855	0.135
49a	anti	235.	118.	114.
49b	syn	11.9	98.5	17.4
50	?	1000.	1000.	1000.

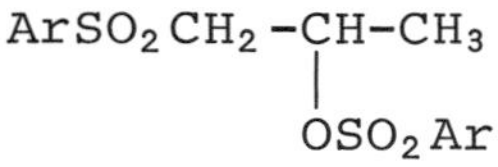

ArSO2CH2-CH-CH3
OSO2Ar

50

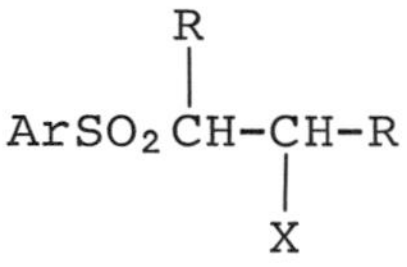

R
ArSO2CH-CH-R
X

51

methyl-methyl interaction in the transition state for anti elimination, compared to an arylsulfonyl-methyl interaction for the threo isomer). Stereospecific elimination (presumably anti, though this was not proved) is also observed with 51 (R = Me, Ar = Ph, X = I).[75] This stereospecificity with open-chain reactants provides evidence that the reactions are really concerted, since it would not be likely in a carbanion process. In fact, sufficient activation does destroy the stereospecificity and, presumably, change the mechanism to E1cB. Both the threo and erythro isomers of 51 (R = Ph, Ar = p-tolyl, X = Cl) react with sodium hydroxide in ethanol to give the more stable of the two possible isomeric products (the one with phenyl groups cis).[76] The isomers of the product are not interconverted under the reaction conditions, hence, it must have been a primary product.

Cyclohexyl derivatives with substituents that fix (or at least strongly favor) a particular conformation have been used to explore conformational effects in sulfonyl-activated eliminations. Compounds 52-56 react with piperidine in dimethylformamide at rates that are

H
OMs
Me
SO2Ar
Me
H

52

OMs
H
Me
SO2Ar
Me
H

53

H
OMs
t-Bu
SO2Ar
H
H

54

H
OMs
t-Bu
H
H
SO2Ar

55

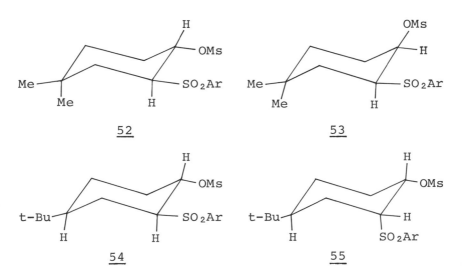

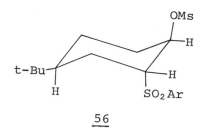

<u>56</u>

recorded in Table 3.[77] The anti-periplanar elimination

TABLE 3   Rates of Reaction of Various Substituted Cyclohexyl Methanesulfonates with Piperidine in Dimethylformamide at 30°

Compound	Conformation	$k \times 10^3$ 1/(mole) (sec)
<u>52</u>	H ax, OMs eq	1.5
<u>53</u>	H ax, OMs ax	28.
<u>54</u>	H ax, OMs eq	0.62
<u>55</u>	H eq, OMs eq	1.9
<u>56</u>	H eq, OMs ax	8.4

is clearly the most favored, but none of the rate ratios is especially large. The other major generalization that can be drawn is that a syn-clinal elimination
with hydrogen equatorial and methanesulfonate axial is
clearly preferred over one with hydrogen axial and methanesulfonate equatorial. The former would be expected to be accelerated by relief of strain caused by
the axial conformation of the arylsulfonyl group, and
the latter hindered by the difficulty in abstracting
an axial proton from a 1,2-diequatorial substituted
cyclohexane.[68,69]

Activation by a β-nitro group is still stronger
than by a β-sulfonyl group, and might be expected to
show a still lower stereochemical preference. The
rates of reaction of nitro acetates <u>57</u> and <u>58</u> with
methoxide ion in methanol are virtually the same, and
there is strong evidence that the reactions occur by
the ElcB mechanism.[78,79] This similarity of syn and
anti rates in an ElcB reaction need not always hold
true, however, because there may be differences in rate
of abstraction of conformationally different protons.

The rate of exchange with sodium methoxide of 59 is 350 times the rate of 60, for example.[68,69]

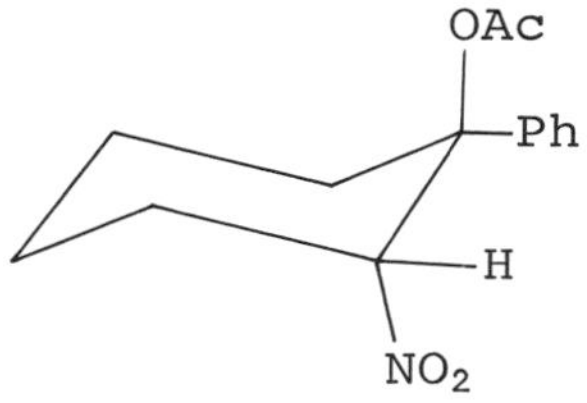

<u>57</u>

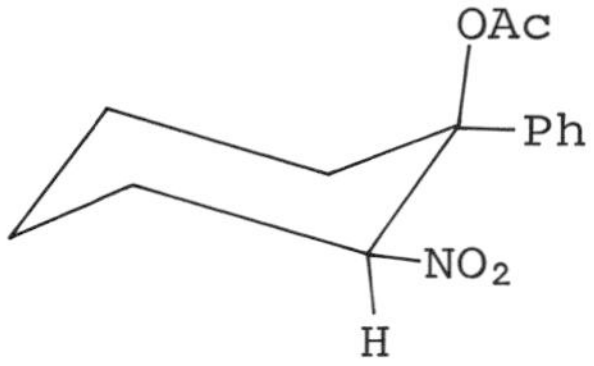

<u>58</u>

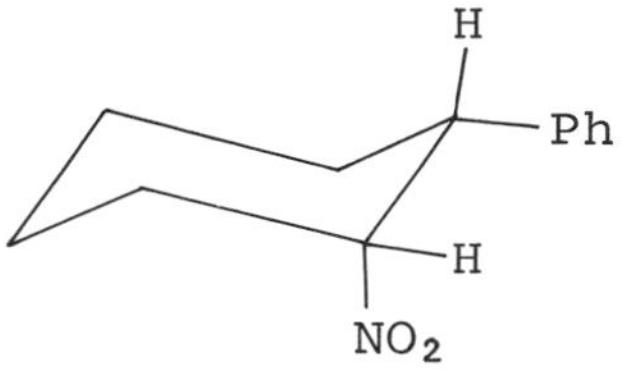

<u>59</u>

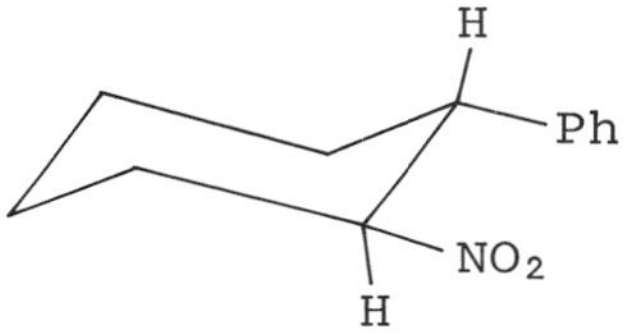

<u>60</u>

3 Small-Ring Compounds

The smaller-ring compounds (up to seven carbon atoms) usually prefer anti elimination if there is no activation at the β-position or if they are not incorporated in a rigid bicyclic system. Cycloalkyltrimethylammonium ions, however, can depart markedly from this behavior (see also Section III.D.2). By using reactants labeled with cis deuterium in the β-position, Cooke and Coke studied the stereochemistry of elimination from cyclobutyl- through cycloheptyl-trimethylammonium hydroxides under Hofmann-type conditions.[15] From the extent of deuterium loss and an independently-determined isotope effect for the syn elimination, they calculated the stereochemistry of the reaction in the absence of deuterium. The results are given in Table 4. The cyclohexyl system is the only one that give nearly or entirely stereospecific anti elimination (the deuterium content is actually within experimental error of that of the starting material, so the small decrease may or may not be real). The others represent major departures from the anti rule, including one case of marked preference for syn elimination. That such

TABLE 4   Stereochemistry of Elimination from Cycloalkyltrimethylammonium Hydroxides

Ring Size	Conditions[a]	% d₁ Olefin	Calculated % syn[b]
4	Dry, 50°	69	90
5	Wet, 110°	86	46
6	Wet, 110°	99	4
7	Dry, 50°	89	37
	Wet, 110°	91	31

[a]The "wet" method involved heating a solution of the
quaternary hydroxide under reduced pressure to concentrate it and then bring about reaction; the "dry" method involved prior concentration under high vacuum at
low temperatures followed by heating to about 50°.
[b]The estimated error in these figures is about ±5%.

departures were not observed sooner can be attributed
to the use of cyclohexyl derivatives in most of the
earlier investigations of stereochemistry. In addition, quaternary ammonium salts are the only common
substrates with a marked propensity for syn elimination, and even then strongly basic conditions such as
the Hofmann pyrolysis are usually required.
    The latter point is well illustrated by a study of
stereochemistry of elimination from cyclopentyl- and
3,3-dimethylcyclopentyltrimethylammonium ions under
various conditions.[80] In Table 5 we see that the importance of syn elimination increases as the basicity
of the medium[81,82] increases. The medium for the

TABLE 5   Stereochemistry of Elimination from Cyclo-
pentyl- and 3,3-Dimethylcyclopentyl-trimethylammonium
Salts

Base/Solvent, °C	Cyclopentyl, % Syn	Dimethylcyclo pentyl, % Syn
NaOH/H₂O, 190	4 ± 3	10 ± 1
NaOH/H₂O-DMSO,[a] 130	1 ± 3	52 ± 2
t-BuOK/t-BuOH, 70	17 ± 2	63 ± 5
t-BuOK/t-BuOH-DMSO,[b] 70	45 ± 1	72 ± 4
Hofmann[c]	46	76

[a]Water containing 60 mole % dimethyl sulfoxide.

[b]t-Butyl alcohol containing 50 mole % dimethyl sulfoxide.
[c]See Table 4 and Ref. 15.

Hofmann elimination is a highly concentrated syrup of
the quaternary hydroxide that would be expected to be
very basic because there are probably not enough water
molecules for full solvation of the hydroxide ions.
The differences between the cyclopentyl and dimethylcy-
clopentyl systems also show that alkyl substitution can
have an important effect on stereochemistry of elimination. In this case, steric interference between the
trimethylammonio group and the methyl group cis to it
probably hinder attainment of an anti-periplanar transition state.[15]

4   Electrostatic Factors

We have already seen a number of cases where electrostatic attraction or repulsion between the base and the
leaving group may affect the stereochemistry of elimination. For example, the anti/syn rate ratio was different for the reaction of hydroxide ion and amine
bases with cyclic β-sulfonyl tosylates (Section III.
D.2), and quaternary ammonium salts show less preference for anti elimination than do the corresponding
bromides or tosylates. Usually it is not possible to
isolate the electrostatic factor from other effects
such as base strength or the steric requirements of the
base.
     A relatively clear-cut example of the operation of
an electrostatic effect is afforded by the elimination
reactions of the dl and meso diastereomers of dichloro-
succinic acid.[83] A preference for the production of
chlorofumaric acid, the trans isomer, is expected because interactions between the carboxyl groups are minimized. In a solution sufficiently basic to ionize
both carboxyl groups, a strong electrostatic repulsion
between them will be added to the normal steric repulsion. The dl acid can give chlorofumaric acid by an
anti elimination, and does so in basic solution, in
water alone, and in acidic solution. The meso acid, on
the other hand, gives mixtures of chlorofumaric and
chloromaleic acids (syn + anti elimination) in aqueous
or aqueous acidic solution, but only chlorofumaric
acid (syn elimination) in basic solution. Electrostatic repulsion between the carboxylate groups clearly
is exerting a powerful effect, but the absence of
measurable amounts of chloromaleic acid makes it difficult to estimate the magnitude of the effect. The

preference for anti elimination that it must overcome
is not especially strong, for the dl isomer reacts only 14 times faster than the meso in alkaline solution
and 2.7 times faster than the meso in acidic solution.
It is not certain whether the mechanism of the syn e-
limination is E2 or ElcB.   There was about 0.5% uptake
of deuterium from deuterated solvent.   Hughes and
Maynard[83] considered this insufficient to justify postulation of an ElcB mechanism, but it is consistent
with an ElcB mechanism having much faster loss of chloride than uptake of a proton by the carbanion.   On the
other hand, formation of a carbanion in the base-promoted reaction would require the generation of a third
negative charge.   Finally, the aqueous and aqueous a-
cidic reactions could involve solvolysis rather than E2
attack by solvent, though solvolysis would be more
likely to yield substitution products.

5   Medium-Ring Compounds

The medium rings have provided a particularly fruitful
area for investigation of steric and stereochemical
factors in chemical reactions because the ring structures help to stabilize certain conformations that are
unlike those commonly found in either small-ring or
open-chain compounds.   The stereochemistry of medium
rings has been thoroughly reviewed by Sicher,[20] and
only those facts necessary for the understanding of
elimination reactions in these systems will be repeated
here.
     The relevant conformational features of the med-
ium rings are best illustrated by the cyclodecane
structure, which possesses a single, low-energy conformation.   Although the other members of the series are
not conformationally homogeneous, and are consequently
more difficult to discuss, their basic features are
very similar to those of cyclodecane.   The two drawings
of the cyclodecane ring, 61 and 62, illustrate the most

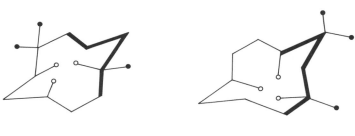

61   62

important of these features. The five-carbon segment
marked by a heavy line in 61 contains syn-clinal conformations about both of the central carbon-carbon
bonds; the segment marked by a heavy line in 62 contains a syn-clinal conformation about one of the central carbon-carbon bonds and an anti-periplanar conformation about the other. These conformations, in
different combinations, are characteristic of all medium-ring compounds. In both structures, the hydrogen
atoms indicated by open circles are in intra-annular
positions, and approach to them is severely hindered
by the ring structure and by the other hydrogen atoms.
In contrast, the hydrogen atoms indicated by closed
circles are in much more accessible extra-annular positions. They are not all equivalent (some are in positions similar to the axial and others in positions
similar to the equatorial hydrogens of the cyclohexane
system), but they are clearly far less hindered than
the intra-annular hydrogens. The hydrogens in other
medium-ring systems can also be divided into intra-
annular and extra-annular hydrogens. Substituents
other than hydrogen atoms will be constrained, by vir-
tue of their greater bulk, to occupy extra-annular positions. Furthermore, conformational homogeneity in a
medium-ring compound can often be attained by placing
gem-dimethyl substituents on the ring, which effectively prohibits conformations that would place either
methyl group of the pair in an intra-annular position
(compare Section III.C.1 for methods of achieving conformational homogeneity in the cyclohexane system).
        Considering the olefins obtained from elimination
reactions in medium-ring systems, we notice the important feature is that the rings are large and flexible
enough that trans as well as cis olefins are possible,
in contrast to the small rings. The cis olefin is
still preferred up through cyclodecane, and for cyclo-
undecane and cyclododecane there is only a 2:1 preference for trans olefin, but the important point is
that it is possible to prepare stable trans olefins in
all ring systems from cyclooctane on up. The preferences quoted are from determinations of the cis/trans
equilibrium by Cope, Moore, and Moore.[84] The extensive
and elegant investigations by Sicher and his group of
the stereochemistry of elimination in medium-ring systems, which will constitute most of the material
covered in this section, were reported to have been
prompted by the curious fact that there was generally
a strong preference for trans olefin in these reactions, although thermodynamic stabilities could, at
best, account for only a modest preference.[7] Although
Prelog had postulated that the medium rings should

prefer a conformation leading via anti elimination to trans rather than cis olefin,[85] Sicher showed that the correct explanation of the phenomenon involved a departure from anti elimination.

Before we describe the evidence, a brief discussion of Sicher's most important tool, the rate profile, is desirable. As we have already mentioned in general terms (Section III.C.3), the use of the rate profile is based on the assumption that reactions of similar stereochemistry will show a similar dependence of rate on ring size. Sicher proposed that reactions be divided into single-constraint and double-constraint processes for the purpose of discussing stereochemistry.[20] The single-constraint process refers to cases where a stereochemical change at a single ring atom is occurring, and includes reactions such as S_N1 and S_N2 processes on cycloalkyl halides or tosylates. The double-constraint process includes reactions that require definite orientations of the valencies of two ring atoms, and would include closures of small rings fused to the medium ring, or elimination reactions. This category includes two distinct types of processes, those requiring syn-periplanar and those requiring anti-periplanar orientation of the valencies on two adjacent ring atoms. To determine the stereochemistry of an elimination reaction from the rate profile, then, one should have available rate profiles for reactions of known stereochemistry that belong to the double-constraint category. The stereochemistry of the elimination reaction is presumed to be the same as the stereochemistry of the model reaction with the more closely similar rate profile. This method is capable of giving only qualitative information on the predominant stereochemical path of elimination, and requires that the syn and anti pathways have distinctly different rate profiles. We will see that this requirement is usually fulfilled, and that the results of this approach are confirmed by the results of more rigorous methods.

Rate profiles for a reaction of known syn stereochemistry, the thermal elimination from cycloalkyldimethylamine oxides (Section VIII.E.2), and for the reaction of cycloalkyltrimethylammonium chlorides with potassium t-butoxide in t-butyl alcohol are shown in Figure 1.[16,86,87] The rates of the amine oxide reactions for cis and trans olefin formation give very similar rate profiles with a clear maximum at cyclooctyl for the cis olefins and at cyclodecyl for the trans olefins. While the two profiles could not a priori have been expected to be so similar in view of the

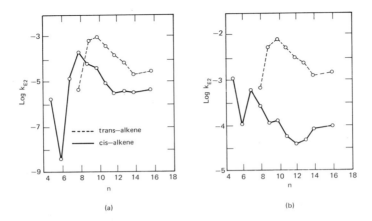

(a)   (b)

<u>Figure 1</u> Effect of ring size on rates of cis- and trans-cycloalkene formation (a) from cycloalkyldimethylamine oxides in t-butyl alcohol at 70.6°, and (b) from cycloalkyltrimethylammonium chlorides and potassium t-butoxide in t-butyl alcohol at 55° (reproduced by permission of the authors[18] and Tetrahedron Letters).

different transition-state conformations involved, the similarity is useful evidence for the validity of the rate profile method. The rate profile for formation of the trans olefin from the cycloalkyltrimethylammonium salt is closely similar to that for its formation from the amine oxide, suggesting that it is formed in both cases by a syn elimination. The rate profiles for the formation of the cis olefins in the two reactions are not at all similar, however, suggesting that the two reactions are of dissimilar stereochemistry. The rates of formation of the cis olefin from the cycloalkyltrimethylammonium salts in fact correlate well with the rates of a known anti elimination, the reaction of potassium iodide with 1,2-dibromides in acetone (Section VII.C.2.a). Interestingly, they also correlate well with the rates of a single-constraint process, the S_N2 reaction of potassium iodide in acetone with the corresponding monobromides.[88] While this correlation can be rationalized in several ways,[86] it is best regarded as an empirical fact of as yet unknown generality.

     The important point about these investigations is the evidence that they provide for an extremely interesting and unexpected conclusion: that the cis and

trans olefins are formed by reactions of different
stereochemistry. Specifically, the cis olefin is formed by an anti elimination, and the trans olefin by a
syn elimination. Sicher uses the term "syn-anti dichotomy" to describe this behavior. In subsequent papers
its generality was demonstrated. For example, Figure
2 shows the results in three different solvent-base

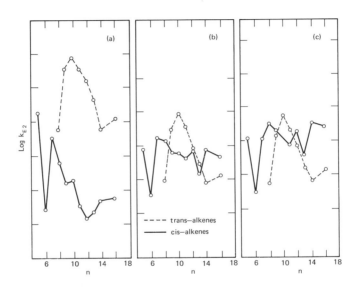

Figure 2 The effect of ring size on rates of formation of cis- and trans-cycloalkenes from cycloalkyltrimethylammonium chlorides and (a) potassium t-butoxide in t-butyl alcohol, (b) potassium ethoxide in
ethanol, and (c) potassium methoxide in methanol
(reproduced by permission of the authors[89] and the
collection of Czechoslovak Chemical Communications).

systems.[89] The rate profiles for formation of the
trans olefin are very similar in all three. Those for
formation of the cis olefin are more variable in shape,
but in no case do they resemble the profile for the
syn elimination to give cis olefin from the amine oxides (Figure 1). Later results in nine additional solvent-base systems gave the same qualitative picture.[7]
Thus the syn-anti dichotomy seems to be the main or exclusive mode of reaction under all of the conditions
studied.
     To test the reliability of the conclusions from
the rate profiles, the technique of stereospecific

deuterium labeling was applied by using 1,1,4,4-tetra-
methyl-7-cyclodecyltrimethylammonium chloride (63a)
and its cis-8-d (63b) and trans-8-d (63c) derivatives.
The methyl groups served as "markers" so that the olefins from elimination into and away from the deuterated
branch would be chemically different, and they also
helped to enforce conformational homogeneity. The olefins were not analyzed for deuterium. Instead,

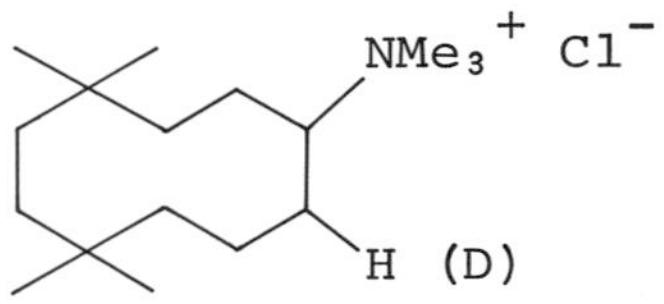

63a (no deuterium)

63b (cis-8-d)

63c (trans-8-d)

apparent isotope effects were calculated from the effect of deuterium on product proportions (Section III.
C.2). The result was that 63b showed no appreciable
isotope effect in the formation of either cis or trans
olefin, while 63c showed isotope effects in the formation of both.[96] The same results were obtained in two
solvent-base systems, potassium methoxide in methanol,
and potassium t-butoxide in dimethylformamide. While
this approach is only semiquantitative and cannot show
that the syn-anti dichotomy is the exclusive path, it
is certainly at least the major one.

That the syn-anti dichotomy is not limited to reactions of 'onium salts was shown by investigations of
tosylates and bromides. Deuterium labeled derivatives
were used in the tosylate investigations.[90,91] The reactions of 1,1,4,4-tetramethyl-7-cyclodecyl-8-d (64)
and 1,1,4,4-tetramethyl-8-cyclododecyl-9-d (65) tosylates with potassium t-butoxide in dimethylformamide,

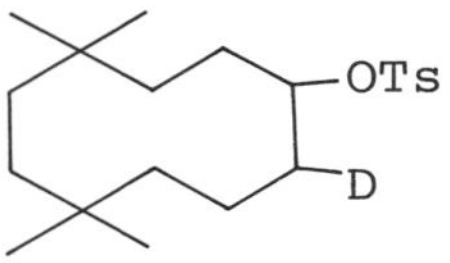

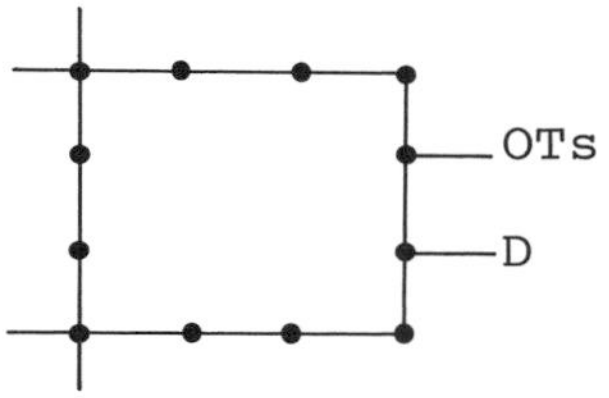

64

65

t-butyl alcohol, and benzene were examined. The stereochemistry of elimination into the deuterated branch
was determined for 64 from apparent isotope effects
just as with the corresponding quaternary ammonium
salt (63), and for 65 by isolation and deuterium analysis of the trans-8,9-olefin. The results given in
Table 6 show that the cyclodecyl derivative follows the
syn-anti dichotomy exclusively, or nearly so. Part of
the trans olefin from the cyclododecyl derivative,

TABLE 6   Stereochemistry of Elimination from 1,1,4,4-
Tetramethyl-7-cyclodecyl-8-d and 1,1,4,4-Tetramethyl-
8-cyclododecyl-9-d Tosylates

Base/Solvent, °C	Ring Size	% Syn-anti Dichotomy
t-BuOK/DMF, 50	10	>95[a]
t-BuOK/DMF, 50	12	30[b]
t-BuOK/t-BuOH, 100	10	>95[a]
t-BuOK/t-BuOH, 100	12	85[b]
t-BuOK/benzene, 130	10	>95[a]
t-BuOK/benzene, 130	12	95[b]

[a]Estimated from apparent isotope effects and from deuterium analysis of the total olefin product.
[b]Calculated from deuterium content of the trans-8,9-
olefin and the assumption that isotope effects for the
syn and anti eliminations are the same. The cis olefin
is assumed by analogy to result entirely from anti elimination.

however, is formed by an anti elimination. The tendency of the solvent to promote syn-anti dichotomy runs in
the order DMF < t-butyl alcohol < benzene, which is the
order of decreasing polarity or decreasing tendency to
promote ionization. Although the medium-ring tosylates
will follow the syn-anti dichotomy under favorable conditions, the propensity does not appear to be so strong
as with the quaternary ammonium salts. The syn-anti
dichotomy is also seen to be particularly favored in
the ten-membered rings.
     With the cycloalkyl bromides, the rate profile
approach was again used.[92] Profiles for the reactions
with t-butoxide in t-butyl alcohol and ethoxide in
ethanol are shown in Figure 3. Rates of formation of
both the trans and cis olefins give rate profiles very
similar to those found with the cycloalkyltrimethylammonium salts (Figure 1) when the reagent is

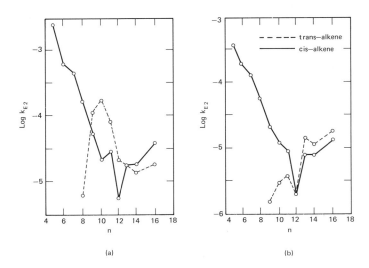

(a)   (b)

Figure 3   The effect of ring size on rates of formation
of cis- and trans-cycloalkenes from cycloalkyl bromides
and (a) Potassium t-butoxide in t-butyl alcohol at
82.5°, and (b) Potassium ethoxide in ethanol at 58.5°
(reproduced by permission of the authors[92] and the
Collection of Czechoslovak Chemical Communications).

potassium t-butoxide in t-butyl alcohol. The only discordant feature is the relatively higher rate of formation of cyclohexene, and this can be explained by the
fact that bromine can much more readily take up the
axial position necessary for anti elimination from a
cyclohexane derivative than can trimethylammonium.
Thus the syn-anti dichotomy seems to be the major or
exclusive mode of elimination under these conditions.
With ethoxide in ethanol, the rate profile for formation of the trans olefin is completely different, while
that for formation of the cis olefin is essentially unchanged. The cis olefin is presumably still formed by
an anti mechanism, and the change in the rate profile
for formation of the trans olefin suggests that it is
arising by a mechanism of different stereochemistry
than in the t-butoxide reaction. More specifically,
the rates correlate well with those of the S_N2 reactions of the cycloalkyl bromides with potassium iodide
in acetone,[88] and we have seen above that such a correlation is characteristic of anti eliminations. In this
case, both olefins are formed by a mechanism with

mainly or exclusively anti stereochemistry. Of two
other solvent systems investigated, potassium t-butoxide in benzene gave results indicating a syn-anti mechanism, while lithium t-butoxide in dimethylformamide
appeared to give all anti elimination. Again, the tendency toward the syn-anti dichotomy is greater in the
less polar and less ionizing solvents. It also appears
that the bromides show still less tendency toward the
syn-anti dichotomy than the tosylates, although definite conclusions are difficult because the tosylate
work (Table 6) used tetramethyl-substituted reactants
and deuterium labeling rather than the rate profile to
determine stereochemistry.

Finally, there is indirect evidence that the positive counter ion to the alkoxide base can exert a decided effect on stereochemistry. The action of potassium t-butoxide in t-butyl alcohol on cyclodecyl bromide gives a ratio of trans- to cis-cyclodecene of 5:1,
while tetramethylammonium t-butoxide gives 1:4 for the
same ratio. Since syn elimination favors trans olefin,
the potassium alkoxide evidently gives considerably
more syn elimination than the tetramethylammonium alkoxide.

Sicher's papers contain a considerable amount of
discussion of possible causes of the syn-anti dichotomy, but no firm conclusions. He points out that it
is most important in solvents of low ionizing ability,
so that ion pairing as in 66 for neutral leaving
groups, or as in 67 for positive leaving groups, would

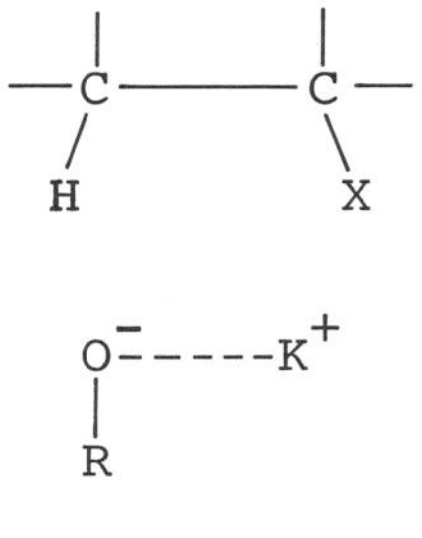

66 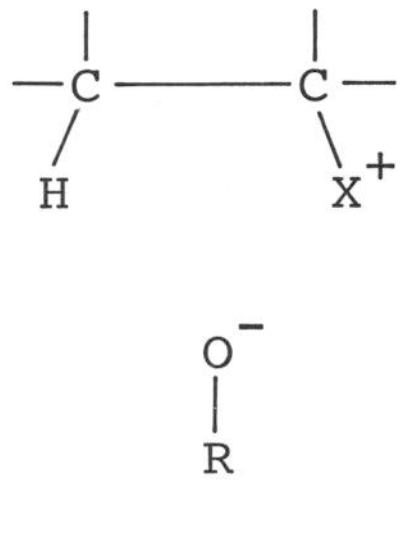   67

encourage syn elimination. While these effects may
well play a role, they do not explain why only one of
the two products is formed by syn elimination. The
double inversion mechanism for the syn elimination,
after a suggestion by Ingold,[6] suffers from the same
defect. In this mechanism it is supposed that the
very strong bases that promote syn elimination cause

the β hydrogen-carbon bond to be so stretched in the
transition state that a well-developed lobe of the p-
orbital which is forming can appear opposite the carbon-hydrogen bond, and that the electrons in this lobe
effect a backside displacement of the leaving group
(see Section III.B. and 6). An additional problem with
this mechanism is that both reliable evidence and theoretical principles indicate exactly the opposite effect
of increasing base strength: a decrease in the extent
of proton transfer (see Section II.B, II.C.3, and the
later discussion in this chapter).

     The most promising approach to explaining the di-
chotomy was based on conformational analysis. As noted
above, the favored equilibrium conformation of a cyclodecane derivative will have the substituent X in an
extra-annular position (68). The anti-β-hydrogen that

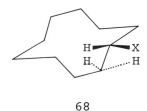

68

would have to be removed to give trans olefin by anti
elimination is then in a very hindered intra-annular
position.[7] Consequently, it is not suprising that the
syn-β-hydrogen, which is in an extra-annular position,
is the one that is attacked. Formation of cis olefin
by anti elimination is easier, since it only requires
that the ring structure open up or unfold to the point
where the leaving group and the β-hydrogen are at
least approximately anti to each other when the two
segments of the chain are cis. Under these circumstances, neither the leaving group nor the β-hydrogen will
be in especially hindered positions. At the time he
proposed this explanation, Sicher expressed doubt that
it could be "the sole or even the main cause responsible for the dichotomy."[7] The reason for this skepti-
cism was that examples of the dichotomy had been found
in open-chain compounds,[17] and he felt that this steric
explanation could not be applicable to these cases.
As we see shortly (Section III.F), there are good
grounds for believing that the steric explanation need
not be dismissed. Sicher pointed out finally that the
the two β-hydrogens are diastereotopic (diastereomeric
by internal comparison[93]), and might be expected to

show different reactivities. While this observation is quite reasonable, it is more a description than a solution of the problem, and it leaves unexplained the very large difference in reactivity.

6 Open-Chain Compounds

As noted above (Section III.D.2), simple open-chain reactants show a distinct preference for anti elimination. For the simplest system in which stereochemistry of elimination has been determined, 2-butyl-2-d, there are no conditions under which appreciable syn elimination has been observed. Among simple alkyl derivatives with longer chains, however, departures from anti elimination begin to appear. The first such example was the quaternary ammonium salt 69, reported by Pánková, Sicher, and Závada.[17] They calculated apparent isotope

$$CH_3-\overset{\overset{\displaystyle CH_3}{|}}{\underset{\underset{\displaystyle CH_3}{|}}{C}}-CH_2CHCHD(CH_2)_3CH_3$$
$$\underset{NMe_3}{|}{}^+$$

69

effects (Section III.C.2) from product proportions for elimination from the threo and erythro isomers and the undeuterated analog of 69. Elimination into the deuterated branch proceeds with isotope effects near unity for the erythro isomer, but with substantial isotope effects (2.3-4.7) for the threo isomer. This pattern held with a variety of reaction media; methoxide in methanol, t-butoxide in t-butyl alcohol, t-butoxide in dimethyl sulfoxide, and pyrolysis of the quaternary hydroxide. This result was checked by deuterium analysis of the total olefinic product, and the percentage of deuterium found agreed well with that calculated, assuming the operation of the syn-anti dichotomy.

Although one could argue that 69 is a sterically congested molecule that might give results atypical of alkyltrimethylammonium ions in general, a similar, though less pronounced, propensity toward the syn-anti dichotomy was soon found with 3-hexyl-4-d-trimethyl-ammonium ion.[94] In this case, the method of apparent isotope effects showed that the syn-anti dichotomy predominated with t-butoxide in t-butyl alcohol and with t-pentoxide in t-pentyl alcohol, but that anti elimination predominated with methoxide in methanol and with n-butoxide in n-butanol. Intermediate stereochemistry was found with s-butoxide in s-butyl alcohol.

A more quantitative study of 5-decyl-6-d-trimeth-ylammonium ion revealed a similar pattern.[95,96] Though the 4- and 5-decene produced could not be separated completely, they could be resolved into two fractions, one containing the cis and the other the trans isomers of each. These fractions were ozonized, and the alco-hols obtained on reduction of the ozonolysis products were separated and analyzed for deuterium. The method should give results equivalent to those from deuterium analysis of the separated olefins provided that there is no isotopic fractionation in the degradation pro-cedure. The calculated proportions of syn elimination in the formation of both the cis and trans-5-decene are recorded in Table 7. For comparison, data on 5-decyl-6-d tosylate are included.[48,96]

TABLE 7   Stereochemistry of Elimination from 5-Decyl-6-d-trimethylammonium Ion and 5-Decyl-6-d Tosylate

Leaving Group, X	Base/Solvent	% Syn/ Trans	% Syn/ cis	% Syn Overall
NMe_3^+	t-BuOK/Benzene	91.6	18.0	84.2
NMe_3^+	t-BuOK/DMSO	93.4	6.0	75.9
NMe_3^+	t-BuOK/t-BuOH	88.9	4.7	64.5
NMe_3^+	Pyrolysis	95.5	4.1	69.0
NMe_3^+	MeOK/MeOH	24.3	6.3	10.1
OTs	t-BuOK/Benzene	27.0	6.6	16.0
OTs	t-BuOK/t-BuOH	14.5	3.9	7.0
OTs	t-BuOK/DMF	3.7	5.8	4.2

The observation made on the medium-ring compounds, that syn elimination is favored by strong bases and by weakly ionizing solvents, is seen to apply to the open-chain compounds as well. There is a lower tendency to-ward the syn-anti dichotomy with the open-chain quater-nary ammonium salts, and a much lower tendency with the open-chain tosylates. Furthermore, such syn elimina-tion as there is with the tosylates is less selective than is the case with the quaternary ammonium salts; in dimethylformamide, the syn→cis route is actually slightly more important than the syn→trans. Benzene is an especially favorable solvent for syn elimination, leading to relatively high proportions of the syn→cis as well as syn→trans path with both leaving groups.

The effect of alkyl structure on the tendency

toward syn elimination has been explored in several studies. From the information already given, the branched alkyl chain of 69 causes adherence to the syn-anti dichotomy under a wider range of conditions than the straight chains of the 3-hexyl and 5-decyl systems. A systematic variation was undertaken by Bailey and Saunders,[97] who determined the stereochemistry of elimination for 3-hexyl-4-d, 3-hexyl-2-d- and 2-hexyl-3-d-trimethylammonium ions. Table 8 gives their results. The tendency toward syn elimination decreases in the order 3-hexyl→3-hexene > 3-hexyl→2-hexene > 2-hexyl→

TABLE 8   Stereochemistry of Elimination from Hexyltrimethylammonium Iodides[a]

Base/Solvent	Reaction[b]	% Syn Trans
n-BuOK/n-BuOH	3 ⟶ 3-ene	17
n-BuOK/n-BuOH	3 ⟶ 2-ene	10
n-BuOK/n-BuOH	2 ⟶ 2-ene	0[c]
t-PeOK/t-PeOH	3 ⟶ 3-ene	83
t-PeOK/t-PeOH	3 ⟶ 2-ene	70
NaOH/H_2O	3 ⟶ 3-ene	9
NaOH/H_2O	3 ⟶ 2-ene	2
Pyrolysis	3 ⟶ 3-ene	60
Pyrolysis	3 ⟶ 2-ene	44
NaOH/62 mole % DMSO in H_2O	3 ⟶ 3-ene	53
MeOK/MeOH	3 ⟶ 3-ene	20[c]
s-BuOK/s-BuOH	3 ⟶ 3-ene	68
t-BuOK/t-BuOH	3 ⟶ 3-ene	80[c]
t-BuOK/t-BuOH	2 ⟶ 2-ene	15[c]

[a]Unless otherwise noted, the result was calculated from deuterium analyses on partially-separated olefin fractions (conditions for complete separation could not be found) plus the assumption that the syn→cis path was negligible. Assumed isotope effects for anti and syn elimination were used to convert to the figure that would be observed in the absence of deuterium. See Ref. 97 for details.
[b]3 ⟶ 3-ene refers to 3-hexyl derivative giving 3-hexene, and so on.
[c]Estimated from apparent isotope effects calculated from product proportions (Section III.C.2).

2-hexene. The closer the functional group of the reactant and the double bond of the product to the center

of the chain, the greater the tendency toward syn elimination. On the other hand, further increase in chain
length beyond hexyl does not have a very marked effect.
Comparison of results for the three sets of conditions
common to Tables 7 and 8 show that the percentage of
syn elimination is markedly higher for 5-decyl than
for 3-hexyl (to give 3-hexene) only in the case of the
pyrolysis, a reaction of rather ill-defined and variable conditions. Table 8 also confirms the previous
conclusion[94-96] that strong bases and poorly ionizing
solvents promote syn elimination.

   When a trisubstituted rather than a 1,2-disubsti-
tuted olefin is the product of an elimination reaction,
the tendency toward syn elimination is distinctly lower
under most conditions. This can be seen from data on
the erythro- and threo-5-methyl-6-decyl derivatives
given in Table 9.[98] Here the situation is analogous

TABLE 9   Stereochemistry of Elimination from 5-Methyl-
6-decyl Derivatives

| | | 5-Methyl-5-decene | | |
Reactant	Base/Solvent	% cis	% trans	% syn
erythro-NMe$_3^+$	MeOK/MeOH	42.1	2.2	5.0
threo-NMe$_3^+$	MeOK/MeOH	0.2	53.3	0.4
erythro-NMe$_3^+$	t-BuOK/t-BuOH	1.8	1.2	40.0
threo-NMe$_3^+$	t-BuOK/t-BuOH	0.2	4.4	4.4
erythro-NMe$_3^+$	t-BuOK/DMSO	0.7	0.4	36.0
threo-NMe$_3^+$	t-BuOK/DMSO	0.2	2.3	8.0
erythro-NMe$_3^+$	t-BuOK/Benzene	0.2	4.5	96.0
threo-NMe$_3^+$	t-BuOK/Benzene	3.2	0.4	89.0
erythro-NMe$_3^+$	Pyrolysis	5.0	0.9	15.0
threo-NMe$_3^+$	Pyrolysis	0.4	15.7	2.5
erythro-OTs	t-BuOK/DMF	38.3	0.2	0.5
threo-OTs	t-BuOK/DMF	0.2	41.6	0.5
erythro-OTs	t-BuOK/t-BuOH	51.3	2.4	4.5
threo-OTs	t-BuOK/t-BuOH	3.7	35.5	9.5
erythro-OTs	t-BuOK/Benzene	35.0	2.0	5.4
threo-OTs	t-BuOK/Benzene	0.8	15.6	4.8

to that in the syn elimination to give cis olefin from
unbranched substrates, for an eclipsing interaction between two alkyl groups cannot be avoided in the transition state leading to either of the two products. With

the quaternary ammonium salts there is consistently more syn elimination from the erythro isomer, which yields trans-5-methyl-5-decene by syn elimination, than from the threo isomer, which yields cis by syn elimination. This fact provides further evidence for the role of eclipsing effects. The small amount of syn elimination from the tosylates, however, does not follow this pattern. These data, and data for 5-decyl-6-d tosylates in Table 7, indicate that syn elimination from tosylates is less dependent on conformational factors than is the case with quaternary ammonium salts.

F   THEORETICAL APPROACHES TO THE SYN-ANTI DICHOTOMY

We have already described (Section III.E.5 and 6) the ways in which the syn-anti dichotomy depends on structural and environmental influences, and we have considered some possible explanations of these patterns. We first describe a theory that attempts to provide a unified basis for the dichotomy in both medium-ring and open-chain compounds.[94,97]

Since there is both the theoretical (Section III. B) and experimental (Section III.D) evidence for a preference for anti elimination, our first step should be to ask why this preference is cverridden in reactions that follow the dichotomy. The fact that strong bases favor the dichotomy provides a possible answer to this question. As noted in the previous chapter (Section II.C.3.a), the hydrogen isotope effect in a proton-transfer reaction is largest when the proton is half transferred in the transition state, and smaller when it is either less or more than half transferred.[99] There is ample evidence that changing the strength of the attacking base affects the extent of proton transfer in elimination reactions, a particularly good example of which is the reaction of 2-phenylethyldimethylsulfonium ion with hydroxide ion in mixtures of water and dimethyl sulfoxide.[100] The deuterium isotope effect at first increases and then decreases as the proportion of dimethyl sulfoxide in the reaction medium increases. While these data do not tell the direction in which the extent of proton transfer is changing, the Hammond postulate and Thornton's theory of the effect of substituents on transition-state geometry (Section II.B.1,2) predict that increasing the strength of the attacking base should result in a decrease in the extent of proton transfer.[100,102] Experimental evidence for the correctness of this prediction is the fact that the deuterium isotope effect for the reaction

of 2-phenylethyltrimethylammonium ion with alkoxide ion
in the corresponding alcohol increases when the base
changes from ethoxide to t-butoxide, and at the same
time the Hammett rho value decreases.[103]   Both reactions are mainly or entirely anti eliminations, so
there is no change in mechanism to complicate the picture.[104]   For anti elimination, at least, the transition state appears to become more like reactant as base
strength increases.   Since most of the explanations of
the anti rule assume a transition state that is elec-
tronically and/or geometrically similar to the product
olefin, it seems reasonable to assume that a reactantlike transition state would show a diminished preference for anti elimination.   Under these circumstances
a relatively small effect, acting either to favor syn
elimination or further disfavor anti elimination, could
produce a change to a syn mechanism.   It should be emphasized, incidentally, that the above evidence does
not necessarily say that the extent of proton transfer
in the syn elimination must also be small, since there
will be numerous steric and electronic differences between the syn and anti transition states.[80]   This point
is discussed more fully in the preceding chapter (Section II.C.3.b).

Examination of models of alkyltrimethylammonium
ions reveals a steric effect that should hinder the
anti⟶trans route for elimination.   In the hexyltrimethylammonium ions, for example, the bulky trimethylammonio group forces the ends of the alkyl chain as
far away from itself as possible.   Specifically, the
groups attached to the β'- and γ-carbon atoms (the un-
primed letter refers to the branch into which elimination occurs) are forced into positions in which they
hinder approach to the anti-β-hydrogen when the molecule is in a conformation suitable for anti elimination.   This situation is depicted in 70, where R_1
is the β' and R_2 the γ group, for the anti⟶trans transition state, and in 71 for the anti⟶cis transition
state.   In accord with our conclusion that the transition state is reactantlike, bond angles similar to
those in the reactant are assumed.   The anti-β-hydrogen
in 70 is effectively protected from approach in almost
any direction.   That in 71 is hindered on one side
only, and is relatively open to approach from the
other.   The steric situation in 70 is more readily
appreciated from the projection shown in 72, from which
it is evident that a β' methyl group introduces 1,4-
interactions, and a γ methyl group 1,3-interactions,
with the anti-β-hydrogen.   These interactions "enclose"
the hydrogen in a manner that makes access by base

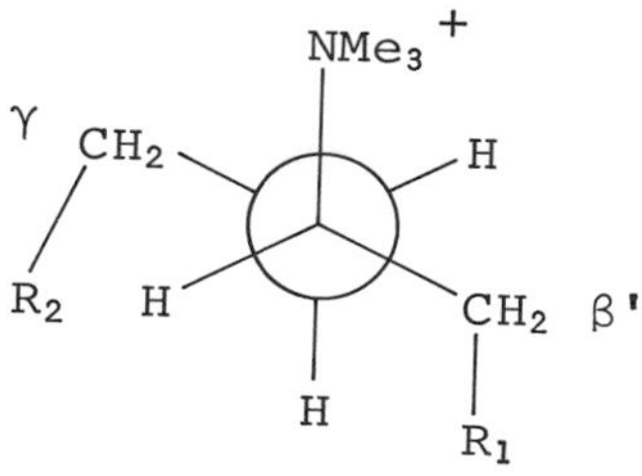

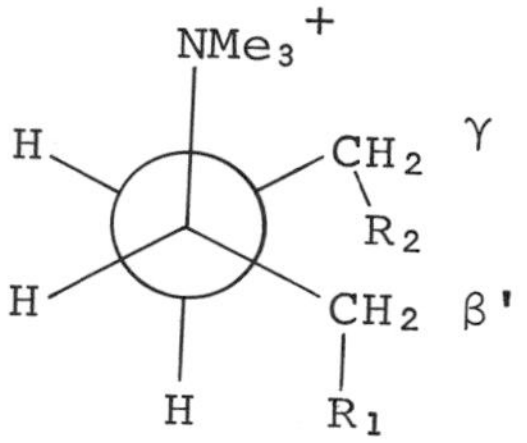

70   71

a: $R_1 = R_2 = CH_3$

b: $R_1 = C_2H_5$, $R_2 = H$

c: R_1 H, $R_2 = C_2H_5$

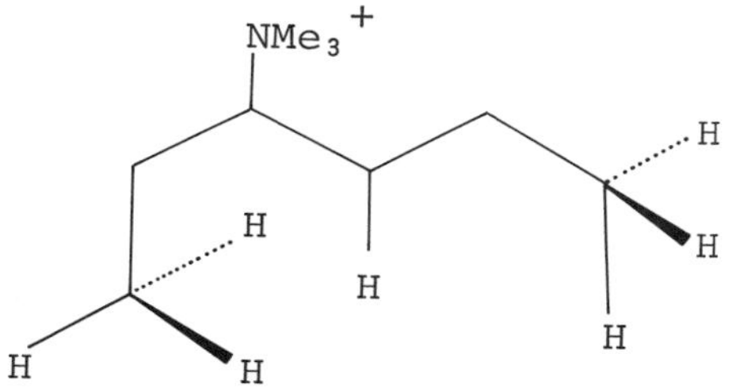

72

difficult. While the same interactions are present in
71, they do not enclose the hydrogen in the same manner. Thus one would expect the anti→cis path to be
easier than the anti→trans. In fact, those eliminations from quaternary ammonium salts that proceed
mainly or entirely by the anti mechanism do give higher yields of cis than trans olefin.[50,95-97] Under conditions that weaken the "normal" preference for anti
elimination, the syn mechanism should first appear as
a competitor to the anti→trans path.
    A variant of this explanation that leads to the
same conclusions has been suggested by Felkin and
Sicher.[110] According to them, the gauche interaction
between the γ-carbon and the trimethylammonio group
in 70 forces the transition state out of a strictly
anti-periplanar conformation, and thereby forces the
anti-β-hydrogen closer to the β-substituent R_1. The

analogous interaction in 71 would force the anti-β-hydrogen away from R_1.

It is still necessary to explain why the syn$\rightarrow$cis path is negligible in almost all eliminations from quaternary ammonium salts. The simplest hypothesis is that this is an ordinary eclipsing effect. While the transition state for syn elimination appears to be relatively reactant like, any double-bond character at all would introduce a preference for syn-periplanar elimination because of the increased orbital overlap it affords. The eclipsing effects that result will evidently be much worse in the transition state leading to cis olefin (73) than in the transition state leading to trans (74). Dissection of product ratios into those

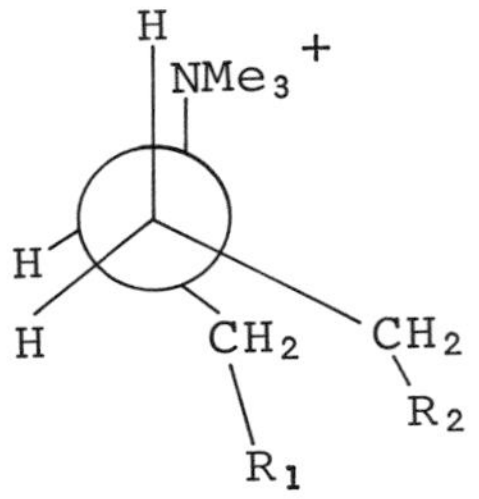

73 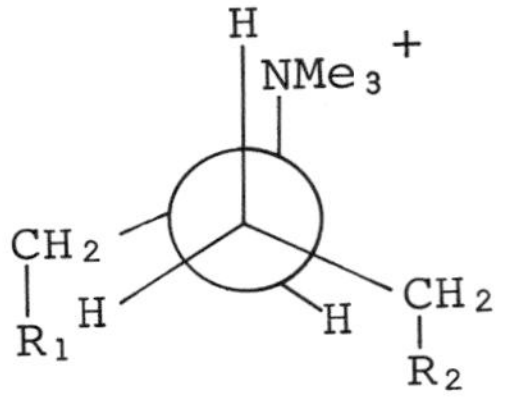   74

produced by syn and those produced by anti elimination reveals that the preference for trans olefin in the syn path can be imposingly large, as high as 50-60 times the yield of cis.[96,105] This is well above the ratio of thermodynamic stabilities of the olefins. Rotational barriers in simple alkanes, however, run 1.0 to 2.7 kcal/mole,[106] so that a syn-periplanar, reactant-like transition state might well give a difference in rate between the syn$\rightarrow$trans and syn$\rightarrow$cis paths approaching two powers of ten.

These basic postulates account qualitatively for the occurrence of the syn-anti dichotomy. The next question to be considered is why it is more important with some reactants than with others. As for the effect of the alkyl structure, closer examination of models and of 72 leads one to conclude that the 1,4-interactions with the β' substituent are more effective in hindering approach to the anti-β-hydrogen than the 1,3-interactions with the γ substituent. Further lengthening of the chain at either end is expected to have a relatively minor effect, for methylene groups beyond the γ' and δ positions are free to adopt conformations that will not further increase hindrance to the anti-β-hydrogen. Branching at the γ' or δ positions will, however, provide additional hindrance. Eclipsing effects will also be increased by chain branching, but will be increased relatively little by chain lengthening.

The data of Bailey and Saunders[97] afford a good test of these ideas. Their 3-hexyl⟶3-hexene reaction corresponds to 70a, where both the β' and γ groups are methyl. The 3-hexyl⟶2-hexene reaction corresponds to 70b, where the β' group is ethyl and the γ group is hydrogen, and the 2-hexyl⟶2-hexene reaction corresponds to 70c, where the β' group is hydrogen and the γ group is ethyl. Hindrance to attack at the anti-β-hydrogen should decrease along this series, and an examination of Table 8 shows that the importance of syn elimination also decreases. The relatively minor effect of further chain lengthening is shown by a comparison of Table 7 and 8, which shows that the tendency for syn elimination in the 5-decyl is not much greater than in the 3-hexyl system. Finally, the greatest tendency toward the syn-anti dichotomy so far observed in an open-chain system is found with 69, where the β' group is the very bulky t-butyl.

Both the open-chain and medium-ring compounds can be treated by this theory. Comparison of the cyclodecyl derivative 75 with the 3-hexyltrimethylammonium ion 76 (which is simply a redrawn version of 72) shows the close conformational similarities between the portions of each molecule directly involved in the

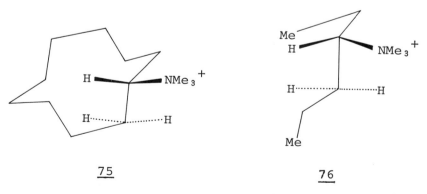

75   76

elimination reaction. Hindrance of the anti-β-hydrogen should be more effective in 75 than in 76, and the dichotomy is more cleanly followed by medium-ring than by unbranched open-chain quaternary ammonium salts. In

addition, a conformation such as 76 will be effectively
enforced only with a bulky leaving group such as trimethylammonium, while the conformation in 75 is enforced
ced to a major extent by the ring structure and to only
a minor extent by the size of the leaving group. Consequently, one would expect the dichotomy to be observed
ved with a wider range of leaving groups in medium-ring
than in open-chain systems. This expectation is also
in accord with the facts, since tosylates and bromides
show little tendency toward the dichotomy in open-chain
compounds, but a marked tendency in the medium rings
(though not so much so as the 'onium salts). This
lesser tendency for the tosylates and bromides is
doubtless associated not only with their smaller size
but also with their tendency to react via more product-
like transition states (Section II.C.2.3). It is evident that the theory accounts for the occurrence of the
dichotomy in both medium-ring and open-chain compounds
and also for the differences between these two classes.
Thus there is good reason for believing that Sicher's
original conformational explanation is more widely
applicable than he thought, and that his doubts over
its validity for open-chain systems were unnecessary.[7]

The steric theory in its present state is doubtless oversimplified. Perhaps its most serious defect
is that it makes no predictions about the effect of
structural and environmental changes on the rate of
syn elimination. Instead, syn elimination is regarded
as the passive beneficiary of hindrance to anti elimination. When the rates of elimination from open-chain
quaternary ammonium salts are dissected into partial
rates of syn and anti elimination, however, structural
variations produce important changes in the rate of
the syn process.[107]

The theory also does not make any explicit reference to ion pairing or electrostatic factors, yet
the evidence given earlier in this chapter (Section
III.E) on a variety of syn eliminations indicates
that they must be considered. A number of recent
studies afford more direct evidence. The importance
of syn elimination from 1,1,4,4-tetramethyl-7-cyclo-
decyl tosylates (64), as judged from the trans/cis
ratio of the product, increases with increasing base
concentration.[108] This is also the expected order of
increasing ion pairing of the base. Complexing the
cation of the base (potassium t-butoxide) with a cyclic
polyether ("crown ether") dramatically reduces the
trans/cis ratio of the olefins from 1,1,4,4-tetrameth-
yl-7-cyclodecyl bromide and tosylate, indicating a
large decrease in the propensity for syn elimination.[109]

Thus an ion-paired metal alkoxide seems much more effective than free alkoxide ion in promoting syn elimination, as suggested earlier by Sicher and his co-
workers.[91,92] A possible mode of operation of the effect is shown in 66 (Section III.E.5).

The effect of ion pairing on eliminations from
'onium salts is more difficult to establish, because
both the base and substrate are ionic, and one must
consider equilibria between free ions and the various
possible ion pairs. It appears that quaternary ammonium salts are largely dissociated, while metal alkoxides are largely ion paired, in the alcoholic media
used for most elimination reactions.[103] Thus the equilibrium of Eq. 4 probably describes well the states of
the ions in the reaction mixtures.[103] If the free RO⁻

$$RNMe_3^+ \; + \; X^- \; + \; RO^-M^+ \; \rightleftharpoons$$

$$RNMe_3^+ \; + \; RO^- \; + \; M^+X^- \qquad (4)$$

is present in higher concentration or is the more reactive base, syn elimination should be favored by
electrostatic attraction between the positively-charged
leaving group and the negatively-charged base.[7,89]
This attraction should be much less for the ion-paired
base, RO⁻M⁺, and the tendency for syn elimination
should be lower.

The evidence available thus far is in accord with
this picture. In eliminations from 3-hexyl-4-d-tri-
methylammonium iodide promoted by potassium phenoxide,
added tetramethylammonium iodide, which should shift
the equilibrium of Eq. 4 to the right, increases sharply the proportion of syn elimination.[111] The proportion of syn elimination with different alkali phenoxides increases in the order K < Na < Li. The tendency
toward ion pairing by the metal cation also increases
in this order, which suggests that increasing stability
of M⁺X⁻ in Eq. 4 is shifting the equilibrium toward the
right in the order K < Na < Li.[111]

Very recent evidence indicates that a leaving
group of large steric requirements is not a necessary
condition for the syn-anti dichotomy, even in open-
chain systems. 5-Decyl fluoride gives 90%, and 5-decyl
chloride gives 62% syn → trans elimination with potassium t-butoxide in benzene, but only 20% and 6% syn →
trans elimination, respectively, with potassium t-
butoxide in dimethyl sulfoxide.[112] Under similar conditions, the corresponding quaternary ammonium salt
gives more than 90% syn → trans elimination in both

solvents.[96]   Similarly, 3-hexyl fluoride gives 70%
syn→trans elimination with potassium t-butoxide in
t-butyl alcohol,[113] conditions under which the corresponding quaternary ammonium salt gives 80% syn→trans
elimination.[97]   In all cases the syn→cis path is minor
for the fluoride and chloride, just as with the quaternary ammonium salts.

Although fluorides and chlorides seem not to be
quite so prone to syn elimination as quaternary ammonium salts, it is clearly no longer tenable to maintain
that a large leaving group is required for the operation of the syn-anti dichotomy in open-chain systems.
The considerable differences in the amounts of syn
elimination between benzene and dimethyl sulfoxide for
the fluoride and chloride suggest that ion pairing
or ion aggregation (which is much more important in
benzene) is a major cause of syn elimination with these
compounds. By contrast, the large amounts of syn elimination from the quaternary ammonium salt in both solvents suggests that the steric effect described earlier
may still be the primary factor with bulky leaving
groups.

The general conclusion one can draw (and it should
be emphasized that the field is still developing rapidly at the time of this writing) is that no single
cause seems capable of accounting for all instances of
the syn-anti dichotomy. The one constant thread that
does seem to run through all the cases is an eclipsing
effect that keeps the syn→cis path minor even when
the syn→trans path is virtually the only route to
trans olefin. Thus the dichotomy follows naturally
when any factor favoring syn elimination is brought
into play, and is not the mysterious phenomenon it
first appeared to be.

The factors favoring syn elimination, which have
thus far been recognized, should be summarized in conclusion. Strong bases favor reactantlike transition
states that have lowered preference for anti elimination. Poor leaving groups also favor reactantlike
transition states. A large leaving group or a ring
structure may enforce transition-state conformations
that hinder approach to the anti-β-hydrogen. Alkyl
substitution at the β' and γ positions promotes syn
elimination, either by providing the shielding of the
anti-β-hydrogen mentioned in the preceding sentence,
or via some as yet undiscovered steric or electronic
mechanism. Finally, ion-pairing effects can favor
syn elimination by promoting electrostatic attraction
between the base and the leaving group.

REFERENCES

1. W. Klyne and V. Prelog, Exp., 16, 521 (1960).
2. W. Hückel, W. Tappe, and G. Legutke, Ann., 543, 191 (1940).
3. S. J. Cristol, N. L. Hause, and J. S. Meek, J. Am. Chem. Soc., 73, 674 (1951).
4. E. L. Eliel, N. L. Allinger, S. J. Angyal, and G. A. Morrison, "Conformational Analysis," Wiley, New York, 1965, p. 482.
5. K. Fukui and H. Fujimoto, Tetrahedron Lett., 4303 (1965).
6. C. K. Ingold, Proc. Chem. Soc., 265 (1962).
7. J. Sicher and J. Závada, Coll. Czech. Chem. Commun., 33, 1278 (1968).
8. W. T. Dixon, Chem. Commun., 402 (1967); Tetrahedron Lett., 2531 (1967).
9. J. P. Lowe, J. Am. Chem. Soc., 94, 3718 (1972).
10. J. Hine, J. Org. Chem., 31, 1236 (1966).
11. F. O. Rice and E. Teller, J. Chem. Phys., 6, 489 (1938); 7, 199 (1939).
12. J. Hine, J. Am. Chem. Soc., 88, 5525 (1966).
13. O. S. Tee, J. Am. Chem. Soc., 91, 7144 (1969).
14. D. J. McLennan, Qu. Rev. (London), 21, 490 (1967).
15. M. P. Cooke, Jr. and J. L. Coke, J. Am. Chem. Soc., 90, 5556 (1968).
16. H. C. Brown, "Hydroboration," W. A. Benjamin, New York, 1962.
17. M. Pánková, J. Sicher, and J. Závada, Chem. Commun., 394 (1967).
18. J. Sicher, J. Závada and J. Krupička, Tetrahedron Lett., 1619 (1966).
19. H. C. Brown and K. Ichikawa, Tetrahedron, 1, 221 (1957).
20. J. Sicher, "The Stereochemistry of Many-Membered Rings," in P. B. D. de la Mare and W. Klyne, Eds., "Progress in Stereochemistry," Vol. 3, Butterworths, London, 1962, Chap. 6.
21. P. Pfeiffer, Z. phys. Chem. (Leipzig), 48, 40 (1904).
22. A. Michael, J. prakt. Chem., 52, 308 (1895).
23. G. Chavanne, Bull. Soc. Chim. Bel., 26, 287 (1912).
24. P. F. Frankland, J. Chem. Soc., 654 (1912).
25. E. D. Hughes, C. K. Ingold, and J. B. Rose, J. Chem. Soc., 3839 (1953).
26. E. D. Hughes and J. Wilby, J. Chem. Soc., 4094 (1960).
27. S. J. Cristol, J. Am. Chem. Soc., 69, 338 (1947).
28. S. J. Cristol, N. L. Hause, and J. S. Meek,

J. Am. Chem. Soc., 73, 674 (1951).

29. S. J. Cristol and W. Barasch, J. Am. Chem. Soc., 74, 1658 (1952).

30. S. J. Cristol and D. D. Fix, J. Am. Chem. Soc., 75, 2647 (1953).

31. E. D. Hughes, C. K. Ingold, and R. Pasternak, J. Chem. Soc., 3832 (1953).

32. P. B. D. de la Mare, N. V. Klassen and R. Konigsberger, J. Chem. Soc., 5285 (1961).

33. W. Hückel and M. Hanack, Angew. Chem., 79, 555 (1967); Angew. Chem. Int. Ed. Engl., 6, 534 (1967).

34. T. H. Brownlee and W. H. Saunders, Jr., Proc. Chem. Soc., 314 (1961).

35. D. V. Banthorpe, A. Louden, and F. D. Waller, J. Chem. Soc., (B), 509 (1967).

36. S. Winstein and N. J. Holness, J. Am. Chem. Soc., 77, 5562 (1955).

37. D. Y. Curtin, R. D. Stolow, and W. Maya, J. Am. Chem. Soc., 81, 3330 (1959).

38. D. Y. Curtin, Rec. Chem. Prog., 15, 111 (1954).

39. R. D. Haworth, J. McKenna, and R. G. Powell, J. Chem. Soc., 1110 (1953).

40. B. B. Gent and J. McKenna, J. Chem. Soc., 573 (1956).

41. R. Ledger, J. McKenna, and P. B. Smith, Tetrahedron Lett., 1433 (1963).

42. D. J. Cram, J. Am. Chem. Soc., 74, 2149 (1952).

43. D. Y. Curtin and D. B. Kellom, J. Am. Chem. Soc., 75, 6011 (1953).

44. D. J. Cram, F. D. Greene, and C. H. DePuy, J. Am. Chem. Soc., 78, 790 (1956).

44b. J. Borchardt and W. H. Saunders, Jr., unpublished results.

45. P. S. Skell and R. G. Allen, J. Am. Chem. Soc., 81, 5383 (1959).

46. R. A. Bartsch, Tetrahedron Lett., 297 (1970); J. Am. Chem. Soc., 93, 3683 (1971).

47. D. H. Froemsdorf, W. Dowd, and W. A. Gifford, Chem. Commun., 449 (1968).

48. J. Závada, M. Pánková, and J. Sicher, Chem. Commun., 1145 (1968).

49. D. H. Froemsdorf, H. R. Pinnick, Jr., and S. Meyerson, Chem. Commun., 1600 (1968).

50. D. S. Bailey, F. C. Montgomery, G. W. Chodak, and W. H. Saunders, Jr., J. Am. Chem. Soc., 92, 6911 (1970).

51. S. J. Cristol and N. L. Hause, J. Am. Chem. Soc., 74, 2193 (1952).

52. S. J. Cristol, F. R. Stermitz, and P. S. Ramey, J. Am. Chem. Soc., 78, 4939 (1956).

53. S. J. Cristol and E. F. Hoegger, J. Am. Chem.

Soc., <u>79</u>, 3438 (1957).

54. N. A. LeBel, P. D. Beirne, E. R. Karger, J. C. Powers, and P. M. Subramanian, J. Am. Chem. Soc., <u>85</u>, 3199 (1963).

55. N. A. LeBel, P. D. Beirne, and P. M. Subramanian, J. Am. Chem. Soc., <u>86</u>, 4144 (1964).

56. H. Kwart, T. Takeshita, and J. L. Nyce, J. Am. Chem. Soc., <u>81</u>, 2606 (1964).

57. H. C. Brown and K. T. Liu, J. Am. Chem. Soc., <u>92</u>, 200 (1970).

58. J. K. Stille, F. M. Sonnenberg, and T. H. Kinstle, J. Am. Chem. Soc., <u>88</u>, 4922 (1966).

59. J. L. Coke and M. P. Cooke, Jr., J. Am. Chem. Soc., <u>89</u>, 670 (1967).

60. C. W. Bird, R. C. Cookson, J. Hudec, and R. O. Williams, J. Chem. Soc., 410 (1963).

61. R. T. Arnold and P. N. Richardson, J. Am. Chem. Soc., <u>76</u>, 3649 (1954).

62. J. Weinstock and F. G. Bordwell, J. Am. Chem. Soc., <u>77</u>, 6706 (1955).

63. A. C. Cope, G. A. Berchtold, and D. L. Ross, J. Am. Chem. Soc., <u>83</u>, 3859 (1961).

64. S. J. Cristol and D. I. Davies, J. Org. Chem., <u>27</u>, 293 (1962).

65. S. J. Cristol and F. R. Stermitz, J. Am. Chem. Soc., <u>82</u>, 4692 (1960).

66. G. Ayrey, E. Buncel, and A. N. Bourns, Proc. Chem. Soc., 458 (1961).

67. C. H. DePuy, G. F. Morris, J. S. Smith, and R. J. Smat, J. Am. Chem. Soc., <u>87</u>, 2421 (1965).

68. F. G. Bordwell and M. M. Vestling, J. Am. Chem. Soc., <u>89</u>, 3906 (1967).

69. F. G. Bordwell and K. C. Yee, J. Am. Chem. Soc., <u>92</u>, 5933 (1970).

70. C. H. DePuy and C. A. Bishop, J. Am. Chem. Soc., <u>82</u>, 2532 (1960).

71. J. Banger, A. F. Cockerill, and G. L. O. Davies, J. Chem. Soc., <u>B</u>, 498 (1971).

72. J. Weinstock, R. G. Pearson, and F. G. Bordwell, J. Am. Chem. Soc., <u>78</u>, 3468 (1956).

73. J. Weinstock, R. G. Pearson and F. G. Bordwell, J. Am. Chem. Soc., <u>78</u>, 3473 (1956).

74. F. G. Bordwell and P. S. Landis, J. Am. Chem. Soc., <u>79</u>, 1593 (1957).

75. P. S. Skell and J. H. McNamara, J. Am. Chem. Soc., <u>79</u>, 85 (1957).

76. S. J. Cristol and P. Pappas, J. Org. Chem., <u>28</u>, 2066 (1963).

77. W. M. Jones, T. G. Squires, and M. Lynn, J. Am. Chem. Soc., <u>89</u>, 318 (1967).

78. F. G. Bordwell, R. A. Arnold, and J. B. Biranowski, J. Org. Chem., 28, 2496 (1963).

79. F. G. Bordwell, M. M. Vestling, and K. C. Yee, J. Am. Chem. Soc., 92, 5950 (1970).

80. K. C. Brown and W. H. Saunders, Jr., J. Am. Chem. Soc., 92, 4292 (1970).

81. K. Bowden, Chem. Rev., 66, 119 (1966).

82. D. Bethell and A. F. Cockerill, J. Chem. Soc., B, 913 (1966).

83. E. D. Hughes and J. C. Maynard, J. Chem. Soc., 4087 (1960).

84. A. C. Cope, P. T. Moore, and W. R. Moore, J. Am. Chem. Soc., 81, 3153 (1959).

85. V. Prelog in "Perspectives in Organic Chemistry," A. R. Todd, Ed., Interscience Publishers, London, 1956, p. 94.

86. J. Sicher and J. Závada, Coll. Czech. Chem. Commun., 32, 2122 (1967).

87. J. Závada, J. Krupička, and J. Sicher. Coll. Czech. Chem. Commun., 31, 4273 (1966).

88. L. Schotsmans, P. J. C. Fierens, and T. Verlie, Bull. Soc. Chim. Belg., 68, 580 (1959).

89. J. Závada and J. Sicher, Coll. Czech. Chem. Commun., 32, 3701 (1967).

90. J. Závada, M. Svoboda, and J. Sicher, Tetrahedron Lett., 1627 (1966).

91. M. Svoboda, J. Závada, and J. Sicher, Coll. Czech. Chem. Commun., 33, 1415 (1968).

92. J. Závada, J. Krupička, and J. Sicher, Coll. Czech. Chem. Commun., 33, 1393 (1968).

93. K. Mislow, "Introduction to Stereochemistry," W. A. Benjamin, New York, 1965.

94. D. S. Bailey and W. H. Saunders, Jr., Chem. Commun., 1598 (1968).

95. M. Pánková, J. Závada, and J. Sicher, Chem. Commun., 1142 (1968).

96. J. Sicher, J. Závada, and M. Pánková, Coll. Czech. Chem. Commun., 36, 3140 (1971).

97. D. S. Bailey and W. H. Saunders, Jr., J. Am. Chem. Soc., 92, 6904 (1970).

98. J. Sicher, M. Svoboda, M. Pánková, and J. Závada, Coll. Czech. Chem. Commun. In press.

99. F. H. Westheimer, Chem. Rev., 61, 265 (1961); L. Melander, "Isotope Effects on Reaction Rates," Ronald Press, New York, 1960, pp. 24-32.

100. A. F. Cockerill, J. Chem. Soc., B, 964 (1967).

101. G. S. Hammond, J. Am. Chem. Soc., 77, 334 (1955).

102. E. R. Thornton, J. Am. Chem. Soc., 89, 2915 (1967); L. J. Steffa and E. R. Thornton, J. Am. Chem. Soc., 89, 6149 (1967).

103. W. H. Saunders, Jr., D. G. Bushman, and A. F.
      Cockerill, J. Am. Chem. Soc., 90, 1775 (1968).
104. A. N. Bourns and A. C. Frosst, Can. J. Chem.,
      48, 133 (1970).
105. J. Sicher, J. Závada, and M. Pánková, Chem.
      Commun., 1147 (1968).
106. Ref. 4, Chap. 1.
107. J. Závada, private communication.
108. J. Závada and J. Svoboda, Tetrahedron Lett.,
      23 (1972).
109. M. Svoboda, J. Hapala, and J. Závada, Tetrahedron
      Lett., 265 (1972).
110. H. Felkin, as quoted by J. Sicher, Pure. Appl.
      Chem., 25, 655 (1971).
111. J. K. Borchardt and W. H. Saunders, Jr.,
      Tetrahedron Lett., 3439 (1972).
112. M. Pánková, M. Svoboda, and J. Závada,
      Tetrahedron Lett., 2965 (1972).
113. J. Swanson, J. K. Borchardt, and W. H. Saunders,
      Jr., unpublished results.

IV   ORIENTATION IN E2 REACTIONS

A   DEFINITIONS AND THE ORIENTATION RULES

1   Types of Orientation

Whenever a single reactant can yield two or more
different olefins in an elimination reaction, the problem of orientation is presented.   There are two main
categories of orientation.   When the position of the
double bond in the product molecule is under discussion, one speaks of positional orientation.   An example
is the reaction of 2-butyl bromide with base (Eq. 1),
which can give either 2-butene or 1-butene.   As we will

$$CH_3CHCH_2CH_3 + EtO^- \longrightarrow CH_3CH=CHCH_3 + CH_3CH_2CH=CH_2$$
$$\mid$$
$$Br \qquad\qquad\qquad 81\% \qquad\qquad 19\%$$

(1)

165

notice from the figures under the products, there is a
decided preference for 2-butene.[1] The preference for
one product over the other can often be quite large,
and one of the major requirements for theories of elimination reactions is that they satisfactorily explain
these preferences. A different type of positional
orientation is found with molecules that are capable of
losing two or more different alkyl groups to give olefins. An example is given in Eq. 2. The actual prod-

$$CH_3CH_2 \overset{CH_3}{\underset{CH_3}{\overset{|}{\underset{|}{-N^+-}}}} CH_2CH_2CH_3 \; OH^- \longrightarrow CH_2 = CH_2 \; + \; \overset{CH_3}{\underset{CH_3}{\overset{|}{\underset{|}{N-}}}} CH_2CH_2CH_3$$

(2)

uct is almost exclusively ethylene, as indicated, but
the reaction could in principle also give propylene.
This second type of positional orientation may not follow precisely the same rules as the first, for the permissible range of structure variation is greater. The
α-position of the reactant in Eq. 1 is common to both
of the paths of reaction, whereas one can change the
degree of α-substitution of one alkyl group in Eq. 2
without affecting the others.

The second main category of orientation is geometrical orientation. For example, the 2-butene in Eq. 1
will, in general, be a mixture of trans-2-butene and
cis-2-butene. Mixtures of geometrical isomers are to be
expected in any case except those where one of the isomers is unstable, such as the trans isomer of a small-
ring olefin. As we show in the subsequent discussion,
however, there is otherwise no necessary relation between product energies and product proportions, either
with respect to positional or geometrical orientation.

2 Orientation and Mechanism

Before attempting to interpret or explain any example
of orientation, we must make certain that all of the
products are formed by the same mechanism. In this
chapter we discuss E2 reactions, which can be distinguished kinetically from the E1 or cyclic mechanisms
(Chapter I). Sometimes formal determination of the
kinetics can be avoided if simple qualitative criteria
give a clear answer. If the substrate does not react
at all in the absence of base there can be no competing
unimolecular reaction. Even if it does, lack of

variation of product proportions over a range of base concentrations indicates that the unimolecular reaction either does not compete with the bimolecular at the base concentrations used, or else gives the same product proportions.

Distinguishing the E2 process from other kinetically bimolecular reactions is more difficult. We have already discussed this problem in Chapter I and will only call attention to it here. The major competitor of this sort is likely to be an E2 reaction of different stereochemistry from the main reaction. In Chapter III we show that many elimination reactions can occur by both anti and syn E2 mechanisms. Any effort to explain product proportions in a reaction of mixed stereochemistry must take this factor into account, since the response of the two modes of elimination to structural and environmental changes may be quite different. We pointed out earlier (Section III.E.6) that anti elimination favors cis olefin and that syn elimination favors trans olefin in E2 reactions of quaternary ammonium salts. We discuss later specific examples of the relationship between orientation and stereochemistry of elimination.

3 Significance of Orientation Data

From a synethetic point of view, orientation data and orientation rules are important because they aid in the choice of conditions that afford the maximum yield of the desired isomer of a product. From the theoretical point of view, however, orientation is important because of the information it affords on relative reactivities. Olefin mixtures are often easily analyzed by gas chromatography, permitting a great deal of data to be obtained in a short time. For this reason, it is important to consider the disadvantages as well as the advantages of orientation data as background material for theories of reactivity.

The most important point to remember is that the same ground state is common to all the paths of reaction of a given reactant, so that the product proportions reflect the free energies of the various transition states. This is an advantage when one discusses the various paths available to that reactant under the specific reaction conditions employed, but it also makes difficult the comparison of the results with those obtained for a different reactant, or for the same reactant under different conditions. For example, an increase in the proportion of 1-butene from 2-butyl bromide on changing to a different base or solvent

might result from an increased rate of formation of 1-
butene, a decreased rate of formation of 2-butene, or
an increase or decrease in both at the same time, but
to different extents. The ambiguity of interpretation
can only be removed by combining orientation and overall rate data so that one can calculate partial rates
of formation of each of the products. Studies including overall rates as well as orientation data are the
exception rather than the rule, however, and we must
keep in mind that the interpretations presented in subsequent sections require unproved assumptions about the
rate changes responsible for the observed changes in
orientation.

4   The Hofmann Rule

This earliest generalization about orientation in elimination reactions was advanced by Hofmann[2] well over
a hundred years ago. He pointed out that quaternary
ammonium hydroxides when heated lost an ethyl group as
ethylene in preference to larger alkyl groups, as illustrated in Eq. 2. Hanhart and Ingold[3] provided further
examples of the rule, and generalized it to the prediction that the olefin having the smallest number of
alkyl groups about the double bond would be preferred.
They also advanced an explanation that we consider
shortly after we present more facts.
    The rule is followed by trialkylsulfonium as well
as tetraalkylammonium hydroxides.[4] In addition to predicting the relative ease of loss of different alkyl
groups, it can also predict the direction of elimination within a single alkyl group. For example, 2-
butyldimethylsulfonium ion reacts with ethoxide ion in
ethanol to give 74% 1-butene and 26% 2-butene,[5] and
t-amyldimethylsulfonium ion under the same conditions
gives 86% of 2-methyl-1-butene and 14% of 2-methyl-2-
butene.[6] That this disparity was specifically due to
a slower rate of elimination into the branch with a
β-alkyl group was shown by combining rate and product
data to calculate partial rates of elimination.[5] The
results for the isopropyl and 2-butyl systems are shown
in 1 and 2, where the figures in parentheses are relative partial rates.

(100) CH_3
               $>CHS(CH_3)_2^+$
(100) CH_3

(36) CH_3CH_2
               $>CHS(CH_3)_2^+$
(98) CH_3

1   2

Adherence to the rule seems to be firmer for ammonium than for sulfonium salts. The pyrolysis of 2-butyltrimethylammonium hydroxide yields 95% of 1-butene and only 5% of 2-butene.[7] The data in Table 1 on the products of elimination from a number of quaternary ammonium hydroxides permit a number of additional conclusions.[7] The first β-methyl group (from ethyl to n-propyl) causes a striking decrease in relative rate of elimination compared to the rather minor effect of a second β-methyl group (n-propyl to isobutyl), or of lengthening the alkyl chain (n-propyl to n-butyl). Smith and Frank[8] reported similar results, though they did find that the β-t-butylethyl group was lost considerably more slowly than other alkyl groups. They also found that relative rates of loss of alkyl groups calculated from one set of substrates could not always predict accurately the product proportions from other substrates. The second major generalization is that α-alkyl substitution does not have the same effect as β-alkyl. The isopropyl and t-butyl groups are both lost more readily than the ethyl group.

TABLE 1 Products from Thermal Decomposition of $R_1 R_2 NMe_2^+ OH^-$

| | | Percent of Olefin from | |
| | | --- | --- |
R_1	R_2	R_1	R_2
Et	n-Pr	97.6	2.4
Et	n-Bu	98.4	1.6
n-Pr	n-Bu	59.8	40.2
n-Pr	i-Bu	72.9	27.1
Et	i-Pr	41.2	55.6
Et	t-Bu	7.2	92.8

Rate results with sulfonium salts show a pattern similar but not quite identical to that for ammonium salts. Substitution in the α-position again has an accelerating effect, for the rate of reaction of the alkyldimethylsulfonium ion with ethoxide ion in ethanol increases along the series ethyl < i-propyl < t-butyl.[5,6,9] The rates for sulfonium and ammonium salts in Table 2 show that the first β-methyl group does not have the unusually large effect with sulfonium salts; the first and second β-methyl groups produce very

TABLE 2   Relative Elimination Rates for $R_1 R_2 CHCH_2 X$ with Ethoxide Ion in Ethanol

R_1	R_2	X, Temperature	
		$\overset{+}{S}Me_2$, 64°	$\overset{+}{N}Me_3$, 104°
H	H	100	100
Me	H	36.7	7.2
Me	Me	12.7	2.4
Et	H	26.6	4.0
i-Pr	H	20.2	1.5
t-Bu	H	0.54	0.12

similar changes in rate.[10]

A brief summary of possible causes of Hofmann-rule behavior is in order at this point, although we discuss the subject more thoroughly when we consider the effect of the leaving group on orientation later in this chapter. The first explanation was that of Hanhart and Ingold,[3] who argued that the positively charged leaving group increased the acidity of the β-hydrogens by an inductive effect, but that β-alkyl substitution partially counteracted this effect by an electron-repelling inductive effect. A similar electronic effect on transition-state stability was first suggested by Cram,[11] and was further elaborated by a number of workers (Section II.A). According to this argument, the trimethylammonio and, to a lesser extent, the dimethylsulfonio groups are relatively poor leaving groups, and a considerable amount of negative charge must be built up on the β-carbon before the transition state is attained. The transition state thus has considerable carbanion character and little double-bond character. Substitution in the β-position by an electron-repelling alkyl group destabilizes the transition state by making it more difficult to build up negative charge on the β-carbon.

An alternate explanation that ascribes the Hofmann rule to steric effects has been advanced by Brown.[12,13] According to this explanation, there are steric interactions between the leaving group X and a β-alkyl group R in the transition state for anti elimination (3), and these interactions are particularly important for large leaving groups such as trimethylammonio. There seems to be little doubt that steric effects

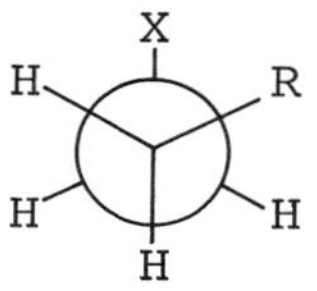

$$\underline{3}$$

contribute to some degree to Hofmann-rule behavior. The large effect of a β-t-butyl group on rate of elimination (Table 2) can hardly be anything but a steric effect. The magnitude of the steric effect in less clear-but cases is hard to assess. Its mode of operation cannot be as simple as described above, for we now know that eliminations from quaternary ammonium salts usually occur with mixed anti and syn stereochemistry (Section III.E.5,6). There is good evidence, however, that the incursion of syn elimination is, at least, in part the result of steric effects (Section III.F). We discuss later (Section IV.C.3) the relationship between stereochemistry of elimination and orientation of the products.

5   The Saytzev Rule

In 1875, Saytzev[14] noted certain regularities in elimination reactions of alkyl halides. When elimination into different branches of an alkyl group is possible, the preferred product will be the one bearing the greater number of alkyl groups on the double bond. The rule was suggested long before anything was known about the mechanisms of the elimination reactions, but subsequent studies by Hughes, Ingold, and their collaborators[15,16] defined the range of mechanistic applicability. They found the rule to apply to both E1 and E2 reactions of alkyl halides, but only to E1 reactions of sulfonium salts. We consider only the E2 reactions in this chapter and discuss orientation in E1 reactions later (Section V.C.3).

An example of the operation of the Saytzev rule has already been given in Eq. 1, where 2-butyl bromide with ethoxide ion gives 2-butene in preference to 1-butene.[1]   In $\underline{4}$ and $\underline{5}$, the figures in parentheses are

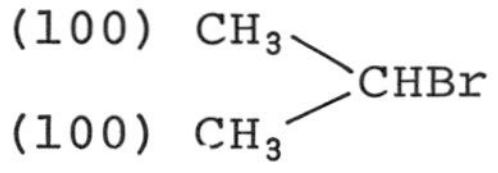

(100) CH$_3$
            $>$CHBr
(100) CH$_3$

$$\underline{4}$$

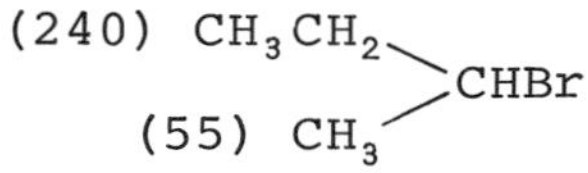

(240) CH$_3$CH$_2$
              $>$CHBr
(55) CH$_3$

$$\underline{5}$$

the relative rates of elimination into the various
branches. It is evident that the β-alkyl group markedly accelerates elimination into the branch bearing it,
an effect directly opposite to that observed with reactions following the Hofmann rule. A similar orientation pattern is observed with t-amyl bromide and ethoxide ion, which yields 2-methyl-2-butene in preference
to 2-methyl-1-butene, again mainly because of an increased rate of elimination into the branch bearing the
β-methyl group.[17] Numerous other examples of the
Saytzev rule could be given, but they are best deferred
until later in the chapter when the influences of various structural and environmental factors on orientation
are discussed.

Of two or more isomeric products of an elimination
reaction, the one bearing the greater number of alkyl
groups about the double bond is almost always thermodynamically more stable than the others. This was
noted by Hughes and Ingold,[16] who suggested that the
alkyl group was capable of stabilizing the partial
double bond in the transition state by hyperconjugation.
This effect is illustrated in 6. Whether hyperconjugation does play a significant role in stabilizing

$$H-C-C\text{-----}C- \quad \overset{X}{\underset{B}{}}$$

6

alkylated olefins or in the transition states leading
to them has been the subject of considerable controversy.[18-22] It is not necessary to take a position in
this controversy to discuss the Saytzev rule, since the
important point is that compounds following the rule
yield preferentially the more stable olefin, not the
precise mechanism by which the stabilization occurs.
One can simply say that the transition states for reactions following the Saytzev rule possess sufficient
double-bond character that they are stabilized by the
same factors that stabilize the product olefins
(Section II.A.4).

Much of our evidence on olefin stabilities comes
from heats of hydrogenation. Turner[18] has summarized
the extensive results in the vapor phase obtained by
Kistiakowsky's group, and he points out that the

number of alkyl groups about the double bond has an
important effect on stability, but that the nature of
the alkyl groups does not have much effect. This finding argues against carbon-hydrogen hyperconjugation,
which predicts that the number of α-hydrogens possessed
by the alkyl group should be important. In any event,
there is no doubt that the 2-ene, with its two alkyl
groups, is normally more stable than the 1-ene, with
its single alkyl group, when considering any isomeric
pair. In one case of violation of the Saytzev rule,
elimination from 2-bromo-2,4,4-trimethylpentane,[23] the
1-ene (7) has been shown to be more stable than the
2-ene (8),[25,26] presumably because the latter possesses

$(CH_3)_3CCH_2$ $\diagdown$ $C=CH_2$
CH_3 $\diagup$

$(CH_3)_3C$ $\diagdown$ $C=C$ $\diagup$ CH_3
H $\diagup$ $\diagdown$ CH_3

7   8

an eclipsing interaction between a methyl and a t-butyl
group. If one defines Saytzev-rule behavior as resulting in the more stable olefin, this result is no longer a violation. Except with t-alkyl groups, the effect
of the size of the alkyl group on the trans/cis ratio
for 2-enes seems small. Turner et al.,[25] have shown
that the relative stabilities of the cis and trans
isomers are almost the same for 2-butene (methyl versus methyl eclipsing) and 4-methyl-2-pentene (methyl
versus isopropyl eclipsing).
     One could argue that stabilities measured by gas-
phase hydrogenation may not be reliable measures of
stabilities in solution under the conditions of elimination reactions. Comparisons that have been made in
the literature do not support this view. The relative
stabilities of 7 and 8 are essentially the same whether
measured by hydrogenation[25] or by direct equilibration.[24] In another case, the relative stabilities of
the 2-methylpentenes measured by equilibration in dimethyl sulfoxide agreed well with gas-phase values.[26]
While subtle differences might exist between solution
and gas-phase results, it seems quite safe to conclude
that the major effects are the same. One should also
keep firmly in mind that the Saytzev rule refers to
effects of substituents on transition-state stability.
It is unlikely that precisely the same blend of substituent effects will operate in the transition state
as in the product, hence, olefin stabilities are at

best a qualitative guide to the product proportions to be expected in elimination reactions.

B THE EFFECT OF STRUCTURE AND REACTION CONDITIONS ON ORIENTATION

1 General Comments

The Hofmann and Saytzev rules are generalizations that cover a wide range of elimination reactions, but they are far from being without exception. Particularly in recent years, a great many intermediate cases and formal violations of the rules have been observed. In this section we examine some of the details of the effects of structure and environment on orientation in elimination reactions, and consider the extent to which the various effects fit into our current theoretical picture of the E2 reaction.

2 The Leaving Group

As we note above (Section IV.A.4,5), the Hofmann rule originally applied to quaternary ammonium salts and the Saytzev rule to alkyl halides. Early mechanistic investigations added other groups to each category, notably sulfonium salts to the first and sulfonate esters to the second. Still more recently, it has been shown that the difference between the two rules does not arise from the charge type of the leaving group. As we see below, there are compounds with neutral leaving groups that follow the Hofmann rule, and compounds with positive leaving groups that follow the Saytzev rule. There is nothing surprising about these observations in the light of the variable transition state theory, which ascribes the major effect of the leaving group to the ease of heterolytic cleavage of the bond between the leaving group and the α-carbon atom, a property that bears no necessary relationship to charge type. The harder the bond is to break, the more negative charge will be built up on the β-carbon atom in the transition state, and the less double-bond character the transition state will possess (Section IV.A.4). Thus one expects a decrease in the rate of formation of the more alkylated relative to the less alkylated product, as the ease of heterolytic cleavage of the bond decreases. Before we discuss in any detail how well this expectation is fulfilled, it is advisable to consider whether there are other explanations that predict the same trend.

Brown and Wheeler[12] have argued that the

proportion of the less-substituted olefin correlates very well with the expected steric requirements of the leaving group. As evidence, they give the product proportions for eliminations from 2-pentyl derivatives with potassium ethoxide in ethanol. The results are shown in Table 3. They point out that the least amount of 1-olefin results with the halides, where the leaving group is monoatomic, next with the sulfonate ester, where the leaving group has one branch, next with the sulfonium salt, where the leaving group has two branches, and most with the sulfone and the quaternary ammonium salt, where the leaving groups have three branches. They had not expected the close similarity of the results with the bromide and iodide, but suggested it probably arose from a balance between the larger Van der Waals' radius and the larger covalent radius of iodine; that is, iodine is larger than bromine, but is also farther away from the site of steric interference. Otherwise, the results are certainly consistent with Brown's steric theory, but the order is also approximately that to be expected from the variable transition state theory. Both the steric requirements of the leaving group and the difficulty of heterolytic cleavage of the bond to the leaving group increase with increasing percentage of 1-pentene.

In an effort to distinguish between the two theories, product proportions from the 2-pentyl and t-amyl halides were determined. The results are given in Table 4.[27] The expected order for the steric requirements of the leaving group was F < Cl < Br < I, while the difficulty of heterolytic cleavage of the bond to the leaving group should run in the opposite direction. The results are evidently in the order predicted by the variable transition state theory, with the fluoride actually showing Hofmann-rule behavior. The predictions of the steric theory are not, however, so simple as they appear to be at first sight. An investigation of the equatorial-axial equilibrium in halocyclohexanes found that the proportion of equatorial isomer at equilibrium runs F < I < Br < Cl.[28] This unexpected order can be accounted for by the argument of Brown and Wheeler[12] that increasing Van der Waals' radius is balanced by increasing covalent radius along the series Cl, Br, I (see above). Results with the 3,3-dimethyl-halocyclohexanes indicate a larger preference for equatorial halogen.[28b] There was no measurable axial halogen at equilibrium for chlorine, bromine, and iodine, but calculations gave a steric preference for the equatorial position that ran I > Br > Cl > F.[28b] The magnitude of this methyl-hydrogen 1,3-diaxial interaction would seem the better measure of the type of

TABLE 3 Products from Reactions of 2-Pentyl Derivatives with Potassium Ethoxide in Ethanol

Leaving Group	% 1-Pentene	% cis-2-Pentene	% trans-2-Pentene
Br	31	18	51
I	30	16	54
OTs	48	18	34
SMe_2	87	5	8
SO_2Me	89	2.5	9
NMe_3	98	1	1

TABLE 4   Products from Reactions of Pentyl Halides with Ethoxide Ion in Ethanol

Alkyl Halide	% 1-ene	trans/cis 2-Pentene
2-Pentyl fluoride	82	2.6
2-Pentyl chloride	35	3.5
2-Pentyl bromide	25	3.8
2-Pentyl iodide	20	4.1
t-Amyl fluoride	71	–
t-Amyl chloride	45	–
t-Amyl bromide	38	–

interference found in the E2 transition state. Whatever the case, the fluoride is clearly the smallest halogen by either criterion, however, and the difference in percent 1-pentene between the fluoride and the other halides is inconsistent with the steric theory. Similar results have been reported for the reactions of the 2-hexyl halides with methoxide ion in methanol,[29] and for the 2-butyl and t-amyl halides (except for the fluoride) with t-butoxide in t-butyl alcohol.[30]   Brown and Klimisch[30] make the interesting suggestion that fluoride may have considerably greater steric requirements in hydroxylic than in aprotic solvents because of solvation. It is difficult to assess this factor in the absence of independent evidence for it.

Another way of distinguishing steric from electronic effects is to choose a series of reactions in which one effect, usually the former, is held constant. This was done by Colter and Johnson,[31] who studied the E2 reactions of sodium ethoxide in ethanol with 2-pentyl benzenesulfonates with substituents in the m- and p-positions. The results were not very conclusive, however, for the change from p-amino to p-nitro was only 1.37 to 1.46 for the Saytzev/Hofmann product ratio, and only 1.98 to 2.14 for the trans/cis 2-pentene ratio. Although these results are in the order expected for the variable transition-state theory, the change is barely beyond experimental error, especially when one considers that a correction for a competing E1 process had to be applied. In an effort to obtain more decisive results, Colter and McKelvey[32] studied eliminations from 2-methyl-3-pentyl arenesulfonates brought about by t-butoxide ion in t-butyl alcohol mixed with dioxane or dimethyl sulfoxide. The overall change in orientation was larger, though the trend was

somewhat irregular. With 50% t-butyl alcohol --
dioxane, for example, most of the results were close
together, but the p-nitrobenzene sulfonate gave substantially more Saytzev product than any of the other
reactants. The total range in the Saytzev/Hofmann ratio in this medium was from 0.96 for p-methoxy to 1.45
for p-nitro. The corresponding variation in the trans/
cis ratio was from 1.48 to 3.21. The overall rate increased by about 60-fold over the same range (the rate
increase from the slowest reactant, the p-dimethyl-
aminobenzenesulfonate, to the p-nitrobenzenesulfonate
was about 200-fold, but the p-dimethylaminobenzenesul-
fonate gave somewhat more Saytzev product than did the
p-methoxybenzenesulfonate). It is interesting to compare these numbers to those for the t-amyl halides with
t-butoxide in t-butyl alcohol, where the Saytzev/
Hofmann ratio changes from 0.28 to 0.66 while the overall reactivity changes by about 350-fold. The change
in the Saytzev/Hofmann ratio for the benzenesulfonates
can be seen to be not so small after all, when viewed
in the light of the relatively restricted range of
overall reactivities.

Colter's results permit another conclusion concerning the mode of operation of the leaving-group
effect on orientation. In the original explanation of
the Hofmann rule, it was considered that the leaving
group (in that case, the positively charged trimethylammonio group) increased the acidity of the β-hydrogens
by an inductive effect, and that the β-alkyl group
slowed elimination by partly counteracting this inductive effect.[3] With the benzenesulfonates, electron-
withdrawing groups favor Saytzev-rule product, so that
the inductive effect of the leaving group on acidity
of the β-hydrogen cannot be an important factor. It
seems unlikely to be important with uncharged leaving
groups, in general, though one cannot exclude it for
charged leaving groups.

Additional valuable evidence on the relationship
between orientation and reactivity has been reported by
Bartsch and Bunnett.[29,33,34] They determined rates and
products of elimination from 2-hexyl derivatives with
a wide variety of leaving groups, including a number of
unusual and less reactive ones such as phenoxy, thiophenoxy, selenophenoxy, and trimethylbenzoyloxy, as
well as the more common ones such as the halogens and
benzenesulfonates. In an effort to uncover linear
free-energy relationships, they plotted a "reactivity
index" that was the logarithm of the rate of formation
of 1-hexene (or an estimate thereof) against the logarithm of the ratio of yields of 2-hexene and 1-hexene,

or of the trans/cis 2-hexene ratio. The plots were
linear for the halogens, the 2-ene/1-ene and trans/cis
ratios decreasing regularly with increasing reactivity.
Certain other groups tended to fall off the lines. In
particular, bromobenzenesulfonate and phenylsulfonyl
fell consistently below the line (Figure 1). With t-
butoxide in t-butyl alcohol, V-shaped plots were obtained (Figure 2). Four of the least reactive groups,
phenoxy, thiophenoxy, selenophenoxy, and trimethylbenzoyloxy, showed an increasing 2-ene/1-ene ratio with
decreasing reactivity. A linear relationship, though
with considerable scatter, is again found with t-butoxide in dimethyl sulfoxide (Figure 3). The results

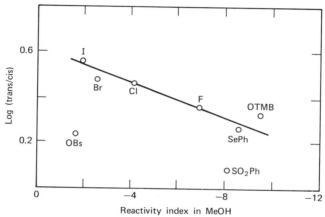

Figure 1   Relationship between log (trans/cis-2-hexene)
and the reactivity index for reactions of 2-hexyl derivatives with sodium methoxide in methanol. The leaving
groups are indicated, OTMB = 2,4,6-trimethylbenzoyloxy
(reproduced by permission of the authors[33] and the
Journal of the American Chemical Society).

demonstrate that a relationship between orientation and
reactivity does indeed exist, but it is perturbed by
other factors unless one keeps to a very limited range
of substrates. The other factors are doubtless in part
steric. The deviations of bromobenzenesulfonate and
phenylsulfonyl seem particularly clear examples of this
category. On the other hand, another bulky group, trimethylbenzoyloxy, tends to deviate in the opposite
sense. The V-shaped plots in t-butyl alcohol suggest
a change in mechanism and/or stereochemistry, although
a simple change in transition-state structure within
the E2 mechanism cannot be excluded.
     Some trans/cis ratios for reactions producing

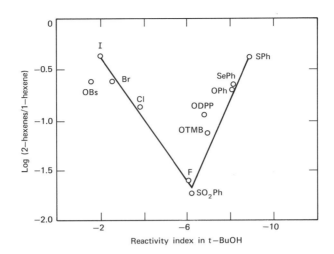

Figure 2   Relationship between log(2-hexene/1-hexene) and the reactivity index for reactions of 2-hexyl derivatives with potassium t-butoxide in t-butyl alcohol. The leaving groups are indicated; OTMB - 2,4,6-trimethylbenzoyloxy, ODPP = OPO(OPh)$_2$ (reproduced by permission of the authors[33] and the Journal of the American Chemical Society).

1,2-disubstituted olefins have already been mentioned. The most common pattern of behavior is for the more reactive leaving groups to give higher trans/cis ratios.[12,27,29-34]   Such a trend is easily explained by the variable transition-state theory, where a better leaving group gives a transition state with more double-bond character and, hence, with greater susceptibility to eclipsing effects.   In principle, one would expect a reversal in this trend as the leaving group becomes still better, for the transition state would then become E1-like and the degree of double-bond character would be consequently less.   In practice, there appear to be no good examples of this type of behavior.   We will also see that there are many instances where simple eclipsing effects cannot account for observed trans/cis ratios.   Quaternary ammonium salts under some conditions give trans/cis ratios below unity (Section III.F and IV.C.3), as do also alkyl tosylates (Section IV.B.4).   Simple eclipsing effects can obviously not explain these ratios.   There are sometimes changes in trans/cis ratios without concomitant changes in positional orientation,[32] so that the factors affecting the two are not always the same.   Most

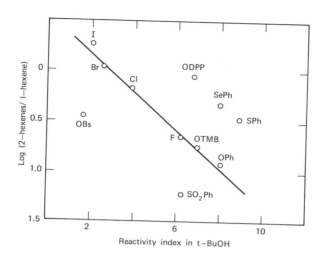

Figure 3   Relationship between log (2-hexene/1-hexene) and the reactivity index in t-butyl alcohol for reactions of 2-hexyl derivatives with potassium t-butoxide in dimethyl sulfoxide. The leaving groups are indicated; OTMB = 2,4,6-trimethylbenzoyloxy, ODPP - OPO(OPh)$_2$ (reproduced by permission of the authors[34] and the Journal of the American Chemical Society).

of the peculiarities in trans/cis ratios are associated with effects of base or solvent (Section IV.B.4) or with variations in mechanism (Section IV.C), and they will be discussed in the indicated sections rather than here.

3   Alkyl Structure

The effect of alkyl structure on orientation are discussed to some extent under the Hofmann and Saytzev rules (Section IV.A.4,5). There we find that the retarding effect of a β-substituent in Hofmann-rule reactions has been explained as the result of an inductive or steric effect, and that the accelerating effect of a β-substituent on Saytzev-rule reactions has been explained as the stabilizing effect of alkyl substitution on the developing double bond in the transition state. In this section, we attempt to discuss more thoroughly the effect of substitution on both positional and trans/cis orientation.

Electronic effects of alkyl substituents fall into two categories. They can exert an inductive effect if there is charge development in the transition state, or they can stabilize developing unsaturation in the

transition state. Steric effects can operate in numerous ways. The α- and β-alkyl groups can interact with
each other (eclipsing effects), or either can interact
with the leaving group or the attacking base. There
can also be various types of "cooperative" steric
effects. One example has already been mentioned (Section III.F); a bulky leaving group can cause the alkyl
chain to adopt a conformation that hinders approach of
the base to the β-proton. We consider others in the
course of this section. Furthermore, it is entirely
possible that two or more effects may be operating at
the same time in a given elimination reaction. The numerous effects available make it a relatively simple
matter to explain experimental results. It is more
difficult to test the validity of the explanation, and
still more difficult to make predictions. An understanding of the characteristics of the various effects
and of the circumstances under which they can be expected to operate is, however, essential if we wish to
make any serious attempt to predict orientation.

The operation of electronic effects, particularly
the stabilizing effect of alkyl substitution on developing double bonds, is sufficiently clear and well
demonstrated that we devote our major attention to steric effects. Examination of the conformations of even
so simple a molecule as 2-butyl bromide reveals some
of the steric effects that may operate. The most
stable conformation is 9, but the conformation must be
either 10 or 11 in the transition state for anti elimination to give 2-butene. On the other hand, a methyl

9   10   11

hydrogen can be anti to bromine without disturbing conformation 9. The effect is, in this case, evidently
insufficient to overcome the stabilizing effect of alkyl groups, for 2-butene is the preferred product (Section IV.A.1). Of the two conformations 10 and 11, the
former has fewer nonbonded interactions, and this is
doubtless why trans olefin is usually favored over cis.
The transition state need not, of course, have

the same geometry as the reactant. Where there is a
very productlike transition state, the α- and β-carbon
atoms will be nearly sp^2-hybridized (12 and 13). The
effect of this change in geometry will be to minimize

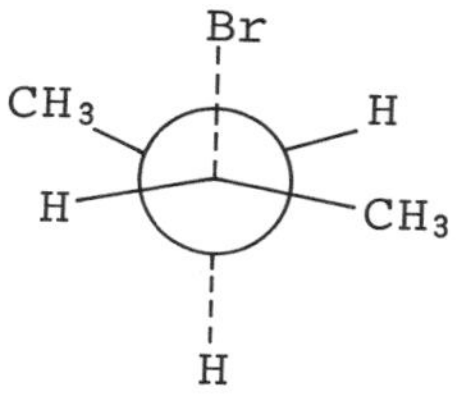

12 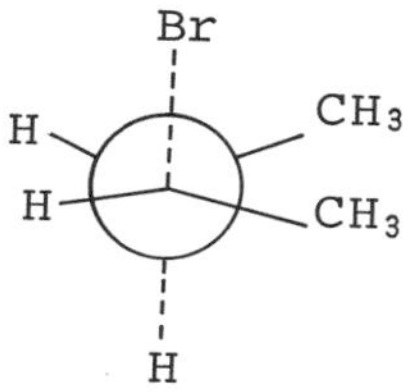   13

interactions between the leaving group and the alkyl
groups, but to accentuate the alkyl-alkyl interactions
in 13 over those in 11. They are already present in
11, however, so a trans/cis ratio does not have to go
below unity to be considered "abnormal"; even a value
close to unity signals the operation of some factor
working in opposition to the eclipsing effect.
      Eclipsing can indirectly affect orientation if the
only way to relieve an alkyl-alkyl interaction is a
change in orientation. A probable example of this
effect is given by Brown, Moritani, and Nakagawa,[35]
who found that the proportion of 1-ene from 14 increased as R increased in bulk from methyl to ethyl
to isopropyl to t-butyl. In this case, the 2-ene, 15,
obviously cannot avoid a methyl-R interaction, but the

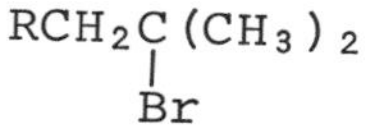

14 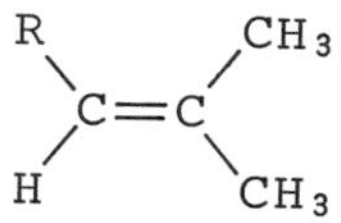   15

1-ene can. It is not certain that all of this trend is
steric (2-pentyl bromide, where eclipsing can be relieved by formation of trans-2-pentene, still gives
more 1-ene than does 2-butyl bromide[36]), but the 86% of
1-ene obtained when R is t-butyl could hardly be the
result of an electronic effect on product stability.
      When the leaving group is not symmetric, the
effect of its conformation on other groups in the

transition state must be considered. An arylsulfonate
leaving group, for example, is expected to prefer a
conformation where the sulfur is anti to an α-alkyl
group, as in 16 and 17. Under these circumstances,
there may be interactions of the groups attached to

          16   17

sulfur (aryl and the two oxygens) with the alkyl group
in the β-position, and there may even be some restrictions on free rotation of the α-alkyl, if that is a
more complex group than a methyl. Interaction of the
leaving group with the β-alkyl group is avoided in 17.
This effect would favor formation of cis over trans
olefin, a phenomenon that we discuss in more detail in
the next section (Section IV.B.4). It could also explain the observation that 2-pentyl tosylate consistently gives 2-pentene with a lower trans/cis ratio
than does 3-pentyl tosylate, since the β-alkyl group
is larger in the 2-pentyl system.[37] A similar effect
has been proposed for sulfonium-salt eliminations.[38]
That leaving-group conformation might play a role in
modifying trans/cis ratios was first suggested by
Brown and Klimisch,[39] and Froemsdorf, Dowd, and
Leimer,[40] although the explanations differed somewhat
in detail from that given above.
    Since steric effects depend closely on conforma-
tion, a change in stereochemistry of elimination may
result in quite different steric effects. To the extent that the bonds to the leaving group and to the
β-hydrogen need to be coplanar, one would expect a very
strong preference for trans over cis olefin in syn eliminations, where the α- and β-alkyl groups will be
eclipsed even in a reactantlike transition state (18
versus 19). The trans/cis ratios from syn eliminations
can, in fact, markedly exceed those predicted from
relative thermodynamic stabilities of the product olefins.[41] Such eclipsing interactions can affect orientation by discouraging elimination into branches where
eclipsing would be important. The relation between

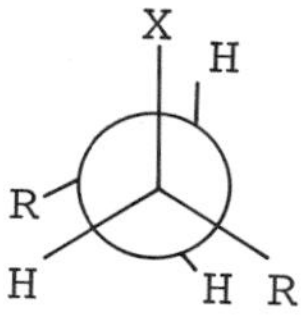

18 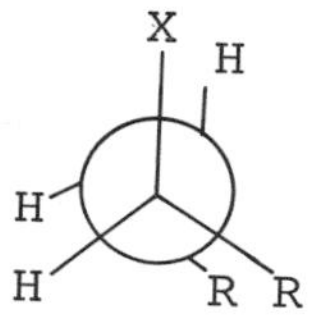   19

stereochemistry and orientation is discussed in detail later (Section IV.C.3).

Most orientation problems involve a choice between elimination into one or another alkyl group, hence, nothing has been said thus far about other substituents. In general, any β-situated functional group that can stabilize either a developing double bond or a developing negative charge will very strongly encourage elimination into that branch. In Section III.E.2, for example, we see that a phenyl, arylsulfonyl, or nitro group can direct elimination exclusively into the branch bearing it, even though a change to syn stereochemistry is involved. The preference is not invariably so overwhelming. Benzyldimethylcarbinyl chloride and methoxide in methanol give a ratio of conjugated to unconjugated product of only 4:1.[42,43] Here the transition state possesses little carbanion character, and methyl-phenyl eclipsing interactions discourage production of the conjugated olefin. Eclipsing interactions involving large groups such as phenyl can be quite important. The proportion of trans-stilbene in the elimination products from 1,2-diarylethyl chlorides is, at least, 99%.[44]

4   Base and Solvent

As we discuss more fully elsewhere in this volume (Section II.C.4), the reactivity of a base in an elimination reaction, or in other proton-transfer reactions, depends strongly on the solvent. For this reason alone, it is seldom possible to separate effects on orientation that are caused by intrinsic properties of the base from those that are the result of solvent effects. A further complication is that a metal alkoxide in the corresponding alcohol is the most common reagent for bringing about elimination reactions, so that much of the data in the literature are on the results of simultaneous changes in the base and the solvent.

Interest in the role of base and solvent in determining orientation developed rapidly after the discovery by Brown, Moritani, and Okamoto[36] that branched alkoxide-alcohol reagents gave with alkyl bromides quite different product ratios from the commonly used ethoxide-ethanol. The results in Table 5 show that

TABLE 5   Olefin Compositions from the Reaction of Alkyl Bromides with RO^- in ROH

RBr	% 1-ene in the Product When RO^- Is			
	ETO^-	$t-BuO^-$	Me_2EtCO^-	Et_3CO^-
2-BuBr	19	53	–	–
2-PeBr	29	66	–	–
i-Pr(Me)$_2$CBr	21	73	81	92
t-AmBr	30	72	77	88
neo-Pe(Me)$_2$CBr	86	98	–	97

t-alkoxides give so much more of the terminal olefin that product distributions characteristic of the Hofmann rule are obtained. From the point of view of synthetic utility, in fact, it is probably more convenient under most circumstances to prepare terminal olefins from the alkyl bromide with a t-alkoxide rather than to prepare the corresponding ammonium or sulfonium salt first. Acharaya and Brown[45] have used triethylcarbinyloxide in the synthesis of β-cedrene and other methylenecycloalkanes. Numerous other examples of preference for the terminal olefin with branched alkoxides are now available in the literature. They now include alkyl tosylates[37,39,40] and a number of compounds with less common neutral leaving groups.[33] Brown and Nakagawa[46] also report a similar, though considerably less pronounced, trend toward terminal olefin in the reaction of alkyl halides with pyridine bases (where the base is the solvent) along the series pyridine, 2-methylpyridine, 2,6-dimethylpyridine.

Although there is no doubt that the phenomenon is general, there is still considerable disagreement over the factor or factors responsible for it. Brown et al.[36,46] proposed originally that the greater steric requirements of the branched alkoxides made it more difficult for them to attack β-hydrogens at nonterminal positions, and that the resulting hindrance to production of 2-ene enabled the 1-ene to predominate. Others have attributed the changes in orientation primarily to changes in electron distribution in the transition state as a result of changing solvent or base strength. In particular, it has been argued that increasing the base strength increases the extent of carbon-hydrogen stretching in the transition state and hence the carbanion character. This explanation has been favored by Froemsdorf et al.[47-49] and by Bunnett.[50] Although the argument is plausible from an intuitive point of view, it is not really in accord with theoretical expectations. Perhaps the simplest way of explaining the discrepancy is to remember that a more reactive reagent is expected to give a transition state that is less productlike, not more.[51] Thornton's theory of the effects of perturbations on transition-state structure likewise predicts that increasing the strength of the attacking base should lead to less complete proton transfer in the transition state (Section II.B.2).[52,53] He explains the increasing yield of 1-ene with increasing base strength as arising from a more reactantlike transition state. Such a transition state would have less double-bond character and, hence, less propensity to adhere to the Saytzev rule. In its simplest form, this theory predicts that the ratio of product olefins should approach a statistical value reflecting complete loss of preference, a prediction incompatible with the observed strong preference for 1-ene in some cases (see, for example, Table 5). A partial reconciliation between this theory and the one advanced by both Froemsdorf and Bunnet is possible if one assumes that increasing base strength does decrease the extent of carbon-hydrogen stretching but decreases the extent of carbon-leaving group stretching even more, thereby giving a transition state of greater carbanion character. Alternatively, one could combine Thornton's and Brown's views and say that steric effects of the base become important when decreasing double-bond character has weakened the normal Saytzev-rule preference.

It is evident from the experimental results that no single property of either base or solvent can correlate successfully all of the variations in olefin proportions. That base strength plays a role is shown by the results of Froemsdorf and Robbins[49] in Table 6. There is apparently a much better correlation between base strength and proportion of 1-butene than there is between steric requirements of the base and proportion of 1-butene, even allowing for the possibility that basicities in dimethyl sulfoxide may not parallel those

TABLE 6   Product Proportions in Eliminations from 2-Butyl Tosylate at 55°

Base	Solvent	% 1-ene	trans/cis 2-ene
t-BuOK	DMSO	61	2.53
EtOK	DMSO	54	2.34
PhOK	DMSO	31	1.96
4-MeOC$_6$H$_4$OK	DMSO	33	1.97
4-NO$_2$C$_6$H$_4$OK	DMSO	16	1.81
2-NO$_2$C$_6$H$_4$OK	DMSO	16	1.85
t-BuOK	t-BuOH	64	0.58
EtOK[a]	t-BuOH	54	0.80
KOH	t-BuOH	50	0.83
PhOK	t-BuOH	34	1.24

[a]Prepared by adding a little ethanol to t-BuOK in t-BuOH. See text for comments.

in the hydroxylic solvents for which pK values are available. The relation between percent 1-ene and base strength is the same in both solvents, but the trans/cis ratios change in opposite directions with increasing base strength in the two solvents. Evidently some factor other than base strength affects trans/cis ratios, perhaps the lesser tendency toward ion pairing of the base in dimethyl sulfoxide. While the steric requirements of the base may well play a role in many cases, they do not seem to be the controlling factor in these experiments. The result for "EtOK" in t-butyl alcohol should be treated with reservation. More recent work (see below) indicates that this solution must have contained significant amounts of t-butoxide ion. Although comparison of results for the two solvents in Table 6 would suggest that the effect of changing solvent is minor, such a conclusion is not generally justified. Solvent changes do affect product proportions in other reactions of alkyl tosylates[34,47,49] and bromides.[34,48] Bartsch and Bunnett[33,34] find a general trend with a variety of 2-hexyl derivatives toward a greater proportion of 2-hexene with potassium t-butoxide in dimethyl sulfoxide than in t-butyl alcohol.

Ratios of trans to cis olefin are also affected by changes in solvent and base. The pattern observed with many substrates, which we will call the "normal" pattern,[37,54] is that an increase in Hofmann-rule product is associated with a decrease in trans-cis ratio. A simple eclipsing effect undoubtedly contributes to this relationship, for less eclipsing of α- and β-alkyl groups is expected the less the double-bond character in the transition state. This cannot be the whole explanation, however, for the trans/cis ratios often go below unity. Brown and Klimisch[39] and Froemsdorf, Dowd, and Leimer[40] found that a number of 2- and 3-alkyl tosylates gave trans/cis ratios around 0.5 with potassium t-butoxide in t-butyl alcohol instead of the nearly 2:1 ratios found with potassium ethoxide in ethanol. Brown and Klimisch suggested that 20 was a lower-energy conformation than 21 for the transition state when the alkyl group of the alkoxide was bulky.

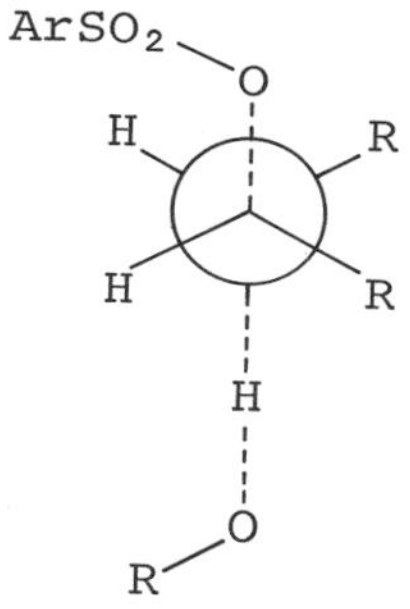

20 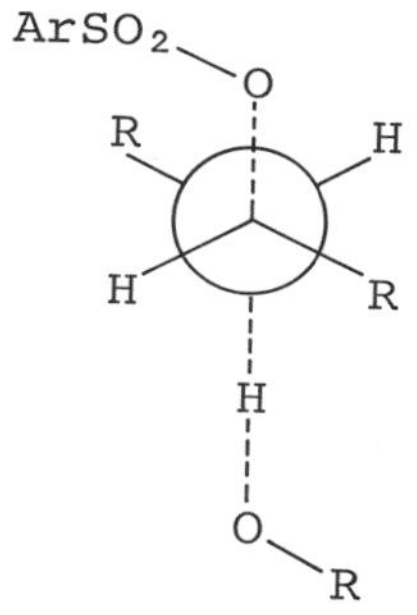   21

Froemsdorf, Dowd, and Leimer argued that the preference arose because the bulky $ArSO_2$ group preferred to be anti to both the α- and β-alkyl groups in the transition state, and that the role of t-butoxide was to cause the transition state to be more reactantlike, thus shortening the carbon-oxygen bond and increasing steric interactions between the sulfonyl and alkyl groups.

Neither the strength nor the steric requirements of the base can account for all changes in trans/cis ratios, since both trans/cis and 2-ene/1-ene ratios tend to be considerably higher in dimethyl sulfoxide than in alcohol solvents (Table 6).[33,34,49] Bases are generally considered to be stronger in dimethyl sulfoxide than in the corresponding alcohols (Section II. C.4),[72] and that predicts the wrong tend for both. Recent results indicate that ion-pairing of the base plays an important role in trans/cis ratios.[73] The trans/cis 2-butene ratio from potassium t-butoxide in t-butyl alcohol at 50° is 0.40, but this rises to 1.9 when dicyclohexyl-18-crown-6 ether (a cyclic polyether

that strongly complexes cations[74]) is added to break up
the potassium t-butoxide ion pairs. It is also high
(2.2) in dimethyl sulfoxide, which effectively solvates
cations and thereby weakens ion pairing. Similarly, an
increase in concentration of potassium t-butoxide
(which should increase ion pairing) lowers the trans/
cis 2-butene ratio from 2-butyl bromide in t-butyl alcohol. Similar observations have been made by Závada,
Svoboda, and Pankova.[73] It is not clear why ion pairing should lower trans/cis ratios so much. Perhaps
the effect is associated with the greater steric requirements of the ion-paired base, or with differing
electrostatic interactions of the free and ion-paired
bases with substrate.

Finally, there are instances, mainly found with
quaternary ammonium salts, where the trans/cis ratios
increase while the 2-ene/1-ene ratios are decreasing.
This "abnormal" pattern appears to be associated with
a change in stereochemistry of elimination and is discussed in the next section (Section IV.C.3).

As we mention at the beginning of this section,
the major problem in interpreting orientation data in
alkoxide-alcohol media lies in assessing the contributions of the solvent and the base to the observed effects. In a recent effort to tackle this problem,
Zwolinski and Griffith[55] used an indicator method to
determine the equilibrium constant for the reaction of
a small amount of ethanol with a potassium alkoxide in
the corresponding alcohol (Eq. 3). The results were

$$EtOH + RO^-K^+ \rightleftharpoons EtO^-K^+ + ROH \qquad (3)$$

rather surprising. The equilibrium constant was barely
above unity in isopropanol and ranged from 2.9 to 3.6
in a number of tertiary alcohols. Addition of a small
amount of ethanol to a tertiary alkoxide in the corresponding alcohol obviously does not, as had previously
been supposed,[49] suffice to convert all of the alkoxide
to ethoxide. Zwolinski and Griffith measured rates and
product proportions for the reactions of 2-pentyl bromide both with the alkoxide-ethoxide mixtures and with
the alkoxide alone. On the assumption that the small
amount of ethanol did not appreciably change the solvent, they dissected the rates and product proportions
into contributions from alkoxide and ethoxide. Although a dissection of this kind must be treated with
caution because it magnifies both experimental error
and deficiences in the assumptions, a clear trend from

ethanol to isopropanol to t-butyl alcohol was noted.
In both instances the proportion of 1-pentene increased, but less so for ethoxide than for the alkoxide.   This result would seem to suggest that the solvent effect, rather than any intrinsic property of the
alkoxide, has the main effect on orientation.   The solvent effect can, of course, operate in a number of
ways.   Changing the solvent changes the solvation of
both the ground and transition states.   In particular,
the strength of the base and its tendency to pair with
cations will change when the solvent changes, even if
the base remains the same.   It seems probable that the
observed effects are primarily the result of a solvent
effect on base strength, with perhaps some steric effect of the base superimposed in the case of the alkoxide.

5   Cyclic Systems

In discussing orientation in eliminations from cyclic
reactants, one must be careful to differentiate intrinsic orientation effects from those caused simply by
stereochemical restraints.   A trans-2-alkylcyclohexyl
derivative, for example, can yield only 3-alkylcyclo-
hexene in an anti elimination (Section III.C.1).   The
cis isomer, on the other hand, can yield either 1- or
3-alkylcyclohexene by an anti elimination, and the
product ratio in this case does give a measure of intrinsic preference.   To avoid uncertainty on this
point, we limit ourselves in this discussion to cis
isomers of systems that can reasonably be supposed to
react largely or entirely by an anti path.
     Halides and sulfonate esters show qualitatively
the same behavior as their open-chain analogs.   Neomenthyl chloride (22, X = Cl) gives 78% 3-menthene (23)
and 22% 2-menthene (24) on treatment with ethoxide ion

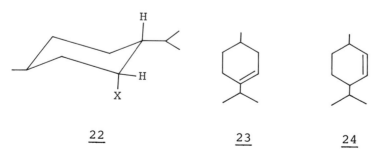

22   23   24

in ethanol under E2 conditions.[56]   2-Methylcyclohexyl

tosylate gives 76% 1- and 24% 3-methylcyclohexene with
the same base.[57] With t-butoxide ion in t-butyl alcohol the 3-methylcyclohexene yield rises to 60%, the
same trend found with open-chain systems (Section IV.
B.4). The dependence of orientation on ring size has
not been thoroughly explored, but it has been reported
that cis-2-methylcyclohexyl and cis-2-methylcycloheptyl
tosylates with sodium isopropoxide in isopropanol give
35 and 63%, respectively, of the 3-methylcycloalkene.[58]
     Quaternary ammonium salts show marked differences
in behavior from their open-chain analogs. Early reports that heating neomenthyltrimethylammonium hydroxide (22, X = NMe$_3^+$OH$^-$) gave mainly the Saytzev-rule
product, 3-menthene (23)[59,60] were confirmed by Hughes
and Wilby.[61] Although the reaction shows the unusual
feature for an elimination from a quaternary ammonium
salt of possessing an El component, the product of the
E2 process still shows an 88:12 predominance of 3-men-
thene. Similar results are found with cis-2-methyl-
cyclohexyltrimethylammonium ion with a variety of
bases, the yields of 1-methylcyclohexene running 94 to
97% of the olefinic product.[57] Hofmann elimination
from a number of cis-2-alkylcyclohexyltrimethylammonium
hydroxides yields olefinic products containing more
than 90% of the 1-alkylcyclohexene in all cases.[62] As
further evidence for the generality of the phenomenon,
6-β-chloestanyltrimethylammonium hydroxide yields 5-
cholestene, the Saytzev-rule product, and no detectable
amount of 6-cholestene, the Hofmann-rule product.[63]
     A plausible explanation[57,61] for this unusual
orientation effect suggests that the difficulty of
forcing the bulky trimethylammonio group into an axial
conformation (25) for anti elimination results in a

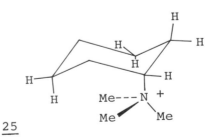

25

transition state with considerable stretching of the
carbon-nitrogen bond, and hence much more double-bond
character than is usual in Hofmann eliminations. Deuterium isotope effects in eliminations from alicyclic
quaternary ammonium salts are consistent with this explanation.[64] It has been challenged on the grounds

that the proportions of 1-alkylcyclohexene from various cis-2-alkylcyclohexyltrimethylammonium hydroxides do not vary as expected from the number of α-hydrogens available for hyperconjugation in the alkyl groups attached to the double bond of the 1-alkylcyclohexene.[62] The percentage of 1-alkylcyclohexene actually varies in a random manner between about 91 and 99%, a range sufficient to conceal the rather small trend to be expected from changing the number of α-hydrogens but not the number of alkyl groups, so that the evidence cannot be regarded as decisive. In any event, closer examination of the facts leaves doubt as to whether the phenomenon is so unusual or so specific to small rings that an explanation based on conformations of ring substituents is needed. The cases presented above all involve competition between elimination into branches of the substrate with one versus two β-alkyl groups. Even in open-chain systems, the preference for the Hofmann-rule product is much less in these cases than where the competition is between branches with no versus one β-alkyl groups.[7,54] Evidence presented later (Section IV.C.3) indicates that the anti component of the competition between branches with one and two β-alkyl groups may generally follow the Saytzev rule.

C   VARIATIONS ON THE E2 MECHANISM

1   General Comments

Although this chapter is intended to deal only with orientation in E2 reactions, there are several variants on the E2 mechanism that should be treated separately. These are the E2C and syn-E2 mechanisms. Like the E2 mechanism, they show second-order kinetics and appear to proceed in a single step without any detectable intermediates. The syn-E2 mechanism can be distinguished from the more common anti-E2 mechanism by use of appropriately designed reactants (Section III.C.1-3). We see below (Section IV.C.3) that the evidence thus far available suggests that syn- and anti-E2 reactions may show quite different orientation patterns. The other variant, the E2C mechanism, differs in a more subtle fashion from the normal E2 mechanism. While there may be a difference in transition-state geometry, a difference has not been unequivocally demonstrated, and it is possible that the difference lies merely in electron distribution and the timing of the bond-making and bond-breaking processes (Section II.A.5). In such a case, the E2C mechanism would more properly be characterized as part of the spectrum of possible E2 mechanisms rather than as a distinctively different mechanism. Nonetheless, there are enough distinctive features of reactions that appear to fall in this category that a separate treatment is desirable to point up these features.

2 The E2C Mechanism

The general features of this mechanism have already been discussed (Section II.C.2.c), and we limit ourselves here to the results of orientation studies and the conclusions that can be drawn from them. Table 7 lists some typical results for simple alkyl derivatives.[65] Two of the bases, tetrabutylammonium bromide and chloride, are typical E2C bases, but the third, sodium ethoxide in ethanol, is typical of the bases used in "normal" E2 (or E2H[65]) reactions. The last column of the table lists the percent 1-olefin found under equilibrating conditions. It is evident that E2C reactions show a very strong preference for the Saytzev-rule product, the observed olefin proportions being close to those found at equilibrium. This pattern of results is not limited to reactions in aprotic solvents, for thiolate ions in alcoholic solvents give similar, though less extreme, orientation. For example, t-amyl bromide with thiophenoxide in ethanol gives 26% of 1-olefin, compared to 37% with ethoxide.[27] The Saytzev-rule product is also strongly favored in E2C-like eliminations from alicyclic systems. Neomenthyl chloride (22, X = Cl) gives 97% of 3-menthene (23) with tetrabutylammonium bromide in acetone, compared to 78% with ethoxide in ethanol.[66] cis-2-Phenylcyclopentyl and cis-2-phenylcyclohexyl brosylates give well over 99% of the 1-phenylcycloalkene on treatment with tetrabutylammonium chloride in acetone.[67]

The high proportion of Saytzev-rule olefin, and particularly the fact that this proportion is close to the one found at equilibrium, suggests a very product-like transition state for E2C reactions.[65] If the double bond is nearly completely formed, then both the leaving group and the β-hydrogen must be very loosely attached. The base is probably loosely attached also, since the effect of changing the structure of the base on orientation is small. For example, t-amyl chloride gives 26-29% of 2-methyl-1-butene with thiolate ions in ethanol, there being no clear trend as the base changes from thiophenoxide to substituted thiophenoxide, and from n- to s- to t-butylthiolate.[68] The question of whether the base is interacting with the α-carbon as well as the β-hydrogen in the transition

TABLE 7 Olefin Compositions in E2C-like and E2H-like Reactions

Substrate	% 1-Olefin with			
	NBu_4Br Me_2CO, 75°	NBu_4Cl Me_2CO, 50°	EtONa EtOH, 50°	Equilibrium
Br C–C–C–C OTs	4	–	19	9
C–C–C–C C	0.3	0.8	18	0.2
Br C–C–C–C C	–	9	37	8.2
Br C–C–C–Ph	0.5	–	–	0.03

state, as suggested by Parker, Winstein, and their
collaborators,[65-67] cannot be answered from the orientation data. Certainly the failure of the considerable
changes in thiolate structure to exert any steric or
electronic effects on orientation indicates that interaction of this kind must be slight.

Apart from the interesting mechanistic questions
they raise, the E2C-type reactions add a useful synthetic tool to those already available for the preparation of olefins. They afford high yields of the more
substituted olefins under mild conditions; where cis-
trans isomerism is possible, the trans-olefin is
strongly preferred.[65] In this respect they complement
nicely the reactions with t-alkoxides in the corresponding alcohols, which give preferentially the
Hofmann-rule product and the cis-olefin (Section IV.
B.4). It is thus possible to obtain completely different product proportions from a single substrate
simply by changing the base and solvent.

3 Syn Eliminations

Until recently, the only examples of syn eliminations
in E2 reactions involved substrates where the question
of orientation did not arise, or where the preference
for one product was so strong in both syn and anti
eliminations that no differences could be discerned
(Section III.E.1,2). Consequently, the literature contains very little data on, and very little discussion
of, the relation between orientation and stereochemistry. Feit and Saunders[54] pointed out that conditions
that favor syn elimination should also favor the
Hofmann-rule olefin with substrates of the type 26,
where the choice is between elimination toward a

$$R_2-\overset{\overset{\displaystyle R_1}{|}}{C}H-CH-\underset{\underset{\displaystyle NMe_3^+}{|}}{C}H_2-R_3$$

26

β-carbon bearing two alkyl groups versus a β-carbon
bearing one. The reason was simply that syn → cis
elimination was known to be difficult, and that the
olefin with two β-alkyl groups necessarily had one of
them cis to the α-alkyl group, while the other olefin
could have its α- and β-alkyl groups trans.

We pointed out previously that the normal pattern

in anti eliminations is for an increase in Saytzev-rule product to be accompanied by an increase in trans/cis ratio, the cause presumably being the greater eclipsing effect with greater double-bond character of the transition state (Section IV.B.2). An abnormal pattern is observed with quaternary ammonium salts, however, where the trans/cis ratio increases with increasing Hofmann-rule product.[54] This pattern has been shown to be the result of increasing importance of syn elimination.[69] Elimination reactions of 2- and 3-hexyltrimethylammonium ions under various conditions show an excellent correlation between the stereochemistry of elimination and the trans/cis ratio of the 2- and 3-hexene produced.[69] Under conditions where the elimination is mainly anti, the cis olefin is preferred by about 3:1, but this ratio is reversed under conditions where 80 to 95% of the trans olefin is produced by syn elimination (the syn-anti dichotomy, see Section III.E.5-6). In fact, it has been reported that the syn component of eliminations of mixed stereochemistry is characterized by extremely high trans/cis ratios.[41] The probable reasons for this behavior are discussed in detail elsewhere (Section III.F). Here we simply point out that the existence of the relationship between stereochemistry and trans/cis ratio enables one to draw qualitative conclusions about stereochemistry from product analyses. The application of this approach to elimination reactions of some simple alkyltrimethylammonium salts has been discussed.[69] It was concluded, for example, that the 2-butyltrimethylammonium ion eliminated mainly or entirely by an anti path under all of the conditions studied because the percentage of cis-2-butene remained nearly constant at 60 to 70%.

Similar reasoning was employed in a suggestion that 2- and 3-pentyldimethylsulfonium ions may undergo syn elimination with t-butoxide ion in t-butyl alcohol.[38] With most of the base-solvent systems the trans/cis ratio is below unity, but it jumps to 3.5 for the 2- and 5.0 for the 3-pentyldimethylsulfonium ions with t-butoxide. It was later found in deuterium-labeling experiments that about two thirds of the elimination in t-butyl alcohol proceeds by the α'-β (Section I.B.4) mechanism.[75] Eclipsing effects in both the α'-β and syn-E2 mechanisms should favor trans olefin. It should be emphasized that acyclic quaternary ammonium salts, unlike the sulfonium salts, appear to undergo syn elimination by a normal E2 mechanism.[71]

The increase in Hofmann-rule product with increasing trans/cis ratio must mean that syn elimination favors the Hofmann-rule product more strongly than does

anti elimination. Consequently, it would be of interest to dissect observed Hofmann/Saytzev ratios into
syn and anti components to see how great this difference is. This can be done in a few cases. The stereochemistry of elimination from the erythro and threo
isomers of 5-methyl-6-decyltrimethylammonium ion (27)
has been investigated by Sicher, Svoboda, Pánková, and
Závada.[70] The stereochemistry of formation of the
Saytzev product (28) can be deduced readily if the configuration of the reactant is known; that of the
Hofmann product (29) was estimated from experiments
with related systems to follow mainly the syn-anti

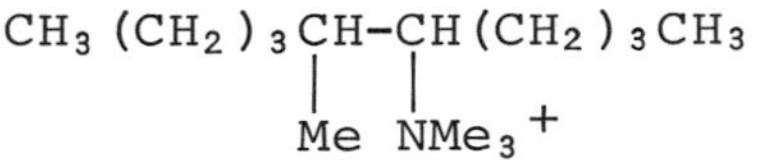

$$CH_3(CH_2)_3CH\text{-}CH(CH_2)_3CH_3 \qquad BuC=CHBu \qquad BuCHCH=CHPr$$
$$\qquad\quad |\quad\ \ | \qquad\qquad\qquad\quad | \qquad\qquad\qquad |$$
$$\qquad\quad Me\ \ NMe_3{}^+ \qquad\qquad\qquad Me \qquad\qquad\qquad Me$$

27 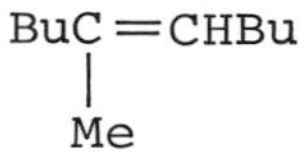   28 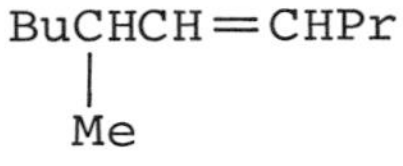   29

dichotomy (trans olefin from syn elimination and cis
olefin from anti elimination). From this information
the Hofmann/Saytzev ratios listed in Table 8 were calculated. In this instance the choice is between

TABLE 8   Ratios of Hofmann-Rule to Saytzev-Rule Products (H/S Ratios) in Eliminations from Some Quaternary
Ammonium Salts

		H/S Ratio for	
Reactant	Conditions	Syn-route	Anti-route
erythro-27	pyrolysis	102	0.38
threo-27	pyrolysis	206	0.10
erythro-27	MeOK/MeOH	22.4	0.15
threo-27	MeOK/MeOH	192	0.15
erythro-27	t-BuOK/t-BuOH	80	0.55
threo-27	t-BuOK/t-BuOH	470	0.34
erythro-27	t-BuOK/C$_6$H$_6$	21.2	<< 0.5
threo-27	t-BuOK/C$_6$H$_6$	30.1	0.25
erythro-27	t-BuOK/DMSO	247	0.43
threo-27	t-BuOK/DMSO	487	<< 0.5
30	t-BuOK/t-BuOH	762	10
30	n-BuOK/n-BuOH	240	23

elimination into a branch bearing two β-alkyl groups versus a branch bearing one. For comparison, a similar approximate calculation may be made where the choice is between one β-alkyl group and none. The stereochemistry of elimination from 2-hexyltrimethylammonium ion (30) to give the Saytzev product, 2-hexene, is known.[71]

$$CH_3CH_2CH_2CH_2CHCH_3$$
$$|$$
$$NMe_3{}^+$$

30

The stereochemistry of formation of the Hofmann product, 1-hexene, cannot be determined by any feasible method, but a reasonable guess may be made. Bailey and Saunders[71] showed that the major structural factor leading to syn elimination in open-chain systems was whether there was an alkyl group at the β'-position (the branch other than the one into which elimination is occurring). We can thus approximate the proportion of syn elimination in 2-hexyl→1-hexene (β'-substituent n-propyl) by that in the case of 3-hexyl→2-hexene (β'-substituent ethyl). Since 30 lacks a γ-substituent in the right-hand branch, this probably slightly overstates the proportion of syn elimination.

The syn elimination shows in all cases a very strong preference for the Hofmann-rule product, in agreement with evidence that the transition state for syn elimination possesses a large amount of carbanion character. The surprising finding is that anti elimination from 27, but not from 30, favors the Saytzev-rule product. Their findings led Sicher and his co-workers[70] to suggest that the preference for the Saytzev-rule product observed in eliminations from cis-2-alkylcyclohexyltrimethylammonium ions might well be the normal outcome of anti eliminations from quaternary ammonium salts, and did not require any special explanation regarding conformational effects in the cyclic system (see Section IV.B.5). The Hofmann rule still holds true, however, for anti eliminations from 30, and no reasonable margin of error in our estimation of stereochemistry of elimination to give 1-hexene would lead to reversal of the figures. It is evident from these results that the same factor or factors cannot be involved in the choice between elimination into branches bearing one or no β-alkyl groups, on the one hand, and two or one, on the other.

REFERENCES

1. M. L. Dhar, E. D. Hughes, and C. K. Ingold, J. Chem. Soc., 2058 (1948).
2. A. W. Hofmann, Ann. Chem., 78, 253 (1851); 79, 11 (1851).
3. W. Hanhart and C. K. Ingold, J. Chem. Soc., 997 (1927).
4. C. K. Ingold, J. A. Jessop, K. I. Kuriyan, and A. M. M. Mandour, J. Chem. Soc., 533 (1933).
5. E. D. Hughes, C. K. Ingold, G. A. Maw, and L. I. Woolf, J. Chem. Soc., 2077 (1948).
6. E. D. Hughes, C. K. Ingold, and L. I. Woolf, J. Chem. Soc., 2084 (1948).
7. A. C. Cope, N. A. LeBel, H. H. Lee, and W. R. Moore, J. Am. Chem. Soc., 79, 4720 (1957).
8. P. A. S. Smith and S. Frank, J. Am. Chem. Soc., 74, 509 (1952).
9. E. D. Hughes, C. K. Ingold, and G. A. Maw, J. Chem. Soc., 2072 (1948).
10. D. V. Banthorpe, E. D. Hughes, and C. K. Ingold, J. Chem. Soc., 4054 (1960).
11. D. J. Cram, F. D. Greene, and C. H. DePuy, J. Am. Chem. Soc., 78, 790 (1956).
12. H. C. Brown and O. H. Wheeler, J. Am. Chem. Soc., 78, 2199 (1856).
13. H. C. Brown and I. Moritani, J. Am. Chem. Soc., 78, 2203 (1956).
14. A. Saytzev, Ann. Chem., 179, 296 (1875).
15. E. D. Hughes and C. K. Ingold, Trans. Far. Soc., 37, 657 (1941).
16. M. L. Dhar, E. D. Hughes, C. K. Ingold, A. M. M. Mandour, G. A. Maw, and L. I. Woolf, J. Chem. Soc., 2093 (1948).
17. M. L. Dhar, E. D. Hughes, and C. K. Ingold, J. Chem. Soc., 2065 (1948).
18. R. B. Turner, Tetrahedron, 5, 127 (1959).
19. M. J. S. Dewar and H. N. Schmeising, Tetrahedron, 5, 166 (1959).
20. M. J. S. Dewar and H. N. Schmeising, Tetrahedron, 11, 96 (1960).
21. M. M. Kreevoy, Tetrahedron, 5, 233 (1959).
22. R. S. Mulliken, Tetrahedron, 5, 253 (1959).
23. H. C. Brown, I. Moritani, and Y. Okamoto, J. Am. Chem. Soc., 78, 2193 (1956).
24. E. D. Hughes, C. K. Ingold, and V. J. Shiner, Jr., J. Chem. Soc., 3827 (1953).
25. R. B. Turner, D. E. Nettleton, Jr., and M. Perelman, J. Am. Chem. Soc., 80, 1430 (1958).
26. A. Schriesheim and C. A. Rowe, Jr., J. Am. Chem.

Soc., $\underline{84}$, 3160 (1962).

27. W. H. Saunders, Jr., S. R. Fahrenholtz, E. A. Caress, J. P. Lowe, and M. Schreiber, J. Am. Chem. Soc., $\underline{87}$, 3401 (1965).

28. A. J. Berlin and F. R. Jensen, Chem. Ind. (London), 998 (1960).

28b. D. S. Bailey, J. A. Walder, and J. B. Lambert, J. Am. Chem. Soc., $\underline{94}$, 177 (1972).

29. R. A. Bartsch and J. F. Bunnett, J. Am. Chem. Soc., $\underline{90}$, 408 (1968).

30. H. C. Brown and R. L. Klimisch, J. Am. Chem. Soc., $\underline{88}$, 1425 (1966).

31. A. K. Colter and R. D. Johnson, J. Am. Chem. Soc., $\underline{84}$, 3289 (1962).

32. A. K. Colter and D. R. McKelvey, Can. J. Chem., $\underline{43}$, 1282 (1965).

33. R. A. Bartsch and J. F. Bunnett, J. Am. Chem. Soc., $\underline{91}$, 1376 (1969).

34. R. A. Bartsch and J. F. Bunnett, J. Am. Chem. Soc., $\underline{91}$, 1382 (1969).

35. H. C. Brown, I. Moritani, and M. Nakagawa, J. Am. Chem. Soc., $\underline{78}$, 2190 (1956).

36. H. C. Brown, I. Moritani, and Y. Okamoto, J. Am. Chem. Soc., $\underline{78}$, 2193 (1956).

37. I. N. Feit and W. H. Saunders, Jr., J. Am. Chem. Soc., $\underline{92}$, 1630 (1970).

38. I. N. Feit, F. Schadt, J. Lubinkowski, and W. H. Saunders, Jr., J. Am. Chem. Soc., $\underline{93}$, 6606 (1971).

39. H. C. Brown and R. L. Klimisch, J. Am. Chem. Soc., $\underline{87}$, 5517 (1965).

40. D. H. Froemsdorf, W. Dowd, and K. E. Leimer, J. Am. Chem. Soc., $\underline{88}$, 2345 (1966).

41. J. Sicher, J. Závada, and M. Pánková, Chem. Commun., 1147 (1968).

42. J. F. Bunnett, G. T. Davis, and H. Tanida, J. Am. Chem. Soc. $\underline{84}$, 1606 (1962).

43. L. F. Blackwell, A. Fischer, and J. Vaughan, J. Chem. Soc., $\underline{B}$, 1084 (1967).

44. J. Griepenburg, Ph.D. thesis, University of Rochester.

45. S. P. Acharaya and H. C. Brown, Chem. Commun., 305 (1968).

46. H. C. Brown and M. Nakagawa, J. Am. Chem. Soc., $\underline{78}$, 2197 (1956).

47. D. H. Froemsdorf and M. E. McCain, J. Am. Chem. Soc., $\underline{87}$, 3983 (1965).

48. D. H. Froemsdorf, M. E. McCain, and W. W. Wilkison, J. Am. Chem. Soc., $\underline{87}$, 3984 (1965).

49. D. H. Froemsdorf and M. D. Robbins, J. Am. Chem. Soc., $\underline{89}$, 1737 (1967).

50. J. F. Bunnett, "Olefin-Forming Elimination
    Reactions," in A. F. Scott, ed., "Survey of
    Progress in Chemistry," Vol. 5, Academic Press,
    New York, 1969.
51. G. S. Hammond, J. Am. Chem. Soc., 77, 334 (1955).
52. L. J. Steffa and E. R. Thornton, J. Am. Chem.
    Soc., 89, 6149 (1967).
53. E. R. Thornton, J. Am. Chem. Soc., 89, 2915
    (1967).
54. I. N. Feit and W. H. Saunders, Jr., J. Am. Chem.
    Soc., 92, 5615 (1970).
55. G. K. Zwolinski and D. L. Griffith, unpublished
    results; J. Hine and M. Hine, J. Am. Chem. Soc.,
    74, 5266 (1952).
56. E. D. Hughes, C. K. Ingold, and J. B. Rose, J.
    Chem. Soc., 3839 (1953).
57. T. H. Brownlee and W. H. Saunders, Jr., Proc.
    Chem. Soc., 314 (1961).
58. W. Hückel and M. Hanack, Angew. Chem., 79, 555
    (1967); Int. Ed. Engl., 6, 534 (1967).
59. W. Hückel, H. Tappe, and G. Legutke, Ann., 543,
    191 (1940).
60. N. L. McNiven and J. Read, J. Chem. Soc., 153
    (1952).
61. E. D. Hughes and J. Wilby, J. Chem. Soc., 4094
    (1960).
62. H. Booth, N. C. Franklin, and G. C. Gidley, J.
    Chem. Soc., C, 1891 (1968).
63. B. B. Gent and J. McKenna, J. Chem. Soc., 137
    (1959).
64. W. H. Saunders, Jr. and T. A. Ashe, J. Am. Chem.
    Soc., 91, 4473 (1969).
65. G. Biale, D. Cook, D. J. Lloyd, A. J. Parker,
    I. D. R. Stevens, J. Takahashi, and S. Winstein,
    J. Am. Chem. Soc., 93, 4735 (1971).
66. G. Biale, A. J. Parker, S. G. Smith, I. D. R.
    Stevens, and S. Winstein, J. Am. Chem. Soc., 92,
    115 (1970).
67. P. Beltrame, G. Biale, D. J. Lloyd, A. J. Parker,
    M. Ruane, and S. Winstein, J. Am. Chem. Soc.,
    93, 4735 (1971).
68. D. S. Bailey and W. H. Saunders, Jr., unpublished
    results.
69. D. S. Bailey, F. C. Montgomery, G. W. Chodak, and
    W. H. Saunders, Jr., J. Am. Chem. Soc., 92, 6911
    (1970).
70. J. Sicher, M. Svoboda, M. Pánková, and J. Závada,
    Coll. Czech. Chem. Commun., 36, 3633 (1971).
71. D. S. Bailey and W. H. Saunders, Jr., J. Am.
    Chem. Soc., 92, 6904 (1970).

72. K. Bowden, Chem. Rev., $\underline{66}$, 119 (1966).
73. R. A. Bartsch, G. M. Pruss, R. L. Buswell, and
     B. A. Bushaw, Tetrahedron Lett., 2621 (1972);
     J. Závada, M. Svoboda, and M. Pánková, Tetrahedron
     Lett., 711 (1972).
74. C. J. Pedersen, J. Am. Chem. Soc., $\underline{89}$, 7017
     (1967).
75. J. K. Borchardt, R. Hargreaves, and W. H.
     Saunders, Jr., Tetrahedron Lett., 2307 (1972).

V THE E1 MECHANISM

A MAIN FEATURES OF MECHANISM

1 The Hughes-Ingold Mechanism

This mechanism has already been mentioned, and its
scope discussed briefly, in Chapter I (Section I.A.2,3).
It is shown again in Eqs. 1 and 2. The second step
very probably involves abstraction of a proton by a

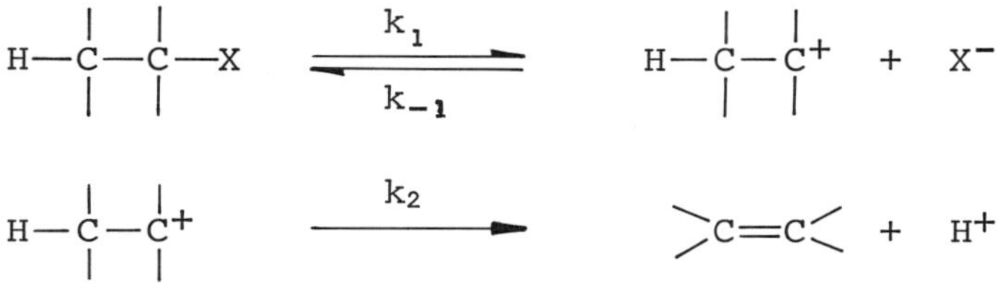

$$H-\underset{|}{\overset{|}{C}}-\underset{|}{\overset{|}{C}}-X \; \underset{k_{-1}}{\overset{k_1}{\rightleftharpoons}} \; H-\underset{|}{\overset{|}{C}}-\underset{|}{\overset{|}{C}}^+ \; + \; X^- \qquad (1)$$

$$H-\underset{|}{\overset{|}{C}}-\underset{|}{\overset{|}{C}}^+ \; \overset{k_2}{\longrightarrow} \; {>}C{=}C{<} \; + \; H^+ \qquad (2)$$

solvent molecule, but this is not shown explicitly for
the sake of simplicity. The mechanism was first

proposed in 1935 by Hughes.[1] It has since been reviewed a number of times by Hughes and Ingold.[2-4] Application of the steady-state treatment to the mechanism gives Eq. 3 for the rate. If $k_2 > k_{-1}(RX)$, we
get the familiar Eq. 4, which is the kinetic form most
commonly encountered in the E1 reaction. If the

$$\text{Rate} \quad = \quad \frac{k_1 k_2 (RX)}{k_{-1}(X^-) + k_2} \tag{3}$$

$$\text{Rate} \quad = \quad k_1 (RX) \tag{4}$$

carbonium ion is formed rapidly and then loses a proton
slowly, $k_{-1}(X^-) > k_2$, and Eq. 5 results. This situation is encountered only when the carbonium-ion intermediate is unusually stable. Returning to Eq. 4, one

$$\text{Rate} \quad = \quad \frac{k_1 k_2 (RX)}{k_{-1}(X^-)} \tag{5}$$

must consider the possibility that an E2 mechanism with
the solvent as base could give the same rate law.
While this possibility is hard to disprove rigorously,
one can be reasonably confident that it is not occurring if small amounts of the conjugate base of the solvent do not affect the rate. If this more powerful
nucleophile does not enter into the rate-determining
step, it is unlikely that the weakly nucleophilic solvent molecule is doing so. It should be noted at this
point that the carbonium ion produced in Equation 1
can also react with solvent to form substitution product by the S_N1 mechanism. The competition between the
E1 and S_N1 processes will be discussed below (Section
V.C.2).

    2   The Ion-Pair Modification

The mechanism of Eqs. 1 and 2 is incomplete in that it
shows the reaction as proceeding via dissociated ions,
whereas there is abundant evidence that many carbonium-
ion reactions involve ion-pair intermediates.[5] Scheme
1 gives an expanded version of the mechanism that incorporates both intimate (1) and solvent-separated (2)
ion-pair intermediates.[5] Although the evidence to
date on elimination reactions does not permit one to

Scheme 1

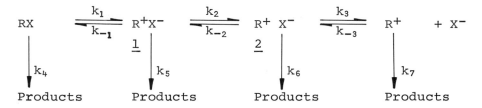

decide whether one or both types of ion pairs are involved in the product-forming step, both are included
for completeness. The step represented by k_4 is not
really part of the E1 process, since it refers to the
situation where an E2 reaction of substrate with a solvent molecule as base occurs.

The best evidence for ion-pair intermediates was
obtained by Cocivera and Winstein, who studied the competition between substitution and elimination in solvolyses of various t-butyl and t-amyl derivatives.[6] If
the product-forming step involved only free carbonium
ions, the intermediate would be the same regardless of
the leaving group in the reactant, and the substitu-
tion/elimination ratio should be independent of the
leaving group. This prediction was found to hold true
for reactions in water, a strongly ionizing solvent,
but not for reactions in ethanol or acetic acid. In
these latter solvents, the proportion of olefin was low
when the reactant was the t-butyldimethylsulfonium ion
(which presumably cannot give an ion-pair intermediate)
but much higher when the reactant was t-butyl bromide
or chloride. Evidence that the composition of a mixture of isomeric olefins produced in an E1 reaction
may depend on the leaving group is afforded by a study
of solvolysis of 2-phenyl-2-butyl derivatives in acetic acid.[7] A considerable variation in proportions of
2-phenyl-1-butene and the cis- and trans-2-phenyl-2-
butenes was observed, though partial equilibration of
some of the product mixtures renders quantitative comparisons difficult. Kinetic evidence for a common ion-
pair intermediate for E1, S_N1, E2, and S_N2 reactions
of 1-phenylethyl bromide has been offered by Sneen and
Robbins (Section I.B.2).[8] Arguments for the ion-pair
mechanism have also been based on the stereochemistry
of the elimination reaction. These arguments are discussed shortly (Section V.B.1,2).

It should be noted at this point that the present
chapter is restricted mainly to solvolytic E1 processes. Other eliminations that appear to proceed via

carbonium-ion intermediates, such as alcohol dehydration, deamination of aliphatic amines, and the deoxidation process are treated in subsequent Chapters (Sections VI.B and VII.A,B).

B   STEREOCHEMISTRY

1   Cyclic Systems

If the intermediate in El reactions were always a free
carbonium ion that survived long enough to equilibrate
rotationally, one would expect no difference between
the products from two diastereomeric reactants. On the
other hand, an intimate ion-pair intermediate should
show considerable stereoselectivity and, perhaps, even
stereospecificity in its reactions. The actual situation in El reactions is often somewhere between these
two extremes. For example, menthyl chloride (3) on
solvolysis in 80% ethanol gives 68% 3-menthene (4)
and 32% 2-menthene (5), which shows that appreciable
syn elimination can occur in this reaction. Neomenthyl
chloride (6), which can eliminate anti in either direction, prefers the anti elimination toward the isopropyl

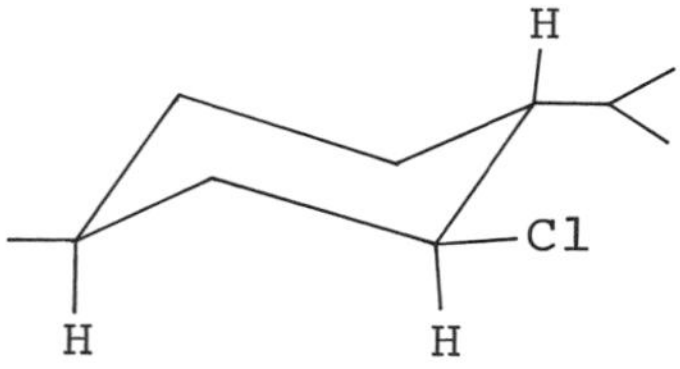

3

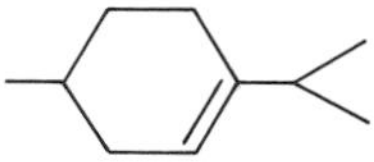

4

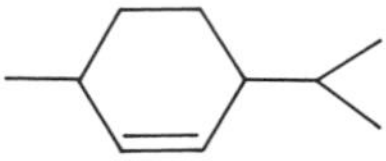

5

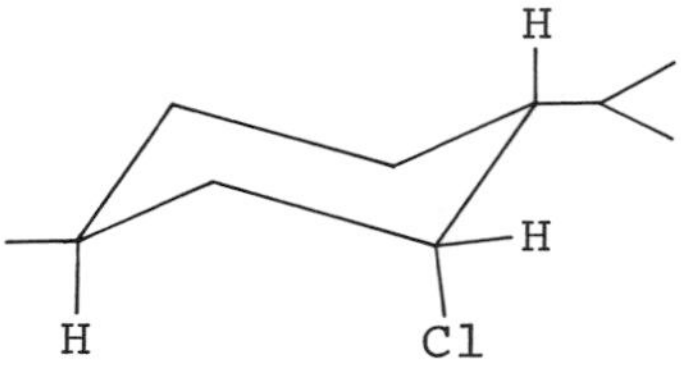

6

group very strongly, giving 99% of 4 and only 1% of
5.[9],[10]   Similar results are obtained with the neomenthyl tosylate[9] and trialkylammonium salt[11] under E1
conditions.   Thus anti elimination seems to be rather
strongly preferred in the cyclohexane system when a
transition-state conformation that places both the
leaving group and the β-hydrogen in axial positions
(anti-periplanar) is favored.   When it is not, there
can be substantial syn elimination.   One cannot draw
any conclusions about intrinsic tendency toward syn
elimination because of the effect of the 2-alkyl group
on product stability.   In the solvolysis of the 4-t-
butylcyclohexyl tosylates, the trans isomer (equatorial OTs) gives 67 to 76% olefin, while the cis isomer
(axial OTs) gives 83 to 87% olefin.[12]   Thus an axial
orientation of the leaving group is not required for
appreciable olefin yields, although the presence of
products of hydrogen shift[12] and large isotope effects
on the rate-determining step with both β-deuterium and
tosylate axial indicate that a hydrogen anti-periplanar
to the leaving group is substantially involved in the
rate-determining as well as the product-determining
step.[13]

     The situation seems to be rather different with
medium-ring compounds.   In the solvolysis of a number
of β-deuterated cyclodecyl tosylates, both the cis and
trans cyclodecenes produced resulted predominantly or
exclusively from syn elimination.[14]   This is in contrast to the E2 reactions with cyclodecyl derivatives,
where the trans olefin results from syn and the cis
results from anti elimination (Section III.E.5).   The
most attractive explanation is that the negative ion of
an ion-pair intermediate abstracts the syn-β-hydrogen.
The solvents used (acetic acid, dimethylformamide, and
pyridine) should promote ion-pair formation.   Unfortunately, no strongly dissociating solvents, such as
aqueous ethanol, were used.   Thus we do not know what
the stereochemical outcome would be under conditions
favoring free carbonium ions.   Ion pairs have also
been suggested as intermediates in deamination reactions (Section VII.A.1.f).

     2   Open-Chain Systems

There has been much less work done on the stereochemistry of the E1 reaction in open-chain than in cyclic
systems, but the one available investigation by Skell
and Hall[15] nicely illustrates the main points made
above.   Solvolysis of 2-butyl-3-d tosylate in the dissociating solvents acetamide and 80% aqueous ethanol
gives predominantly anti elimination (66 to 91%,

uncorrected for the isotope effect on proton loss),
while solvolysis in the nondissociating solvents nitrobenzene and acetic acid gives 65 to 99% syn elimination
(again, uncorrected for the isotope effect). No consistent differences in stereochemistry between reactions leading to trans- and cis-2-butene were noted.
The preference for syn elimination in solvents promoting ion-pair formation is, as indicated above, consistent with proton abstraction by the negative ion of an
ion pair. Although we cannot say rigorously what type
of ion pair is involved, it seems reasonable to suppose
that an intimate ion pair (1) could more easily pre-
serve configuration than could a solvent-separated ion
pair (2). The behavior in dissociating solvents suggests that a free carbonium ion may not have time to
equilibrate rotationally between the time of its formation and the time it loses a proton, or else that the
negative ion is still close enough to exert some influence. Perhaps a solvent-separated ion pair (2) is involved to, at least, some degree in these media.

C   REACTIVITY AND ORIENTATION

1   Conditions Favoring the E1 Mechanism

There are three main factors that favor the E1 mechanism: a substrate that gives a relatively stable carbonium ion, an ionizing solvent, and the absence of
strong bases or nucleophiles. The first factor means
that reactivity with simple alkyl derivatives runs in
the order tertiary > secondary > primary. The trend
is quite marked, since simple primary alkyl derivatives
seldom if ever react by the E1 mechanism, while the E1
mechanism is often difficult to suppress with tertiary
derivatives even under conditions favoring the E2 mechanism.[3] As for the solvent effect, tertiary alkyl
halides are sufficiently prone to solvolysis that it
cannot be completely avoided in solvents such as methanol or ethanol even at quite high concentrations of the
conjugate base of the solvent.[16,17] On the other hand,
the same compounds react with t-butoxide in t-butyl alcohol almost entirely by the E2 mechanism.[16] Solvolysis of t-butyldimethylsulfonium ion in water competes
to an important extent with the E2 reaction even at
high hydroxide-ion concentrations, but the base-promoted reaction is much faster than the solvolysis in 97%
ethanol.[18,19] Tosylates are somewhat more prone to
solvolyze than are halides. Even some secondary tosylates give appreciable E1 reaction in alkoxide/

alcohol media.[20,21] Finally, the role of strong bases
in diverting reaction from the E1 to the E2 pathway is
obvious, but it has recently been noticed that strong
nucleophiles, especially in dipolar aprotic solvents,
can also effectively promote the E2 mechanism. These
reactions are discussed elsewhere in this book (Section II.A.5).

2   Elimination Versus Substitution

The E1 reaction has never gained favor as a method of
synthesis of olefins because the ratio of elimination
to substitution products is often substantially lower
than in the E2 reaction. For example, t-butyl bromide
yields only 19% olefin on solvolysis in ethanol at 25°,
while the reaction in the presence of 2M ethoxide ion
yields 93% olefin.[22] The corresponding figures for
t-amyl bromide are 36% and 99%, respectively.[22] The
actual proportion of elimination for pure E2 reactions
may run even higher.

   An increase in temperature produces a modest increase in the proportion of olefin in E1-S_N1 reactions,
just as in E2-S_N2 reactions.[23] A similarly modest increase is found on going from aqueous ethanol to anhydrous ethanol.[24] Although this may be simply a solvent
effect on the product-forming reaction arising from the
greater dispersal of charge in the E1 than in the S_N1
transition state, it may also reflect the incursion of
ion-pair intermediates in the less polar media (Section V.A.2). Under conditions favoring ion pairing,
higher olefin yields are obtained from tertiary halides.[6]

   Perhaps the major factor affecting olefin yield
in E1 reactions is the structure of the alkyl group.
Unbranched secondary halides give quite poor olefin
yields, ranging from 5% for isopropyl to 15% for 3-
pentyl bromide in 60% aqueous ethanol at 80°.[25] With
tertiary reactants, the yields run higher, around 25 to
35% for t-butyl and 40 to 50% for t-amyl derivatives in
80% ethanol in the range 50 to 65°.[23,26] With a series
of t-alkyl chlorides, the proportion of olefin obtained
on solvolysis in 80% ethanol rises as the alkyl groups
attached to the central carbon atom become larger and
more highly branched.[27] Ingold and his co-workers
observed that higher olefin yields resulted from those
reactants which could yield the more highly substituted olefins, which in turn were known to be more stable
than less substituted olefins.[3] The reasons for this
stabilization are discussed more fully below (Section
IV.C.3). For the present, the argument can be simply

stated by saying that the transition state leading to
olefin will possess double-bond character and will be
stabilized by the same factors that stabilize the olefin.

Brown and Fletcher disagreed with this analysis,
arguing that steric effects were mainly responsible
for the increase in olefin yield.[27] According to them,
the sp^2-hybridized carbonium ion is less strained than
the sp^3-hybridized reactant or substitution product, so
that the larger the alkyl groups the lower the rate of
substitution. They also argued that the more strained
the carbonium ion, the more readily it should expel a
proton to give olefin. Hughes, Ingold and Shiner reit-
erated and presented in further detail the product-
stability explanation, claiming that steric effects
were significant in few if any E1 reactions.[28] It
seems likely that there is truth in both points of
view. It is reasonable to suppose that the more stable
olefin will be formed more rapidly, but one cannot neglect changes in rate of the substitution reaction in
doing so. Certainly steric strain must play an important role with the highly branched alkyl groups. When
all three alkyl groups attached to the central carbon
atom are unbranched, however, the variations in olefin
yield are too small and irregular to be assigned to a
simple steric effect, but they can be explained quite
well in terms of product stability.

3 Orientation Effects - the Saytzev Rule

We turn now from the competition between elimination
and substitution to the competition involved in the
formation of two or more isomeric olefins from the
same reactant. Orientation in E2 reactions is discussed earlier (Chapter IV), and many of the same considerations apply to orientation in E1 reactions. As
we see below, nearly all E1 reactions follow the
Saytzev rule (Section IV.A.5), which predicts that
formation of olefins bearing the greater number of
alkyl groups on the double bond will be favored. Since
alkyl substitution almost always increases olefin
stability, the rule can be rephrased to predict preference for the thermodynamically more stable product
(or products).

The Saytzev-rule product is preferred regardless
of the nature of the leaving group, as would be expected for a carbonium-ion intermediate. For example,
solvolysis of t-amyl bromide in ethanol gives 82% 2-
methyl-2-butene, while solvolysis of t-amyldimethylsulfonium ion gives 87%.[26] It is not clear whether these

two figures are significantly different. There is some
other evidence for different product compositions with
different leaving groups in E1 reactions, but the area
has not been explored systematically enough to allow
any reliable generalizations (Section IV.A.2).[7] It
does not seem very likely that the negative ion of an
ion pair would have a large effect on orientation.

The preference for the more stable olefin is quite
marked, generally significantly more so than in E2 reactions that follow the Saytzev rule. Dhar, Hughes,
and Ingold report that t-amyl bromide gives 82% 2-
methyl-2-butene on solvolysis in ethanol, but gives
only 71% on reaction with ethoxide ion in ethanol.[22]
Later work has given a somewhat lower figure for the
E2 reaction (62%), thereby widening the gap further.[17]
There are many examples of this phenomenon in the literature, most recently from the work of Feit and
Saunders.[21] They determined product proportions in the
solvolysis of several secondary alkyl tosylates in n-,
s- and t-butyl alcohol, and in the E2 reactions with
the conjugate bases of the alcohols in the same solvents. In all cases the proportion of the Saytzev-rule
product, and the ratio of trans to cis olefin in that
product, were much higher in the E1 than in the E2 reaction. The product proportions did not vary significantly with solvent in the E1 reaction, apart from one
case where the yield of a rearranged product did
(Section V.D.2).

The question of why alkyl substitution increases
the thermodynamic stability of an olefin is discussed
earlier (Section IV.A.5), and that discussion will not
be repeated here. Evidence that steric effects play a
role in determining product proportions in E1 reactions
has been adduced by Brown and his co-workers in a series of papers.[29-33] An illustrative set of results is
shown in Table 1.[30] Here the number of alkyl groups
about the double bond remains constant, but the size of

TABLE 1 Products of Solvolysis of $RCH_2CBr(CH_3)_2$ in
85% n-Butyl Cellosolve at 25°

R	% Olefin Yield	% 1-ene	% 2-ene
Me	27	21	79
Et	32	29	71
i-Pr	46	41	59
t-Bu	57	81	19

one of them increases. There is a clear increase in
the proportion of the 1-ene with increasing size of
the group R, doubtless arising from the fact that
eclipsing interactions between the R group and the
methyl groups in the 2-ene become increasingly severe
as R becomes larger. The last member of the series
gives only 19% of the 2-ene but, as noted previously
(Section IV.A.5), the 1-ene is here more stable than
the 2-ene. Thus the Saytzev rule is not violated if
it is expressed as a preference for the more stable
product. A similar but less pronounced trend is noted
with the products of acetolysis of $\underline{7}$, where the pro-

$$RCH_2CH(OBs)CH_3$$

$$\underline{7}$$

portion of 1-ene and, especially, the trans/cis ratio
of the 2-ene increase with increasing bulk of R.

    4   The Transition States of the Rate Determining
       and Product-Determining Steps

The strong preference for the more stable olefin and
the pronounced eclipsing effects suggest that the transition state of the product-determining step is quite
close to olefin in structure. Reconciliation of this
conclusion with the Hammond postulate, which predicts
that a high-energy intermediate should decompose via a
transition state which resembles the intermediate more
than the products, is not entirely easy.[34] Deuterium
isotope effects on proton loss of up to 3.1 have been
observed.[35] This less-than-maximum effect is consistent with a proton transfer to solvent which is either
distinctly less or distinctly more than half complete
in the transition state (Section II.C.3.a), so that
the experimental data all fit a productlike transition
state.

    If the carbonium-ion intermediate already pos-
sessed some of the characteristics of the product, a
productlike transition state for its decomposition
would be more readily understandable. Even in simple
solvolyses there is evidence for some loosening of the
β-carbon-hydrogen bonds in the rate-determining step,
for deuterium substitution lowers the solvolysis rate
by 10 to 30% per deuterium atom.[36,37] Where the
β-deuterium is in a tertiary position, or is fixed
anti-periplanar to the leaving group, k_H/k_D values in

excess of 2.0 have been observed, suggesting considerable weakening of the β-carbon-hydrogen bond in the transition state.[13,38] Although we generally consider that the products are determined during the decomposition rather than the formation of the carbonium ion in an El reaction, the direction of elimination may already be fixed during the rate determining step when there is such marked interaction of the β-hydrogen with the developing carbonium-ion center during ionization. There is some evidence that trans/cis ratios depend on the structure of the original reactant in El reactions, since they are consistently higher in the formation of 2-enes from 3-alkyl than from 2-alkyl tosylates.[21] This could result either from the sort of "predetermination" discussed above, or from the tosylate ion in an ion pair exerting some influence in the product-determining step. Yet another source of this phenomenon could be failure of the carbonium ion to equilibrate rotationally before reaction.

D   REARRANGEMENTS

1   Bridged Versus Open Ions

The role of bridged ions in carbonium-ion reactions has been an active field of research and the subject of much controversy for more than twenty years. It is not within the scope of this book to review the field or even to summarize the arguments of those who have taken various positions in the controversy. We restrict ourselves to the possible intermediacy of bridged ions in the formation of elimination products, where the picture appears to be considerably less complex and controversial. In brief, we know of no evidence that requires aryl- or alkyl-bridged carbonium ions to be precursors of the olefinic products in El reactions of simple reactants. This is not to say that they can always be definitely excluded, for the olefinic products often are not examined or do not permit conclusions about the stereochemistry of their precursors.

Acetolysis of the diastereomeric 3-phenyl-2-butyl tosylates has been thoroughly studied, and evidence points to the intermediacy of a phenyl-bridged (phenonium) ion in the formation of substitution product.[39] Whether the best or only interpretation of the evidence is a phenonium ion is part of the controversy that we wish to avoid. Suffice it to say here that a phenonium-ion intermediate is consistent with the

stereochemical evidence on the substitution product.
It does not, however, seem to be consistent with the
stereochemical evidence on the elimination product.[40]
The phenonium ion, 8, from the threo tosylate possesses
a plane of symmetry, but the olefinic product is not
optically inactive. Its component having an asymmetric
center, 9, thus could not have arisen from 8. Prior

$$\underset{\underset{CH_3}{\diagdown}}{\overset{\overset{H}{\diagup}}{C}} \overset{\overset{Ph}{\diagdown}}{\underset{+}{=\!=\!=\!=\!=\!=\!=}} \underset{\underset{CH_3}{\diagdown}}{\overset{\overset{H}{\diagup}}{C}}$$

$$\overset{\overset{Ph}{|}}{CH_3 - \overline{C}H - CH = CH_2}$$

$$\underline{8} \hspace{5cm} \underline{9}$$

racemization of the starting material can explain such
loss of activity as does occur in the formation of 9.
Otherwise, the olefins are formed partly by anti-elimination and partly by nonstereospecific proton loss.
Cram suggested the hydrogen-bridged ion 10 as precursor
of the anti-elimination product, and the open ions 11
and 12 as precursors of the nonstereospecifically-

$$\underset{\underset{CH_3}{|}}{Ph-\overset{\overset{/\overset{H}{+}\diagdown}{}}{C}\!-\!-\!-\!-CH-CH_3}$$

$$\overset{\overset{Ph}{|}}{CH_3\,CH-CHCH_3} \\ +$$

$$\overset{\overset{Ph}{|}}{CH_3-C-CH_2\,CH_3} \\ +$$

$$\underline{10} \hspace{3.5cm} \underline{11} \hspace{3.5cm} \underline{12}$$

formed product.[40]   The anti-elimination product could
also result from tight ion pairs.

    2 Conditions Favoring Rearrangements

Acid generated by a solvolysis reaction may cause isomerization of olefinic products to isomers not directly
obtainable from the reactant, thereby giving apparent
rearrangement. Rearrangement products from this source
can usually be avoided by adding a base to neutralize
the acid as it is formed. The base must be too weak
to cause an E2 reaction. A hindered base such as 2,6-
lutidine in alcoholic solvents, or the conjugate base

of the solvent in acetic or formic acid may be used for this purpose.

In irreversible E1 reactions, rearrangement products are seldom formed unless the initially formed carbonium ion can be converted to a more stable carbonium ion by a simple 1,2-shift. An exception to this rule is cis-4-t-butylcyclohexyl tosylate, 13, whose axial leaving group apparently confers on it an enhanced tendency to rearrange.[12] 3-Methyl-2-butyl tosylate, 14, is an excellent example of the sort of system that is likely to give rearrangement.[38] The olefinic product from solvolysis contains little 15, which

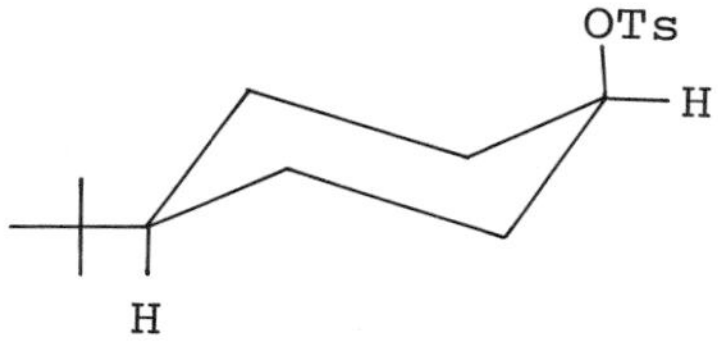

OTs

—H

H

13

OTs

$(CH_3)_2CH-CHCH_3$

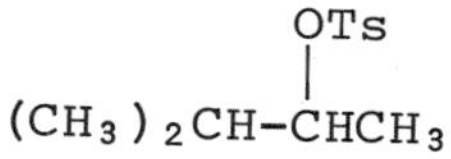

14

must come from the unrearranged secondary carbonium ion. It consists mainly of 16, which can come either from unrearranged secondary or rearranged tertiary carbonium ion, and 17, which can come only from rearranged tertiary carbonium ion.

$(CH_3)_2CHCH{=}CH_2$

$(CH_3)_2C{=}CHCH_3$

$CH_2{=}C-CH_2CH_3$

CH_3

15

16

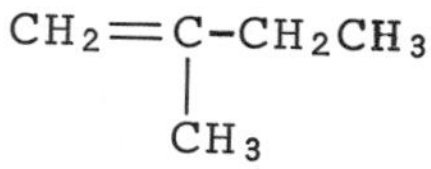

17

The effect of solvent on the tendency to give rearranged products is not entirely clear. One generally expects the greater tendency toward rearrangement in the more highly ionizing solvents, and many examples of this trend exist where rearrangement or neighboring-group participation by aryl or alkyl groups is involved.[41] It is not the general rule for hydrogen migrations. cis-4-t-Butylcyclohexyl tosylate, 13, gives increasing amounts of products of hydrogen shift along the series 60% acetone < acetic acid < formic acid,[12] but a different result is obtained for the secondary →tertiary carbonium ion rearrangement in simple aliphatic systems.   3-Methyl-2-butyl tosylate, 14, gives more 17 in acetolysis than in formolysis,[42] and increasing amounts of 17 along the series n-butyl alcohol < s-butyl alcohol < t-butyl alcohol.[21]   The lack of entirely consistent trend with solvent ionizing power for hydrogen migration reactions suggests that, at least, two solvent properties are affecting the tendency toward rearrangement.   Perhaps the more basic solvents can aid migration of hydrogen by nucleophilic solvation.

REFERENCES

1.   E. D. Hughes, J. Am. Chem. Soc., 57, 708 (1935).
2.   E. D. Hughes and C. K. Ingold, Trans. Far. Soc., 37, 657 (1941).
3.   M. L. Dhar, E. D. Hughes, C. K. Ingold, A. M. M. Mandour, G. A. Maw and L. I. Woolf, J. Chem. Soc., 2093 (1948).
4.   C. K. Ingold, "Structure and Mechanism in Organic Chemistry," Cornell University Press, Ithaca, Chap. 8.   First ed., 1953; second ed., 1969.
5.   S. Winstein, E. Clippinger, A. H. Fainberg, R. Heck, and G. C. Robinson, J. Am. Chem. Soc., 78, 328 (1956); and subsequent papers.
6.   M. Cocivera and S. Winstein, J. Am. Chem. Soc., 85, 1702 (1963).
7.   D. J. Cram and M. R. V. Sahyun, J. Am. Chem. Soc., 85, 1257 (1963).
8.   R. A. Sneen and H. M. Robbins, J. Am. Chem. Soc., 91, 3100 (1969).
9.   W. Hückel, W. Tappe and G. Legutke, Ann., 543, 191 (1940).
10.   E. D. Hughes, C. K. Ingold, and J. B. Rose, J. Chem. Soc., 3839 (1953).
11.   E. D. Hughes and J. Wilby, J. Chem. Soc., 4094 (1960).
12.   S. Winstein and N. J. Holness, J. Am. Chem. Soc., 77, 5562 (1955).
13.   V. J. Shiner, Jr. and J. G. Jewett, J. Am. Chem. Soc., 87, 1382 (1965).
14.   M. Svoboda, J. Závada and J. Sicher, Coll. Czech. Chem. Commun., 32, 2104 (1967).
15.   P. S. Skell and W. L. Hall, J. Am. Chem. Soc., 85, 2851 (1963).
16.   H. C. Brown and I. Moritani, J. Am. Chem. Soc., 76, 455 (1954).
17.   W. H. Saunders, Jr., S. R. Fahrenholtz, E. A. Caress, J. P. Lowe, and M. Schreiber, J. Am.

Chem. Soc., <u>87</u>, 3401 (1956).

18. W. H. Saunders, Jr. and S. E. Zimmerman, J. Am. Chem. Soc., <u>86</u>, 3789 (1964).

19. E. D. Hughes, C. K. Ingold, and L. I. Woolf, J. Chem. Soc., 2084 (1948).

20. A. K. Colter and D. R. McKelvey, Can. J. Chem., <u>43</u>, 1282 (1965).

21. I. N. Feit and W. H. Saunders, Jr., J. Am. Chem. Soc., <u>92</u>, 1630 (1970).

22. M. L. Dhar, E. D. Hughes, and C. K. Ingold, J. Chem. Soc., 2065 (1948).

23. K. A. Cooper, E. D. Hughes, C. K. Ingold, G. A. Maw, and B. J. MacNulty, J. Chem. Soc., 2049 (1948).

24. K. A. Cooper, M. L. Dhar, E. D. Hughes, C. K. Ingold, B. J. MacNulty, and L. I. Woolf, J. Chem. Soc., 2043 (1948).

25. M. L. Dhar, E. D. Hughes, and C. K. Ingold, J. Chem. Soc., 2058 (1948).

26. K. A. Cooper, E. D. Hughes, C. K. Ingold, and B. J. MacNulty, J. Chem. Soc., 2038 (1948).

27. H. C. Brown and R. S. Fletcher, J. Am. Chem. Soc., <u>72</u>, 1223 (1950).

28. E. D. Hughes, C. K. Ingold, and V. J. Shiner, Jr., J. Chem. Soc., 3827 (1953).

29. H. C. Brown and I. Moritani, J. Am. Chem. Soc., <u>77</u>, 3607 (1955).

30. H. C. Brown and M. Nakagawa, J. Am. Chem. Soc., <u>77</u>, 3610 (1955).

31. H. C. Brown and M. Nakagawa, J. Am. Chem. Soc., <u>77</u>, 3614 (1955).

32. H. C. Brown and Y. Okamoto, J. Am. Chem. Soc., <u>77</u>, 3619 (1955).

33. H. C. Brown and I. Moritani, J. Am. Chem. Soc., <u>77</u>, 3623 (1955).

34. G. S. Hammond, J. Am. Chem. Soc., <u>77</u>, 334 (1955).

35. M. S. Silver, J. Am. Chem. Soc., <u>83</u>, 3487 (1961).

36. E. S. Lewis and C. E. Boozer, J. Am. Chem. Soc., <u>74</u>, 6306 (1952); <u>76</u>, 791, 794 (1954).

37. V. J. Shiner, Jr., J. Am. Chem. Soc., <u>75</u>, 2925 (1953).

38. S. Winstein and J. Takahashi, Tetrahedron, <u>2</u>, 316 (1958).

39. D. J. Cram, J. Am. Chem. Soc., <u>75</u>, 2129 (1952), and preceding papers.

40. D. J. Cram, J. Am. Chem. Soc., <u>75</u>, 2137 (1952).

41. S. Winstein, M. Brown, K. C. Schreiber, and A. H. Schlessinger, J. Am. Chem. Soc., <u>74</u>, 1140 (1952).

42. A. J. Finlayson and C. C. Lee, Can. J. Chem., <u>38</u>, 787 (1960).

VI   DEHYDRATION OF ALCOHOLS

A   INTRODUCTORY COMMENTS

The dehydration of alcohols represents one of the most

common methods for the synthesis of alkenes. Under
heterogeneous conditions, dehydration can be effected
by passing the alcohol vapors over heated catalysts
such as alumina or silica. For homogeneous reactions,
dehydration occurs under strongly basic conditions only
if the β-hydrogen is activated by an adjacent carbonyl
or carbon-carbon double bond. However, a range of
acidic reagents can be used; strong acids and iodine
convert the hydroxyl function into a better leaving
group (e.g., $-\overset{+}{O}H_2$ or $-\overset{+}{O}HI$), and the subsequent decompositions often involve carbonium ion intermediates.
Acids such as phosphoric and boric, and halogenating
agents such as thionyl chloride may esterify the alcohol and, depending on the reaction conditions, the resulting ester can decompose by E2, E1, or cyclic Ei
mechanisms. It is only possible within the space
available in this chapter to discuss a portion of the
many papers concerned with the mechanism of dehydration, particularly over solid catalysts, and the reader
is referred to the more detailed reviews on this subject that have appeared in recent years.[1-4]

B   DEHYDRATION WITH STRONG MINERAL ACIDS

1   General Observations

Most alcohols dehydrate in aqueous solutions containing
either Brönsted or Lewis acids to give the thermodynamically more stable alkene.[5,6] This apparent adherence to the Saytzev rule should not be considered as
evidence for a carbonium-ion mechanism unless it has
been shown that the Saytzev olefin is a primary and
not a secondary reaction product, for under the reaction conditions the less stable Hofmann alkene is
often rapidly isomerized. Such thermodynamic control
does limit the synthetic utility of dehydration reactions, but the lability of the alkenes to the reaction
conditions can be put to use kinetically. By studying
the reverse reaction, the hydration of alkenes, and
applying the principle of microscopic reversibility,*

*The general form of this principle states that if
there are many reaction paths from given reactants to
given products, then, in equilibrium, as many molecular
systems will pass forward and backward in each individual path.[7] Here we are concerned with the assumption
that the same intermediates will be involved in the
lowest energy path for both the forward and backward
reaction.

we can obtain more complete understanding of the dehydration mechanism.

For some time, carbonium ions have been accepted
as intermediates for both the dehydration and hydration
reactions. However, the concept of the types of ions
involved and their relative energetics has been modified as a greater understanding of the kinetic methods,
on which our mechanistic deductions depend, has been
developed.

2   Isotope Studies

(a) Exchange and Elimination. In the alkyl series,
the rate of dehydration decreases along the order -
tertiary alcohol > secondary > primary. This trend is
analogous to the reactivity order observed in the El
reactions of alkyl halides and sulfonate esters in
solution (Section V.C.1). Skeletal rearrangements,
typical of carbonium-ion intermediates, have also been
observed in dehydration reactions.[8] For example, in
85% phosphoric acid, trans-2-phenylcyclohexanol eliminates to give a significant proportion of five-mem-
bered-ring products (Scheme 1), while the cis-isomer,
which can eliminate with anti-stereospecificity from
the most stable conformation with the hydroxyl group
axial, dehydrates to give exclusively only the expected
β-elimination products (Eq. 1).[9] Certainly, this evidence suggests that carbonium-ion intermediates are involved in the dehydration of the trans-isomer, but the
products of reaction of the cis-isomer accord with the
operation of an E2-dehydration, or with elimination
from the initially formed phosphate ester, or with an
El mechanism in which the hydrogens situated trans to
the developing carbonium-ion orbital are correctly
aligned for effective orbital overlap. When the leaving group and the β-hydrogen are axial, El reactions
of alkyl halides and esters are markedly facilitated
(Section V.B.1). Labeling experiments with the trans-
isomer show that about 25% of the 1-phenylcyclohexene
arises via a phenonium ion and that isomerization of
3-phenylcyclohexene occurs under the reaction conditions.

A number of authors have shown that the rate of
olefin formation is slower than that of hydroxyl exchange with labeled solvent ($^{18}OH_2$).[10-15] Exchange for

Scheme 1

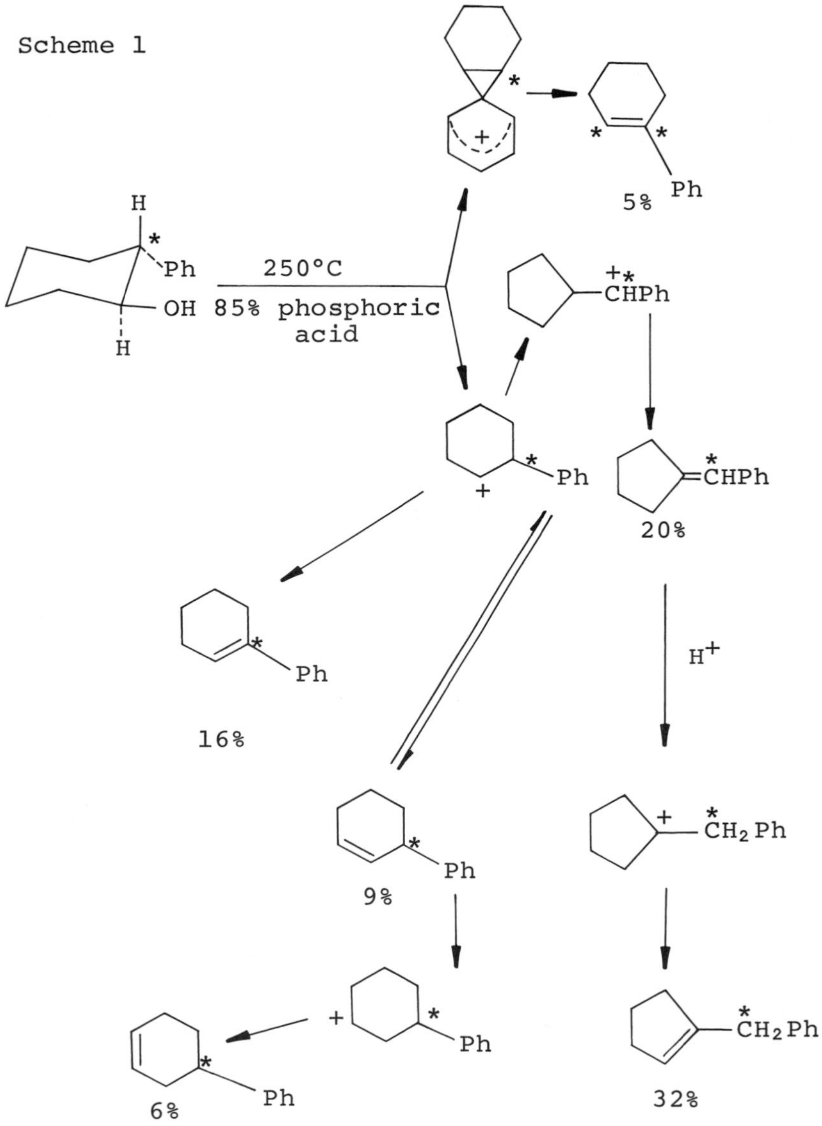

5%

250°C
85% phosphoric
acid

+*
cyclopentyl—CHPh

=*CHPh

20%

+*Ph

*Ph

16%

H⁺

*Ph

9%

+
*CH₂Ph

+ *Ph

*CH₂Ph

6% Ph

32%

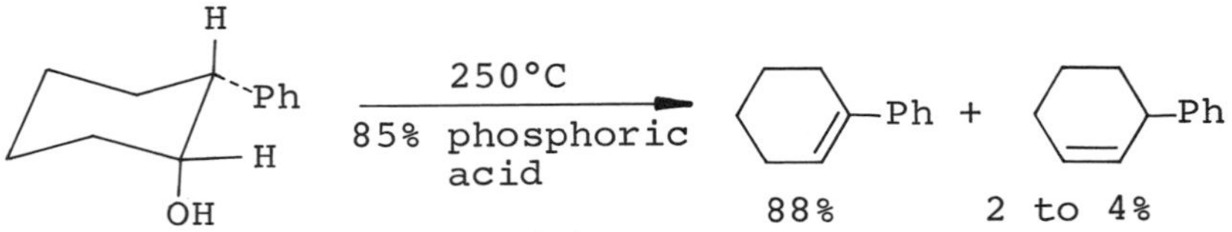

250°C
85% phosphoric
acid

Ph + Ph

88%   2 to 4%

(1)

224

tertiary alcohols is proposed to occur via a "loosely" solvated or "encumbered" carbonium ion* (cf. k_{-2}, Eq. 2), while a symmetrically solvated intermediate or a direct displacement reaction are preferred to account for the exchange of primary hydroxyl groups (cf. k_{-2}, Eq. 3), thus avoiding the intermediacy of the much less stable primary carbonium ions.[16] The nature of the exchange mechanism for secondary alcohols depends on

$$ROH \underset{k_{-1}}{\overset{k_1}{\rightleftharpoons}} ROH_2^+ \underset{k_{-2}}{\overset{k_2}{\rightleftharpoons}} R...OH_2^+ \overset{k_3}{\longrightarrow} alkene \qquad (2)$$

$$RCH_2OH \underset{k_{-1}}{\overset{k_1}{\rightleftharpoons}} RCH_2OH_2^+ \underset{k_{-2}}{\overset{k_2}{\rightleftharpoons}} \left[R-\overset{\overset{OH_2}{|}}{\underset{\underset{OH_2}{|}}{C}}\overset{H}{\underset{H}{\big\langle}} \right]^+ \overset{k_3}{\longrightarrow} alkene$$

$$(3)$$

the stability of the potential carbonium ion. Alcohols bearing an α-phenyl substituent should simulate the behavior of t-alcohols, while simple s-alkyl alcohols probably utilize the mechanism suggested for the primary alcohols. The exchange reaction of s-butanol falls into this latter category,[14] as the racemization of optically active material occurs twice as fast as the rate of hydroxyl exchange.[13] As required for a symmetrically solvated intermediate, each act of exchange is accompanied by an inversion of configuration.

Manassen and Klein also reported that the rate of exchange is twice as fast as that of dehydration of s-butanol. For the same reaction conditions (0.55N $HClO_4$ at 100°C), the dehydration if 12.5 times slower than the exchange reaction for t-butyl alcohol.[14] To explain the difference in selectivity between these isomeric carbonium ions, they suggested that exchange from a solvated carbonium ion would require a greater

*The term "encumbered carbonium ion" is used to describe intermediates that possess substantial carbonium-ion character and involve measurable interactions between the cationic center and the leaving or entering groups that may be uncharged or charged.[32]

activation energy than from an encumbered ion, but for elimination, the activation energy should be lower for the solvated ion as the water molecules should be able to participate more effectively in the breaking of the C_β-H bond (cf $\underline{1}$). For t-butyl alcohol, the activation energy for the exchange reaction is about 30 kcal/mole, and the ratio of exchange to elimination is in close agreement with the values found for E1 reactions of t-butyl halides under solvolytic conditions.[12] Banthorpe[5] has attributed the unimportance of E2 mechanisms involving $R\overset{+}{O}H_2$ to the lack of a sufficiently active base in the aqueous acidic media.

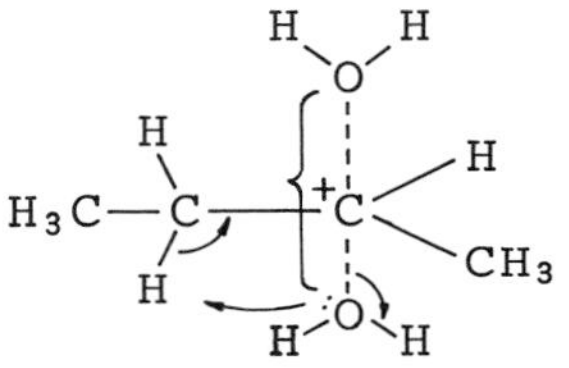

$\underline{1}$

In aqueous sulfuric acid, the values of k(racemization)/k(elimination) for 1,2-diphenylethanol (58),[17] and 1-phenylethanol (> 100),[12,18] are much larger than the values found for the simple alkyl series, reflecting the stronger stabilizing influence of the α-aryl groups. However, for 1-arylethanol, k(exchange)/k (racemization) is 0.82, consistent with exchange occurring at a partially shielded carbonium ion.[12]

(b) <u>Isotope Effects in Elimination and Hydration</u>. Substitution of deuterium for β-hydrogen should reduce the rate of dehydration if proton transfer from the carbonium ion to the reaction medium occurs in the rate-determining step. Noyce, Hartter, and Pollack[17] in fact found that 2-deutero-1,2-diphenylethanol dehydrates to trans-stilbene 1.83 times more slowly than does 1,2-diphenylethanol at 25°C in 5%-ethanol-water containing 53 to 59% sulfuric acid. Only 78.5% of the original deuterium was retained in the trans-stilbene, and transformation of the product isotope effect into a primary kinetic isotope effect gives a value of 1.57. This small but significant value demonstrates that rate-determining proton transfer from the carbonium ion involves a highly unsymmetrical transition state,

presumably more like the carbonium ion than the alkene
in view of the reactivity of the former. For the reverse reaction, the hydration of an alkene, considering
the principle of microscopic reversibility, proton
transfer to the alkene should be extensive in the transition state. The Brönsted component, α, of 0.9 for
the hydration of isobutene by aqueous perchloric acid
confirms this view.[19]

    Noyce and his co-workers have also measured the
solvent isotope effect (k_{D_2O}/k_{H_2O}) for the dehydration
of 1,2-diphenylethanol with sulfuric acid. They found
that the reaction occurred 1.62 times more rapidly in
the deuterated medium [at 25°C, 5% ethanol-H_2O(D_2O),
51% H_2SO_4(D_2SO_4)].[17,20] In this case, isotopic substitution influences the position of the pre-equilibria
which precede the rate-determining step. As $D_3\overset{+}{O}$ is a
stronger acid than $H_3\overset{+}{O}$, the alcohol is protonated to a
greater extent in the deuterated medium. On the other
hand, in the acid catalysed isomerization of cis-stilbene under similar reaction conditions, k_{H_2O}/k_{D_2O} =
2.59. In this case, the rate-determining step involves
transfer of the isotopically substituted atom to the
alkene and a normal isotope effect is observed.

    The rate of formation of trans-stilbene in the
isomerization of cis-stilbene exhibits an induction
period, attributed by Noyce, Hartter, and Miles[20] to
the buildup of a steady-state concentration of 1,2-
diphenylethanol, a deduction that is verified by consideration of the rate-acidity profile for the dehydration reaction.[17] Two schemes accord with an induction
period and the kinetics of appearance of trans-stilbene; the first (Eq. 4) involves 1,2-diphenylethanol
as an intermediate, and the second (Eq. 5) as a storage
form of the intermediate trans-R^+ ion. Certainly the
equilibrium between the alcohol and the trans-R^+ ion
is rapid as optically active 1,2-diphenylethanol race-
mizes much more rapidly than it dehydrates. In deuterated media, the trans-stilbene formed in the isomerization reaction contains 0.8 atom of deuterium per
molecule, a value almost identical with that found[17]
in the dehydration of 1,2-diphenylethanol. This result
suggests that product composition in both the dehydration and isomerization is controlled by the same step.

$$\text{cis-stilbene} \xrightarrow{\text{slow}} \text{1,2-diphenylethanol} \longrightarrow$$
$$\text{trans-stilbene} \tag{4}$$

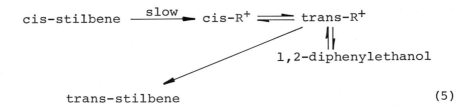

$$\text{cis-stilbene} \xrightarrow{\text{slow}} \text{cis-R}^+ \rightleftharpoons \text{trans-R}^+$$

1,2-diphenylethanol

trans-stilbene   (5)

By combining the above facts with the activation parameters for both the dehydration and isomerization reactions, Noyce, Hartter, and Miles constructed a free energy diagram relating both the stilbenes, 1,2-diphenylethanol and the carbonium ions (Figure 1). They estimate the difference in stability between the cis-R$^+$ and trans-R$^+$ ions to be about 4 kcal/mole.[20],[21]

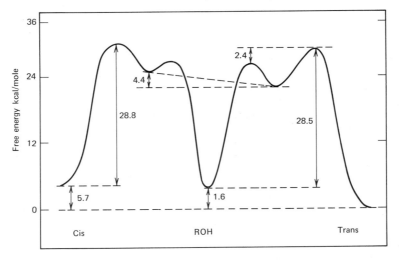

Figure 1   Reaction coordinate diagram for the isomerization of cis-stilbene (reproduced by permission of the authors[20] and the Journal of the American Chemical Society).

Kieboom and van Bekkum reported that the dehydration of 2-phenyl-3-methyl-2-butanol by 20% sulfuric acid-acetic acid at 50°C gives mainly the Hofmann alkene, 3-methyl-2-phenyl-1-butene, which then subsequently isomerizes under the reaction conditions to give the more stable Saytzev product (Eq. 6).[22] The rate of dehydration exceeds that of isomerization by about 100 times. As they anticipated for a reaction involving a carbonium-ion intermediate, the rate of

dehydration is markedly increased by electron-releasing
aryl substituents. The authors attribute the unexpected preference for formation of the less stable alkene
in the initial dehydration to steric consequences.
Proton loss accompanied by double-bond formation, de-
mands a maximum overlap of the sp^3-orbitals of the

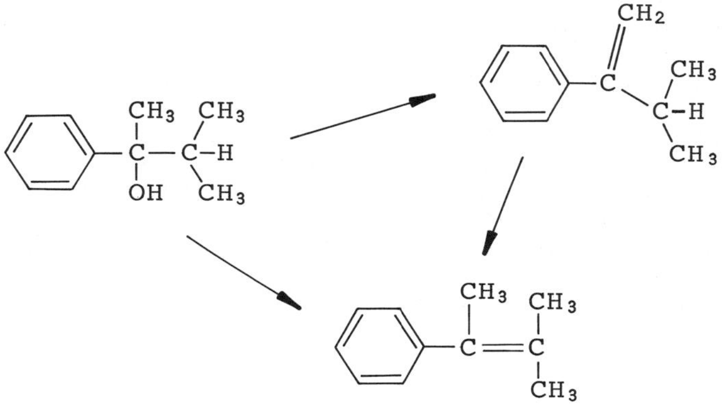

(6)

carbon-hydrogen bond with the empty orbital of the carbonium ion. Van der Waals forces of repulsion between
the 3-methyl group and the ortho-hydrogens of the aryl
ring in 3 make this conformation less favored than 2.
In 4, Van der Waals forces involving the aryl group
are the same for both the Hofmann and Saytzev product
formation and 2-phenyl-2-butanol dehydrates to give
mainly cis-2-phenyl-2-butene, reflecting thermodynamic
rather than steric control in the collapse of the carbonium ion.

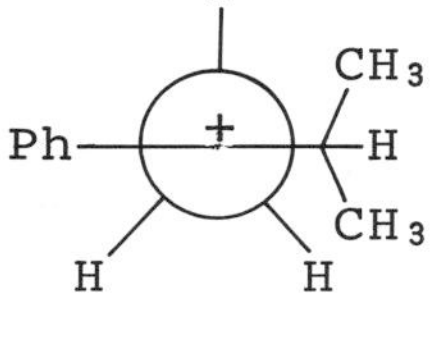

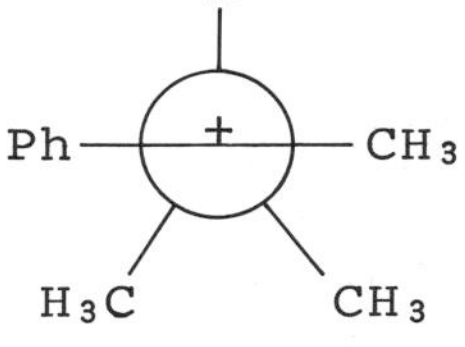

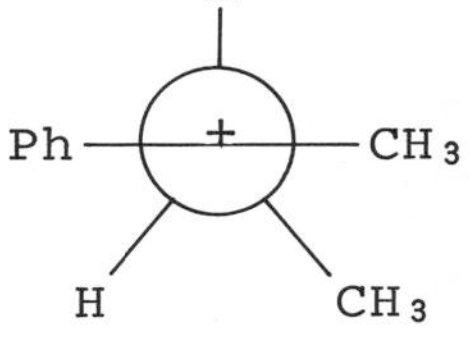

A solvent isotope effect, k_{AcOH}/k_{AcOD}, of 2.38 at 25°C is found in the isomerization of 3-methyl-2-phenyl-1-butene. 3-Methyl-2-phenyl-3-d-1-butene isomerizes 2.34 times more slowly than the undeuterated material. Thus both a solvent and a kinetic isotope effect are observed and they, when considered in conjunction with the kinetics of the dehydration reaction, accord with the mechanism of Scheme 2. From the selectivity of the dehydration reaction, k_2/k_3 varies between 4 to 10 and $k_4 \ll k_1$, k_2, and k_3 from the olefin composition at equilibrium. Consequently by using the steady-state approximation on the intermediate carbonium ion, one can express the rate of appearance of the Saytzev olefin by Eq. 7. As k_3 makes a smaller contribution than k_2 to the denominator, both a solvent (k_1) and a kinetic isotope effect (k_3) are expected. As the authors note, assignment of the rate-determining step in the acid catalyzed isomerization of cis-stilbene by Noyce and his co-workers, only on the basis of solvent isotope measurements, may be rather uncertain.[20]

Scheme 2

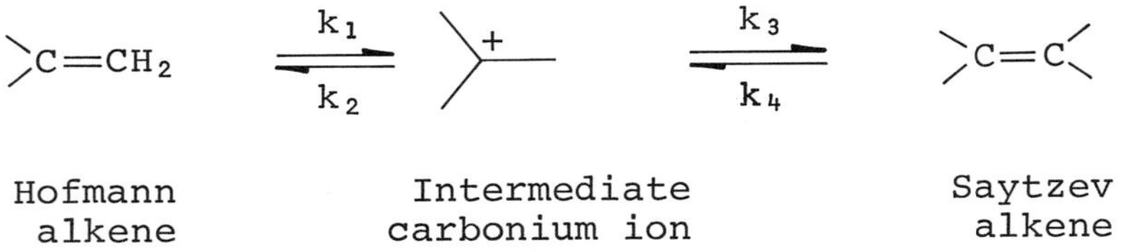

$$>C=CH_2 \quad \underset{k_2}{\overset{k_1}{\rightleftharpoons}} \quad >\!\!+\!\!\!\!\!- \quad \underset{k_4}{\overset{k_3}{\rightleftharpoons}} \quad >C=C<$$

| Hofmann alkene | Intermediate carbonium ion | Saytzev alkene |

$$\frac{-d\left[>C=CH_2\right]}{dt} = \frac{d\left[>C=C<\right]}{dt} = \frac{k_1 \ k_3 \left[>C=CH_2\right]}{k_2 + k_3} \quad (7)$$

3 Measurement of Acidity in Concentrated Acidic Solution

In concentrated acidic solution, the rates of acid catalyzed reactions such as dehydration or alkene isomerization, increase more rapidly than stoicheiometric acid concentration. For example, the dehydration of 1-p-anisyl-2-phenylethanol in 5% ethanol-water at 25°C proceeds 10^5 times faster with a sulfuric acid concentration of 55.4% than with one of 8.4%.[23] The increase in acid concentration is counterbalanced by a decrease in the water concentration and, consequently, as the acid concentration is increased, the extent of solvation of the proton decreases and not suprisingly its activity increases dramatically [e.g., $\overset{+}{H}(H_2O)_4 \longrightarrow$ $\overset{+}{H}(H_2O)_3 \longrightarrow \overset{+}{H}(H_2O)_2$ etc]. Thus, to explain the effect of acidity on rate or equilibria in concentrated acidic solution, it is necessary to have a measure of the activity and not just the simple stoicheiometric concentration of the acidic species.

Hammett and Deyrup were the first to attempt to construct an empirical measure of acidity in aqueous sulfuric acid solutions.[24] They measured the extent of ionization of a number of weak bases (Eqs. 8 to 10), and termed the expression

$$-\log \frac{a_{\overset{+}{H}}\ \gamma_A}{\gamma_{\overset{+}{AH}}}$$

as H_O, which equivalent to pH in dilute aqueous solution, when the activity coefficients approach unity. H_O was regarded as a measure of the ability of the medium to protonate an uncharged indicator and was assumed to be independent of the indicator used in the measurement. By making use of the relationship between the ionization ratios of successive indicators of similar basicity ($\Delta pK \leqslant 2$, ideally, Eq. 11), and by anchoring the scale in dilute aqueous solution, they established an acidity function using a number of nitro-anilines and carbonyl compounds (Figure 2).

$$A + \overset{+}{H} \rightleftharpoons \overset{+}{AH} \quad \text{(A is a weak base)} \tag{8}$$

$$K_a = \frac{[A]}{[\overset{+}{AH}]}\ \frac{a_{\overset{+}{H}}\ \gamma_A}{\gamma_{\overset{+}{AH}}} \tag{9}$$

$$H_O = -\log \frac{a_{\overset{+}{H}}\ \gamma_A}{\gamma_{\overset{+}{AH}}} = pK_a + \log\frac{[A]}{[\overset{+}{AH}]} \tag{10}$$

If γ_A/γ_{AH^+} varies in the same way as γ_C/γ_{CH^+} with changes in the reaction medium, Eq. 11 will hold (A and C are weak bases).

$$\Delta pK = pK_a - pK_c = \log \frac{[A]}{[AH^+]} - \log \frac{[C]}{[CH^+]} + \log \frac{\gamma_A \gamma_{CH^+}}{\gamma_{AH^+} \gamma_C}$$

$$= \log \frac{[A]}{[AH^+]} - \log \frac{[C]}{[CH^+]}$$

(11)

Subsequently, Jorgenson and Hartter[25] developed an acidity function for aqueous sulfuric acid using only nitroaniline indicators. They found their scale differed from that of Hammett and Deyrup in the region where carbonyl compounds had been used previously (Figure 2). This discrepancy revealed a weakness in the acidity function approach, namely that the function is not independent of the structure of the indicator. The variation was attributed to the tendency of the activity coefficients ratio γ_A/γ_{AH^+} changing in a specific manner for each type of weak base and its conjugate acid. For the most part, variations in the extent of solvation account for the different behavior of each kind of indicator. For example, anilinium ions, which possess a highly localized charge, are likely to bond with the solvent much more strongly than are highly delocalized carbonium ions.[26] The term H_O is now reserved for the acidity function based on the ionization of only nitroanilines.

Of particular relevance to the dehydration of alcohols are the acidity functions H_R and $H_{R'}$, generated from the protonation of carbinols[27] and alkenes,[28] respectively. The H_R function, derived by using triarylmethanols (Eqs. 12 and 13), rises much more rapidly than H_O with increasing acid concentration (Figure 2) and empirically should relate to H_O and the water activity if the activity coefficient term was indicator invariant (i.e., if γ_{ROH}/γ_R^+ varies as does γ_A/γ_{AH^+} with changes in the reaction medium). However, the failure to adhere to the relationship of Eq. 14 again illustrates the complication arising from the activity coefficient term. On the other hand, the term γ_R/γ_R^+ for the alkene equilibria varies in the same manner as γ_{ROH}/γ_R^+ as the acidity function $H_{R'}$ relates to H_R and $\log a_{H_2O}$.

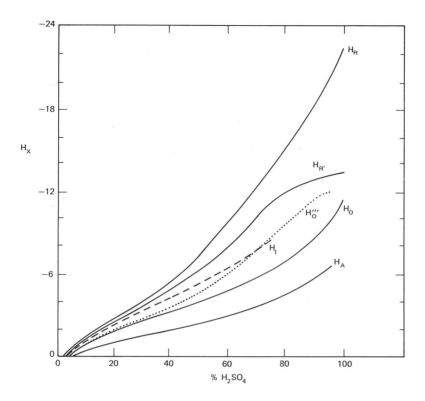

Figure 2   Acidity functions for aqueous sulfur acid
H_R (triarylmethanols); $H_{R'}$ (olefins); H'''_O (tertiary
anilines); H_I (indoles); H_O (nitroanilines); H_A (amides).

$$ROH \; + \; \overset{+}{H} \; \underset{\longleftarrow}{\overset{K_R}{\rightleftharpoons}} \; \overset{+}{R} \; + \; H_2O \qquad (12)$$

(ROH is a triarylmethanol)

$$H_R \; = \; -\log \; \frac{a_H^+ \, \gamma_{ROH}}{a_{H_2O} \, \gamma_{R^+}} \; = \; pK_R \; + \; \log \; \frac{[ROH]}{[R^+]} \quad (13)$$

$$H_R \; = \; H_O \; + \; \log a_{H_2O} \qquad (14)$$

Some other acidity functions derived from the protonation of amides,[29] indoles,[30] and tertiary anilines[31] are also shown in Figure 2. Clearly, the
apparent variation in medium acidity as defined by
these scales limits their application as a precise
criterion on which to base a reaction mechanism. However, there is no doubt that they all constitute a
better measure of medium acidity than simple acid concentration. Over a limited range, all the scales run
parallel to each other, so whichever acidity function
is related to rates of an acid catalyzed reaction, a
linear response is expected, but the magnitude of the
slope of the log k versus H_x plot should be treated
with caution.

4   Correlation of Rate and Medium Acidity

In aqueous sulfuric acid a linear relationship, with
unit slope, is observed for a plot of log k against H_O
for the following reactions: (a) the dehydration of
t-butyl alcohol and t-pentyl alcohol,[32] (b) the hydration of trimethylethylene,[32] (c) the isomerization of
cis-cinnamic acid,[33] and (d) the dehydration of 1-
methylcyclohexanol (in acetic acid-sulfuric acid-
water).[34] Slopes greater than unity are observed for
the following related reactions: (a) the dehydration
of 1,2-diphenylethanol (5% ethanol-water), slope
1.32,[17] (b) the racemization of (+) β-phenyl-β-hydroxy-
propionic acid, 1.17,[35] (c) the hydration of styrene,
1.26,[36] (d) the exchange reaction ($^{18}OH_2$) of t-butyl
alcohol, 1.20,[32] and (e) the isomerization of cis-
stilbene, 1.25.[20] Clearly, there is not an apparent
division in structure between the reactions giving the
unit slope correlations and those that do not. Most
of the substrates giving the unit slope correlation are
simple alkyl compounds, which presumably react via
carbonium ions in which the charge is poorly delocalized and interact strongly with the aqueous solvent
molecules. Thus the anilinium indicators used for the
H_O function are probably appropriate model compounds
for describing the behavior of the activity coefficients of these substrates and their charged transition states. On the other hand, the presence of an
α-phenyl substituent (e.g., 1,2-diphenylethanol) enables more effective delocalization of the positive
charge in the carbonium ion, and there is a lesser
tendency for the activated complex to interact with
the solvent molecules. Consequently, with a decrease
in water activity, the free energy of solvation suffers
less in the case of 1,2-diphenylethanol than for

t-butyl alcohol and the rate of dehydration rises more rapidly with increasing acidity. An allowance of one water molecule difference between the activated complex and the solvated anilinium ion appears to rectify the position as a plot of log k against $-(H_O + \log a_{H_2O})$ for the dehydration of 1,2-diphenylethanol has almost unit slope.[17] However, this approach is not recommended for estimating differences in solvation between transition states and model compounds. For example, for a series of substituted cis-stilbenes, the plots of log k (isomerization) against H_O give slopes varying from 0.88 to 1.27. Thus even the change in electron density at the heteroatom centers of the substituents, in passing from reactant into the transition state, can give rise to variations in activity coefficient behavior.

   A plausible explanation is that the rates of dehydration should follow the H_R function and not the H_O function. However, the triarylcarbinols used to establish the H_R scale yield carbonium ions in which the charge is much more effectively delocalized than in the activated complexes formed in the dehydration reactions outlined above. The proximity of a water molecule to the carbonium ion center, in the encumbered ions preferred for the dehydration of t-butyl alcohol, causes the activity coefficients to resemble those of the anilinium rather than the delocalized carbonium ions, and hence the closer adherence of rate change to H_O than H_R.

   5   The π-Complex Hypothesis

Although it has since been shown that slow proton transfers can follow H_O,[37] an earlier disbelief in this view led to the suggestion by Taft[38] of the formation of a π-complex between the olefin and the proton prior to the rate-determining formation of the carbonium ion (Scheme 3). It was suggested that π-complex formation was rate determining in the reverse process, although this is not required by the principle of microscopic reversibility. The intermediacy of a π-complex was also favored to explain the preferential formation of cis-2-butene over trans-2-butene in the dehydration of 2-butanol by sulfuric acid.[14]

   The concept of a π-complex has been challenged on theoretical grounds, and quantum mechanical calculations infer that such a complex should be less stable than the corresponding carbonium ion.[39] The lack of β-hydrogen exchange in the hydration of styrene-

Scheme 3

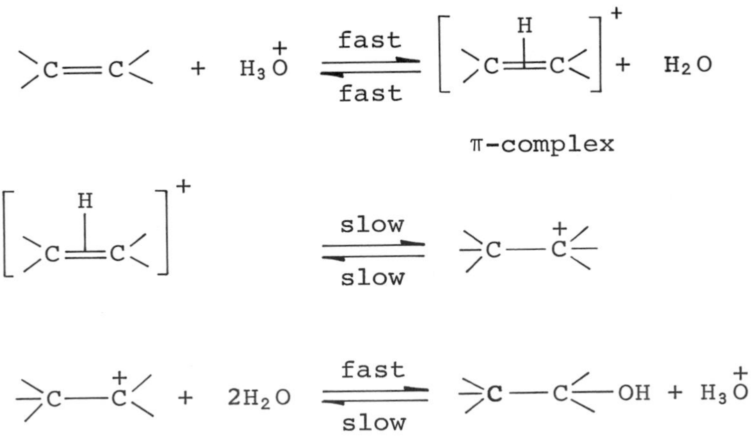

$$\text{>C=C<} + \text{H}_3\overset{+}{\text{O}} \underset{\text{fast}}{\overset{\text{fast}}{\rightleftharpoons}} \left[\overset{\overset{\text{H}}{|}}{\text{>C}\!=\!\text{C<}} \right]^{+} + \text{H}_2\text{O}$$

π-complex

$$\left[\overset{\overset{\text{H}}{|}}{\text{>C}\!=\!\text{C<}} \right]^{+} \underset{\text{slow}}{\overset{\text{slow}}{\rightleftharpoons}} \text{>C}\!-\!\overset{+}{\text{C}}\!<$$

$$\text{>C}\!-\!\overset{+}{\text{C}}\!< + 2\text{H}_2\text{O} \underset{\text{slow}}{\overset{\text{fast}}{\rightleftharpoons}} \text{>C}\!-\!\text{C}\!<\!\!-\text{OH} + \text{H}_3\overset{+}{\text{O}}$$

(the terms "slow" and "fast" refer to the rate of material transfer)

β,β-d_2,[36] and in the dehydration of 1,2-diphenyletha-
nol-1-d,[17] argues against the intermediacy of a π-com-
plex in which hydride rearrangement might be expected.
Noyce and his co-workers have presented evidence
against π-complexes using the Hammett equation, as we
see in the next section.[20,23]

6   Hammett Equation Correlations

Although the Hammett equation has been used with great
success to correlate the effect of meta- and para-
substituents on the rates of reaction of side-chain
derivatives of benzene (Section II.C.2.a),[40] severe
limitations arise when reactivities of benzene derivatives toward electrophiles and carbonium ion mechanisms
are considered. For these reactions, electron-releasing substituents such as methoxyl in the para-position,
cause much greater rate enhancements than are expected
from the magnitude of the substituent constant σ. In
the ionization of the substituted benzoic acids used
to define σ, the role of a substituent such as p-OMe is
confined to an inductive (or field effect) and a
resonance interaction with the π-electron cloud of the
benzene ring. However, for reactions in which an
electron-deficient center is formed, before or in the
rate-determining step, the possibility of resonance

interaction with an appropriate substituent must be considered. To cater for this circumstance, Brown and Okamoto defined a new substituent constant σ^+ from the rates of solvolysis of substituted t-cumyl chlorides in 90% aqueous acetone at 25°C (Eqs. 15 and 16).[41] The value of ρ^+ for this reaction was determined by using

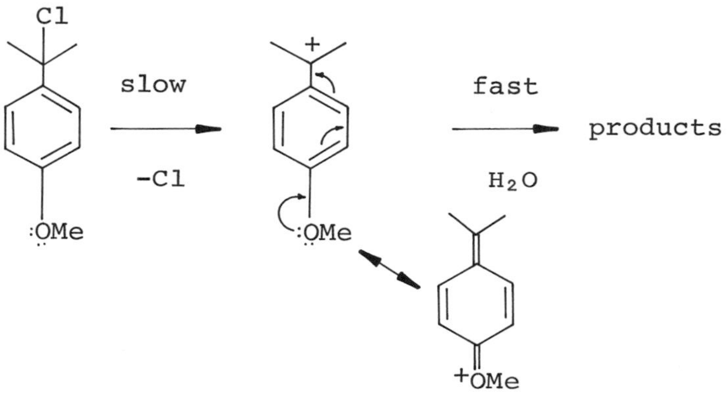

$$\log\ (k_X/k_H)\ =\ \rho^+\ \sigma^+ \tag{16}$$

data only for meta-substituents, which cannot participate in this direct resonance interaction. The σ^+ values only differ significantly from the σ values for those para-substituents which possess an electron-releasing resonance effect (e.g., OMe, OH, NH_2, NMe_2, Me).

Application of Eq. 16 to rate data necessitates the assumption that the degree of resonance interaction in the reaction concerned is the same as in the solvolysis of t-cumyl chloride. In some instances, this assumption may be in considerable error, and to allow for variable resonance interactions, Yukawa and Tsuno introduced the modified Eq. 17, which contains an additional constant, r (r = 1 for the solvolysis of t-cumyl chloride.).[42] Both Eqs. 16 and 17 have been used to explain substituent effects in dehydration and hydration reactions. The need to employ σ^+ constants in a rate correlation implies that a carbonium-ion intermediate or transition state with considerable carbonium-ion character is involved.

If before or in the rate-determining step, the two aryl groups of 1,2-diphenylethanol become equivalent,

$$\log (k_X/k_H) = \rho^+ [\sigma + r(\sigma^+ -\sigma)] \tag{17}$$

substituents in either aryl ring should exert a similar effect on the rate of dehydration. However, if non-equivalence is maintained until after the transition state, the effect of 1-aryl substituents (5) should be described by σ^+ constants and that of the 2-aryl substituents by σ constants. This latter circumstance is, in fact, observed, and the rate of dehydration by sulfuric acid in 5% ethanol-water at 25°C is described by Eq. 18.[23] The large negative ρ^+ value (-3.78) and the

$$\log k_{X,Y} = -3.78(\sigma_X^+ + 0.23 \sigma_Y) - 3.19 \tag{18}$$

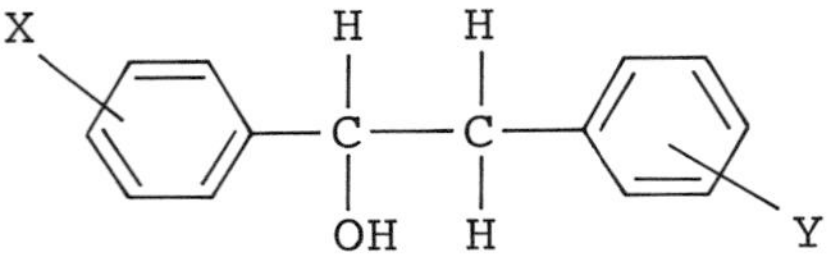

5 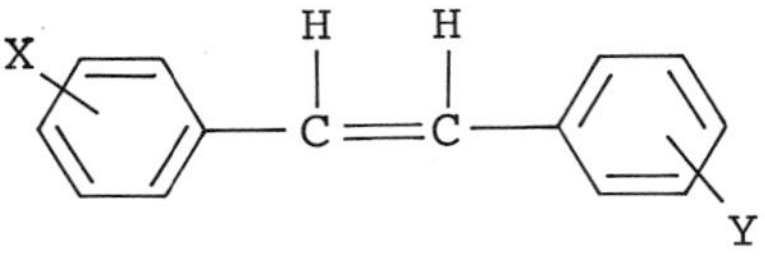   6

nonequivalence of the substituents X and Y accords with the intermediacy of a benzylic carbonium ion, being inconsistent with rate-determining π-complex formation. A similar conclusion is afforded by the effect of substituents on the rates of isomerization of substituted cis-stilbenes (6), the rates being described by Eq. 19 in which X is the greater electron-releasing substituent.[20] For symmetrical stilbenes, X = Y, correlation with the Hammett equation requires a statistical correction for two identical sites of protonation. The intermediacy of a π-complex again seems unlikely, as the symmetrically substituted stilbenes (e.g., X = Y = p-OMe) would be expected to show abnormally rapid isomerization rates.

$$\log k_{X,Y} = -4.46 - 3.30(\sigma_X^+ + 0.29 \sigma_Y)$$

$$\tag{19}$$

Kieboom and van Bekkum[22] have correlated the rates of dehydration, hydration, and isomerization of a number of substituted aryl derivatives with the Yukawa-Tsuno equation (Table 1). All these reactions involve benzylic cations as intermediates as evidenced by the large negative values of ρ^+. The resonance parameters r, vary quite markedly and are smaller for reactions of substrates carrying an additional α-alkyl substituent in which steric interactions prevent coplanarity between the carbonium ion and the aromatic π-system.

TABLE 1 Yukawa-Tsuno Parameters for Reactions Involving Benzylic Cations[22]

Reaction	Medium	ρ	r	Reference
Hydration of styrenes	3.83M HClO$_4$	-3.42	1.0	43
	6.3M H$_2$SO$_4$	-3.87	0.83	44
Isomerization of cis-cinnamic acids	20% H$_2$SO$_4$	-4.3	1.0	45
Dehydration of 1-aryl-2-phenylethanols	{ 5% EtOH-H$_2$O, 55% H$_2$SO$_4$	-4.3	0.91	23
Dehydration of 2-aryl-2-hydroxypropionic acids	60% H$_2$SO$_4$	-4.6	1.0	46
Hydration of 2-arylpropenes	20% H$_2$SO$_4$	-3.36	0.61	44
	20.2% H$_2$SO$_4$	-3.3	0.65	47
Isomerization of 2-aryl-3-methyl-1-butenes	{ 0.435M H$_2$SO$_4$, 99.7% AcOH	-3.81	0.48	22

7   General Reaction Scheme

In the preceding sections, the experimental evidence presented accords with the operation of a spectrum of transition states for acid catalyzed dehydration of aliphatic alcohols.[48]   These vary from the intermediacy of essentially free or encumbered carbonium ions for the elimination from tertiary alcohols (Scheme 4), to the symmetrically solvated carbonium ions which decompose in a concerted manner in the dehydration of primary alcohols.   The transition states for the reactions of secondary alcohols range between these extremes, substituents which lend stability to carbonium ions favoring a shift toward the pathway for tertiary alcohols.

Scheme 4

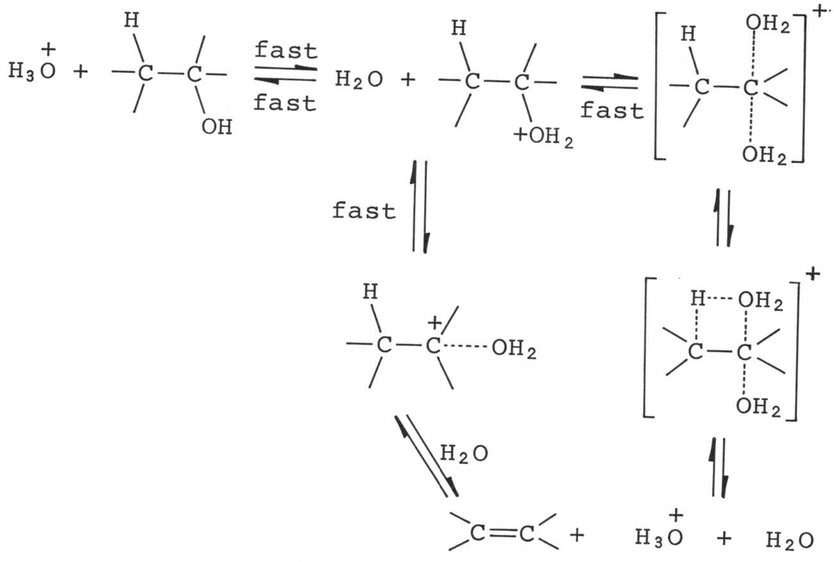

C   DEHYDRATION VIA ESTER INTERMEDIATES

1   General Comments

A number of acidic reagents effect dehydration of alcohols by converting the hydroxyl function into an ester, or a complex, thereby increasing the ionizing ability of the C-O bond. Possibly the best known example of an ester intermediate is ethyl hydrogen sulfate, which partitions to diethyl ether and ethylene in the dehydration of ethanol by sulfuric acid.[49] In this instance, ester formation is more rapid than its decomposition, and its intermediacy can be proved by isolation. However, for more highly substituted alcohols, ester formation may be a slower process than its decomposition to give a moderately stable carbonium ion. In such cases, the intermediacy of an ester cannot be established directly, but it may be inferred if the products of dehydration and decomposition of the ester, prepared by an independent route, are compared, and combined with the knowledge of the rate of esterification of an alcohol of similar structure, which cannot undergo dehydration.

2   Dehydration in Aprotic Solvents

The high affinity of some aprotic solvents for water enables them to act as good dehydrating agents. Tertiary alcohols, excluding t-butyl alcohol, and secondary benzylic alcohols dehydrate on heating in dimethyl sulfoxide (160 to 200°C) to alkenes.[50-52] The orientation of dehydration usually accords with the Saytzev rule, although for sterically encumbered alcohols, Hofmann orientation has been found for reactions under kinetic control.[22] Rearrangement products, characteristic of carbonium ion intermediates, are often found and 1,2-diols partially dehydrate by a pinacol rearrangement (Eq. 20).[51] Considerable ether formation can result if only a slight excess of dimethyl sulfoxide over the alcohol is employed.

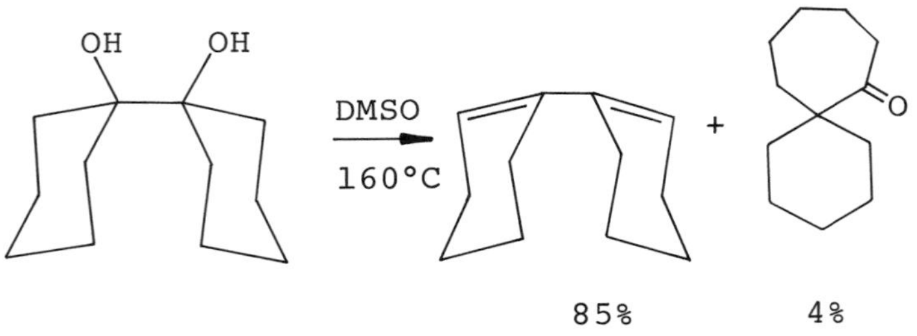

85%   4%

+ other products   (20)

Both threo- and erythro-1,2-diphenyl-1-propanol dehydrate in dimethyl sulfoxide to give an excess of cis-methylstilbene, but a different product distribution. These results exclude the involvement of a common intermediate for the two reactions, and elimination pathways modeled solely on anti-E2 elimination or syn-elimination. Traynelis and Hergenrother[50-52] preferred the intermediacy of a phenonium ion. Kieboom and van Bekkum proposed a cyclic mechanism with considerable carbonium-ion character (7).[22]

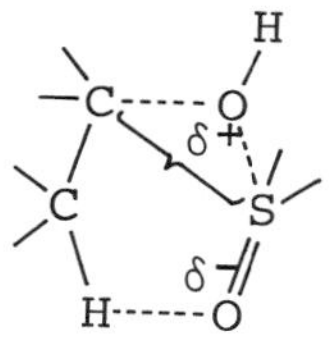

7 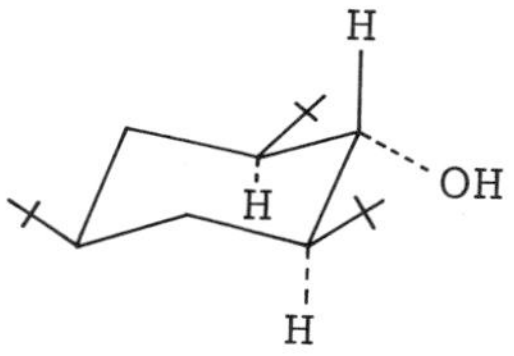   8 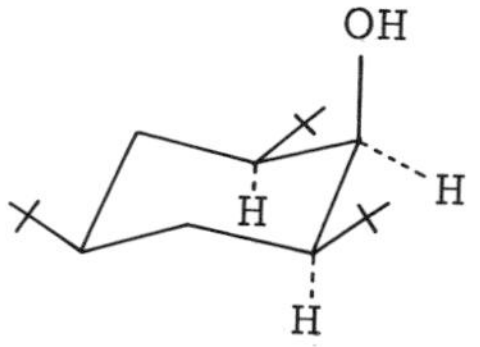   9

The Hofmann orientation in the dehydration of 2-aryl-3-methyl-2-butanols was attributed to eclipsing interactions being greater in the syn-periplanar transition state leading to the Saytzev olefin. The reaction constant of only -2.0 ± 0.5 was cited as too small for a mechanism involving a phenonium ion or a free carbonium ion. However, it should be pointed out that the reaction temperature of 189°C is much greater than that employed for typical carbonium-ion reactions in acidic solution (with which comparisons were made), and the reaction constant usually decreases with increasing temperature.[53]

The greater rate of dehydration from 8 than from 9 supports a cyclic mechanism.[54] Kieboom and van Bekkum also prefer a cyclic mechanism to account for the isomerization of the Hofmann to Saytzev alkene (Eq. 21).[22]

Recently, Monson has reported the use of hexamethylphosphortriamide (HMPT) for the dehydration of alcohols.[55] Slightly greater reaction temperatures (220 to 240°C) are required than for the dimethyl sulfoxide reactions, but simple aliphatic primary and secondary alcohols can be converted into alkenes. Despite reaction temperatures above the boiling point of the alcohols, only the secondary alcohols can be partially

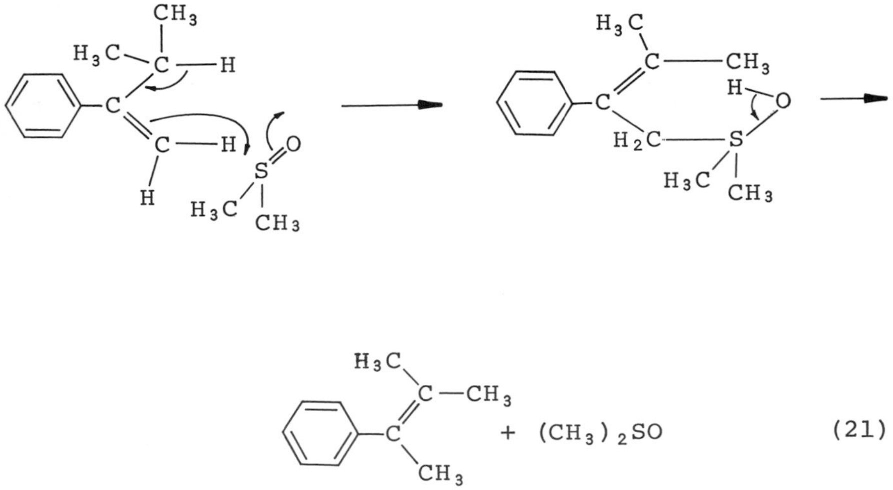

$$+ (CH_3)_2SO \tag{21}$$

distilled from the reaction mixtures. This suggests that primary alcohols are extensively complexed with HMPT, but steric factors reduce the complexing ability of the hydroxyl group in secondary alcohols. Primary alcohols usually decompose to give similar yields of the terminal alkene and 1-N,N-dimethylaminoalkane along with some dimethylamine. Secondary alcohols dehydrate more rapidly and do not give the amino by-products. The mechanism of dehydration, as outlined by Monson, for primary alcohols is shown in Scheme 5.

Scheme 5

$RCH_2CH_2OH + HMPT \rightleftharpoons RCH_2CH_2OH(HMPT)$

$RCH_2CH_2OH(HMPT) \longrightarrow RCH=CH_2 + \bar{O}H(HMPT) + HO\overset{+}{P}(NMe_2)_3$

$\bar{O}H + HO\overset{+}{P}(NMe_2)_3 \longrightarrow H_2O + HMPT$

$H_2O + HMPT \longrightarrow Me_2NH + (Me_2N)_2P(OH)=O$

$$RCH_2CH_2OH(HMPT) \ + \ Me_2NH_2 \longrightarrow$$

$$RCH_2CH_2 \overset{+}{N}HMe_2 \ + \ \bar{O}H(HMPT)$$

$$RCH_2CH_2 \overset{+}{N}HMe_2 \ + \ \bar{O}H \longrightarrow H_2O \ + \ RCH_2CH_2NMe_2$$

(1-octanol $\longrightarrow$ 1-octene(55%) + 1-N,N-dimethylamino-
octane(45%))[55]

3   Halogenating Reagents

A number of halogenating reagents have been used to effect dehydration, including sulfur tetrafluoride,[56]
phosphorous oxychloride-pyridine,[57] and thionyl chlo-
ride-pyridine.[58]   The latter two are well-known chlorinating reagents, and the elimination may occur directly
on the alcohol, via the ester or via an alkyl chloride,
by either unimolecular, base catalyzed, or thermal decompositions.   Thus, the reactions of alcohols with
these reagents offer a formidable mechanistic challenge.
     The esters are too unstable to allow isolation,
but the intermediacy of alkyl chlorides can be excluded
by comparing the products of their decomposition with
those of the alcohol.   Care should be taken with asymmetric and alicyclic alcohols in which the stereochemistry of the hydroxyl relative to the β-hydrogen is
fixed, as chlorination may occur with either retention,
inversion, or racemization.   For instance, although the
reaction of thionyl chloride, via primary alkyl chlorosulfites is a classic example of the S_Ni reaction,
which occurs with retention of configuration (Eq. 22),[59]
secondary chlorosulfites decompose into the corresponding chloride via a solvated carbonium ion, with nearly
complete retention in dioxane but inversion in toluene.[60]   Stille and Sonnenberg have shown that pyridine
encourages carbonium-ion intermediates, resulting in
considerable rearrangement in the chlorination of nor-
borneols.[61]

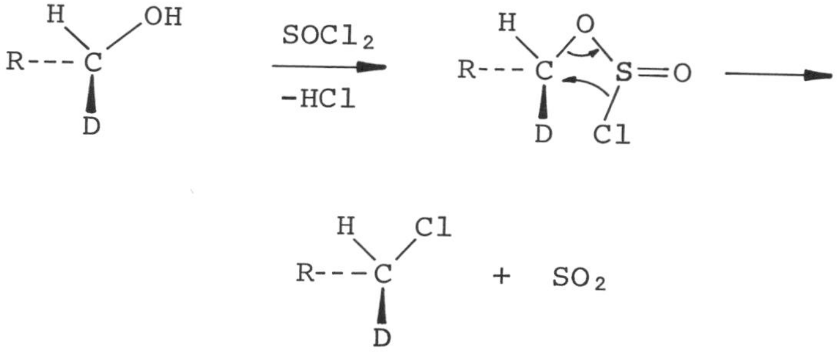

$$(22)$$

3β-Methyl-5α-cholestan-3α-ol (<u>10</u>) is dehydrated by POCl$_3$-pyridine to the Δ^2-endocyclic alkene (<u>11</u>), but the 3-β-ol (<u>12</u>) gives mainly the exocyclic alkene (<u>13</u>).[57] The change in orientation of elimination is explained in terms of an E2-elimination of an intermediate phosphoryl ester, which occurs with anti-stereospecifity. The selectivity of this dehydration has been used in stereochemical assignments of hydroxyl groups in other rigid molecules with success.[62] However, in less rigid systems, departure from such stringent stereochemical control has been noted. Both

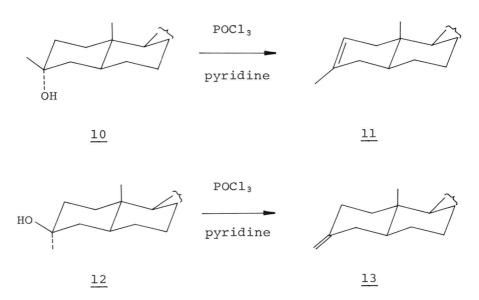

<u>10</u>	<u>11</u>
<u>12</u>	<u>13</u>

isomers of 1-methyl-4-t-butylcyclohexanol dehydrate to give mainly the endocyclic olefin, reflecting thermodynamic control.[63] The greater flexibility of the monocyclic compounds allows the required anti-orientation of the leaving groups to be attained, if elimination from cis-1-methyl-4-t-butylcyclohexanol occurs from a twist-boat conformation (<u>14</u>).

<u>14</u>

In the bridged bicyclic series, Sauers has shown
that both 15 and 16 dehydrate mainly to the more stable
exo alkene, but considering the lesser stability of the
endo product and the inability to attain coplanarity in
the transition state, the yield of the endo alkene is
unexpectedly high (Scheme 6).[64]   Steric inhibition, to
the approach of the methyl hydrogens, by the molecular
framework was suggested to explain the lower yield of
the exo alkene from 14 than from 15.   Bridged structures 17 and 18 are dehydrated by $POCl_3$-pyridine into
different proportions of α- and β-pinene and a number
of fenchyl compounds.[65]   The different product mixtures
exclude a common intermediate for the two reactions,
and the fenchyl products imply that carbonium-ion

Scheme 6

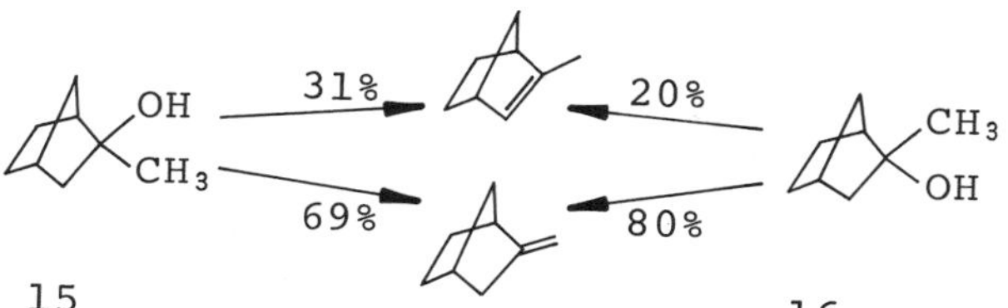

intermediates account for some of the dehydration reaction.   Thus, for compounds that are capable of forming
stabilized carbonium ions, the concept of an E2 reaction for the $POCl_3$-pyridine dehydration is an oversimplification.

Kirk and Shaw[66] have recently shown that the orientation of dehydration, even in rigid molecules, is
determined as much by the leaving group and the base
employed as by the conformation of the hydroxyl function.   Some of their results are listed in Table 2,
along with one of the corresponding dehydrohalogenations.   Clearly, the dehydrohalogenation products

differ markedly from those of dehydration. The authors assumed that the leaving groups were -OSOCl and -OPOCl$_2$ for the SOCl$_2$ and POCl$_3$ reactions, respectively. For the SOCl$_2$ reactions in the 3-yl series, the conformation of the hydroxyl has little effect on the orientation of elimination. In the 12-yl series, the yield of exo alkene is significantly larger when the hydroxyl is equatorial. For the POCl$_3$ reactions, in the 3-yl series, the results obtained earlier by Barton and his co-workers[56] were confirmed, but a change of base from pyridine to collidine markedly alters the product distribution. In the 12-yl series, the exo alkene is the major reaction product for both isomers, but the yield as expected is slightly greater for the equatorial epimer. Thus, in such cases, it is only possible to assign hydroxyl configuration with certainty, from orientation of elimination, if results of dehydration for both epimers are available.

To explain the above variations in product proportions, Kirk and Shaw invoke the variable transition state for conventional E2 reactions (Section II.A.2). They regard -OSOCl as a better leaving group than -OPOCl$_2$ and, consequently, predict a more E1-like transition state for its reactions and, hence, a greater tendency to exhibit Saytzev orientation (Section V.C.).

Previously used to prepare 1,1-di-t-butylethylene from alcohol 19 (Scheme 7, R = H),[58] SOCl$_2$-pyridine has recently proved useful in the synthesis of tri-t-butyl-ethylene from alcohol 19 (Scheme 7, R = t-butyl).[67] Other alkenes arising via Wagner-Meerwein 1,2-methyl shifts were also isolated, and the authors proposed initial formation of an alkyl chlorosulfite, which subsequently decomposed by an E1 process (Scheme 7).

TABLE 2 Olefinic Products from Dehydration of Some Alicyclic Alcohols[66]

Compound	Reagent	% exo-Alkene	% endo-Alkene
3-Methyl-5α-cholestan-3-yl derivatives			
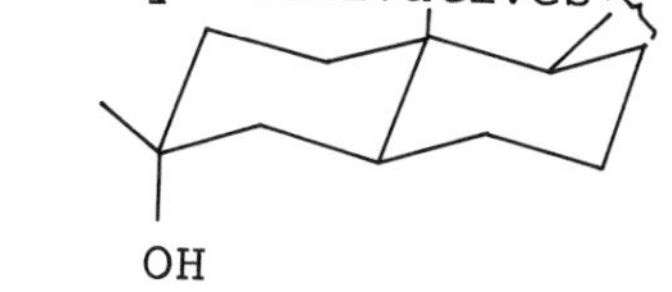 (OH)	SOCl$_2$-pyridine	<1	>99
	POCl$_3$-pyridine	1	99
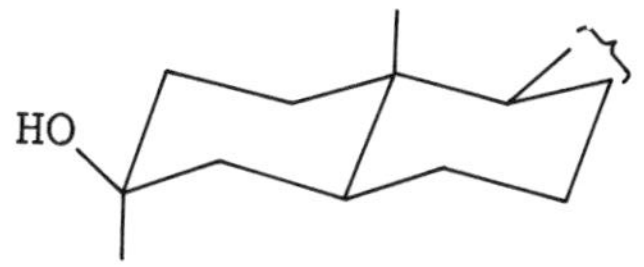 (HO)	SOCl$_2$-pyridine	7	93
	POCl$_3$-pyridine	52	48
	POCl$_3$-collidine	30	70
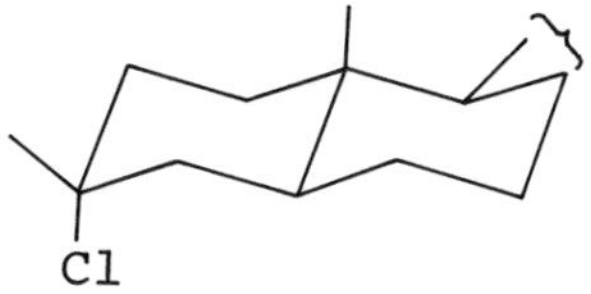 (Cl)	t-BuOH- t-BuOK	49	51
	collidine	46	54

TABLE 2 Continued

Compound	Reagent	% exo-Alkene	% endo-Alkene
(25R)-3β-Acetoxyl-12-methyl-5α-spirostan-12-yl derivatives			
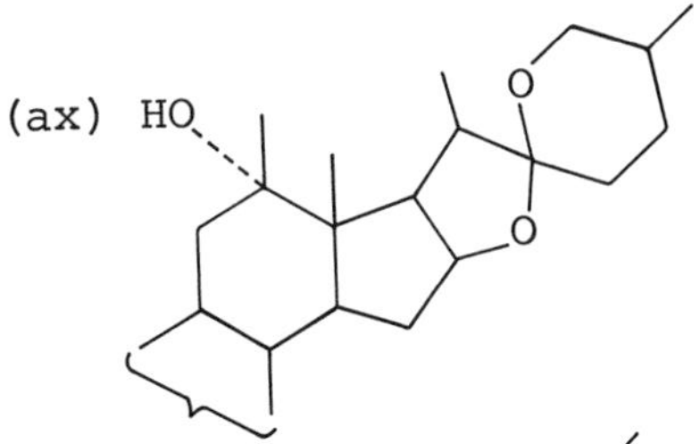 (ax)	SOCl₂-pyridine	36	64
	POCl₃-pyridine (16 hr at 25°C)	74	26
	(2 hr at 100°C)	60	40
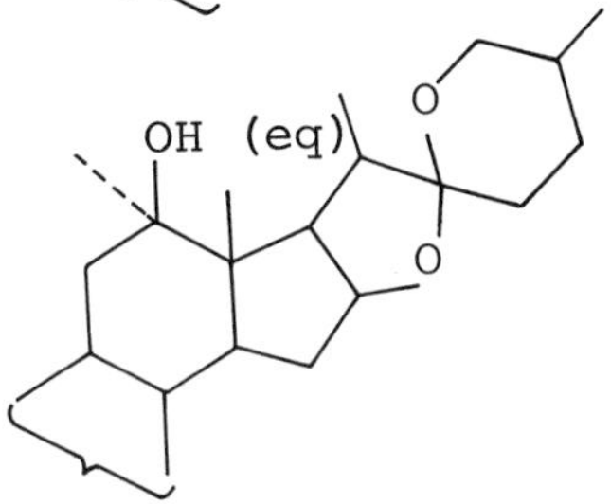 (eq)	SOCl₂-pyridine (30 min at 25°C)	88	12
	(2 hr at 25°C)	87	13
	POCl₃-pyridine (16 hr at 25°C)	95	5
	(24 hr at 25°C)	95	5

Scheme 7

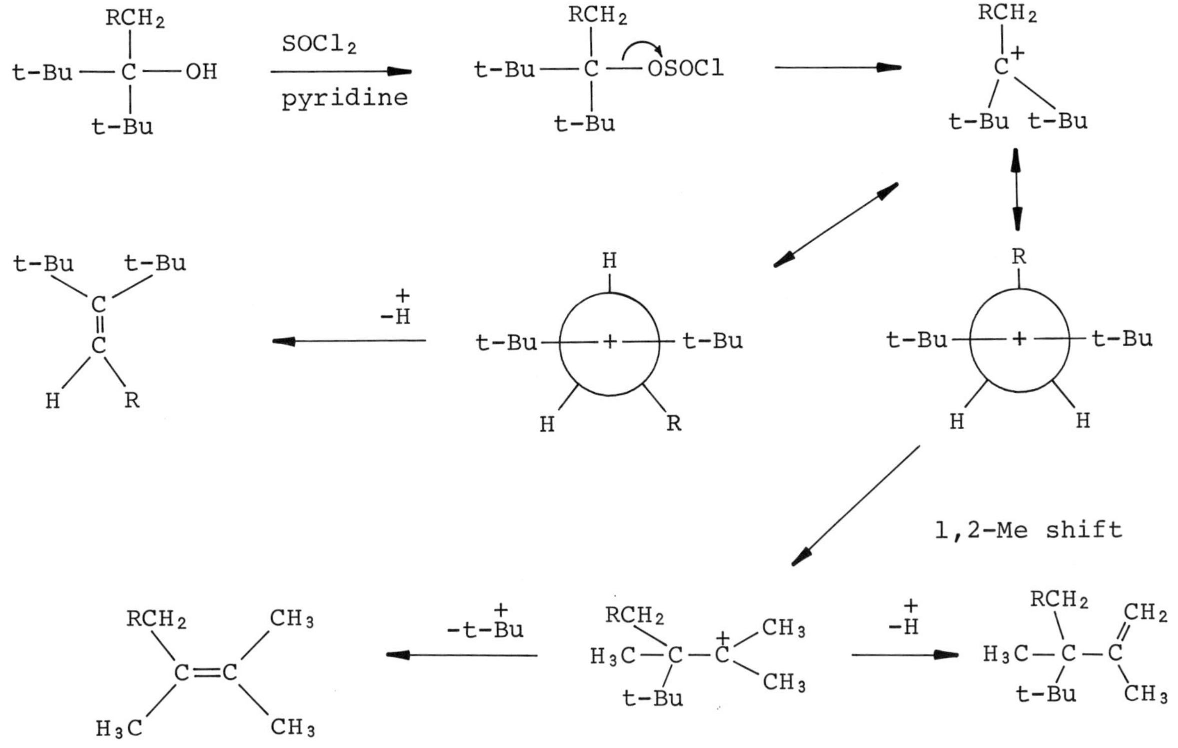

4 Miscellaneous Dehydrating Reagents

Dehydration by boric acid appears to involve pyrolysis of the boronic ester, which decomposes in situ at 350° C, a reaction temperature much lower than required for analogous acetate pyrolyses (Section VIII.C).[68] p-Toluenesulfonic acid and formic acid dehydrations are consistent with El reactions of initially formed esters.[69],[70] The inner salt, 20, induces dehydrations of secondary and tertiary alcohols in accordance with Saytzev orientation in a stereospecific syn elimination, presumably involving an ion-pair (Scheme 8).[71-73] Phenyl isocyanate,[74] iodine (the Hibbert reaction),[75] boron trifluoride,[76] P$_2$O$_5$,[77] and naphthalenesulfonic acid[78] have all been used successfully to effect dehydration on a preparative scale. Recent studies reveal that the Westphalen dehydration[79] using acetic anhydride and sulfuric acid (or KHSO$_4$) is an example of the El mechanism. The precursor to the carbonium ion is the acetyl

Scheme 8

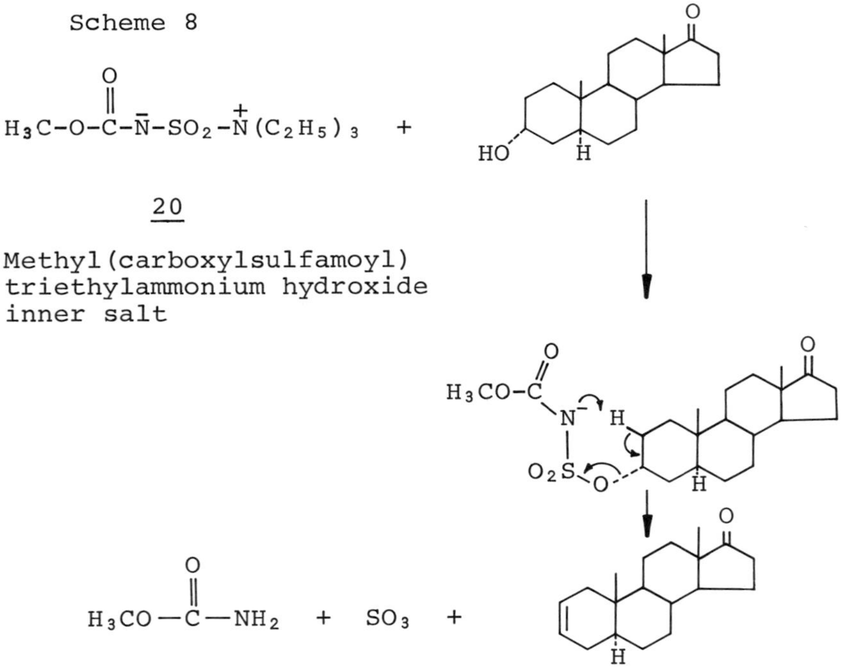

$$H_3C-O-\overset{\overset{\displaystyle O}{\|}}{C}-\bar{N}-SO_2-\overset{+}{N}(C_2H_5)_3 \quad +$$

20

Methyl(carboxylsulfamoyl)
triethylammonium hydroxide
inner salt

$$H_3CO-\overset{\overset{\displaystyle O}{\|}}{C}-NH_2 \quad + \quad SO_3 \quad +$$

sulfonic ester (Scheme 9).[80] The reaction is first order in acetic anhydride, and also first order in the alcohol and sulfuric acid, when the concentrations of these latter two reagents differ significantly from

each other. For changes in substituent Y, (F,Cl,OMe),
a Taft reaction constant, $\rho*$, of -4.8 is observed. It
is interesting that other acidic reagents ($HClO_4$, HCl,
p-toluenesulfonic acid, sulfoacetic acid, and hydro-
fluoroboric acid) catalyze only acetylation of the
alcohol,[80,81] to give the 5-acetate derivative, which
is stable to the Westphalen reaction conditions.[80]

Scheme 9

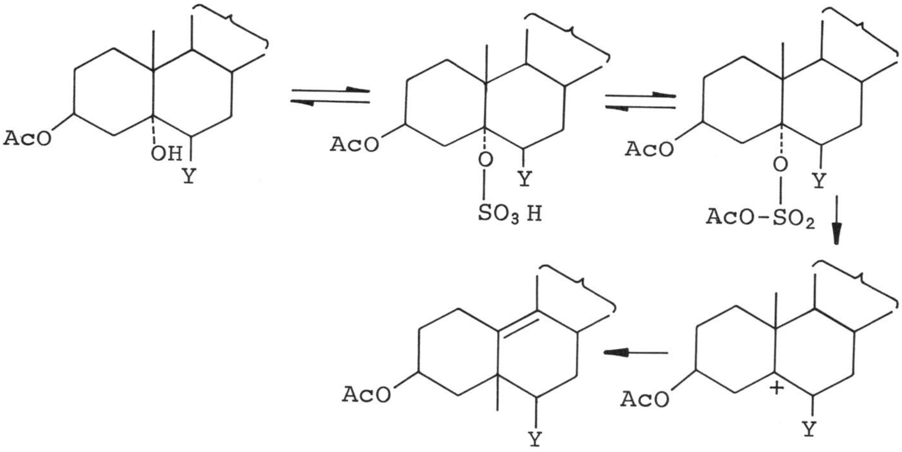

(Y = OAc, F, Cl, OMe, Br, I, $OCOCF_3$)

The importance of dehydration catalyst on determining the reaction pathway is amply illustrated in the
reactions depicted in Schemes 10 and 11. For the ter-
penol, phthalic anhydride dehydration reflects steric
and statistical control, and is presumably an example
of pyrolytic syn elimination of the initially formed
ester (Section VIII. C). These results were obtained
before the advent of gas-liquid chromatography, and
it is likely that mixtures and not single products are
obtained for each reaction.

For the steroid series, the authors proposed E1
reactions with BF_3, H_2SO_4/acetic anhydride and formic
acid, and concerted E2 reactions with $POCl_3$-pyridine
and $KHSO_4$/acetic anhydride.[76]

Scheme 10   Dehydration of Optically Active
             α-Terpineol[82]

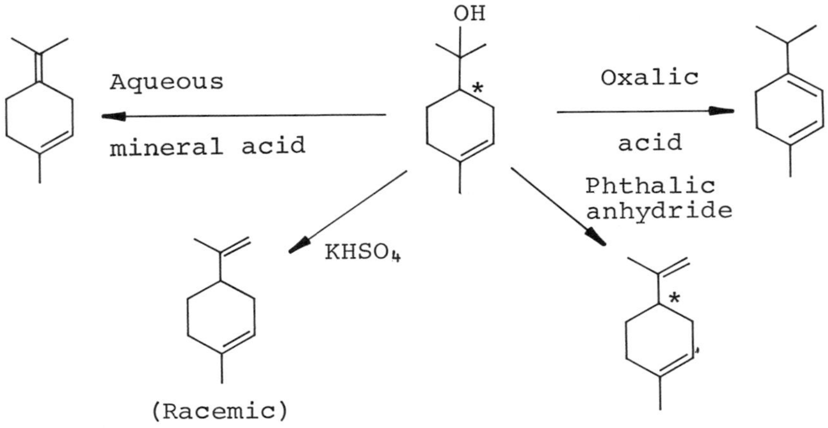

(Racemic)

Scheme 11   Dehydration of an Androstane Derivative
             with various Reagents[76]

(Major Reaction Products Only Indicated)

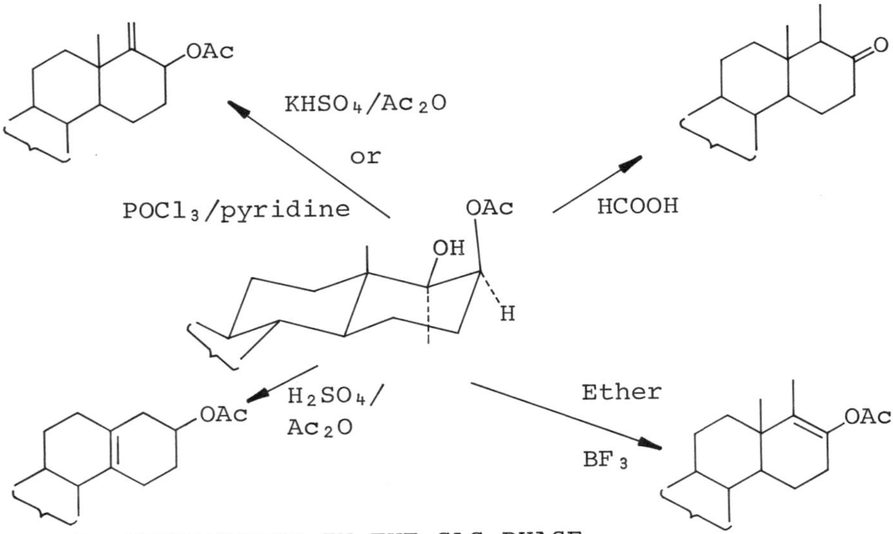

D   DEHYDRATION IN THE GAS PHASE

Simple primary and secondary alcohols such as ethanol,
n-propanol, n-butanol, and i-propanol undergo pyrolytic

decomposition above 500°C to give mainly dehydrogenation products.[83,84]   On the other hand, t-butyl and t-pentyl alcohols dehydrate quantitatively.[85]   In the temperature range 487 to 620°C the rate of dehydration of t-butyl alcohol follows first-order kinetics and is not affected by the addition of water or the radical inhibitor nitric oxide.[86]   The rate of reaction can be expressed as $k = 10^{11.51} e^{-54,500/RT}$ sec^{-1}.   These facts are consistent with the operation of a four-center transition state as depicted in Chapter VIII for alkyl halide pyrolysis, but the activation energy is 12-16 kcal/mole greater than those for the dehydrohalogenations (see Table 1, Section VIII.B.5.a).   Nitric oxide does inhibit the rate of dehydrogenation of the primary and secondary alcohols, and radical chain mechanisms were proposed for these reactions.[83,84]

Maccoll and Stimson and their co-workers[87,88] have investigated the gas phase pyrolysis of aliphatic alcohols in the presence of hydrogen halide catalysts. Unlike the situation for the simple pyrolysis, even isopropyl alcohol undergoes quantitative dehydration. The rate of dehydration follows first-order kinetics in both reactants and is not affected by added cyclohexene, a radical inhibitor.   The activation energies for the acid catalyzed pyrolysis are about 20 kcal/mole lower than for the simple dehydration and are, in fact, lower than those for dehydrohalogenation of the corresponding alkyl halides [e.g., for t-butyl alcohol, added HCl, T = 328 to 454°C, $k_2 = 2 \times 10^{-12} e^{-32,700/RT}$ cc/(mole)(sec)].   Thus the intermediacy of an alkyl halide can be excluded and two possible transition states were suggested: 21, a cyclic six-membered transition state, equivalent to that involved in ester pyrolysis (Section VIII.C), and 22, an ion pair.   For dehydration of t-butyl alcohol, at 320°C, the reactivity of the hydrogen halides is as follows; HCl:HBr:HI = 1:25:200.

21

22

E   DEHYDRATION OVER SOLID CATALYSTS

1   General Comments

By careful selection of the solid catalyst and the reaction temperature, alcohols can be dehydrated unimolecularly to an alkene, bimolecularly to an ether, or
can be dehydrogenated to a carbonyl compound. Proper
preparation of the solid catalyst and selective poisoning of its surface by modification of the porosity are
properties that can be varied to control the direction
of elimination and to minimize the subsequent isomerization of the alkene product. Many of the early workers in this field failed to realize the importance of
the physical state of the catalyst, and some of their
work must be regarded as uncertain in the light of
more recent developments and improvement in analytical
techniques.

2   Nature and Selectivity of the Catalysts

Metal oxides have been used most often as dehydration
catalysts, but sulfides, mineral salts, and ion-exchange resins have also been employed. Tertiary alcohols, which cannot be dehydrogenated without rearrangement, can be dehydrated over metal catalysts. Tabula-
tions of catalysts have been compiled by Sabatier,[89,90]
Winfield,[1] and Knozinger.[3] Pines and Manassen have
reviewed the properties of the various preparations of
alumina.[2] Here we confine our subsequent discussion
to metal oxides.
        The change in selectivity of the catalyst with
gross structure and mode of preparation affords valuable insight into the mechanism of dehydration, dehydrogenation, ether formation, and alkene isomerization.
Dehydrogenation is favored at the expense of dehydration in primary and secondary alcohols by methods that
increase the crystal size, decrease the surface area,
and reduce the catalyst porosity. For example, the
dehydrating activity of alumina is decreased when it
is heated to high temperatures, a process that causes[91]
healing of the irregularities in the crystal lattice.
This observation suggests that dehydrogenation is a
surface reaction but dehydration occurs within the
pores of the catalyst. Water, by forming hydrates with
the catalyst surface, can drastically reduce its dehydration activity, but some water or surface hydroxyls
are required for a catalyst to be active in dehydration, as freshly heated oxides always exhibit an induction period with respect to dehydration activity.[92]

The role of these surface hydroxyls is possibly as H-bridge donors or proton donors.

Szabo[93] noted that dehydrogenation activity increases with the ionic character of the metal-oxygen bond of the catalyst, whereas dehydration activity is enhanced by increased covalent bond character, thus the catalysts can be placed in the following selectivity sequence:

dehydrogenation ◄──────────────────────────

CaO MgO ZnO FeO Fe$_2$O$_3$ Cr$_2$O$_3$ TiO$_2$ Al$_2$O$_3$ SiO$_2$ WO$_3$

──────────────────────────► dehydration

The surfaces of metal oxides expose the following groups: (1) hydroxyl groups, (2) oxygen ions, and (3) incompletely coordinated metal ions, which can act as Lewis acid centers. To understand and control the selectivity of the various catalysts, it is essential to know the function each site plays in the reactions of alcohols and their decomposition products. Alumina has been studied most systematically, and we use its properties to demonstrate the approach.

Pines and Pillai suggested that both a basic and an acidic site are involved in dehydration.[94] Highly purified aluminas usually give rise to substantial alkene isomerization, a process that can be suppressed by selective poisoning of the Lewis acid sites by amines.[94,95] However, the dehydration activity of catalysts modified in this manner is little affected, suggesting that the Lewis acid sites are not required for dehydration, the surface hydroxyl groups acting as the acidic center. The presence of hydroxyl sites alone is insufficient, for Bayerite (aluminum trihydroxide), which exposes only hydroxyl groups at the surface, is devoid of dehydration activity.[96] It appears that the oxygen ions function as the basic site as dehydration activity of alumina is suppressed by poisoning with tetracyanoethylene.[97] Accepting the involvement of both basic and acidic sites in dehydration, one can envisage an elimination process as depicted by 23.

In the dehydration of ethanol over alumina, the rate of the bimolecular elimination, which produces diethyl ether, can be reduced by catalyst poisoning by pyridine.[96] It appears that Lewis acid sites are involved in ether formation, an intermediate surface alkoxide being alkylated by a second molecule of

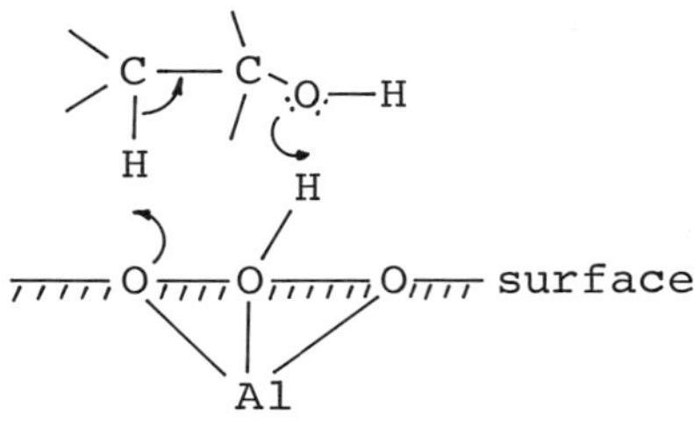

$$\underline{23}$$

ethanol (Eq. 23).[96,98,99] A similar displacement
mechanism, with a minor role for surface hydroxyl
groups in bonding the alcohol molecule, has also been
proposed (Eq. 24).[100] For dehydrogenation, a mechanism

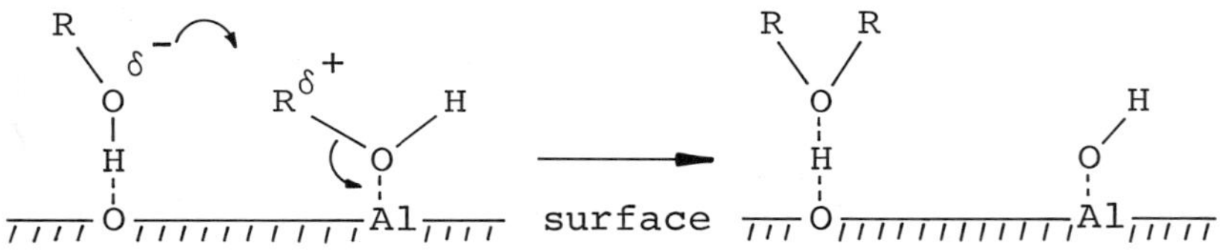

$$(23)$$

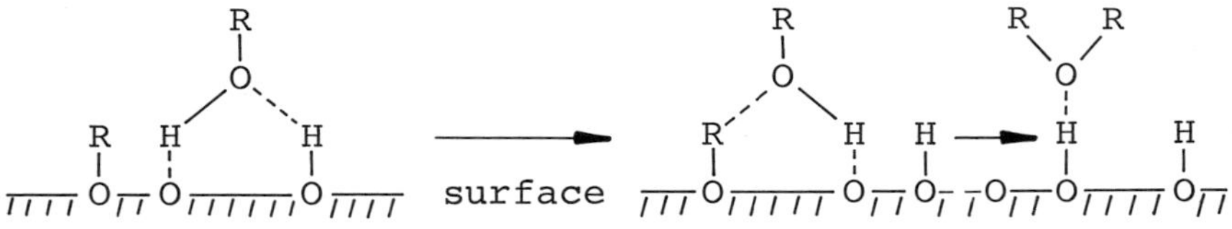

$$(24)$$

involving a hydride transfer to an incompletely coordinated metal atom is probable (24).[101] Thus, of the
four major reactions of alcohols over metal-oxide surfaces, only dehydration does not require the involvement of metal ions. For this reason, aluminas prepared
from sodium aluminate, which retain about 0.1% of their
sodium ions, are good dehydration catalysts, as they
possess a large number of weakly acidic sites but few
strongly acidic (Lewis acid) sites.[102]

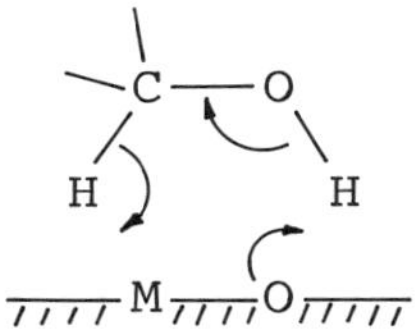

24

3 Measurement of Catalyst Acidity

The acidity of catalyst surfaces has been measured by titration against basic solvents,[103],[104] chemisorption of gaseous bases,[102],[104],[105] and by indicators that form colored complexes.[102],[105] Most of these methods suffer from not enabling a distinction between Lewis and Brönsted acidity, but thermometric titration with dioxane estimates only the former.[106] The number of hydroxyl groups have been determined by reaction with thionyl chloride.[107]

The "extent" and "depth" of isomerization of alkenes can provide information on the acidity of catalysts.[102] To induce arrangements of alkenes via primary carbonium ions, strong acid sites are required, but both strong and weak acid sites can cause rearrangement through secondary and tertiary carbonium ions. Cyclohexene and 3,3-dimethyl-1-butene (Scheme 12) were used as model alkenes.

For dehydration of i-propanol on various chromia-alumina catalysts, relationships between the activation parameters and the number of surface hydroxyl groups have been developed.[108]

Scheme 12

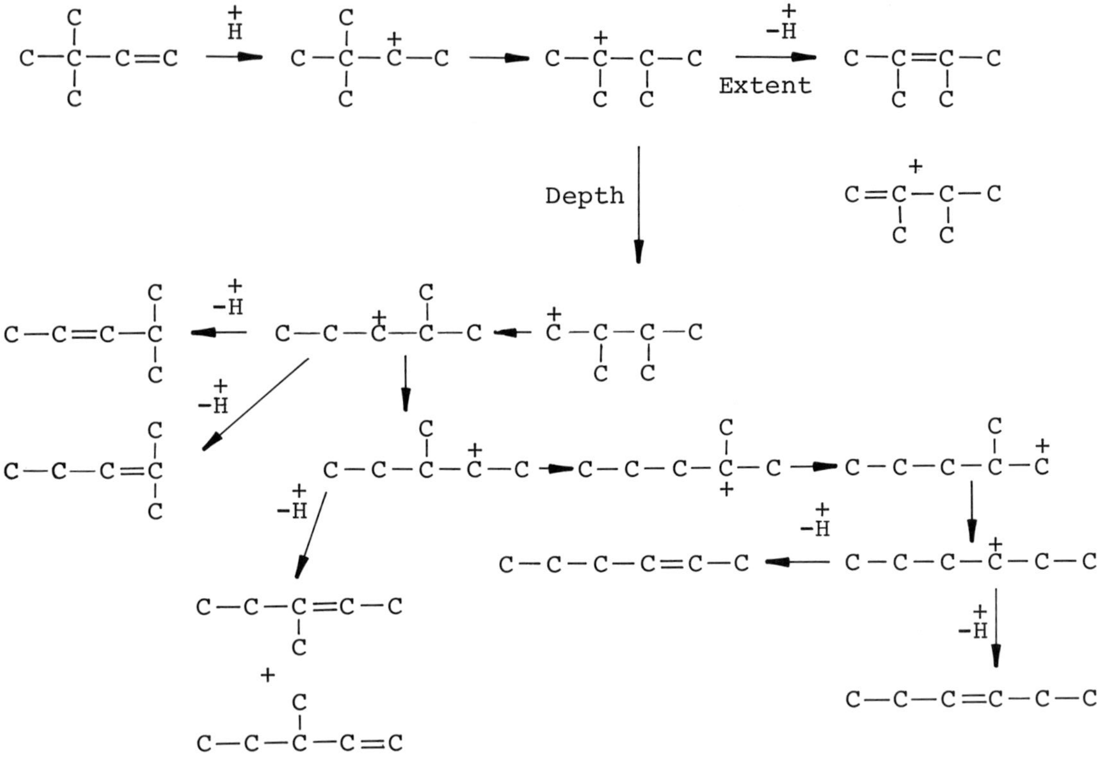

260

4 Relationship between the Dehydration Mechanisms

Contrary to the beliefs of earlier workers, who favored consecutive[109] and parallel schemes[110] for the dehydration of ethanol into diethyl ether and ethylene over alumina, Balaceanu and Jungers proposed a parallel-consecutive reaction sequence (Scheme 13),[111] which has since been verified by a kinetic analysis[112] and also using ^{14}C-labeled ethanol.[113] The same mechanism holds true for the dehydration of the normal aliphatic alcohols up to C_6 [114] and dodecanol.[115] If product

Scheme 13

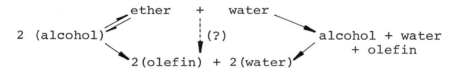

5 Elimination Mechanisms

analysis is to be used as a guide to mechanism, it is essential to distinguish between primary reaction products and secondary products, which arise by subsequent isomerization. Cyclohexanol dehydrates to cyclohexene over alumina containing 0.4% alkali metal ions, but gives large amounts of methylcyclopentenes when dehydrated over high purity alumina prepared from aluminum isopropoxide.[116] The cyclopentenes do not arise directly from cyclohexanol since extrapolation of the changing product proportions to zero reaction time reveals cyclohexene as the sole product. To avoid isomerization, one can employ catalyst poisoning using alkali metal ions, ammonia, or organic bases.

The dehydration of n-butanol over alumina gives mainly the β-elimination product, 1-butene (97.3%), but also small amounts of the 2-butenes of which the cis-isomer predominates by a factor of two. To account for the 2-enes, a γ-elimination with migration was invoked (Eq. 25).[117] Dehydration of 2-methyl-1-propanol over a variety of alumina catalysts gives a mixture of alkenes (Eq. 26).[118] Nonclassical, rather than classical intermediates were preferred (Eq. 27), since the least acidic alumina, which possessed the least dehydration activity, caused the greatest skeletal isomerization.

$$H_3C-\overset{H}{\underset{H}{C}}-\overset{H}{\underset{H}{C}}-\overset{H}{\underset{OH}{C}}-H \longrightarrow \diagdown\!/\!\diagup \;+\; \diagup\!\diagdown\!\diagup \tag{25}$$

$$\overset{}{\underset{B}{\diagup\!\diagup\!\diagup}\underset{}{\diagup\!\diagup\!\diagup\!\diagup\!\diagup\!\diagup}\underset{A}{\diagup\!\diagup\!\diagup}}$$

$$H_3C-\overset{H}{\underset{CH_3}{C}}-CH_2OH \xrightarrow[\Delta]{\text{alumina}} \searrow\!\!=\!\!\swarrow \;+\; \diagup\!\diagdown\!\diagup$$

$$(77 \text{ to } 88\%) \qquad (4 \text{ to } 10\%)$$

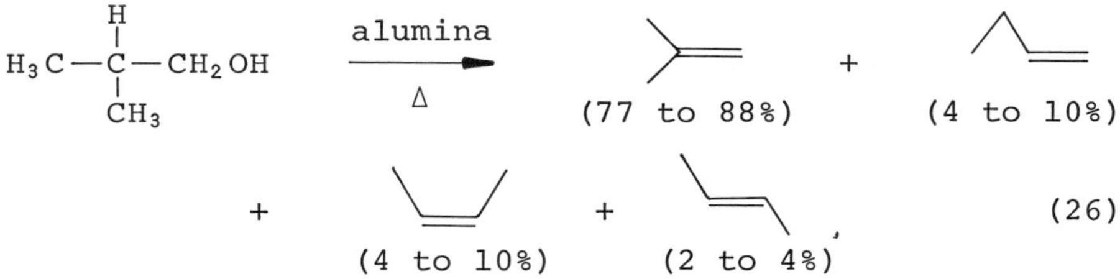

$$+\; \diagdown\!/\!\diagup \;+\; \diagup\!\diagdown\!\diagup \tag{26}$$

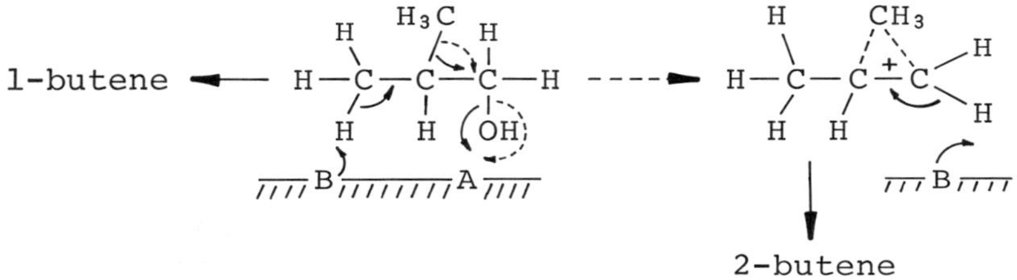

$$(4 \text{ to } 10\%) \qquad (2 \text{ to } 4\%)$$

1-butene

Scheme (27)

2-butene   (27)

The dehydration of 2-phenyl-1-propanol-1-^{14}C on alumina catalysts yields allylbenzene, α-methyl styrene and β-methyl styrene.[119] The allyl benzene is labeled only on the benzylic carbon and, therefore, arises by a γ-hydrogen abstraction accompanied by a concerted phenyl migration (Eq. 28). α-Methyl styrene is equally labeled on both the 1- and 3-carbon atoms, an observation consistent with either an intermediate cyclopropane or a tertiary carbonium ion (Scheme 14). From

$$H-\overset{H}{\underset{H}{C}}-\overset{H}{\underset{Ph}{C}}-\overset{H}{\underset{OH}{\overset{*}{C}}}-H \longrightarrow H_2C=CH-\overset{*}{C}H_2Ph \tag{28}$$

the label distribution in the β-methyl styrene, it can be concluded that phenyl migration occurs about eight times more readily than methyl migration. As is found for carbonium ion reactions in solution (e.g.,

Scheme 14

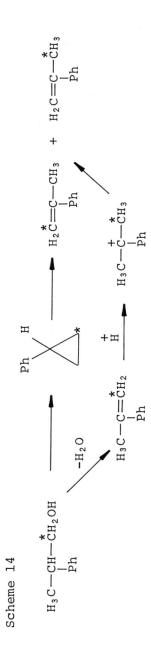

deamination, deoxidation, see Section VII.A.B), aryl
migration is enhanced by electron-releasing substituents. In the 2-arylethyl series, the p-tolyl group
migrates between 8.9 to 18%, while phenyl migration
ranges between 1 to 8.3% with the various aluminas.[120]
    Deuterium-labeling studies reveal that α-elimina-
tion is responsible for about 25% of the dehydration
reaction of CH_3CD_2OH over alumina (380 to 430°C),[121]
and for the formation of 2-butene from 2-deutero-2-
butanol over calcium sulfate.[122]

6 Stereochemistry and Orientation of β-Elimination

Unlike the pyrolytic eliminations in the gas phase
(Section VIII.A), dehydration reactions of secondary
alcohols over alumina usually occur with anti-stereospecificity. Thus menthol yields mainly 2-menthene, and
neomenthol eliminates mainly to 3-menthene (Scheme
15).[117] In both reactions, a small amount of 1-men-
thene, arising by a γ-hydrogen abstraction-β-hydrogen
migration mechanism, is detected (Eq. 29). The products of dehydration of cis- and trans-2-R-cyclohexanol
(R = alkyl,phenyl) differ from each and also accord
with anti elimination.[123] These facts accord with concerted rather than carbonium ion mechanisms. In the
acyclic series, the dehydrations of threo-3-methyl-2-
pentanol[124] and 3-deutero-dl-erythro-2-butanol (and its
threo isomer)[125] are examples of anti elimination.
    To explain the phenomenon of anti elimination,
Schwab and Schwab-Agallidis[91] proposed that dehydration
occurs within the pores of the catalyst, and Pines and
his co-workers have extended this view, assuming that
acidic and basic sites are located on opposite walls
of a cavity.[2,117] In fact, they regard the role of
the catalyst as a pseudo-solvent.
    The orientation of elimination in secondary alco-
hols varies with both the catalyst, and the substrate
and dehydration in favor of either the Hofmann or
Saytzev rules can be observed. Over alumina, the primary dehydration products of 2-butanol are mainly the
Saytzev olefins of which cis-2-butene predominates
over the trans-alkene by a factor of between four and
ten.[116,126] There is no parallel to this product distribution in pyrolytic or base catalyzed eliminations,
although the acid catalyzed dehydration in sulfuric
acid does give a slight preference of the cis-alkene.[14]
Cis-π-complexes are known to be more stable than trans-
π-complexes, and these were invoked to explain the unexpected predominance of the less stable isomer
(Scheme 16).[2,14,116] Preferential cis-alkene formation

Scheme 15

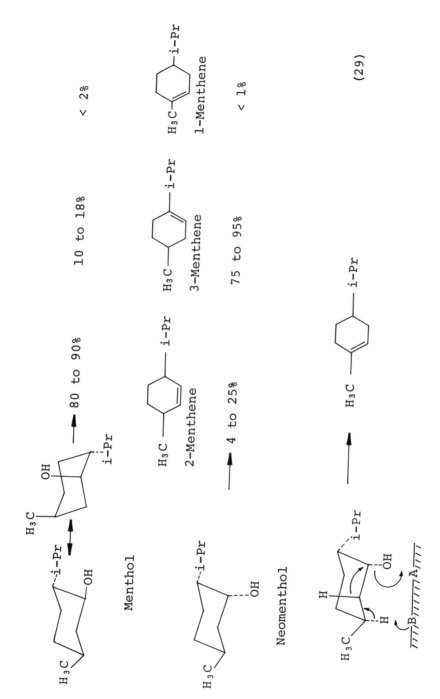

Menthol

Neomenthol

3-Menthene

2-Menthene

1-Menthene

80 to 90%

10 to 18%

< 2%

75 to 95%

4 to 25%

< 1%

(29)

265

is also found in the dehydration of 2- and 3-pent-anol[116],[127] and a number of other aliphatic alcohols.[4]

Scheme 16

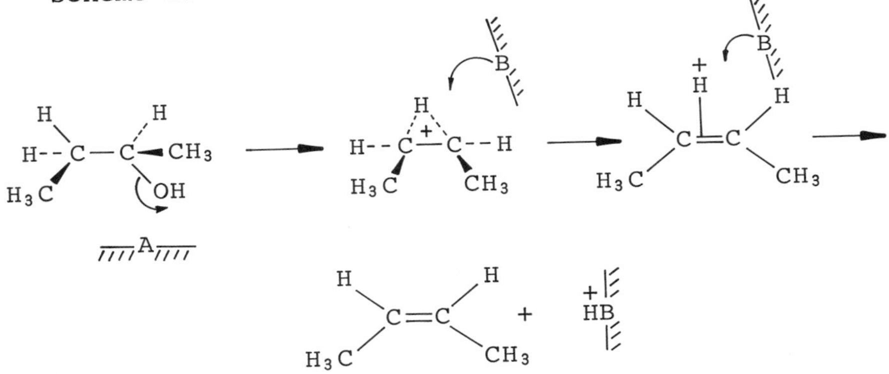

On thoria, the dehydration of threo-3-deutero-2-butanol occurs with anti stereospecificity,[128] but deu-terium analysis of cis- and trans-4-methyl-2-pentenes demonstrates that syn eliminations are involved in the reactions of threo- and erythro-2-methyl-4-deutero-3-pentanols. In striking contrast to the alumina cata-lyzed reactions, dehydrations of 2-hydroxy alkanes over thoria lead to high yields (> 93%) of the Hofmann prod-uct, the terminal alkene.[129] These results are con-sistent with a cyclic elimination with the alcohol bonded to the catalyst surface. Nonbonding interac-tions between the catalyst and adsorbed alcohol are less in the structure 24, required for 1-ene formation, than those leading to the 2-enes (25, 26).

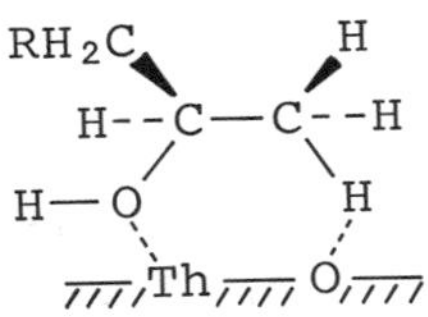

24 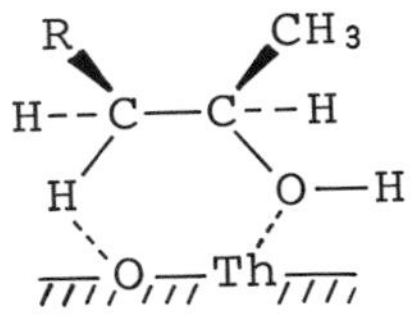   25 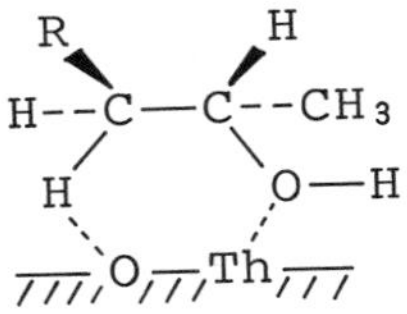   26

7 Energetics and Intermediates in β-Elimination

The activation energies for the dehydration of ethanol, i-propanol, t-butanol over bauxite decrease as follows; 31, 25.5, 20 kcal/mole.[130] These figures parallel those for dehydrohalogenation of the corresponding chlorides over lithium chloride (25, 20, 16 kcal/mole),[131] values which are much smaller than those for homogeneous gas phase pyrolysis (see Section VIII, Table 1) illustrating the pseudo-solvent effect of the catalyst. Both series of figures relate to the heterolytic bond dissociation energy, suggesting that C_α-OH bond breaking runs ahead of C_β-H bond cleavage in the transition state.

The rates of dehydration of a series of aliphatic alcohols [RCH(OH)CHMe$_2$, R = Me, Pr, i-Pr, 3-pentyl] correlate with the Taft equation, but the sign of the reaction constant, ρ^*, varies with the catalyst.[132] Negative reaction constants are found for silica and titania catalysts, indicative of carbonium-ion-like intermediates, whereas on alumina and zirconia, small positive reaction constants, suggestive of an E2-like dehydration, are observed. However, for the same catalysts, large negative, but virtually invariant reaction constants, ρ^+, were observed in the dehydration of substituted 1-arylethanols. In this case, the mechanism is determined by the substrate; the α-aryl group, by increasing the stability of the carbonium ion, encourages an E1-like intermediate.[133,134]

Significant deuterium isotope effects have been reported for the dehydration of t-butyl alcohol over alumina; $k(t-C_4H_9OH)/k(t-C_4D_9OH)$ = 3.35 at 120°C, 1.15 at 260°C.[135] At low reaction temperatures, the isotope effect accords with a rate-determining proton removal, possibly in a concerted E2-like mechanism. At higher temperatures, however, the small value reflects mainly the operation of a secondary isotope effect, and an E1-like mechanism is preferred. Primary isotope effects have also been observed in the alumina-catalyzed dehydration of s-butyl alcohol [$k(C_2H_5CH(OH)CH_3)/k(C_2D_5CH(OH)CH_3)$ = 3.95 at 120°C],[136] and i-butyl alcohol ($k(Me_2CHCH_2OH)/k(Me_2CDCH_2OH)$ = 1.7 at 200°C),[136] consistent with concerted eliminations.

The strict adherence to anti stereospecificity in the surface catalyzed dehydrations of some secondary alcohols infers the operation of concerted E2-like β-eliminations. However, the tendency toward Saytzev orientation and the observation of Wagner-Meerwein rearrangements suggests E1-like transition states and, in some instances, even carbonium ion intermediates are involved in the dehydration of some secondary and tertiary alcohols. Pines and Manassen,[2] with regard to alumina catalysts, have proposed that tertiary alcohols react via free carbonium ions, secondary alcohols react via intermediates possessing considerable carbonium-ion character, and primary alcohols react via concerted eliminations. The more recent kinetic results support these proposals and illustrate that the concept of the variable transition state, which has been well documented for eliminations in solution (see Section II), can be extended to reactions over solid catalysts.

8 Concluding Remarks

Solid catalysts add a vital dimension to the methods available to the synthetic chemist interested in converting alcohols into previously inaccessible alkenes. Factors such as catalyst nature and temperature should be carefully considered when contemplating a synthesis. Solid catalysts, of course, can be used to effect eliminations other than dehydrations, and Noller and his co-workers have recently outlined their value in dehydrohalogenation.[4]

REFERENCES

1. M. E. Winfield, in "Catalysis", Vol. VII, P. H. Emmett, Ed., Reinhold, New York, 1960, p. 93.
2. H. Pines and J. Manassen, Adv. Catal., 16, 49 (1966).
3. H. Knozinger, in "The Chemistry of the Hydroxyl Group," S. Patai, Ed., Interscience Publishers, London, 1971, p. 642.
4. H. Noller, P. Andréu, and M. Hunger, Angew. Chem. Int. Ed., (in English), 10, 172 (1971).
5. D. V. Banthorpe, "Elimination Reactions," Elsevier, Amsterdam, 1963, Chap. VI.
6. R. L. Taber and W. C. Champion, J. Chem. Educ., 44, 620 (1967).
7. C. K. Ingold, "Structure and Mechanism in Organic Chemistry," Cornell University Press, Ithaca, 1953, p. 216.
8. F. C. Whitmore and K. C. Laughlin, J. Am. Chem. Soc., 55, 3732 (1933).
9. H. J. Schaeffer and C. J. Collins, J. Am. Chem. Soc., 78, 124 (1956).
10. I. Dostrovsky and F. S. Klein, J. Chem. Soc., 791 (1955).
11. I. Dostrovsky and F. S. Klein, J. Chem. Soc., 4401 (1955).
12. E. Grunwald, A. Heller, and F. S. Klein, J. Chem. Soc., 2604 (1957).

13. C. A. Bunton and D. R. Llewellyn, J. Chem. Soc., 3402 (1957).
14. J. Manassen and F. S. Klein, J. Chem. Soc., 4203 (1960).
15. R. H. Boyd, R. W. Taft, Jr., A. P. Wolf, and D. R. Christman, J. Am. Chem. Soc., 82, 4729 (1960).
16. D. Bethell and V. Gold, "Carbonium Ions," Academic Press, London, 1967, p. 59.
17. D. S. Noyce, D. R. Hartter, and P. M. Pollock, J. Am. Chem. Soc., 90, 3791 (1968).
18. W. M. Schubert and B. Lamm, J. Am. Chem. Soc., 88, 120 (1966).
19. V. Gold and M. A. Kessick, J. Chem. Soc., 6718 (1965).
20. D. S. Noyce, D. R. Hartter, and F. B. Miles, J. Am. Chem. Soc., 90, 4633 (1968).
21. F. J. Adrian, J. Chem. Phys., 28, 608 (1958).
22. A. P. G. Kieboom and H. van Bekkum, Recl., 88, 1424 (1969).
23. D. S. Noyce, D. R. Hartter, and F. B. Miles, J. Am. Chem. Soc., 90, 3794 (1968).
24. L. P. Hammett and A. J. Deyrup, J. Am. Chem. Soc., 54, 2721 (1932).
25. M. J. Jorgenson and D. R. Hartter, J. Am. Chem. Soc., 85, 878 (1963).
26. N. C. Deno, in "Survey of Progress in Chemistry," Vol. I, A. F. Scott, Ed., Interscience, New York, 1963, p. 155.
27. N. C. Deno, J. J. Jaruzelski, and A. Schreisheim, J. Am. Chem. Soc., 77, 3044 (1965).
28. N. C. Deno, P. T. Groves and G. Saines, J. Am. Chem. Soc., 81, 5790 (1959).
29. K. Yates, J. B. Stevens and A. R. Katritzky, Can. J. Chem., 42, 1957 (1964).
30. R. L. Hinman and J. Lang, J. Am. Chem. Soc., 86, 3796 (1964).
31. E. M. Arnett and G. W. Mach, J. Am. Chem. Soc., 86, 2671 (1964).
32. R. H. Boyd, R. W. Taft, Jr., A. P. Wolf and D. R. Christman, J. Am. Chem. Soc., 82, 4729 (1960), and references theirin.
33. D. S. Noyce, H. S. Avarbock, and W. L. Reed, J. Am. Chem. Soc., 84, 1647 (1962).
34. J. Roček, Coll. Czech. Chem. Comm., 25, 375 (1960).
35. D. S. Noyce and C. A. Lane, J. Am. Chem. Soc., 84, 1635 (1962).
36. W. M. Schubert and B. Lamm, J. Am. Chem. Soc., 88, 120 (1966).
37. L. Melander and P. C. Myhre, Ark. Kem., 13, 507 (1959).

38. R. W. Taft, Jr., J. Am. Chem. Soc., 74, 5372 (1952).
39. I. I. Moiseev and Ju K. Syrkin, Dokl. Akad. Nauk. S.S.S.R., 115, 541 (1957).
40. (a) H. H. Jaffé, Chem. Rev., 53, 191 (1953).
    (b) L. P. Hammett, "Physical Organic Chemistry," McGraw-Hill, New York, 1940.
41. H. C. Brown and Y. Okamoto, J. Am. Chem. Soc., 79, 1913 (1957); 80, 4979 (1958).
42. (a) Y. Yukawa and Y. Tsuno, Bull. Chem. Soc. Jap., 32, 971 (1959).
    (b) Y. Yukawa, Y. Tsuno and M. Sawada, Bull. Chem. Soc. Jap., 39, 2274 (1966).
43. W. M. Schubert, B. Lamm, and J. R. Keeffe, J. Am. Chem. Soc., 86, 4727 (1964).
44. J. P. Durand, M. Davidson, M. Hellin, and F. Coussement, Bull. Soc. Chim. Fr., 33, 52 (1966).
45. D. S. Noyce and H. S. Avarbock, J. Am. Chem. Soc., 84, 1644 (1962).
46. D. S. Noyce, P. A. King, C. A. Lane and W. L. Reed, J. Am. Chem. Soc., 84, 1638 (1962).
47. N. C. Deno, F. A. Kish, and H. J. Peterson, J. Am. Chem. Soc., 87, 2157 (1965).
48. A. F. Cockerill, "Elimination Reactions," in "Comprehensive Chemical Kinetics," Vol. 9, Chap. 3, Elsevier, Amsterdam, 1973.
49. L. F. Fieser and M. Fieser, "Organic Chemistry," Reinhold, London, 3rd ed., 1956, p. 55.
50. V. J. Traynelis, W. L. Hergenrother, J. R. Livingston, and J. A. Valicenti, J. Org. Chem., 27, 2377 (1962).
51. V. J. Traynelis, W. L. Hergenrother, H. T. Hanson, and J. A. Valicenti, J. Org. Chem., 29, 123 (1964).
52. V. J. Traynelis and W. L. Hergenrother, J. Org. Chem., 29, 221 (1964).
53. H. H. Jaffé, Chem. Rev., 53, 191 (1953).
54. H. van Bekkum, H. M. A. Buurmans, G. van Minnen-Pathius, and B. M. Wepster, Recl., 88, 779 (1969).
55. R. S. Monson, Tetrahedron Lett., 567 (1971).
56. R. E. A. Dear, E. E. Gilbert, and J. J. Murray, Tetrahedron, 27, 3345 (1971).
57. D. H. R. Barton, A. da. S. Campos-Neves, and R. C. Cookson, J. Chem. Soc., 3500 (1956).
58. M. S. Newman, A. Arkell, and T. Fukunga, J. Am. Chem. Soc., 82, 2498 (1960).
59. W. A. Cowdrey, E. D. Hughes, C. K. Ingold, S. Masterman, and A. D. Scott, J. Chem. Soc., 1252 (1937).
60. E. S. Lewis and C. E. Boozer, J. Am. Chem. Soc., 74, 308 (1952); 75, 3182 (1953); 76, 794 (1954).

61. J. K. Stille and F. M. Sonnenberg, J. Am. Chem. Soc., 88, 4915 (1966).

62. (a) H. Heusser, N. Wahba, and F. Winternitz, Helv. Chim. Acta, 37, 1052 (1954).
    (b) E. J. Corey and R. R. Sauers, J. Am. Chem. Soc., 81, 1739 (1959).
    (c) J. M. Coxon, M. P. Hartshorn, and D. N. Kirk, Aust. J. Chem., 18, 759 (1965); Tetrahedron, 23, 3511 (1967).

63. B. Cross and G. H. Whitham, J. Chem. Soc., 3892 (1960).

64. R. R. Sauers, J. Am. Chem. Soc., 81, 4873 (1959).

65. R. R. Sauers and J. M. Landesberg, J. Org. Chem., 26, 964 (1961).

66. D. N. Kirk and P. M. Shaw, J. Chem. Soc., C, 182 (1970).

67. J. S. Lomas, D. S. Sagatys, and J. E. Dubois, Tetrahedron Lett., 599 (1971).

68. W. Brandenberg and A. Galat, J. Am. Chem. Soc., 72, 3275 (1950).

69. I. Ho and G. J. Smith, Tetrahedron, 26, 4277 (1970).

70. C. Quannes, M. Dvolaitzky, and J. Jaques, Bull. Soc. Chim. Fr., 776 (1964).

71. G. M. Atkins, Jr., and A. M. Burgess, J. Am. Chem. Soc., 90, 4744 (1968).

72. A. M. Burgess, E. A. Taylor, and H. P. Penton, Jr., 159th National Meeting of the American Chemical Society, Houston, Texas, February 1970, Abstracts No. ORGN-105.

73. P. Crabbé and C. León, J. Org. Chem., 35, 2594 (1970).

74. W. Oroshnik, G. Karmas, and A. D. Mebane, J. Am. Chem. Soc., 74, 295 (1952).

75. F. C. Whitmore and K. C. Laughlin, J. Am. Chem. Soc., 54, 4012 (1932).

76. C. Monneret, C. Tchernatinsky, and Q. Khuong-Hun, Bull. Soc. Chim. Fr., 1520 (1970).

77. J. P. Desvergne, R. Lapouyade, and H. Bouas-Laurent, C. R. Acad. Sci. Paris, 270, 642 (1970).

78. F. C. Whitmore and K. C. Laughlin, J. Am. Chem. Soc., 55, 3732 (1933).

79. T. Westphalen, Chem. Ber., 48, 1064 (1915).

80. J. W. Blunt, A. Fisher, M. P. Hartshorn, F. W. Jones, D. N. Kirk, and S. W. Young, Tetrahedron, 21, 1567 (1965).

81. (a) Z. Hattori, J. Pharm. Soc. Jap., 59, 129 (1939);
    (b) M. Davies and V. Petrow, J. Chem. Soc., 2536 (1949);

(c) A. T. Rowland and H. R. Nace, J. Am. Chem. Soc., 82, 2833 (1960).

82. J. L. Simonsen, "The Terpenes," Vol. 1, Cambridge, 1947, p. 256.

83. J. A. Barnard and H. W. D. Hughes, Trans. Faraday Soc., 56, 55 (1960); 56, 64 (1960).

84. J. A. Barnard, Trans. Faraday Soc., 56, 72 (1960); 53, 1423 (1957).

85. G. B. Kistiakowsky and R. F. Schultz, J. Am. Chem. Soc., 56, 395 (1934).

86. J. A. Barnard, Trans. Faraday Soc., 55, 947 (1959).

87. A. Maccoll and V. R. Stimson, Proc. Chem. Soc., 80 (1958); J. Chem. Soc., 2836 (1960).

88. V. R. Stimson, (and K. G. Lewis), J. Chem. Soc., 3087 (1960); (and R. A. Ross), J. Chem. Soc., 3090 (1960); (and E. J. Watson), J. Chem. Soc., 3920 (1960); 1392 (1961); (and F. L. Failes), J. Chem. Soc., 653 (1962).

89. P. Sabatier, "Die Katalyse in der Organischen Chemie," Akad. Verlagsgesellschaft, m.b.H., Leipzig, 1914.

90. P. H. Emmett, P. Sabatier, and E. E. Reid, "Catalysis, Then and Now," Franklin Pub. Co., Englewood, N. J., 1965.

91. G. M. Schwab and E. Schwab-Agallidis, J. Am. Chem. Soc., 71, 1806 (1949).

92. (a) L. A. Munro and W. R. Horn, Can. J. Res., 12, 707 (1935);
(b) A. Eucken and E. Wicke, Naturwiss., 32, 161 (1945);
(c) K. V. Topchieva, E. N. Rosolovskaya, and O. K. Sharaev, Vestn. Mosk. Univ. Ser. Khim., 14, 217 (1959).

93. Z. G. Szabó, J. Catal., 6, 458 (1966).

94. H. Pines and C. N. Pillai, J. Am. Chem. Soc., 82, 2401 (1960).

95. C. N. Pillai and H. Pines, J. Am. Chem. Soc., 83, 3274 (1961).

96. H. Knozinger, Angew. Chem. Int. Ed., (in English), 7, 791 (1968).

97. F. Figueras Roca, A. Nohl, L. de Mourges, and Y. Trambouze, C. R., 266, 1123 (1968).

98. (a) R. O. Kagel, J. Phys. Chem., 71, 844 (1967); (b) D. Treibmann and A. Simon, Ber. Bunsenges. Phys. Chem., 70, 526 (1966).

99. J. R. Jain and C. N. Pillai, J. Catal., 9, 322 (1967).

100. H. Knozinger, E. Ress, and H. Bühl, Naturwiss., 54, 516 (1967).

101. (a) A. Eucken, Naturwiss., 36, 48 (1949);
     (b) A. Eucken and K. Heuer, Z. Phys. Chem.
     (Leipzig), 196, 40 (1950);
     (c) E. Wicke, Z. Elektrochem., 53, 279 (1949).

102. H. Pines and W. O. Haag, J. Am. Chem. Soc., 82, 2471 (1960).

103. Y. Trambouze, C. R. Acad. Sci., 236, 1261 (1953).

104. A. N. Webb, Ind. Eng. Chem., 49, 261 (1957).

105. E. Eschigoya and T. Shiba, Bull. Tokyo Inst.
     Technol. Ser., B3, 133 (1960); Chem. Abstr., 55, 20577 (1961).

106. Y. Trambouze, M. Perrin and L. de Mourges, Adv.
     Catal., 9, 44 (1957).

107. H. P. Boehm and M. Schneider, Z. Anorg. Allg.
     Chem., 301, 326 (1959).

108. H. Bremer, F. Janiak, and H. Stach, Z. Chem., 4, 397, 466 (1964).

109. (a) R. N. Pease and C. C. Young, J. Am. Chem.
     Soc., 46, 390 (1924);
     (b) A. M. Alvarado, J. Am. Chem. Soc., 50, 790 (1928);
     (c) K. Kearby and S. Swann, Ind. Eng. Chem., 32, 1607 (1940).

110. (a) H. Adkins and B. N. Nissen, J. Am. Chem.
     Soc., 46, 130 (1924).
     (b) H. Adkins and P. P. Perkins, J. Am. Chem.
     Soc., 47, 1163 (1925);
     (c) H. Adkins and F. Bishoff, J. Am. Chem. Soc., 47, 810 (1925).

111. J. C. Balaceanu and J. C. Jungers, Bull. Soc.
     Chim. Belg., 60, 476 (1951).

112. H. J. Solomon, H. Bliss, and J. B. Butt, Ind.
     Eng. Chem. Fundam., 6, 325 (1967).

113. G. V. Isagulyants, A. A. Balandin, E. I. Popov,
     and Yn. I. Derbentsev, Zh. Fiz. Khim., 38, 20 (1964).

114. H. Knozinger and R. Köhne, J. Catal., 5, 264 (1966).

115. R. M. Langer and C. A. Walker, Ind. Eng. Chem., 46, 1299 (1954).

116. H. Pines and W. O. Haag, J. Am. Chem. Soc., 83, 2847 (1961).

117. H. Pines and C. N. Pillai, J. Am. Chem. Soc., 83, 3270 (1961).

118. J. Herling and H. Pines, Chem. Ind., 984 (1963).

119. J. Herling and H. Pines, J. Org. Chem., 31, 4088 (1966).

120. J. Herling, N. C. Sih, and H. Pines, J. Org.
     Chem., 31, 4085 (1966).

121. G. M. Schwab, O. Jenkner, and W. Leitenberger,
     Z. Elektrochem. Ber. Bunsenges. Phys. Chem.,

63, 461 (1959).

122.   P. Bautista, M. Hunger, and H. Noller, Angew. Chem., 80, 150 (1968).

123.   E. J. Blanc and H. Pines, J. Org. Chem., 33, 2035 (1968).

124.   K. Narayanan and C. N. Pillai, Ind. J. Chem., 7, 409 (1969).

125.   W. L. Hall, Diss. Abstr., B27, 754 (1966).

126.   H. Knozinger and H. Bühl, Z. Phys., Chem. (Frankfurt), 63, 199 (1969).

127.   L. Kh. Freidlin, V. Z. Sharf and Z. T. Tukhtamuradov, Neftekhim., 2, 730 (1962).

128.   A. J. Lundeen and W. R. van Hoozer, J. Org. Chem., 32, 3386 (1967).

129.   A. J. Lundeen and W. R. van Hoozer, J. Am. Chem. Soc., 85, 2180 (1963).

130.   H. Dohse, Z. Phys. Chem. (Bodenstein-Festband), 533 (1931).

131.   H. Noller and K. Ostermeier, Z. Elektrochem. Ber. Bunsenges. Phys. Chem., 60, 926 (1956).

132.   H. Kochloefl, M. Kraus, and V. Bažant, presented at the IVth Intern. Congr., Catalysis, Moscow, 1968; quoted in Ref. 3.

133.   M. Kraus and K. Kochloefl, Coll. Czech. Chem. Comm., 32, 2320 (1967).

134.   M. Kraus, unpublished observations, quoted in Ref. 3.

135.   H. Knozinger and A. Scheglila, Z. Phys. Chem. (Frankfurt), 63, 197 (1969).

136.   H. Knozinger and A. Scheglila, J. Catal., 17, 252 (1970).

VII MISCELLANEOUS 1,2-ELIMINATIONS GIVING ALKENES

A   DEAMINATION REACTIONS

It is not feasible to obtain alkenes directly from amines by a base catalyzed elimination as such a reaction would involve the formation of an amide ion, an exceptionally strong base. In an earlier chapter, (see Chapter I), the base catalyzed Hofmann elimination of an amine derivative is described. After successive alkylation of the nitrogen atom, the amine is converted into a quaternary ammonium salt in which the positively charged nitrogen not only activates the β-carbon-hydrogen bond but also acts as a good leaving group, forming the weak trialkylamine bases on reaction with moderate alkali. Nitrogen also possesses a capability of combining with a second nitrogen, leading subsequently to the formation of a stable nitrogen molecule. The reactions that fall into this category, are called deamination reactions.

The oldest,[1] and possibly the most commonly used deamination, involves the reaction of the amine with nitrous acid, generated from sodium nitrite and a mineral acid, in either water or acetic acid as the solvent. From the synthetic standpoint the reaction is of little use because of the multitude of products, which fall into three major types: elimination components, substitution, and rearrangement products (Scheme 1,[2]).

The nitrosoamide decomposition affords an alternative means of deamination. After conversion into an amide, nitrosation is performed with nitrogen peroxide[3] and the nitrosoamide is then thermally decomposed in a solvent (see Scheme 2).[4] For primary aliphatic amines, this reaction constitutes a useful synthesis of esters.[5] In nonpolar solvents the amount of skeletal rearrangement is small and generally much less than in the nitrous acid deamination.

Closely related to the nitrosoamide decomposition are the thermal decompositions of nitroamides,[6,7] nitroso and nitrocarbamates,[6,8] nitrosohydroxylamine derivatives,[5,6,9,10] and nitrososulfamides.[4]

Scheme 1

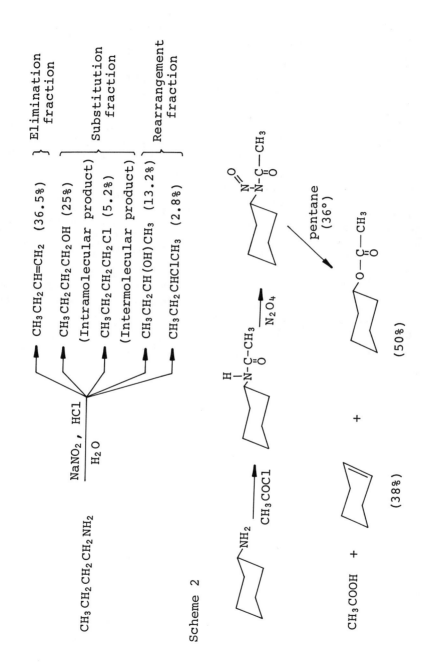

$CH_3CH_2CH_2CH_2NH_2$

$\xrightarrow[\substack{H_2O}]{\text{NaNO}_2, \text{HCl}}$

$CH_3CH_2CH=CH_2$ (36.5%) $\left.\right\}$ Elimination fraction

$CH_3CH_2CH_2CH_2OH$ (25%) (Intramolecular product)

$CH_3CH_2CH_2CH_2Cl$ (5.2%) (Intermolecular product) $\left.\right\rangle$ Substitution fraction

$CH_3CH_2CH(OH)CH_3$ (13.2%)

$CH_3CH_2CHClCH_3$ (2.8%) $\left.\right\}$ Rearrangement fraction

Scheme 2

277

The triazene reaction,[11,12] in which the aliphatic amine is initially treated with a diazonium salt and the resultant triazene is then decomposed under acidic conditions (see Scheme 3), provides a third mode of deamination.

The presence of products of electrophilic substitution (the absence of ethylbenzene among the reaction products even in the presence of thiophenol excludes a radical mechanism)[12] from the triazene reaction and the multitude of products in the other deamination processes suggests that carbonium ion intermediates are involved. Although of little utility in olefin synthesis, it is interesting to compare the reactivity of the intermediate carbonium ions that participate in the deamination reactions with carbonium ions formed during solvolysis of alkyl halides and esters or acid catalyzed dehydration of alcohols. Any mechanism we formulate must be able to accomodate formation of all and not just the elimination fragments, in which we are primarily interested.

On recent occasions, the mechanistic features of deamination reactions have been the subject of a number of reviews.[13-16] An excellent account of the mechanism of the nitrosoamide decomposition has also been recently compiled by White and Woodcock.[17] Consequently, in this chapter, we concentrate on the salient mechanistic features rather than attempt to present a comprehensive survey.

Scheme 3

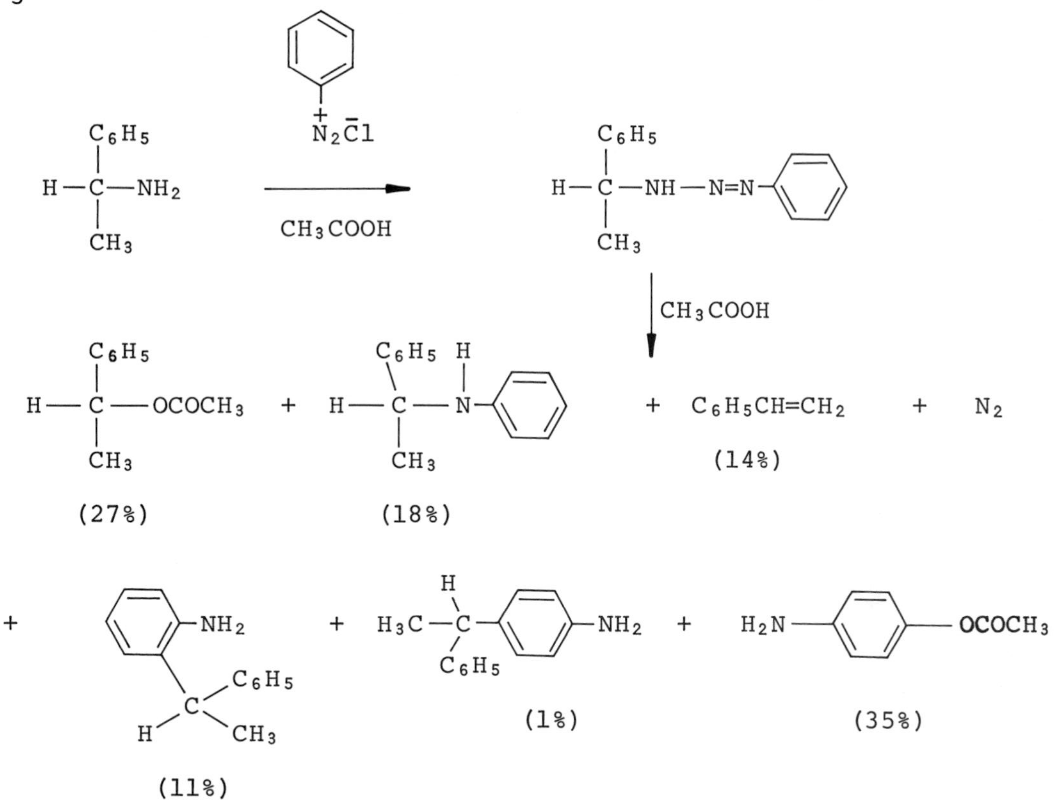

$$C_6H_5-\underset{\underset{CH_3}{|}}{\overset{\overset{H}{|}}{C}}-NH_2 \quad \xrightarrow[CH_3COOH]{C_6H_5-\overset{+}{N_2}\bar{C}l} \quad C_6H_5-\underset{\underset{CH_3}{|}}{\overset{\overset{H}{|}}{C}}-NH-N=N-C_6H_5 \quad \xrightarrow{CH_3COOH}$$

$$C_6H_5-\underset{\underset{CH_3}{|}}{\overset{\overset{H}{|}}{C}}-OCOCH_3 \;+\; C_6H_5-\underset{\underset{CH_3}{|}}{\overset{\overset{H}{|}}{C}}-\underset{\underset{C_6H_5}{|}}{\overset{\overset{H}{|}}{N}} \;+\; C_6H_5CH=CH_2 \;+\; N_2$$

$$(27\%) \qquad\qquad (18\%) \qquad\qquad (14\%)$$

$$+ \quad \text{(2-)} \underset{C_6H_5\ CH_3}{H}\text{C}_6H_4\text{-}NH_2 \;+\; H_3C-\underset{\underset{C_6H_5}{|}}{\overset{\overset{H}{|}}{C}}\text{-}C_6H_4\text{-}NH_2 \;+\; H_2N\text{-}C_6H_4\text{-}OCOCH_3$$

$$(11\%) \qquad\qquad\qquad (1\%) \qquad\qquad\qquad (35\%)$$

1   The Deamination of Amines with Nitrous Acid

The treatment of amines with nitrous acid results in a
rate-determining nitrosation,[18] followed by diazotiza-
tion and subsequent deamination.[13]   As the structure
of the amine is varied, the stage to which the reaction
proceeds, as typified by Scheme 4, is modified.   In
fact, rationale for the reaction scheme is provided by
characterization of the various stable entities of the
different amines.   Tertiary amines generally do not react, but have on very few occasions been reported to
undergo a cleavage reaction, via an ylide-intermediate,
to a secondary amine and an aldehyde.[19]   For secondary
amines, the reaction stops at the nitrosoamine because
of the inability of this latter species to tautomerize.

Scheme 4

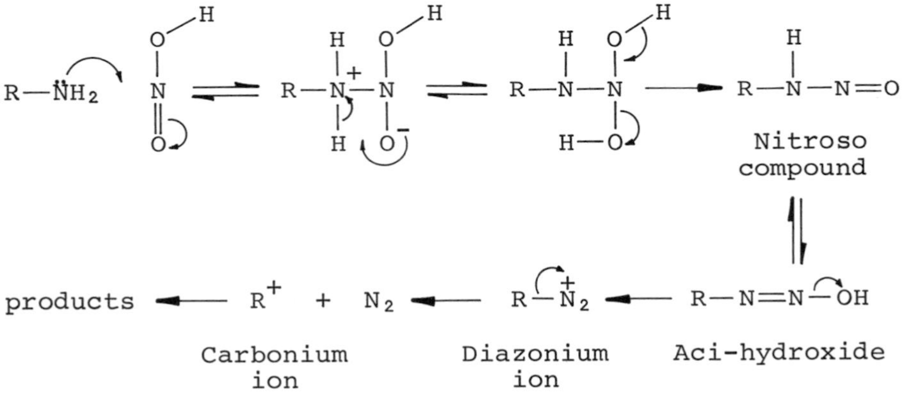

Fast reaction for aromatic amines effectively ceases
at the diazonium ion.   However, the mesomeric stabilization present in aromatic diazonium ions is lacking in
the primary aliphatic amines, and it seems reasonable
to expect a rapid loss of a nitrogen molecule to give
a carbonium ion which can give rise to varying products
depending on the substrate structure and the reaction
conditions.   As nitrosation is the rate-determining
step, evidence for carbonium-ion intermediates has to
come from analysis of the nature and stereochemistry of
the reaction products rather than from kinetic studies.

(a)   Outline of Reaction Products

   (i)   Similarities to other solvolytic reactions.
A number of features in the nitrous acid deamination

have their counterparts in typical solvolytic reactions of alkyl halides and esters. Alcohols or esters of retained configuration are obtained on deamination of α-amino-acids or solvolysis of α-carboxylsulfonates or halides.[20],[21] In both series of reactions the dominating factor is the neighboring group participation of the carboxylate group, which exerts an internal nucleophilic displacement on the leaving group, thus excluding attack by the solvent from the rear side. Transannular 1,5- and 1,6-shifts (see Scheme 5) have been observed in deaminations of eight- to twelve-membered alicyclic amines and also in the solvolyses of the corresponding p-toluenesulfonates.[22],[23] Similar extents of racemized substitution product have been reported in the deamination and solvolysis of the p-nitrobenzenesulfonate of the 1-deutero-1-butyl X structure.[24] The similar proportions of products from different reactions indicates that similar intermediates may be involved.

Scheme 5

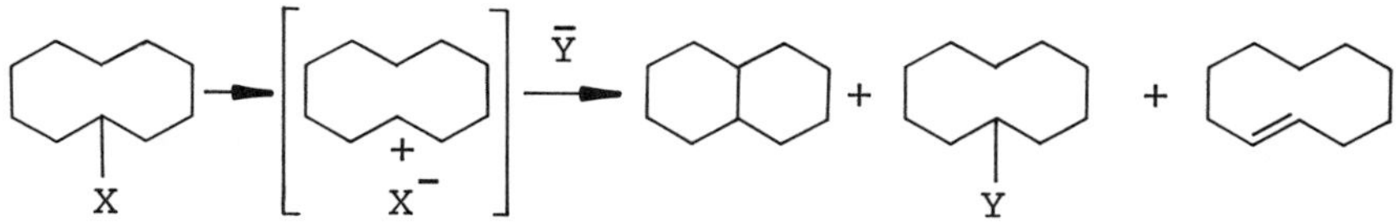

(ii) <u>Contrasting differences with other solvolytic reactions</u>. The main difference between deamination reactions and solvolytic carbonium ion processes is reflected in the greater competitive nature of unimolecular elimination and in particular rearrangement. In solvolytic reactions, rearrangement is only encountered in the presence of nonnucleophilic reagents such as Friedel-Crafts catalysts[25] or under conditions of "limiting" solvolysis in a poorly ionizing solvent of low nucleophilicity such as acetic acid.[26] Under these latter conditions, ionization of the C-X bond is aided by migration of an alkyl or aryl group or a hydride transfer from the adjacent β-carbon. Relative to the straight chain simple primary alkyl halides, deamination of the corresponding amines with nitrous acid gives enhanced yields of alkenes and additional rearranged products. Rearrangement generally decreases with increasing chain length; for example, percentage of rearranged alcohol product - 58(n-propylamine), 34(n-butyl), and 5(n-octyl).[27-29] The olefin mixture obtained in the aqueous deamination of s-butylamine consists of 25% 1-butene, 19% cis-2-butene, and 56%

trans-2-butene,[24] whereas for the acetolysis of the corresponding p-toluenesulfonate, the relative yields are 10, 43, and 46%, respectively.[30] The acetolysis results are explicable in terms of electromeric factors favoring stabilization of the carbonium ion, giving rise to Saytzev orientation (Section V.C), and the two hydrogens which eliminate to give the 2-butene becoming essentially equivalent in a planar intermediate.[24] The olefin proportions for the deamination reaction cannot be explained in these terms.

Isopropyl acetate is formed in only 2.8% yield in the acetolysis of n-propyl tosylate but is formed in 32% yield in the deamination of the corresponding amine in the same solvent.[14] The migratory aptitudes in the alcoholic products of the deaminative and tosylate acetolysis of the 3-phenyl-2-butyl system are quite distinctive (see Table 1).[31] In the solvolytic reaction, the predominant migration is by the phenyl group, with a small amount of hydride shift but no methyl migration. However, in the deamination of the threo-isomer, the methyl group migrates 1.5 times more readily than phenyl with which hydride migration is competitive. In the erythro-amine, phenyl migration exceeds that of hydride by four times and that of methyl by a factor of eight. The alcoholic products were isolated by reducing the crude reaction mixture with lithium aluminium hydride. This procedure results in pooling all the initially formed alcohol with acetate, nitrate, and nitrite esters, and is thus partly open to criticism. However, there is no doubt about the characteristic differences in the migratory aptitudes between the two types of reaction.

By using radioactive-^{14}C as a label, Burr and Cieresko have compared the relative rates of migration of phenyl to p-anisyl in the acid catalyzed alcohol rearrangement[32] to the deamination reaction.[33] When water is the leaving group (Scheme 6, X = OH_2^+), the migration ratio (anisyl/phenyl) is 24, but for the deamination reaction it is only 1.4. Similar findings arise in pinacol studies. The anisyl/phenyl migration

TABLE 1[31]   Alcoholic Products of the Deaminative and Tosylate Acetolysis of the 3-Phenyl-2-Butyl Systems[a,b]

| | 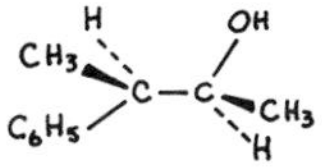 Yield[b] threo, [% | Yield[b] erythro, % | 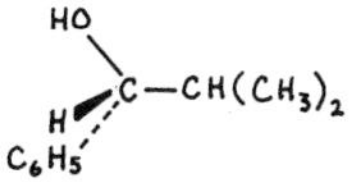 Yield,[b] % | 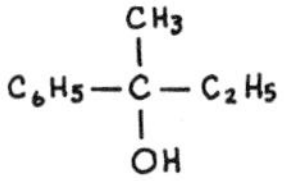 Yield Alcohol Fraction,[c] % |
	Active Racemic	Active Racemic	Active Racemic	%
$\overset{\text{H}}{\underset{\text{C}_6\text{H}_5}{\text{CH}_3-\text{C}}}-\overset{\text{NH}_2}{\underset{\text{H}}{\text{C}-\text{CH}_3}}$ $\xrightarrow[\text{2,LiAlH}_4]{\text{1,HNO}_2\ \text{AcOH}}$	6 19	14 5	5 27	24 50
[L(+)-threo-I] 1,AcOH $\xrightarrow{\text{2,LiAlH}_4}$ L(+)-threo-3-phenyl-2-butyl tosylate	0.5 85	1 2.5	0 0	11 70

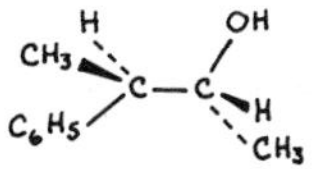

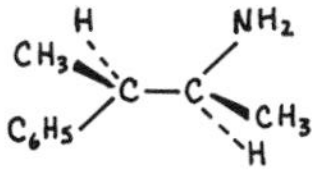

283

TABLE 1[31] Continued

| | 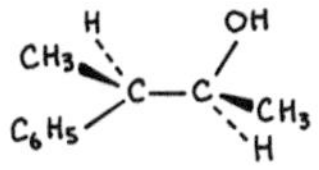 Yield[b] threo, % | | 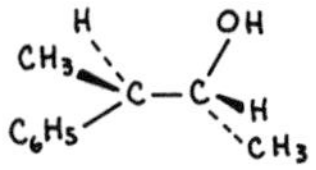 Yield[b] erythro, % | | 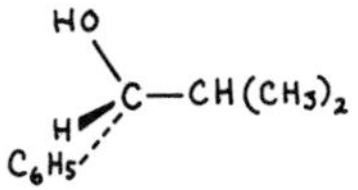 Yield,[b] % | | 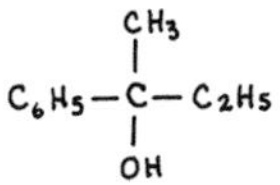 Yield,[b] % | Yield Alcohol Fraction, % |
	CH₃—C—CH₃ / C₆H₅ / H (OH) structure		CH₃—C / C₆H₅ / H / OH—C—CH₃ structure		HO—C—CH(CH₃)₂ / H / C₆H₅ structure		CH₃—C—C₂H₅ / C₆H₅ / OH structure	
	Active	Racemic	Active	Racemic	Active	Racemic	%	%
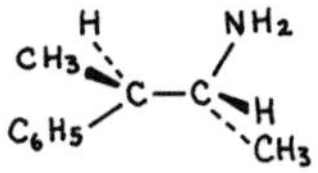 [L(+)-erythro-I] 1,HNO₂ AcOH → 2,LiAlH₄	6	0	68	0	0.2	6	20	50
L(+)-erythro-3-Phenyl-2-butyl tosylate 1,AcOH → 2,LiAlH₄	5	0	89	0	0	0	6	74

[a]The deaminative and tosylate acetolysis were both run at room temperature. However,

TABLE I[31]   Continued

a complete analysis of the secondary alcohol components of tosylate acetolysis was made only on product obtained from runs at 75°. Enough data were obtained at both temperatures to indicate the patterns of results to be almost identical. The ratios of the various secondary alcohols to one another were taken from the runs at 75°.

bTotal alcohol fraction = 100%.

cStarting amine or tosylate = 100%.

Scheme 6

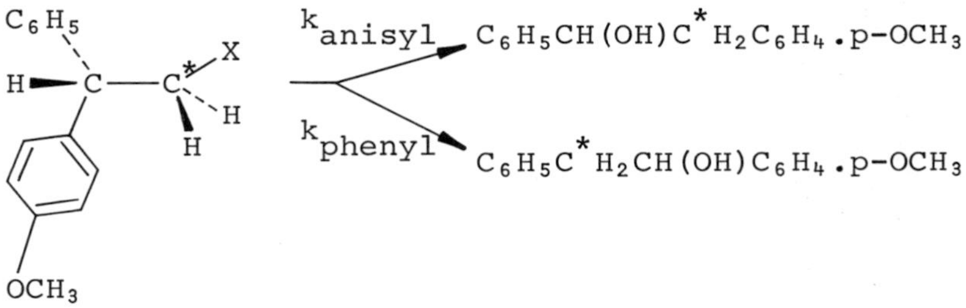

ratio is 500 in the pinacol rearrangement of compound 1,[34] but only 1.56 in the pinacolic deamination of substrate 2.[35]

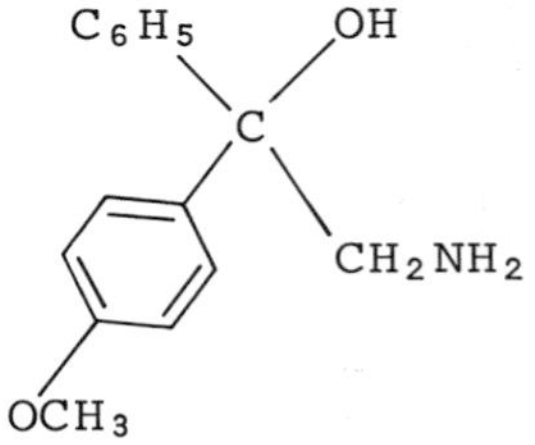

The increase in the competitive nature of eliminations and rearrangements relative to substitutions in
deamination relative to solvolysis reactions and the
general leveling in migratory aptitudes suggests the
intermediates involved in deamination are highly reactive and possess little discriminative potential compared to carbonium ions formed in the solvolytic reactions. These observations indicate that low activation energies are involved in the reactions of the intermediates leading to products. A measure of the magnitude of the energies involved is afforded by the
brilliant stereochemical and carbon labeling study of
Benjamin, Schaeffer, and Collins on the pinacolic deamination of optically active 2-amino-1,1-diphenylpro-
pan-1-ol (see Scheme 7).[36a] The major product 3 is
formed with inversion of configuration and migration
of the labeled phenyl group. However, the other
ketone 4 is formed with retention of configuration and
migration of the unlabeled phenyl group. These

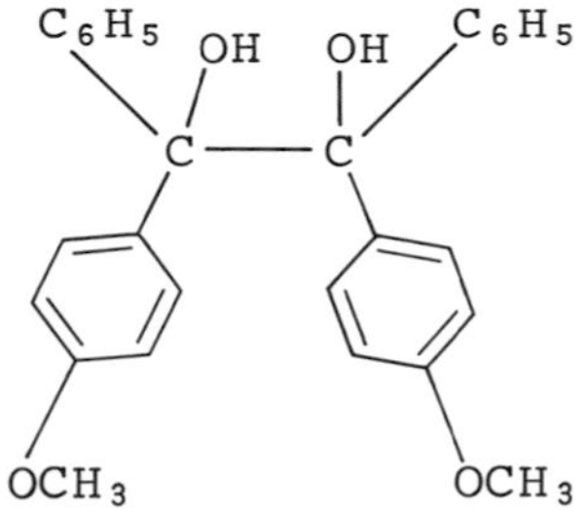

results, which are interpretable in terms of classical carbonium ions, indicate that the lifetime of the intermediate carbonium ion is sufficiently long to allow some rotation to occur, but is too short for complete rotation through 180° to be attained.[36] Consequently, the activation energies for reactions of intermediates leading to products are of the order of energies required for bond rotation, namely approximately 3 to 5 kcal/mole. This figure should be compared to that of 25 to 30 kcal/mole for the activation energies of solvolysis reactions. Since the barrier to reaction in deamination is lower, differences in activation energy between competing modes of reaction are smaller, and rates of competing reactions become more similar.[24]

Scheme 7

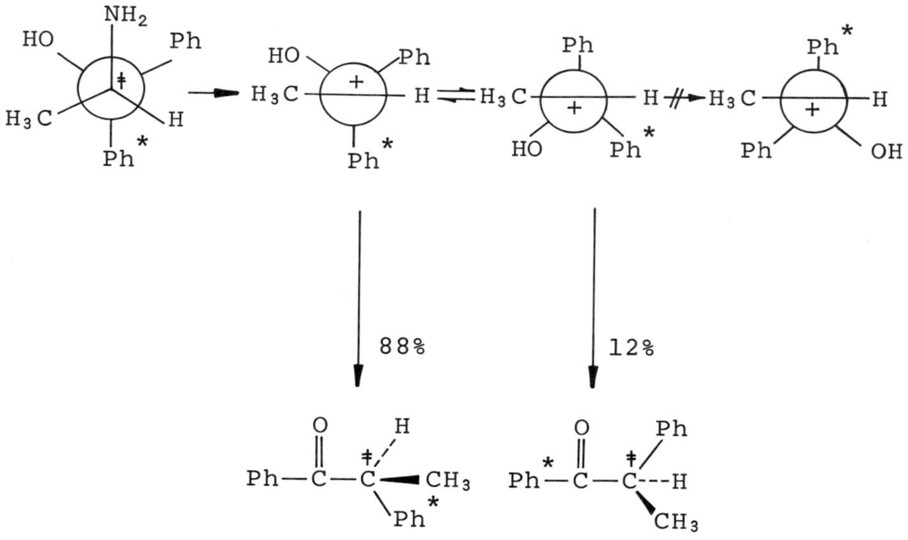

3
(Inversion[†])

4
(Retention[†])

*[14]C-labeled.
[†]Optically active center.

(b) The Possibility of Diazoalkane Intermediates. It is possible that the differences between the products of deamination and solvolysis reactions arise as the former reaction does not involve carbonium-ion intermediates, but occurs via carbenoid reactions of diazoalkanes (see Scheme 8). If diazoalkanes are involved, deuterated products should be formed when the deamination is carried out in deuterated protic solvents. Conversely, amines initially labeled with α-deuterium should yield products lacking the label. These approaches have been used to exclude diazoalkanes as intermediates in deamination of ethylamine and iso-butylamine in polar solvents such as water and acetic acid.[24],[28],[37] However, in the deamination of α-amino-ketones in aprotic solvents, the intermediacy of diazo-alkanes seems probable.[39]

Scheme 8

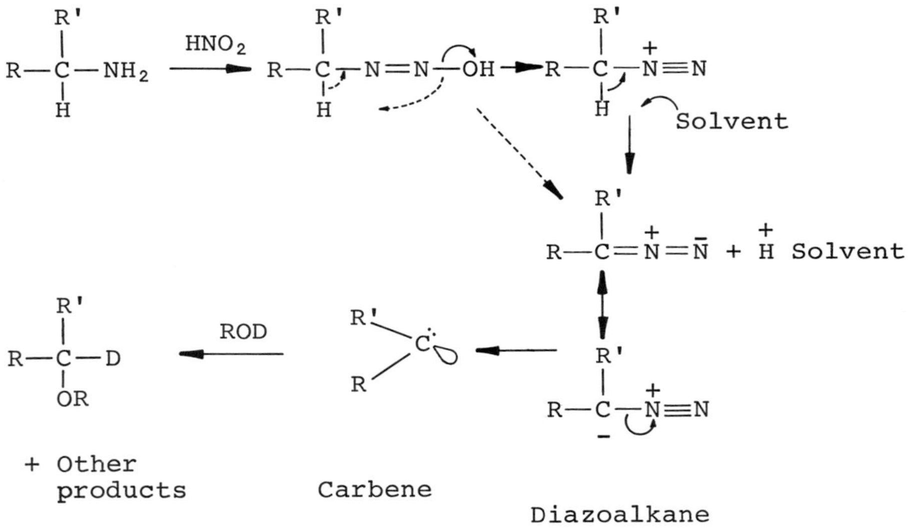

+ Other
products   Carbene

                         Diazoalkane

(c) The Hot Carbonium Ion Hypothesis. In a solvolysis reaction, the formation of an intermediate carbonium ion from a neutral substrate molecule is a highly endo-thermic process. Energy has to be supplied to stabi-lize (via solvation, etc.) the developing ions and to overcome the force of attraction of these two ions for each other. The general increase in solvation may also be unfavorable from entropy considerations. However, in a deamination reaction, the precursor to the car-bonium ion is a highly unstable diazonium ion. On de-composing it forms a neutral, highly stable molecule of nitrogen. There is no force of attraction between ions or dipoles of opposite charge to overcome. Solvation of the departing nitrogen is not required. These factors, coupled with the initial instability of the diazonium ion, make the activation energy for its decomposition very small and, in fact, it may even be an exothermic process.[38b] The loss of nitrogen from the diazonium ion is regarded as such a facile process that neither neighboring group participation nor solvent assistance, normally required in the ionization of more stable carbonium ion precursors, is needed. Thus, the carbonium ion formed is an exceptionally reactive species as it possesses essentially an unsolvated vacant p-orbital and has no counterion in the immediate vicinity. The terms, "hot",[33,40] "unsolvated",[31] and "vibrationally excited",[38b] have been used to describe these highly reactive carbonium ions, which are capable of promoting rearrangements of appropriately situated groups on an adjacent carbon atom. Such rearrangements occur more rapidly than molecular collision as hydride shifts occur intramolecularly, tracer not being incorporated when deamination is performed in labeled solvents.[28]

From consideration of maximum orbital overlap with the electrons of the migrating group, substituents situated trans to the vacant p-orbital should migrate preferentially. As the lifetime of the carbonium ion is shorter than the time for rotation of a single C-C bond, the populations of the various conformations of the aliphatic amine, which to a large extent will be reflected in the diazonium ion, determines the product distribution.[36b] Ground state control invoked in this manner adequately explains the data listed in Table 1. Similarly, the olefin distribution in the deamination of secondary butylamine can be accomodated.[24] The most stable conformation 5 gives rise to trans-2-butene (56%); cis-2-butene arises from the less stable conformation 6 (19%). Methyl migration from conformation 7 does not occur as this would require formation of a less stable primary carbonium ion from a secondary carbonium ion. Rearrangement only occurs to give the same ion (scrambling, see Section VII.A.1.e below), or a more stable ion, namely primary $\longrightarrow$ secondary $\longrightarrow$ tertiary.[41,42] All three conformations can give rise to 1-butene, which is formed in much greater abundance than in E1 reactions of 2-butyl halides or esters.

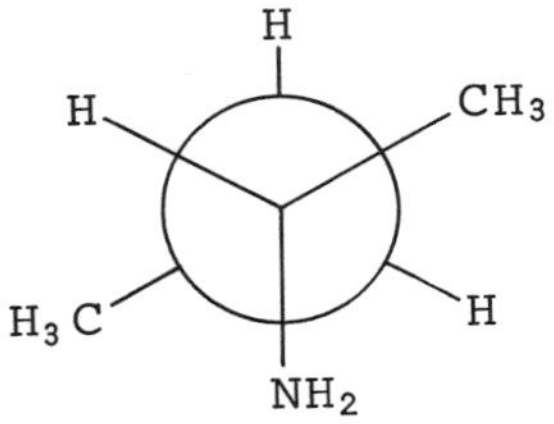

5 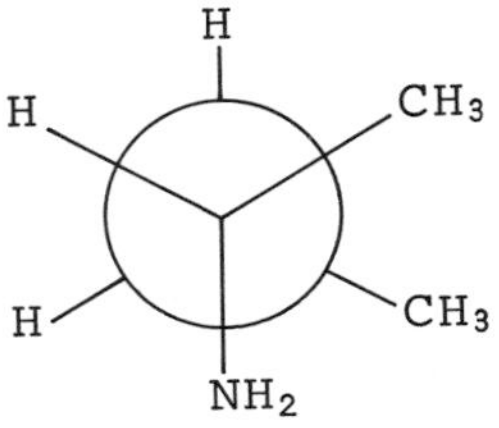   **6** 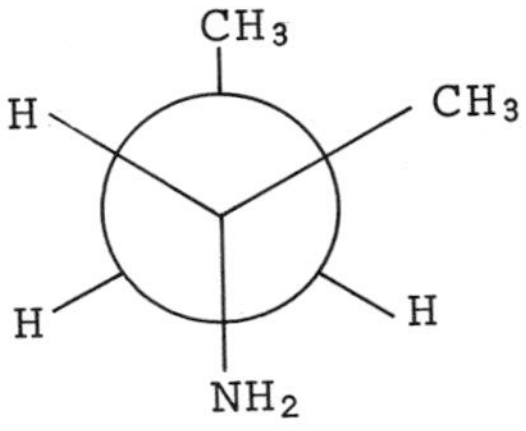   **7**

(d) Concerted Decompositions of Diazonium Ions.
Whereas axial cyclohexylamines undergo extensive elimination (50 to 80%) on deamination, the equatorial isomers show a much lower preference for this mode of decomposition (5 to 25%).[43-46] Similar observations have
been recorded for the pinacolic deamination using a t-
butyl substituent to "fix" the stereochemistry of the
cyclohexyl ring.[47] Compound 8 undergoes only rearrangement to 9, while the axial amine 10 eliminates
quantitatively to the cyclohexanone 11. These results
bear a close relationship to those for the E2 reactions
of the corresponding trialkylammonium ions with basic
reagents. Compound 12, in which the leaving group is

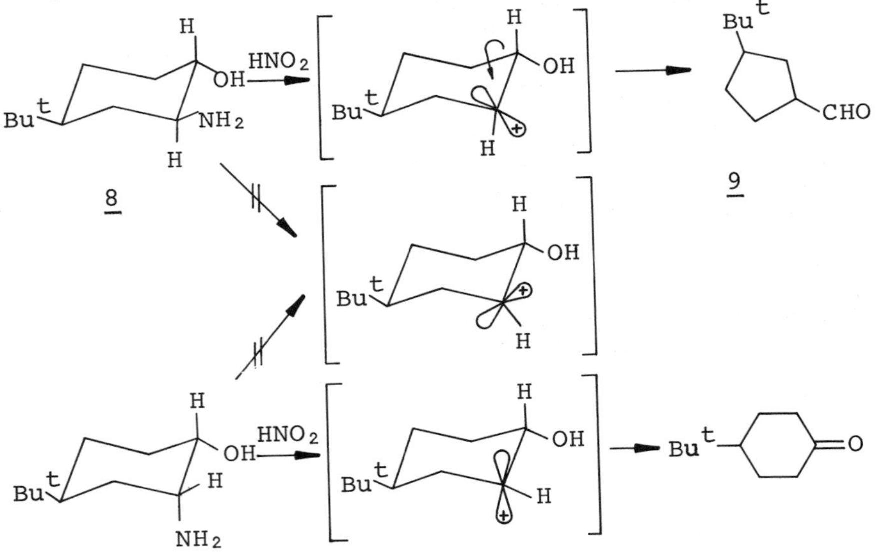

8   **9**

10   **11**

confined to the equatorial position undergoes only de-
alkylation to the corresponding dimethylamine, but the
isomer 13, in which the leaving group is axial in about
half of the molecules, as the t-butyl and trimethylammonium groups have similar spacial requirements,
gives 93% elimination product.[48] The obvious stereochemical analogies between deamination and the Hofmann
elimination led Streitwieser to suggest that the
difference between the products of E1 processes and deamination was attributable to the concerted E2 reaction
of the diazonium ion with the solvent.[14,24] Similar
concerted rearrangements or bimolecular substitution
reactions with the solvent account for the other reaction products.

In some ways, the concerted diazonium hypothesis
differs little from the "hot" carbonium ion theory. In
terms of energetics, if the potential energy well of an
unsolvated carbonium ion has sides of lower energy than
thermal vibrations, the carbonium ion may in effect be
bypassed. Under this condition, the decomposition of
the diazonium ion becomes essentially a one-step
process.[49] However, a number of observations are not
in accord with a concerted decomposition of a diazonium
ion.[50]

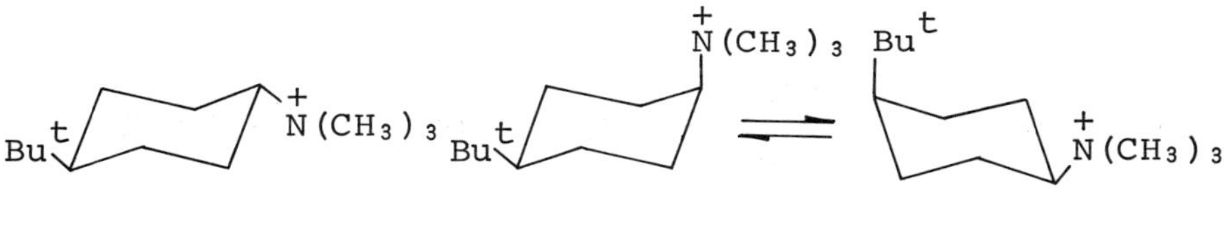

12   13

Let us direct our attention to the substitution
products. On deamination in aqueous solution, axial
cyclohexylamines give the alcohol of inverted configuration, but the equatorial isomers substitute with
retention.[43-46] On the "hot" ion theory these results
can be explained in terms of ground state control and
inhibition to solvent approach from the backside for
the equatorial substituent by the ring system. The
only reasonable concerted process for the formation of
the retention product is an S_Ni-type reaction[51] of the
diazohydroxide (14).

14

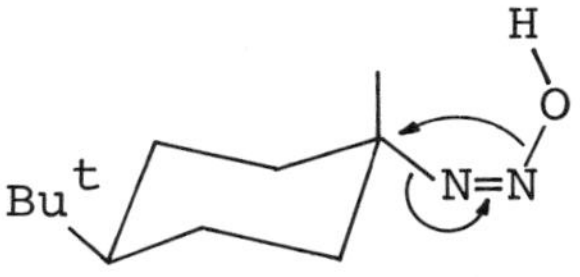

Considering the rearrangements, we find that the migration of the phenyl group in Scheme 7 (product 4) resulting in a product with retention of configuration at the reaction center is equally difficult to accommodate by a concerted reaction.

(e) Hidden Rearrangements. An apparently unrearranged alcohol is one of the products of deamination of 2-(p-anisyl)ethylamine.[52,53] Carbon-labeling experiments indicate that the two carbon atoms of the side chain have become almost equivalent during the course of the reaction, the label initially located on the α-carbon being almost equally distributed in the alcoholic product (see Scheme 9). This result is accomodated in terms of a bridged-ion intermediate, 15, rather than a pair of rapidly equilibrating classical ions, although this preference is a matter of conjecture rather than proven fact.* With the p-anisyl derivative, about 90% of the reaction must proceed through this intermediate, but participation is much less with other substituents, falling to a negligible level along the series; p-anisyl > phenyl > methyl > hydrogen > benzyl.[37,52-55] The trend along this series reflects

Scheme 9

$$RCH_2{}^*CH_2NH_2 \xrightarrow[\text{H}_2\text{O}]{\text{HNO}_2} RCH_2{}^*CH_2OH \ + \ R\overset{*}{C}H_2CH_2OH$$

R	Reference	%	%
p-Anisyl	52, 53	55	45
Phenyl	52, 53	73	27
H	37	98.5	1.5
Benzyl	55	100	0

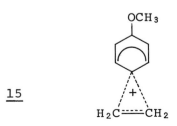

15

*See subsequent discussion on deoxidation reactions.

a decreasing order of stability of the bridged "ethylene onium" ion relative to the classical ion. It is
interesting that this trend is paralleled by the ease
of cleavage of R-Hg by hydrochloric acid.[56] In other
words, there is a close analogy between group transfer
to a proton and to a carbonium-ion center.[55]
     The deamination of l-aminopropane with sodium ni-
trite in aqueous perchloric acid yields four major
products (Scheme 10).[27,57,58] The alcohol product from
the deamination of 2-aminopropane is exclusively 2-pro-
panol. The presence of cyclopropane among the products
raises the possibility that all the n-propanol arises
from attack of water on a protonated cyclopropane.[59]

Scheme 10

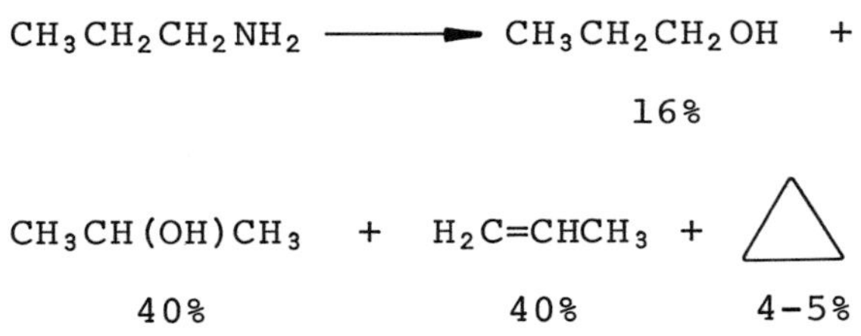

$$CH_3CH_2CH_2NH_2 \longrightarrow CH_3CH_2CH_2OH \ +$$

16%

$$CH_3CH(OH)CH_3 \ + \ H_2C=CHCH_3 \ + \ \triangle$$

40%   40%   4-5%

To test this idea, Karabatsos and Meyerson and their
co-workers studied the deamination of carbon and deuterium labeled substrates.[60] The products were separated by gas chromatography, and after conversion to
the trimethylsilyl ether, the propanol fractions were
analyzed by mass spectometry. 1-Aminopropane-1,1-d_2,
-2,2-d_2, -3,3,3-d_3, and -1-^{13}C gave l-propanols that
were about 4% isotope-position rearranged and 2-pro-
panols that were exclusively isotope-position unrearranged, having arisen from a nominally irreversible
1,2-hydride shift (see Scheme 11). These results,
which are in close agreement with those of previous
workers who used ^{14}C and tritium labels,[61] show that
1-propanol arises mainly directly from solvolysis of a
primary carbonium ion or one of its precursors. The
distribution of the ^{13}C and deuterium label in the rearranged 1-propanols is consistent with the three carbons of the propane system having become equivalent before product formation, but with equilibration of the
hydrogens having progressed to a lesser extent. Explanations involving equilibrating classical ions or
face protonated cyclopropanes require unrealistic
assumptions.[60] However, the intermediacy of edge-pro-
tonated cyclopropanes accommodate the observations
(see Scheme 12). As all three carbons are equilibrated

Scheme 11

$$CH_3CH_2CD_2NH_2 \xrightarrow{40°} CH_3CH_2CD_2OH + C_2H_4DCHDOH + C_2H_3D_2CH_2OH$$
$$96.2\% \qquad 0.8\% \qquad 3.1\%$$

$$CH_3CD_2CH_2NH_2 \xrightarrow{40°} CH_3CH_2CD_2OH + C_2H_4DCHDOH + C_2H_3D_2CH_2OH$$
$$1.1\% \qquad 0.9\% \qquad 97.9\%$$

$$CD_3CH_2CH_2NH_2 \xrightarrow{40°} C_2H_4DCD_2OH + C_2H_3D_2CHDOH + C_2H_2D_3CH_2OH$$
$$1.0\% \qquad 1.2\% \qquad 97.8\%$$

$$CH_3CH_2{}^{13}CH_2NH_2 \xrightarrow{40°} CH_3CH_2{}^{13}CH_2OH + {}^{13}CH_3{}^{13}CH_2CH_2OH$$
$$97\% \qquad 3\%$$

in the product, edge-edge equilibration (e-e) must be faster than nucleophilic attack by the solvent, but hydrogen scrambling (s-s) must be slower and competitive with nucleophilic attack.

These labeling experiments indicate that more than one type of intermediate decomposes to products in deamination.

Scheme 12

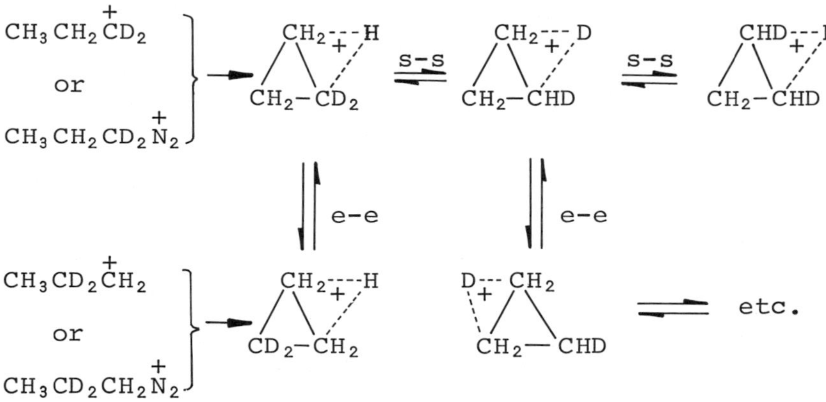

$$CH_3CH_2\overset{+}{C}D_2$$ or $$CH_3CH_2CD_2\overset{+}{N}_2$$
$$\longrightarrow \quad \underset{CH_2-CD_2}{CH_2 \cdots H} \overset{+}{\xrightleftharpoons{s-s}} \quad \underset{CH_2-CHD}{CH_2 \cdots D} \overset{+}{\xrightleftharpoons{s-s}} \quad \underset{CH_2-CHD}{CHD \cdots H}$$

$$\Big\updownarrow e-e \qquad \Big\updownarrow e-e$$

$$CH_3CD_2\overset{+}{C}H_2$$ or $$CH_3CD_2CH_2\overset{+}{N}_2$$
$$\longrightarrow \quad \underset{CD_2-CH_2}{CH_2 \cdots H} \quad \underset{CH_2-CHD}{D \cdots CH_2} \quad \rightleftharpoons \quad etc.$$

(f) Ion-Pair Mechanisms. Our discussion thus far has ignored the influence of the solvent and the counterion on the products of deamination. Recent findings indicate this approach to be an oversimplification.

A change of solvent from water to aqueous dimethyl formamide causes a reduction in the percentage rearranged product from 58 to 31% for the deamination of n-propylamine.[62] In water, deamination of isobutylamine gives iso:sec:tert -products in the ratio 10 : 19 : 71, but in the presence of 7N sodium thiocyanate, this ratio changes to 23 : 21 : 56.[41] Alcohols are formed, in low yield in deamination in acetic acid, usually with retention of configuration.[62-64] The alkyl acetate (solvolysis product) is often formed with overall retention of configuration.[65] Despite the low nucleophilicity of the nitrate ion and its very low concentration in nitrous acid solutions, alkyl nitrates arise as minor products of deamination. Finally, attention is drawn to the observation that axial amines give vastly different product ratios from the corresponding alicyclic equatorial isomers. If planar hot carbonium ions had been involved as intermediates, both isomers should have given the same product distribution.[17]

The observations just described can be accounted for, in part, by assuming that the products arise directly from the diazohydroxide, the diazonium ion, or a hot carbonium ion, the relative contributions of these paths varying with the reaction conditions. However, a more attractive explanation invokes ion-pair intermediates as outlined below.

Cohen and Daniewski have examined the stereochemistry of elimination in deamination from trans-2-decalylamines (see Table 2),[66] using specifically deuterated substrates. The degree of deuteration in the 1-octalin provides an estimate of the extent of deuteration in the reactants. From a measure of the extent of deuteration in the 2-octalin, and the 1-ene/2-ene ratio, the isotope effect* for proton abstraction can be assessed and, hence, the percentage reaction involving syn and anti elimination can be calculated. The results clearly indicate that for the axial amino substrate, 16a, syn elimination accounts for 56 to 61% of the reaction product, and for the equatorial epimer, 16b, for 78% (syn-ae as against syn-ee). Similar results have recently been reported by Levisalles and Mouchard for deamination of the epimeric 3-amino-4,4-dimethylcholestanes in chloroform-acetic acid solutions.[67] These findings clearly cast doubt on the validity of the concerted E2 hypothesis involving the diazonium ion.

There is considerable evidence in solvolysis reactions that the partitioning of the carbonium ion in solvents of low polarity, such as acetic acid, is significantly influenced by the nature of the leaving group.[68,69] The products of elimination in the unimolecular solvolysis of erythro-3-deutero-2-butyl tosylate in nitrobenzene or acetic acid are consistent with a syn elimination.[70] However, in more polar and basic solvents such as water or acetamide, product analysis clearly supports an anti elimination. These observations are explained in terms of the relative basicities of the solvent and the tosylate anion. Solvent participation in proton removal is minimal for the weakly basic solvents and the counterion removes the β-proton giving syn elimination. The more basic solvents simulate the base in an E2 reaction and anti elimination results. The obvious analogy between these solvolyses

---

*As might be expected, the primary isotope effect for decomposition of the carbonium ion or its precursor diazonium ion is small, the transition state for such an exothermic process being very reactantlike (Section II.C.3.a).

TABLE 2 Stereochemical Analysis of the Alkenes Formed during Nitrous Acid Deamination of Trans-2-decylamines in Acetic Acid and Thermal Decomposition of N-Nitrosocarbamates in Cyclohexane[66]

Substrate	% Monodeuterated 1-Octalin[c]	% Monodeuterated 2-Octalin[c]	% Monodeuterated 1-/2-Octalin Isotope Effect		% Syn-elimination
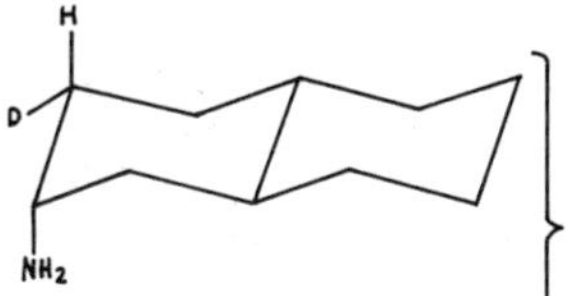 16a	95	54	0.91	1.71	56
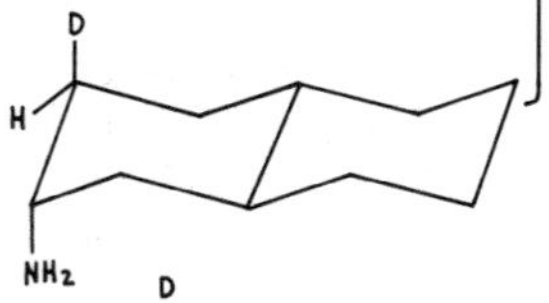	94	67	0.90	1.61	61
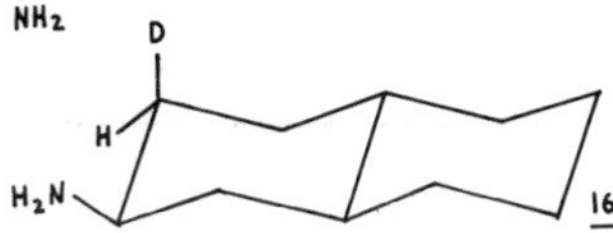 16b	93	30	a	1.66[b]	78

TABLE 2 Continued

Substrate	% Monodeuterated 1-Octalin[c]	% Monodeuterated 2-Octalin[c]	1-/2-Octalin Isotope Effect	% Syn-Elimination
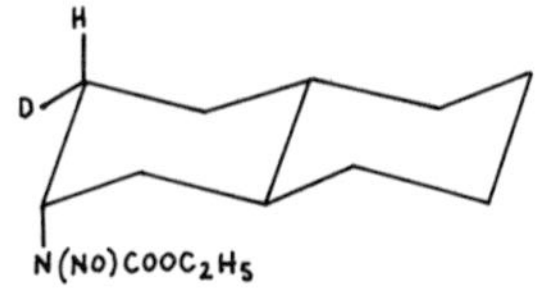	95	9.5	1.2	1.30	92
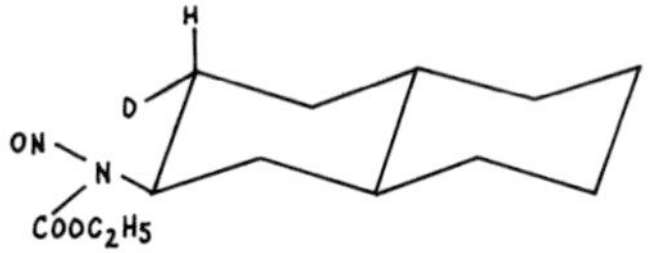	95	83	0.75	1.58	81

[a]Too low for accurate determination.
[b]Assumed value is the average of the values for the axial amines (16a).
[c]Estimated by mass spectrometry at 20eV.

and the deamination of trans-decalylamines suggests that ion pairs are involved in diazonium ion formation and subsequent decomposition.[71,72]

For the axial epimer, the diazonium ion collapses to give 17, in which the hydrated acetate ion maintains the configurational identity of the carbonium ion. As the trans-axial-β-hydrogen atoms are on the opposite side of the molecule, the counterion can only remove a cis-related equatorial proton. This process competes favorably with direct solvent attack on the axial β-hydrogen.

For the equatorial amine, the counterion can effect essentially a syn elimination with either the axial or equatorial β-hydrogens, 18. The reason for

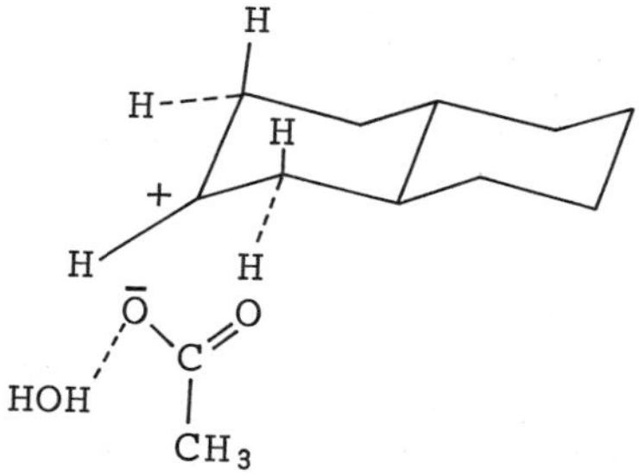

17 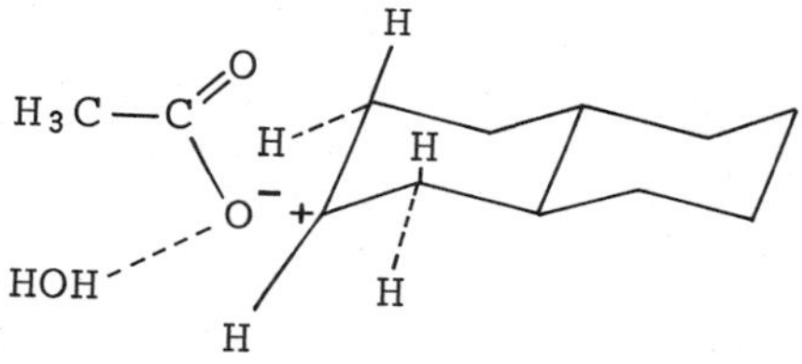   18

the marked preference for ae-elimination, which is also observed in pyrolytic syn elimination of equatorial acetates (Section VIII.C.2,C.3.(c)), is not obvious. It could reflect stereoelectronic control,[66] or could arise as less buckling of the carbon skeleton is required for an ae than an ee elimination.[49] However, considering the small activation energy involved in reactions of the ion-pair, it is surprising that such selectivity is observed.

It should be realized that ion-pairs formed from decomposition of a diazonium ion will be more reactive than those formed in solvolyses. In the diazonium ion-pair, two nitrogen atoms separate the counterion from the carbon center on which the empty orbital will form. The poor solvating ability of nitrogen ensures its rapid and essentially irreversible departure, generating simultaneously an anion and a cation at a separation greater than is achieved in solvolysis reactions.[73,74] Thus relative to the highly solvated or contact ion-pair formed in solvolysis, the entity formed in deamination can be regarded as a "vibrationally excited" ion-pair. Alternatively, it can be considered as an ion-pair separated by a nitrogen molecule.[74] In either case, it is capable of undergoing skeletal rearrangements and hydride shifts on a comparable time scale to deactivation to a solvated, or contact ion-pair.

White and Woodcock[17] have devised a "counter-ion" hypothesis to account for all the products of deamination (see Scheme 13). Their scheme is particularly attractive in explaining the substitution products in the deamination of secondary and tertiary carbamines in acetic acid (see Table 3). The intramolecular alcohol products, formed with predominant retention of configuration, are proposed to arise via predominant collapse on the front side of the carbonium ion ($27 \longrightarrow 29$). Dilution of acetic acid by water (0 to 8 mole%) has no influence on the stereochemical consequences of the products of deamination of 2-phenyl-2-butylamine.[17] This result excludes an S_N2 reaction of water on the diazonium ion being the route for formation of the alcohol of inverted configuration. The separation of ions in the ion-pair enables some rotation of the carbonium ion before it collapses to product ($27 \longrightarrow 31$).

The intermolecular solvolysis product (acetate) is regarded as arising from front side exchange ($21 \longrightarrow 22$) followed by predominant collapse to a product with retention of configuration. Some inverted acetate may arise by this route, but most likely it originates mainly from a solvent cage reaction ($27 \longrightarrow 28$), the solvent exerting a backside attack on the carbonium ion. A direct displacement reaction on the diazonium ion by the solvent can be ignored, as the product distribution and the stereochemistry of the alkyl acetates remain unchanged when the deamination is performed in the presence of varying amounts of acetate ion.[72,76]

The primary carbamine, 1-deutero-1-butylamine undergoes deamination in acetic acid to give solvolysis products with predominant inversion of configuration.[24] This observation is in marked contrast to the results reported in Table 3 for secondary and tertiary carbamines. It could reflect a direct nucleophilic displacement by the solvent on the diazonium ion. Steric interactions in secondary and tertiary substrates may preclude this mode. Alternatively, the lesser stability of the primary carbonium ion may increase the lifetime of its precursor diazonium ion to enable a direct solvolytic displacement to become competitive with unimolecular decomposition.

Scheme 13

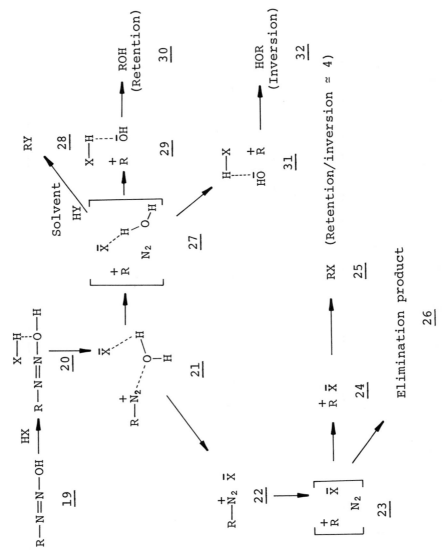

TABLE 3 Stereochemistry of Substitution Products in the Nitrous Acid Deamination of Some Secondary and Tertiary Carbamines in Acetic Acid

$$RNH_2 \longrightarrow ROH + ROAc$$

R	ROH		ROAc		Reference
	% Yield	% Retention	% Yield	% Retention	
ArCHCH₃	2	79	35	54	62
ArC(C₂H₅)CH₃	15	75	41	61	63
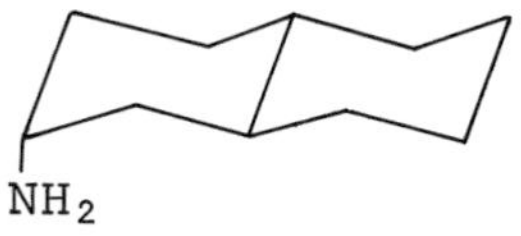 NH₂	7.5	94	18.5	45	72
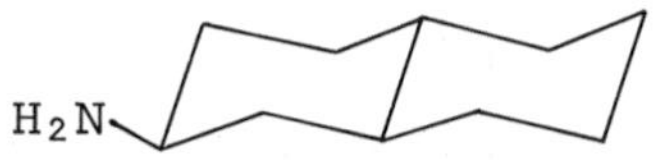 H₂N	–	96	–	75	75

(g) <u>Deamination in Aqueous Solution</u>. In highly polar
media, by analogy with solvolytic reactions, ion-pair
entities should play a very small role.[70] Although
low concentrations of water in acetic acid exert little
influence on the solvolytic products, in more aqueous
conditions marked variations are observed. The deamination results of Cohen and Jankowski amply demonstrate
this point (Table 4).[72]

As the medium becomes progressively more aqueous,
the following observations are apparent; (1) the percentage alcoholic product increases, (2) the axial
amine, <u>33</u>, undergoes solvolysis with a marked increase
toward inversion of configuration, (3) the equatorial
epimer, <u>34</u>, shows a slight increase in the tendency to
undergo solvolysis with retention of configuration relative to the anhydrous media, and (4) the product
proportions from the two epimers approach each other.

The last observation militates against a concerted
reaction of the diazonium ion becoming more important
in the more aqueous solutions. However, it is consistent with a diminishing role for the counterion as it
becomes more solvated, and the carbonium ion therefore
becomes more symmetrically solvated. Inversion becomes
dominating for <u>33</u> as this leads to lesser nonbonded
interactions between the solvent and the axial protons
of the ring system. By comparison, equatorial epimer
<u>34</u> naturally undergoes substitution with predominant
retention of configuration. Finally, observation (4)
suggests that a common intermediate plays a major role
in the decomposition of the two amines in the most
aqueous solutions. Perhaps with a more stable carbonium ion, epimeric amines may give identical products.

There is little doubt that the reaction of nitrous
acid with primary amines is one of the most complex
mechanisms. Deprived of kinetic methods, the investi-
gator must develop his picture entirely from product
analysis. Care should be taken to combine simple product distribution measurements with appropriate labeling techniques. Ion-pairs certainly play a vital role
as intermediates in acetic acid, but their involvement in other solvents is not completely established.
Future investigations will presumably remove this
deficiency. It would also be useful to study the influence of a range of counterions, thereby affording a
more thorough idea of the selectivity of the intermediate carbonium ions. Possibly carbonium ions generated in deamination are more representative of electron deficient carbon species than their more stable,
highly solvated counterparts which are formed during
solvolysis reactions.

TABLE 4 The Influence of Solvent on the Solvolysis Products from Deamination of Trans-2-decalylamines[72]

Mole % Water in Acetic Acid	0	25	50	75	96.6
Substrate Products	% Yields of Total Solvolysis Product				

Substrate: 33 (NH₂)

Products	0	25	50	75	96.6
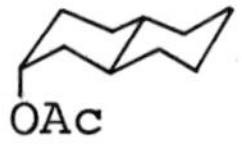 OAC	32	29	25	17	7
 AcO—	39	35	29	24	26
 OH	27	31	33	32	16
 HO—	2	5	13	27	51

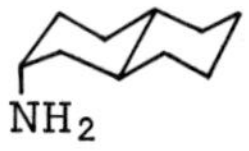

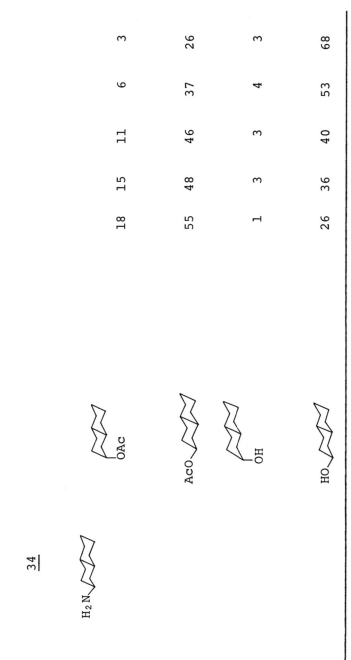

18	15	11	6	3
55	48	46	37	26
1	3	3	4	3
26	36	40	53	68

A great deal of work has been compiled on deamination of bicyclic amines such as norbornylamine, the
decomposition of which can occur via both classical and
nonclassical ions. We regard the scope of these latter
intermediates as outside the field of this chapter and
refer the reader to other sources.[16,17,42]

2   The Nitrosoamide Reaction

The initial step in this reaction is the slow rearrangement of the nitrosoamide, 35, to the diazo ester,
36, which subsequently decomposes rapidly by a number
of routes depending on the reaction conditions.[8,77-79]
At the low temperatures used to initiate the reaction,
the decomposition of the diazo ester appears to be a

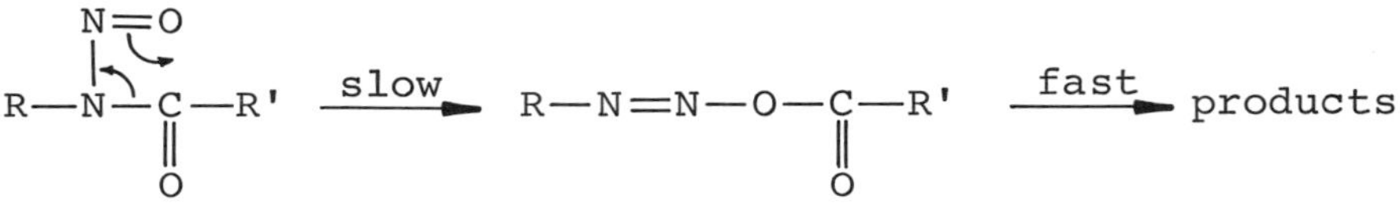

Nitrosoamide   Diazo ester

35   36

heterolytic process. The reaction products are unchanged when the reaction is performed in the presence
of the scavenger nitric oxide,[80] and no polymer is
detected when the reactions are carried out using styrene as the solvent.[81]

(a) Intermediacy of Diazo Compounds. Unlike the nitrous acid deamination, the nitrosoamide and closely related reactions often occur via diazo intermediates
(37), which undergo rapid reactions with carboxylic
acids to yield the corresponding esters. The method of
detecting these diazo compounds is determined by their
stability. For instance, when the alkyl group R in 37
possesses an α-carbonyl group, the diazo compound can

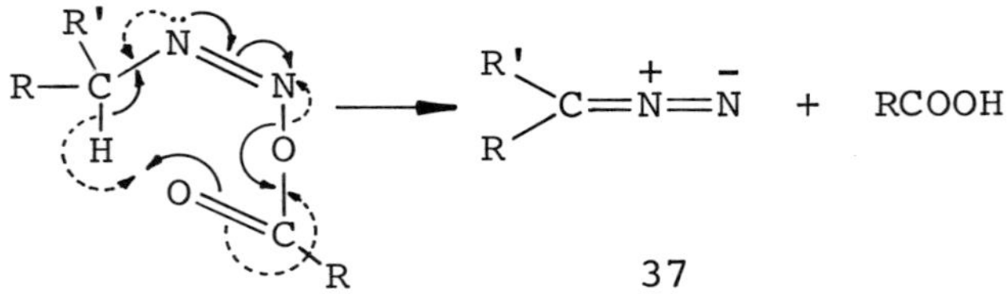

37

(The electrons may move in either direction, ----- or ———.)

actually be isolated.[82]   Similar proof applies to the
thermal decomposition of N-nitrosocarbamates under
basic conditions.[83]   For primary carbamines (37, R' =
H), the instability of primary carbonium ions in nonpolar solvents militates against their formation, and
diazo intermediates can be established by a trapping
procedure. Thus diazobutane has been identified as a
reaction product in the thermal decomposition of N-
(n-butyl)-N-nitroso-trimethylacetamide in pentane in
the presence of excess diazoethane. The diazoethane
rapidly consumes the carboxylic acid (preventing its
decomposing the diazobutane) as it is formed along with
diazobutane.[7]   Optically active N-(1-butyl-1-d)-N-
nitrosoamide decomposes thermally in cyclohexane to
give racemic 1-butyl-1-d-acetate, along with n-butyl
acetate, 1-butyl-1,1-d$_2$-acetate and 1-butene.[84]   These
results, coupled with the observation that 1-butyl-1-d-
acetate is formed when the undeuterated substrate is
thermally decomposed in cyclohexane containing O-deuteroacetic acid, are consistent with a diazo alkane intermediate (see Scheme 14). In benzene, the nitroso
benzamide of benzhydrylamine yields 4% of diphenyl-
diazomethane.[85]   However, in more polar solvents such
as O-deuteroacetic acid, esters lacking deuterium are
formed in the thermal decompositions of the nitrosoamides of 1-phenylethylamine,[5] 3-cholestanylamine[64]
and benzhydrylamine.[86]   Thus it appears that in more
polar solvents, nitrosoamides of secondary and tertiary carbamines do not decompose via diazo alkanes
and, consequently, we confine our subsequent discussion
to reactions meeting these conditions.

Scheme 14

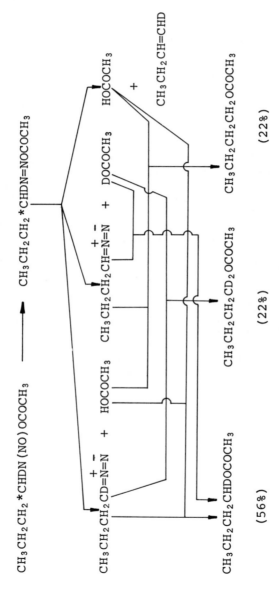

(% yield refers to the ester product only, yield of 1-butene = 16 - 20%).[84]

(b) <u>The Elimination Component</u>. The nitrosoamide and
its related reactions have long been regarded as useful
synthetic methods for the preparation of esters, the
alkene product being considered as an unwanted by-
product, which is always formed when the alkyl group
bears a β-hydrogen. Consequently, product and kinetic
studies concerning the olefinic fragment are rather
sparse. Several years ago, it was proposed that alkenes arose via a cyclic elimination, <u>38</u>, of the diazoester.[4],[84] The recent stereochemical experiments of
Cohen and Daniewski on the thermal decomposition of N-
nitrosocarbamates support this type of syn elimination,
but the authors preferred the intermediacy of an ion-
pair (see Table 2).[66] An analogy was drawn with the
deamination of the corresponding trans-2-decalylamines
by nitrous acid in acetic acid. Whether syn elimination operates when decomposition is effected in more
polar and basic solvents remains to be tested.

<u>38</u>

(c) <u>The Ester Component</u>. Insight into the mechanism
of decomposition of the diazoester into nitrogen and
the ester product has been gained through many interesting stereochemical and labeling techniques.

   (i) <u>Stereochemical Consequences of Ester</u>
<u>Formation</u>. In acetic acid, the nitrosoacetamides of
secondary carbamines in the acyclic or alicyclic series
yield acetates with predominant retention of configuration.[7],[64],[79] The tertiary substrates give products of
almost total retention.[63] Steric factors clearly prevent a displacement (S_N2-type) in the tertiary structures, but as we see subsequently, oxygen labeling
techniques indicate that a displacement reaction pathway plays only a minor role in the secondary series.
   In dioxan, in the presence of 3,5-dinitrobenzoic
acid, optically active N-(s-butyl)-N-nitrosobenzamide
is converted into two esters, the benzoate and the 3,5-
dinitrobenzoate, both formed with about 70% retention
of configuration.[79] Dioxan, the solvent, was regarded

as capable of stabilizing the intermediate carbonium ion (39), thus causing the acid to substitute on the same side from which the leaving group was expelled. When the roles of the two carboxylic acids were reversed, only 3,5-dinitrobenzoate ester was isolated. This rules against a concerted displacement of the diazoester being the mode of formation of the intermolecular* substitution product with inversion of configuration, as benzoic acid should be more nucleophilic than its dinitro derivative. However, this result does accord with formation of an ion-pair preceding ester formation. As dinitrobenzoic acid is a much stronger

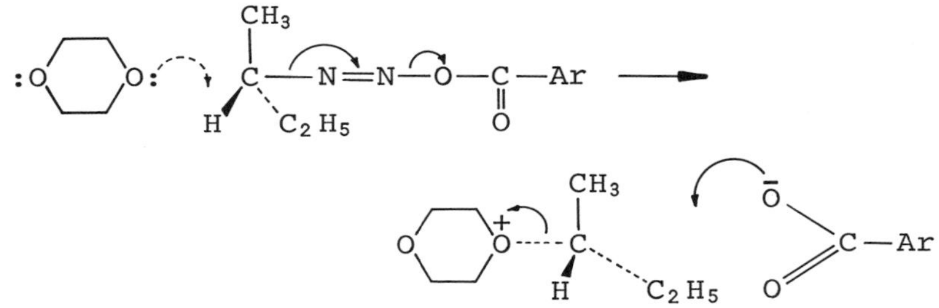

acid, the equilibrium between dinitrobenzoate and benzoate will lie considerably in favor of the former. Thus dinitrobenzoic acid can displace a benzoate ion from an ion-pair (front side exchange[17]), but the reverse process is unlikely (40).

40

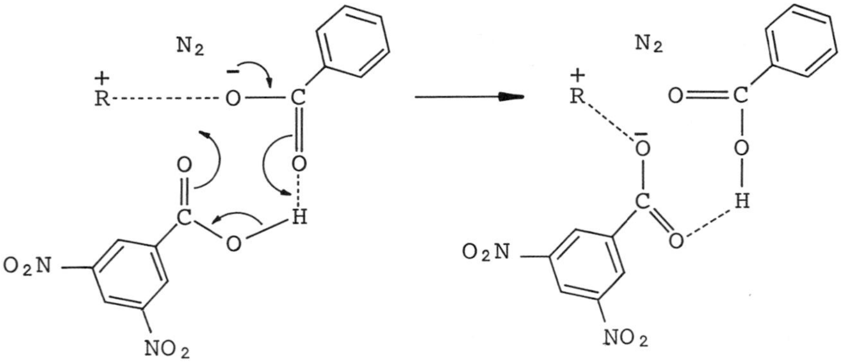

(ii) <u>Labeling techniques</u>. The decomposition of

*See Scheme 1.

nitrosoacetamides in ^{18}O-labeled acetic acid does not produce labeled esters, thereby excluding the involvement of a displacement reaction.[5] The intramolecular* substitution product formed with retention of configuration can arise by collapse of an ion-pair. However, formation of the product of inversion is less easily visualized. White and Aufdermarsh have solved this problem using an elegant combination of ^{18}O-labeling and stereochemical analysis.[7] Optically active N-nitroso-N-(L-1-phenylethyl)-2-naphthamide labeled in the carbonyl group with ^{18}O (41) was decomposed in acetic acid (see Scheme 15). Analysis of the ester showed that 81% was formed with retention and 19% was formed with inversion of configuration. The ester 42 was cleaved with lithium aluminum hydride. The alcohol 43 was treated with phenyl isocyanate, and the carbamates formed were resolved into the L and DL forms. Analysis for ^{18}O showed that both carbamates contained about 31% of the label in the ether oxygen and, consequently, 69% of the label must have resided in the carbonyl group of the ester.

Two conclusions can be drawn from the above results: (1) there are similarities concerning the intermediate leading to both retention and inversion products, and (2) the two oxygens have become only partly equivalent in the intermediate preceding ester formation. These results can be explained in terms of an ion-pair, which collapses at a rate so rapid that product formation is comparable with randomization of the two oxygens by anion rotation and translation. In addition, internal rotation of the cation can occur, leading to the product of inversion, weak bonding between the cation and the anion maintaining the non-equivalence of the oxygen atoms. The configurational change must involve a carbonium ion, but the scrambling stage is less obvious to assign.

For decomposition of the corresponding nitrosoamides in acetic acid, the extent of ^{18}O scrambling in the ester product varies little as the substrate is changed from cyclohexyl, to 1-phenylethyl, to diphenylmethyl.[5,17] These observations do not accord with the scrambling occurring at the carbonium ion-pair stage, leaving the diazonium ion-pair as the most probable source. If this deduction is correct, then factors that destabilize the carbonium ion should prolong the lifetime of the diazonium ion, and greater oxygen scrambling in the product should be observed. Also, the amount of ester formed with inversion of

*See Scheme 1.

Scheme 15

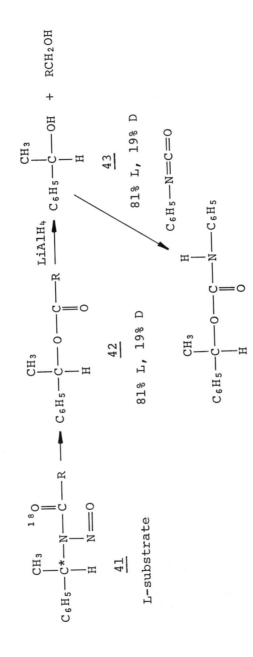

configuration will decline, as ion-pair collapse becomes increasingly more rapid than internal rotation of the carbonium ion in the ion-pair.

Thus, by combining all the data so far discussed the nitrosoamide reaction can be envisaged as occuring along the lines of Scheme 16. The concept that oxygen equilibration in the carboxylate anions is a slower process than reactions of ion-pairs is not unreasonable, an example having been observed in the solvolytic carbonium-ion rearrangement of trans-9-decalyl hydroperoxide.[86] The ratio of the ester product of retention to that of inversion gives a measure of the ease of internal rotation of the carbonium ion. Not suprisingly, this is more facile for secondary than tertiary substrates.

If diazonium ion-pairs are involved in the nitrosoamide reaction, similarities in the reaction products to nitrous acid deaminations in acetic acid are to be expected. White and Woodcock have listed a number of these observations.[17] However, the much greater abundance of rearrangement products in nitrous acid deamination remains to be explained. Perhaps the answer lies in a closer examination of the diazonium ion-pairs involved in the two reactions. In the nitrous acid reaction, the protonated diazohydroxide decomposes to give the diazonium ion and a water molecule, which will solvate the counterion (presumably acetate). Consequently, when the diazonium ion collapses to the carbonium ion-pair, the anion (which is insulated from the carbonium ion center by a nitrogen molecule) will exert a weaker influence on the carbonium ion in deamination owing to the water of solvation, than in the nitrosoamide reaction. The influence of this inherent water molecule should be less in more polar solvents, and there is obvious scope for comparing the nitrosoamide reaction and nitrous acid deamination in highly aqueous solutions.

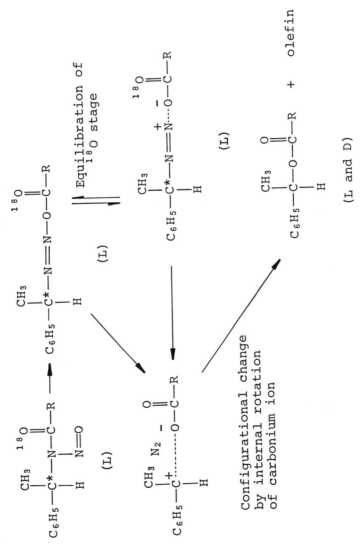

Scheme 16

3   The Triazene Reaction

There are comparatively few mechanistic studies of the triazene reaction.[11],[12],[87] For deamination of optical-ly active 1-phenylethylamine, the mechanism outlined in Scheme 17 has been proposed.[12] The presence of products of electrophilic substitution provides strong evidence for carbonium ion intermediates. The configu-ration of R is predominantly retained in both the products of N-substitution and ring substitution. A change in the structure of the aniline (Y = p-Cl or H) has virtually no effect on the product distribution, indicative of highly reactive intermediates of low selectivity. These results are consistent with the in-volvement of carbonium ion-pairs in which the ions are at a separation greater than in solvolytic reactions or are insulated from each other by a nitrogen molecule. Some rotation of each ion can occur in the ion-pair, accounting for some loss of configuration in group R and the para-substituted aniline.

A displacement pathway for the formation of the secondary amine is unlikely as 1,3-diphenyltriazene does not yield diphenylamine on treatment with hydro-chloric acid, and the decomposition of the p-chloro-triazene in the presence of aniline does not yield any amines derived from aniline.

Maskill, Southam, and Whiting[87] have focused attention on the decomposition of the diazo intermedi-ates, being concerned with whether this is a synchro-nous fragmentation (A, Scheme 18) or a two-step process involving the diazonium ion (B, Scheme 18). In the synchronous fragmentation, the anion $\bar{X}$ is formed in close proximity to the carbonium ion and should exert a controlling influence on the product distribution (cf. Scheme 17). However, in the stepwise route, they argue that diffusion must equate $\bar{X}$ with other anions or nucleophiles in the medium and, consequently, the product distribution should be much less dependent on X. To test this idea, they studied the deamination of 1-octylamine by nitrous acid, by the nitrosoamide reac-tion, and by the triazene method using aniline and p-nitroaniline as the sources of diazonium ion. In all cases acetic acid was the solvent. For all the reac-tions, the percentage yields of the multitude of prod-ucts were almost invariant. This suggests a minor role for the identity of X and a two-step fragmentation involving a diazonium ion. On the other hand, the product distribution for the same reactions of the 4-octyl system was found to vary markedly with X.

The above results strongly suggest that the

Scheme 17

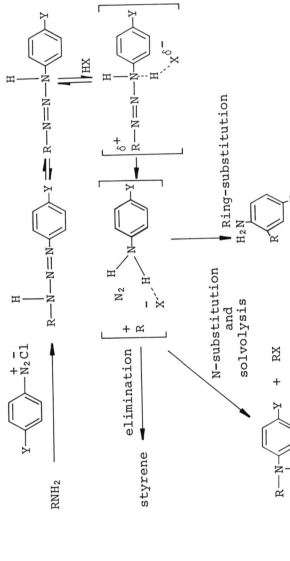

For the relative yields of products, see Scheme 3; R = 1-phenylethyl; Y = Cl or H(extra substitution product, H₂N-C₆H₄-p-R formed); HX = acetic acid or 3,5-dinitrobenzoic acid; solvent = acetic acid at 25°.

Scheme 18

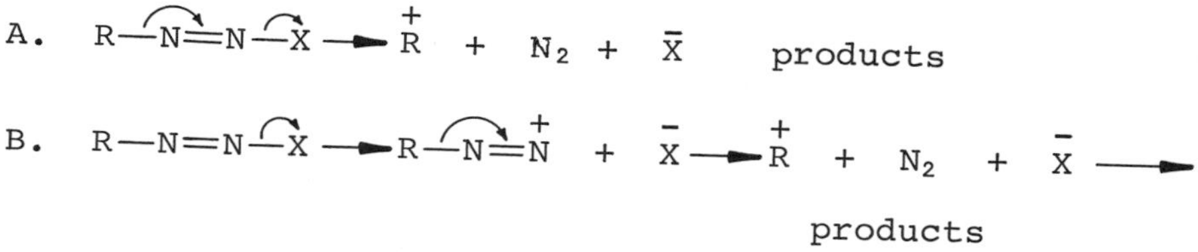

A. $R-N{=}N-X \longrightarrow \overset{+}{R} + N_2 + \bar{X}$ products

B. $R-N{=}N-X \longrightarrow R-\overset{+}{N{\equiv}N} + \bar{X} \longrightarrow \overset{+}{R} + N_2 + \bar{X} \longrightarrow$

products

deaminations described in this chapter, in a common solvent, involve similar intermediates. The lesser stability of primary than secondary carbonium ions allows the intervention of primary diazonium ions to influence the product distribution to a greater extent.

B   DEOXIDATION REACTIONS

Hine and his co-workers elegantly demonstrated that under alkaline reaction conditions, haloforms are converted generally via a rapidly and reversibly formed trihalomethyl carbanion, into dihalomethylene (Scheme 19).[88] In aqueous solution, the final reaction products are carbon monoxide and formate ion.[89] The mechanism of this α-elimination is discussed more fully in Chapter X.

Scheme 19

$$CHCl_3 + \bar{O}H \xrightleftharpoons{\text{fast}} \bar{C}Cl_3 + H_2O$$

$$\bar{C}Cl_3 \xrightarrow{\text{slow}} :CCl_2 + \bar{C}l$$

$$:CCl_2 \xrightarrow[\bar{O}H]{H_2O} CO + HCO\bar{O}$$

When the dihalomethylene is generated in alcoholic media, olefins and cycloalkanes are isolated among the reaction products,[90] and the yield of carbon monoxide is reduced relative to that observed in the reaction in aqueous solution.[91] Dehydrations of aliphatic alcohols performed in this manner were initially termed deoxidation reactions,[90] the name being changed to deoxidation in more recent years.

Skell and Starer investigated the reaction of bromoform with a range of propanols, butanols, and

pentanols containing their respective lyate ions.[91]
Along with carbon monoxide, they isolated the hydrocarbon fragments listed in Table 5. The yield of carbon
monoxide was found to increase from 49 to 94% (based
on $CHBr_3$) as the nature of the alcohol was changed
from primary to secondary to tertiary. The olefin/
carbon monoxide yields ranged from 0.6 to 0.8. The
authors noted that the product distribution, the cis-/
trans-olefin ratios, and the ratio of 2-methyl-1-bu-
tene/2-methyl-2-butene for deoxidation reactions of
n-propanol, n-butanol, i-butanol, and the pentanols,
were almost identical with the products of the nitrous
acid deamination of the corresponding amines.[2,28,92]
Despite the basicity of the medium, these observations
led to the postulation of carbonium ion intermediates.

In deamination, the facile loss of a nitrogen
molecule from the diazonium ion produces a highly reactive carbonium ion, which displays low discrimination
in its reactions. By analogy, the species $R\text{---}O\text{---}\overset{+}{C}$:,
which is isoelectronic with the corresponding dia-
zonium ion, should lose a molecule of carbon monoxide
very easily, thus, providing the necessary driving
force to generate a highly reactive carbonium ion
(Scheme 20). Alternatively, the carbene may attack an
alkoxide molecule to yield species $R \quad O \quad \overset{..}{C} \quad X$, which
undergoes a synchronous fragmentation to the carbonium
ion.

TABLE 5 Hydrocarbon Products from the Reaction of Bromoform with Refluxing Alkoxide / Alcoholic Solutions[91]

Alcohol	Reaction Products	cis-2/trans-2-Butene
n-PrOH	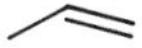 (90%) (10%)	
n-BuOH		0.54
s-BuOH		0.60
i-BuOH	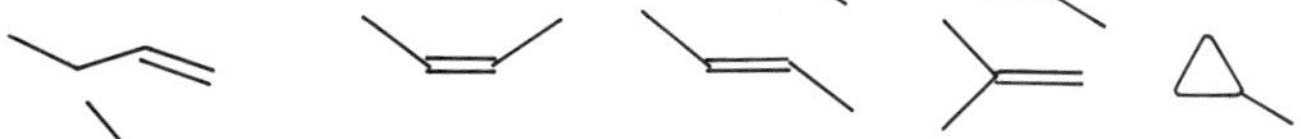	0.55
t-BuOH	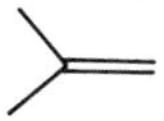	
		2-Methyl-1-/2-Methyl-2-butene
t-Pentanol		4.0
Neopentanol		2.1

319

Scheme 20

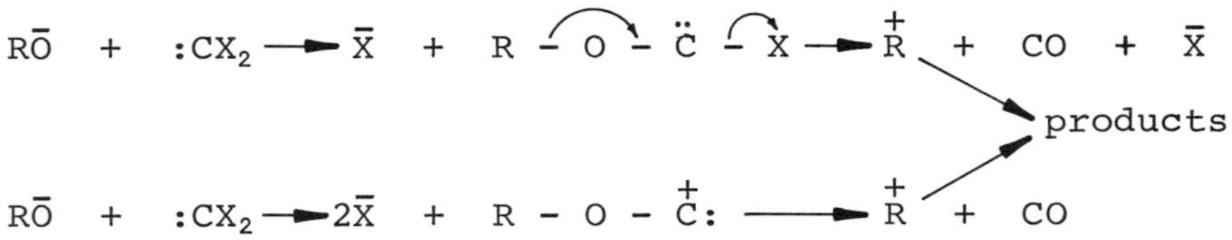

$$\text{R}\bar{\text{O}} + :\text{CX}_2 \longrightarrow \bar{\text{X}} + \text{R} - \overset{\curvearrowright}{\text{O}} - \overset{..}{\text{C}} \overset{\curvearrowright}{-} \text{X} \longrightarrow \overset{+}{\text{R}} + \text{CO} + \bar{\text{X}}$$

$$\searrow \text{products}$$

$$\text{R}\bar{\text{O}} + :\text{CX}_2 \longrightarrow 2\bar{\text{X}} + \text{R} - \text{O} - \overset{+}{\text{C}}: \longrightarrow \overset{+}{\text{R}} \nearrow + \text{CO}$$

The thermolysis of 1-diazopropane in aprotic media gives a similar product distribution to both the nitrous acid deamination of 1-aminopropane and the deoxidation of n-propoxide.[93] This observation suggests the possible involvement of ethyl carbene as an intermediate. To exclude this possibility, Skell and Starer analyzed the products of deoxidation of n-propanol-1,1-d_2.[94] A carbene mechanism should yield monodeuterated cyclopropane, whereas a dideuterated product is expected for a carbonium ion route (Scheme 21). A mass spectral analysis of the cyclopropane fraction revealed it to be 94% dideuterated and only 6% monodeuterated.

Scheme 21

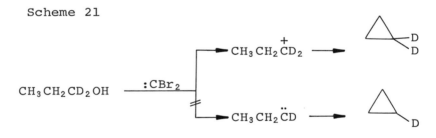

Although the small fraction of monodeuterated product could have arisen by a carbene insertion, the authors preferred a 1,3-hydride shift to give an intermediate-isomeric carbonium ion, which subsequently partitioned to the elimination products. Evidence supporting a 1,3-hydride shift, rather than the normal 1,2-rearrangements, which are usually observed under solvolytic conditions, was provided by the deamination of carbon-labeled n-propylamine.[95] n-Propylamine-1-^{14}C gives n-propanol-1-[14]C (92%) and n-propanol-3-[14]C (8%). If two successive 1,2-shifts had been involved, some isopropanol would have been formed. In addition, deamination of 2-propylamine does not yield n-propanol. Considering all these facts, the deoxidation of n-propanol-1,1-d_2, by dibromocarbene, can be accounted for by Scheme 22.

Scheme 22

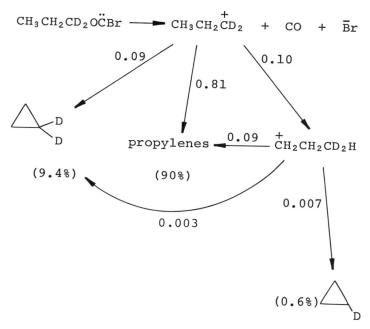

The array of products of the deoxidation of 2-methyl-1-butanol can be explained in terms of carbonium ions undergoing both 1,2- and 1,3- rearrangements (Scheme 23).[96] Two successive 1,2-shifts, leading to 3-methyl-1,1-butene can be excluded, as this would involve conversion of a t-amyl cation into the secondary 3-methyl-2-butyl cation. The t-amyl cation, which is probably involved in the deoxidation of both t-amyl alcohol and neopentyl alcohol, gives rise to only 2-methyl-2-butene and 2-methyl-1-butene, uncontaminated by 3-methyl-1-butene (see Table 5).[91] Thus, assuming equal partitioning of the 3-methyl-2-butyl cation, at least 8% of the products of deoxidation of 2-methyl-1-butanol arise from a 1,3-rearrangement.

Scheme 23

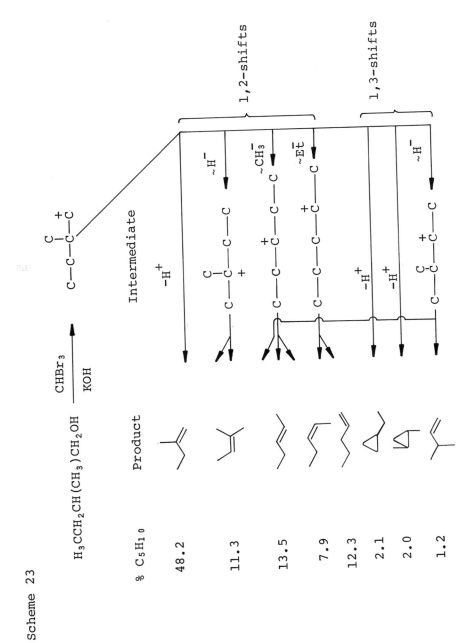

Protonated cyclopropanes offer an attractive alternative to classical alkyl carbonium ions as intermediates, especially as cyclopropane products are often formed in deoxidation reactions.[59] However, the deoxidation of neopentyl alcohol-1,1-d_2 by dibromocarbene is not consistent with this idea (see Scheme 24).[97] If a protonated cyclopropane had been an intermediate, or a transient storage form of the carbonium ion, deuterium should have appeared on both the C-3 and C-4 atoms of both 44 and 45. However, both products were deuterated only in the 3-position.

Scheme 24

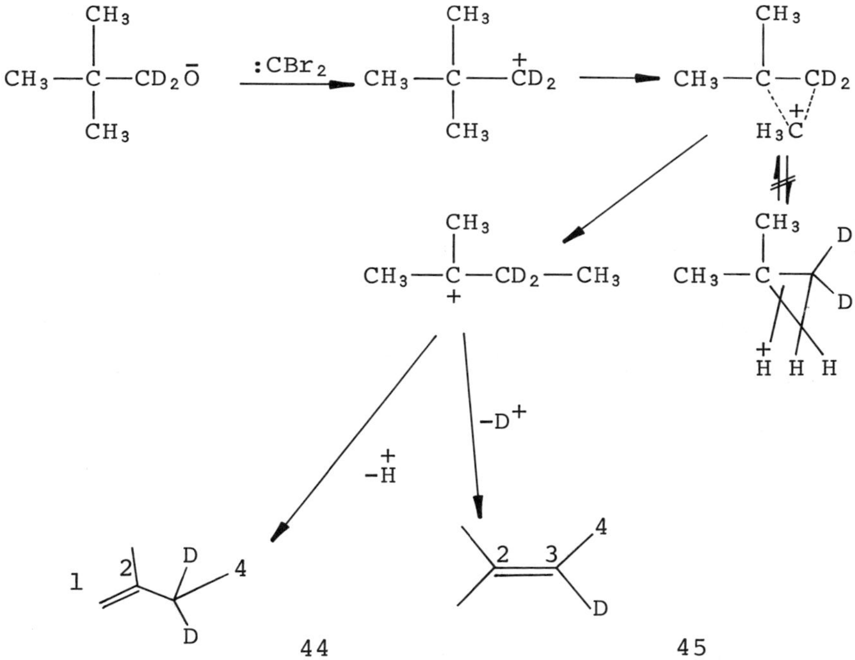

44   45

Sanderson and Mosher[98] prepared optically active S-(+)-neopentyl alcohol-1-d by a fermentation process, and heated it under reflux with bromoform and potassium hydroxide for four hours. Along with unreacted substrate (24%), three major reaction products were isolated; 46 (cf. 44), 31%; 47 (cf. 45), 14%; 48, 3%. Control experiments demonstrated that no interconversion of reaction products occurred and that no racemization of 46 took place during the reaction sequence used to verify its stereochemistry. Two interesting points arise from this work; (1) the 2-methyl-1-butene-3-d (46) is almost wholly the optically active S-(+)-form; (2) the thermodynamically less stable olefin (46) is formed in greater yield than 47. Under equilibrating conditions, dehydration of t-amyl alcohol gives a 2-methyl-1-butene/2-methyl-2-butene ratio of 1:7.[99] Thus the reaction products are those of kinetic control, and the major product is formed with almost complete inversion at the C-1 atom of neopentyl alcohol. Neither of these results can be accomodated by a planar carbonium ion intermediate.

Sanderson and Mosher,[98] considered a number of possible mechanisms, but preferred the concerted proton-abstraction-rearrangement process 49. The 9:2 ratio of γ- to α-hydrogens was cited as the factor determining orientation in favor of the less stable alkene. However, although this process accounts for the stereochemistry of the reaction and formation of 2-methyl-1-butene as the major product, it has a number of deficiencies. The low acidity of the γ-hydrogens, and the mediocre basicity of the reaction medium are not experimental conditions likely to cause much C-H bond breaking in the transition state. The driving force must come from ionization of the C-O bond. The origin of 48 is unlikely to lie in this route, as this would involve effectively an S_N2 reaction at the tertiary carbon atom. In addition, such a mechanism is not conceivable for t-amyl alcohol, yet under a variety of reaction conditions, both neopentyl and t-amyl

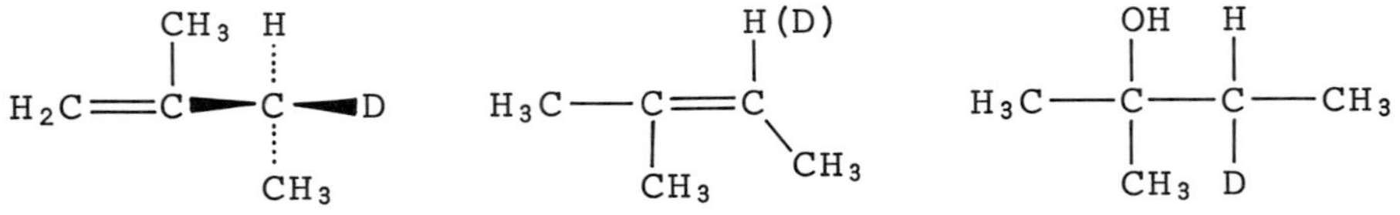

46   47   48

alcohol undergo deoxidation to give almost identical reaction products.[100] This observation strongly suggests a common intermediate or intermediates for the two reactions.

49

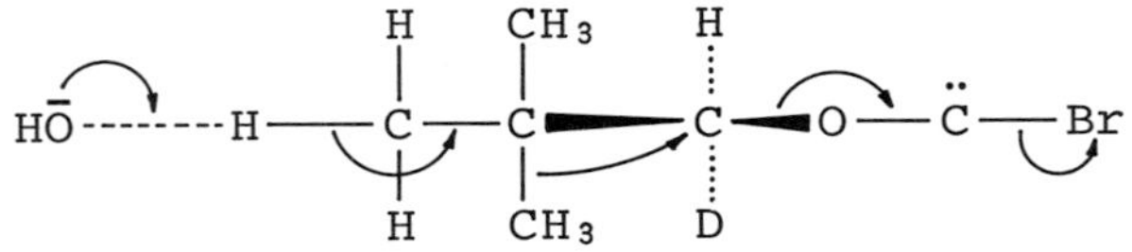

The stereochemistry of the 2-methyl-1-butene product can be explained by a concerted loss of carbon
monoxide and methyl migration, 50, or by a two-step reaction, involving an intermediate ion pair, 51 (or ions
separated by a carbon monoxide molecule). Ion pairs
are likely intermediates in these deoxidation reactions, as the solvent (usually the actual alcohol)
often has a low dielectric constant, and any water
molecules will be largely bound up with the concentrated alkaline solutions that are usually employed.
Whichever route operates, the penultimate intermediate
is most probably the t-amyl cation. The energetics

of this cation then determine the product distribution.
The elimination of a proton from the t-amyl cation is
controlled by two factors; (a) thermodynamic considerations that favor the 2-butene, and (b) statistical influences that favor the 1-butene, there being six
available eliminatable hydrogens leading to this product, but only two leading to 2-butene formation. For
thermodynamic factors to predominate, double-bond
formation in the transition state for the reaction of
the carbonium ion should be extensive; that is, overlap
of the vacant orbital of the carbonium ion with the
developing electron pair of the eliminating hydrogen
should be well advanced. However, the t-amyl cation
formed in deoxidation of t-amyl alcohol or neopentyl
alcohol is formed in a highly exothermic decomposition
in a poor solvating medium. It is, thus, largely unsolvated, and its reactions should be very exothermic,
proceeding through a reactantlike transition state in
which orbital overlap is little developed. Under these
conditions, statistical factors control the orientation
of reaction, and the less stable alkene is formed.
    A possible weakness in the above argument is the
assumption that the energetics and, consequently, the
mode of decomposition of the t-amyl cation formed

directly from t-amyl alcohol or indirectly from neopentyl alcohol will be the same. As the primary neopentyl carbonium ion is likely to be much less stable
than the tertiary amyl cation, it is possible that it
may decompose to give a tertiary ion with similar energetics to that formed directly. Recently Collins,
Raaen, and Eckart have shown that presumably identical
carbonium ions, formed indirectly by deamination of
isomeric amines in acetic acid, exhibit a striking difference in stereoselectivity (Scheme 25).[101] They
attributed the difference to anion control, although it
is tempting to point out the relative position of the
β-hydroxy groups in the two amines; in one case correctly positioned to aid fragmentation of the diazonium
ion by a backside attack.

Scheme 25

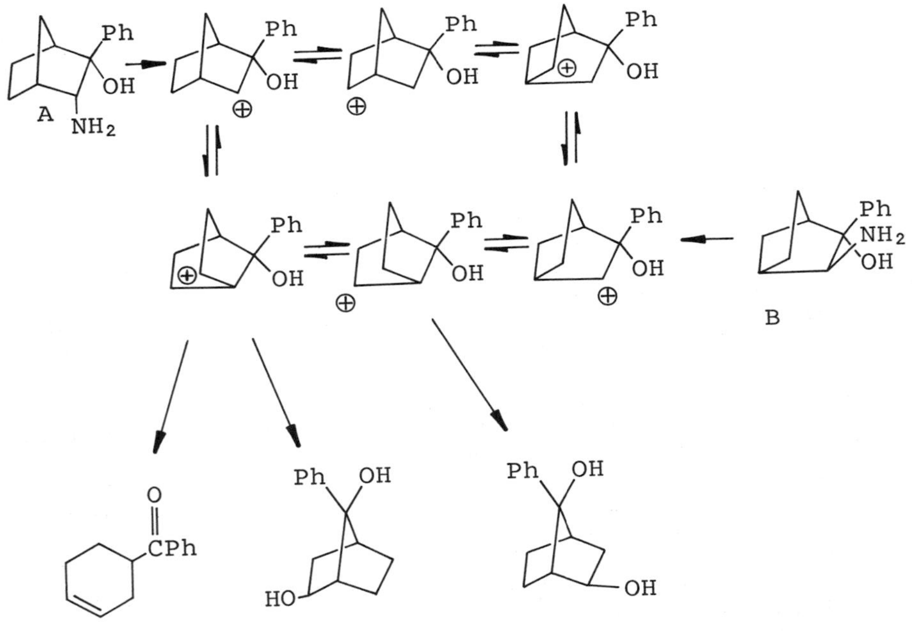

| 18.5 | 22.1 | 9.7 | (% yield from A) |
| 0.4 | 29.2 | 37.2 | (% yield from B) |

Lee and Wan[102] used carbon tracers to elucidate
the mechanism of deoxidation of potassium n-propoxide

in n-propanol by bromoform. From the distribution of label in propene, the relative contributions of the primary and secondary carbonium ions as intermediates can be gauged. The absence of label on the 2-position excludes a protonated cyclopropane as a precursor to the alkene (see Scheme 26). It is of interest to compare these results with those for the deamination of labeled n-propylamine by aqueous nitrous acid (Scheme 27). The fates of the 1-propyl cation are quite distinctive in the two reactions. In deoxidation, rearrangement to the secondary ion is a minor process. Possibly some concerted breakdown of 52, thus, bypassing the primary carbonium ion, aided by the basicity of the medium, occurs in deoxidation but not in deamination. Solvolysis products are much more abundant in deamination, possibly because of the greater proximity of the solvent molecules to the carbonium ion derived from a diazo-hydroxide and diazonium ion than that derived from an alkoxide-carbene adduct.

Scheme 26   The Fate of the 1-Propyl Cation from Deoxidation of K, 1-^{14}C-n-propoxide by $CHBr_3$.[102]

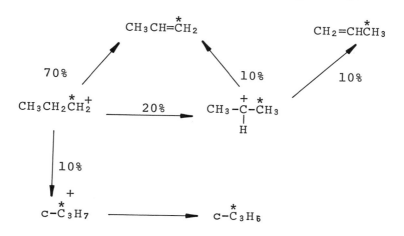

Scheme 27 The Fate of the 1-Propyl Cation from Deamination of n-propylamine-1-¹⁴C[102,103]

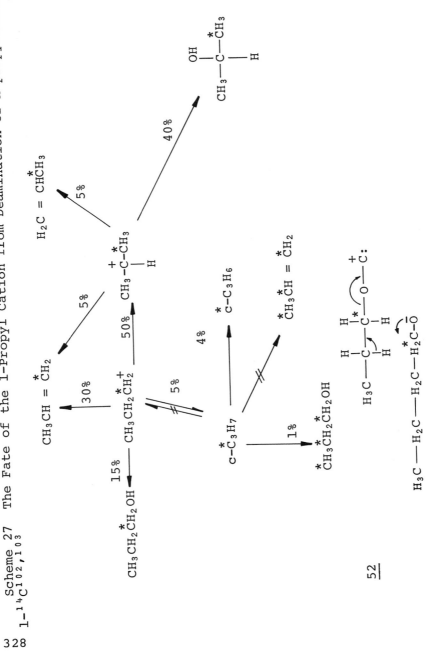

52

2-Phenylethanol-1-^{14}C reacts with bromoform and potassium hydroxide at 100° to give styrene in which the label is 22 to 26% rearranged.[104] This result is in close agreement with the % rearranged substitution product found in the nitrous acid deamination of the corresponding amine.[52,53] The results for both reactions can be ascribed to two simultaneous reaction paths. The <u>rearranged</u> isomer arises via a <u>phenonium ion</u>* and the <u>unrearranged</u> product forms via a reaction of the primary carbonium ion or, more probably, certainly in the deoxidation reaction, by a concerted elimination of type <u>52</u>. Among the reaction products, 2-hydroxy-1[14]C-3-phenylpropane was isolated in 3 to 5% yield. This observation shows that during deoxidation there is time for both rearrangement and subsequent reaction with hydroxide ion to occur.

Our discussion thus far has been limited to the ability of bromoform to convert alkoxide ions to alkenes. Hine and his co-workers[105] have estimated in a semi-quantitative manner, all of the reaction products formed when a variety of haloforms are treated with potassium isopropoxide in isopropanol. Along with propylene and carbon monoxide, bromoform gives rise to small amounts of acetone, methylene bromide, di-isopropyl ether, and tri-isopropyl orthoformate. With dilute base, chloroform yields propylene, carbon monoxide, diisopropyl ether, and tri-isopropyl orthoformate; with more concentrated base it yields, in addition, acetone and methylene chloride. However, dichlorofluoromethane and chlorodifluoromethane gave only tri-isopropyl orthoformate. To explain all these reaction products, Hine and his co-workers suggested a complex reaction

*The relative merits of bridged versus rapidly equilibrating classical ions as intermediates in solvolysis reactions has been a matter of dispute.[107a] In the closely related acetolysis of ^{14}C-labeled 2-phenylethyl tosylate and threo-1,2-dideutero-2-phenylethyl tosylate, Coke and his co-workers have shown that the results accord very well in terms of competing solvent displacement and anchimeric assistance via a bridged ion.[107b] The displacement gives unrearranged product with inversion, and the bridged ion gives rise to both rearranged and unarranged products with retention of configuration. As would be expected, because of less solvent participation in the bridged intermediate, the activation entropy is 7 eu more positive for the anchimeric assistance route. Unreasonable assumptions are required to explain the above facts in terms of rapidly equilibrating classical ions.

scheme (Scheme 28).

The formation of acetone is an example of a hydride transfer reaction. Presumably this occurs by one of the higher energy reactions of the carbene, as it does not appear to make a contribution in the case of difluorocarbene. Of the dihalomethylenes, this is the most stable, and by analogy the most selective in its reactions, because of the ability of fluorine to back-donate its lone pairs into the vacant orbitals of the electron deficient carbon atom. Most often, the reaction of alkoxide ion with dihalomethylene leads to the normal deoxidation reaction. However, when one of the halogens is fluorine, the decomposition of the adduct $R\ddot{O}CF$ does not occur unimolecularly to give the isopropyl cation, but involves addition of isopropoxide and eventual formation of tri-isopropyl orthoformate. The reluctance of the C——F bond to ionize is not surprising as, in typical S_N1 reactions, alkyl fluorides are often 10^6 times less reactive than the corresponding chlorides.[106]

The sensitivity of the reaction of haloforms with alkoxide ions to changes in the nature of the halogen atoms, limits the use of counterion studies on reaction products to gain information on the possible involvement of ion-paired carbonium ions in deoxidation. However, it would be worthwhile attempting to gain a knowledge of the stereochemistry of the olefin-forming step by studying deoxidation of appropriately labeled alkoxide ions in the manner analogous to that used by Cohen and Danweiski[66] in the deamination reaction (see Section A.1.f of this Chapter). The present available data suggest that carbonium ion intermediates are involved in deoxidation of secondary and tertiary alkoxides by bromoform but that concerted reactions of in termediate alkoxyhalocarbenes with alkoxide provide a competitive route for the reactions of normal alkoxides.

Scheme 28

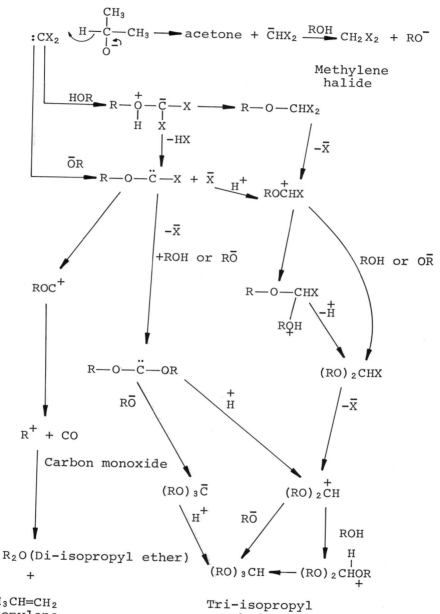

R = (CH₃)₂CH-

C DEHALOGENATIONS AND RELATED β-ELIMINATIONS

1 Introductory Remarks

Of the olefin-forming β-eliminations which do not involve hydrogen as one of the leaving groups, dehalogenations are the most common. With the exception of fluorine, as the leaving group, they can be induced with a variety of reducing agents. In fact, since Perkin initially reported an iodide-promoted elimination of coumarin dibromide,[108] many reductants have been discovered. These include metals (e.g., sodium,[109][110] lithium,[111] zinc,[112]) metal ions (e.g., iron II,[113] titanium III[114]), and a variety of nucleophilic reagents (e.g., thiolate,[115] acetate,[116] phosphite[117]). Recent investigations have revealed that the stereochemistry of the reaction and by implication, the reaction mechanism, is often mainly influenced by the choice of the reductant and the reaction conditions.[118] The overall process can be regarded either as a nucleophilic attack on a <u>positive</u> halogen or as a redox process involving a two-electron reduction of the oxidant dihalide.[119] We divide our subsequent discussion into categories determined by the class of the reagent as follows: (a) nucleophilic reagents, (b) metals, and (c) metal ions. An excellent summary of the reagents used to effect dehalogenation has recently appeared.[120]
    Vicinal dihalides should not be regarded as readily available precursors for alkenes. They are usually prepared by the electrophilic addition reaction of a halogen molecule to an olefinic double bond. However, they can be useful synthetic intermediates. For example, as oxidizing agents attack an olefinic double bond more readily than a hydroxyl group, allyl alcohol cannot be converted directly into acrylic acid. Initial protection of the olefin bond is required. This can be achieved by addition of bromine. By subsequent oxidation, followed by dehalogenation, the desired aim can be achieved (Scheme 29).

Scheme 29

$$H_2C=CH-CH_2OH \xrightarrow{Br_2} BrH_2C-CHBr-CH_2OH \xrightarrow{HNO_3}$$

$$BrH_2C-CHBr-COOH \xrightarrow{Zn} H_2C=CH-COOH$$

In addition to vicinal dihalides, a variety of other β-substituted alkyl halides can be reduced to alkenes. These include β-halogeno-alcohols, ethers, esters, and amines.[119] In general, these substrates only eliminate under more drastic reaction conditions than the corresponding vicinal dihalides. They are also prone to give lower yields of the alkene, reduction of only the carbon-halogen bond becoming more competitive with elimination as a result of the poor nature of the leaving groups. In some cases, both halogens can be replaced; for example, 1,2-ditosylates undergo elimination with iodide ion.[121] The mechanisms of these reactions are similar to those of the vicinal dihalides, and they are conveniently discussed at the same time.

2   Dehalogenations Induced by Nucleophilic Reagents

(a) <u>Iodide Induced Dehalogenations</u>. Of the nucleophilic reagents used to dehalogenate vicinal dibromides and dichlorides, iodide ion has been the most popular. Usually sodium iodide and the substrate are dissolved in acetone or alcoholic solvents, and the reaction temperature is modified to suit the reactivity of the dihalide. Other solvents, which have been used successfully, include methyl cyanide, dimethyl formamide (DMF), dimethyl sulfoxide, and dimethoxyethane (DME).[120] Kinetic and stereochemical studies have been confined mainly to the bromides, which are generally much more reactive than the corresponding chlorides.[122]

(i) <u>General mechanistic features</u>. The rates of dehalogenation of a variety of vicinal dibromides by sodium iodide in acetone or alcoholic solvents follow first-order kinetics in each reactant after an appropriate allowance is made for the iodide ion consumed by the iodine liberated during the course of the reaction.[123-127] The overall change is illustrated in Eq. 1. The stoicheiometry excludes a mechanism involving

$$RCHBrCHBrR \ + \ 3 \ \bar{I} \longrightarrow RHC{=}CHR \ + \ \bar{I}_3 \ + \ 2B\bar{r} \qquad (1)$$

a di-iodo intermediate,[128] but is still consistent with two other reaction pathways: a direct elimination or a rate-determining substitution to give a mono-iodo intermediate, which rapidly eliminates (Scheme 30).[129] The choice between these two routes appears to be dependent mainly on the structure of the vicinal

dibromide. In acyclic systems, when both or one of the
bromine atoms is attached to a primary carbon atom, the
S_N2-E2 mechanism is followed, but for substrates in
which the halogen is situated on secondary and tertiary
carbons, the experimental observations are usually
accommodated by the direct elimination. Kinetic and
stereochemical evidence support this generalization.

Scheme 30

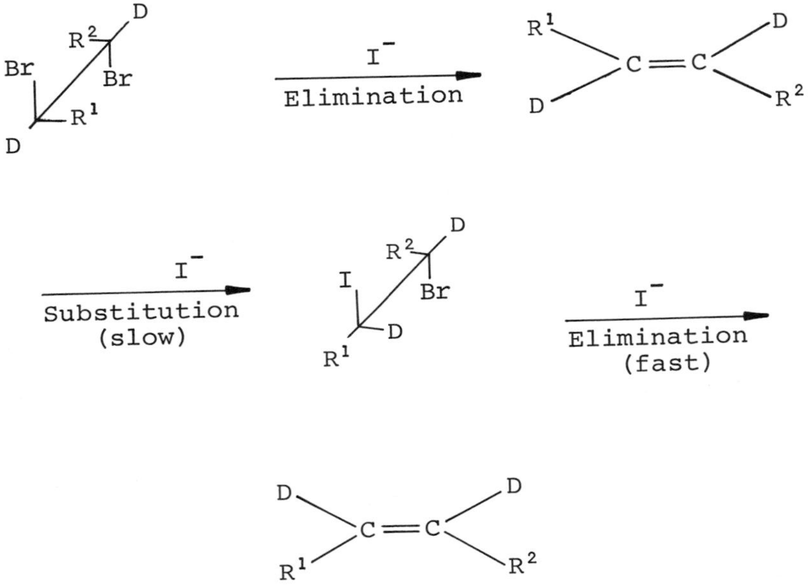

(ii) Stereochemistry. In the 1,2-dihalogeno-
ethanes, the identity of the β-halogen has been shown
to have little influence on the ease of bimolecular
nucleophilic displacement of the alpha-halogen.[130]
Comparison of the rates of substitution and elimination
by iodide ion indicates the dehalogenation occurs by
the S_N2-E2 route.[123] This suggestion is substantiated
by stereochemical studies on meso-dideuterodibromides.
If one (or both) halogen is primary (Scheme 30, R^1 =
R^2 = H, or R^1 = H, R^2 = CH$_3$, C$_2$H$_5$), the alkene is the
product of an apparent syn-elimination.[112] Initial
displacement with inversion of configuration is followed by anti elimination of BrI. However, meso-2,3-
dibromo-2,3-dideuterobutane (Scheme 30, R^1 = R^2 = CH$_3$)
gives the product anticipated for a direct anti

elimination.[124] In this case, the steric environment
of the carbon-halogen bond precludes substitution at
the expense of elimination relative to the primary
series.

A secondary isotope effect (k_H/k_D) of 1.28 at 80°
was observed in the debromination of sym-tetrabromo-
ethane by iodide ion in methanol.[127] On theoretical
grounds, a normal isotope effect is expected for a
direct elimination, but an inverse effect is expected
for a rate-determining substitution.[131] In view of
the similar spatial requirements of a bromine atom and
a methyl group, it is not surprising that the direct
elimination mechanism observed for 2,3-dibromobutane
is also followed by the tetrabromoethane.

So far our arguments have been based on the
assumption that the elimination step, whether it involves IBr or BrBr, occurs with anti-stereospecificity.
Justification for this approach is provided by debrominations of alicyclic substrates. In the rigid tri-
terpenoid system, 11,12-diequatorial dibromides, in
which the dihedral angle between the eliminating bonds
is 60°, are inert to iodide ion. However, diaxial isomers undergo facile anti elimination.[132,133] Cyclo-
hexyl-trans-1,2-dihalides undergo rapid elimination,
but the cis-isomers react more slowly, initial substitution being followed by rapid anti elimination from
the trans-1-iodo-2-halide.[134,135] The reactions of
1-halogeno-2-sulfonate esters exhibit similar stereochemistry. Using labeled bromine, Stevens and
Valicenti[136] demonstrated that trans-addition of bromine to 1-bromocyclohexene is followed by anti elimination with sodium iodide (Eq. 2).

(2)

The rate of appearance of iodine is slower than
that of sodium p-toluenesulfonate in the reaction of
iodide ion with the ditosylate esters of cyclohexene-
1,2-diols.[137] This observation is not consistent with

a direct elimination. The cis-substrate reacts by the
S_N2-E2 mechanism (Eq. 3), but the reaction sequence
for the trans-isomer is more complex (Eq. 4). Initial
bimolecular substitution gives cis-2-iodocyclohexyl
p-toluenesulfonate, which was actually isolated when
the reaction was quenched before completion. (The
corresponding trans-isomer undergoes anti elimination
too rapidly for isolation during the reaction.) A
second displacement yields the trans-di-iodo substrate
which, not suprisingly, eliminates rapidly to cyclohexene. Mechanisms involving mono-[138] and di-iodo[139]

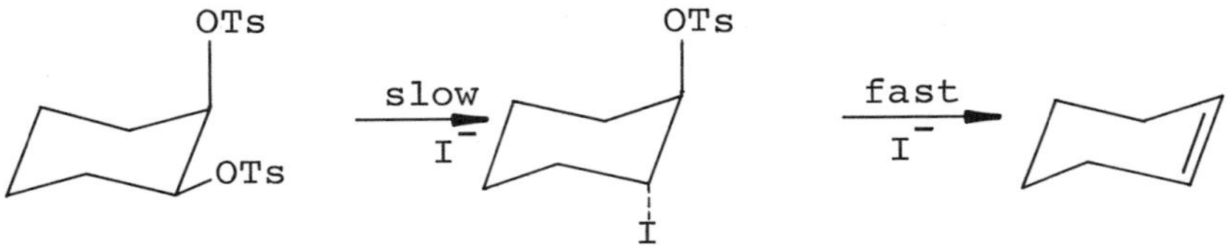

intermediates and even a carbanion (Elcb mechanism,
see Section I.B.1)[121] have been proposed for the
iodide-induced eliminations of tosylates of 1,2-diols
in the alicyclic, steroidal, and carbohydrate series.

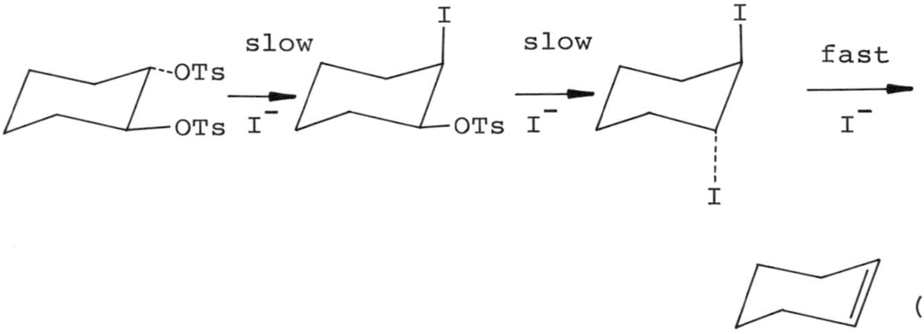

      (iii) Intermediates and transition states. The
experimental evidence we have presented thus far demonstrates that the iodide-induced elimination, whether
it occurs directly or after a rate-determining nucleophilic displacement, proceeds with anti-stereospecificity. We now discuss some of the simple transition
states and intermediates proposed by a number of
authors and devote the subsequent section (iv) to a
more detailed approach recently outlined by Miller and
his co-workers.[126]

On the basis of stereochemistry, a number of
authors have likened dehalogenation to a base-induced
E2 reaction (Section I.A.1) (see 53).[124,140-142] A
slight modification, 54, invoked by others, involves a
bridged structure in which the bromine atom attacked
by the iodide can interact with the α-carbon thereby
aiding C_α-Br ionization.[123,143,144] On the other hand,
Csapilla[145] has suggested "merged" transition states
(Section II.A.5): 55, in which iodide interacts with
C_α and the C_β-Br atom and 56, which allows an additional solvent interaction at C_β. As in the base-induced
eliminations, these latter transition states were proposed to explain substitution accompanying, or replac-
ing, elimination in the rate-determining step.

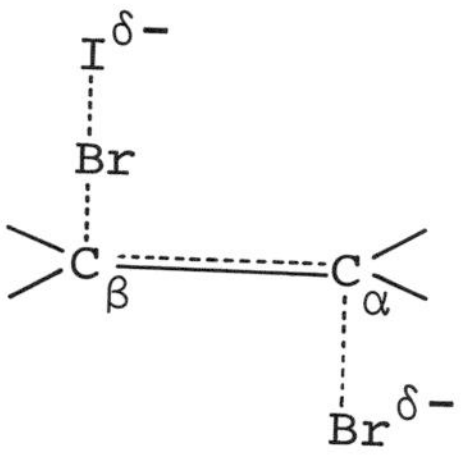

$$\underline{53}$$

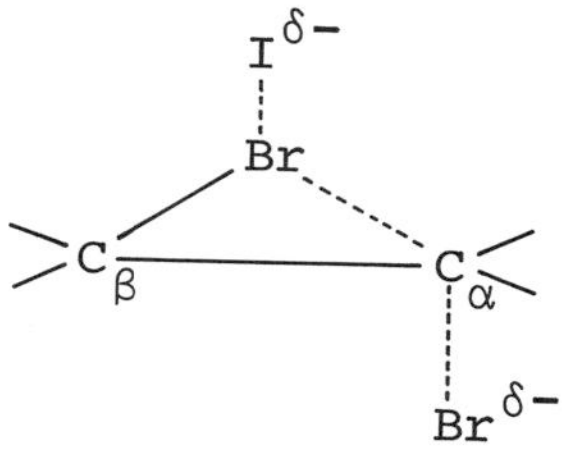

$$\underline{54}$$

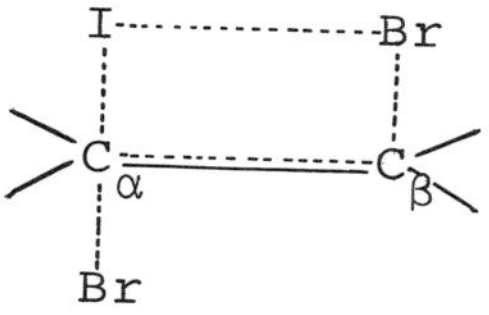

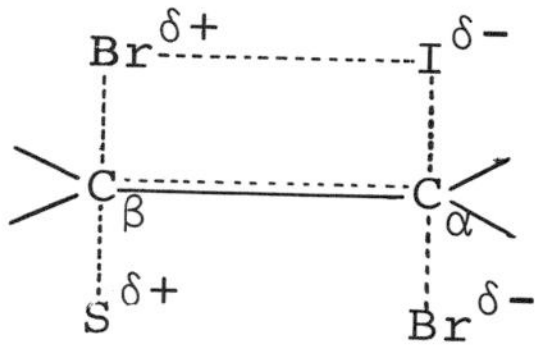

(S = solvent)

$$\underline{55}$$

$$\underline{56}$$

By analogy with base-catalyzed E2 reactions (Section III.D), the strict adherence to anti-stereospecificity suggests the importance of effective overlap
between the developing vacant and occupied orbitals
on the C_α and C_β atoms in the transition state for dehalogenation. Thus double-bond character should be
well advanced in the transition state and, in a series

of reactions, the relative rates should reflect to a certain extent the thermodynamic stabilities of the olefinic products. In part, the results listed in Table 6 support this hypothesis.[125-127,146-149] For example, the dehalogenations of meso- and erythro-1,2-dibromides, which give trans-alkenes almost quantitatively, occur much faster than those of the corresponding dl- and threo-substrates, which undergo anti elimination to give the thermodynamically less stable cis-alkenes in rather varying yields. The larger difference in enthalpies between the stilbenes[150] ($\Delta H_O = 2.3$ kcal/mole) resulting from greater eclipsing interaction of the aryl rings in the cis-alkene, is reflected in the 60-fold difference in the rate of formation of these products by anti elimination from the meso- and dl-substrates, respectively.[126]

TABLE 6   The Relative Reactivities of Some 1,2-Dibromides in Dehalogenation with Iodide Ion in Methanol at 59.6°[125-127,146-149]

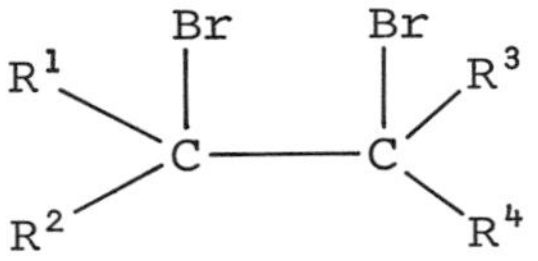

R^1 Br Br R^3 R^2 C—C R^4		meso- or erythro- a	dl- or threo- b	Other
R^1R^2	R^3R^4	$k_{Relative}$	$k_{Relative}$	$k_{Relative}$
Me,Me	Me,Me			23
Ph,H	H,Ph	19	0.3	
H,COOH	H,COOH	4	0.08	
Me,Me	H,Me			2.3
H,H	H,COOH			1.4
Ph,H	H,H			1.3
H,H	H,H			1.0
n-Pr,H	H,Me	0.1	0.03	
n-Pr,H	n-Pr,H	0.09	0.07	
Et,H	H,Me	0.06	0.03	
Me,H	H,Me	0.03		
Br,Br	Br,Br			0.0011

aReaction product trans-alkene.
bReaction product mainly cis-alkene.

The interesting trends in the individual figures and the ratios, k_{meso}/k_{dl} and $k_{erythro}/k_{threo}$ (see Table 6), cannot be regarded solely as a measure of double-bond character in the transition state. Not all the substrates react by the direct elimination route, the S_N2-E2 mechanism being followed by the dibromide of ethylene,[151] favored for that of styrene[152] and a likely possibility for that of acrylic acid. The presence of both geometrical isomers in the elimination fraction from dl-stilbene dibromide raises the possibility that syn elimination or the S_N2-E2 route competes with anti-elimination.[126] Thus we can only interpret certain aspects of the results and even these on only a very cautionary basis.

In the simple alkyl series, the rates of dehalogenation increase with the complexity of the alkyl moeities and the degree of substitution on the carbons bearing the halogen atoms. In fact, the increase in reactivity in the order; $1:10^2:10^3$, as the substitution on the C_α and C_β carbons goes from sec-sec. to tert-sec. to tert-tert. suggests that considerable S_N1 character develops in one of the C-Br bonds in the transition state. S_N2 interactions should decrease in importance along this same series as a result of progressive increases in steric hindrance. These results are, therefore, inconsistent with the merged transition states 55 and 56.

The Hammett equation[153] has been extremely useful in providing information concerning the charge distribution in the transition states of base-induced E2 reactions (Section II.C.2.a). However, although a number of authors have studied the influence of aromatic substituents on the rates of dehalogenation by iodide ion,[142,152,154] their results afford little definitive information on the transition state for elimination. For example, in the chalcone series, 57, both electron-withdrawing (X = p-Cl, p-NO$_2$) and the electron-releasing p-methoxyl substituent increase the rate of debromination relative to that of the unsubstituted compound. The failure of the Hammett equation in this instance is attributable to a number of causes. Changes in the substituent may induce a different mechanism (E S_N2-E2), may result in subtle variations in bond making and bond breaking in the transition state, or may cause the role of the bromine atoms to be interchanged.

$$X-\langle\langle\rangle\rangle-CHBrCHBrCOC_6H_5$$

57

In view of the greater electronegativity of chlorine than bromine, dehalogenation of erythro-1-bromo-2-chloro-1,2-diphenylethane most certainly involves elimination of $C\bar{l}$. Baciocchi, and Schiroli[122] have compared the rate of dehalogenation by iodide ion of the above substrate to that of meso-stilbene dibromide. In both methanol and DMF, both substrates eliminate quantitatively to trans-stilbene. The magnitude of the leaving-group effect [at 50°, k_{Br}/k_{Cl} = 87 (MeOH), 260 (DMF)] indicates a high degree of C-Cl(Br) bond breaking in the transition state (Section IX.B.2.c). As chloride ion is smaller and, hence, the negative charge is more localized, ionization of the C-Cl bond will require more solvent assistance than the C-Br bond. Not surprisingly, the reactivity of the chloro-substrate is depressed relative to the bromide in the poorly anion-solvating dipolar aprotic solvent.[156] This factor acts against the increase in reactivity of the iodide ion in DMF and, consequently, the dehalogenation of both substrates is only slightly faster in this solvent than in methanol. The same authors also showed that both the strongly electron-withdrawing p-nitro- and electron-releasing p-methoxyl-substituents increased the debromination of meso-stilbene dibromide only slightly in both solvents. Clearly, little development of charge on either of the carbons occurs in the transition state. The negative charge introduced into the substrate by $\bar{I}$ resides mainly on the leaving group, and double-bond character is well developed. Baciocchi and Schiroli concluded that their results could be adequately explained in terms of the conventional E2-transition state 53. They excluded the merged intermediates as the dehalogenations were slower with chloride than bromide than iodide, whereas in DMF, if carbon nucleophilicity had been important, the reverse order of reactivity should have been observed.[155] We discuss bromide and chloride-induced dehalogenation in this Chapter (Section C.1.b).

The stereochemical and substituent effects discussed in this section accord equally well with either transition state 53 or 54. The merged proposals 55 and 56 seem less appropriate.

(iv) Dehalogenation of stilbene dibromides. Miller and his co-workers have recently reported the dehalogenations of meso- and dl-stilbene dibromides with a variety of one- and two-electron reductants.[120] More than 50 different reagents have been used to debrominate the meso substrate to trans-stilbene. Some 30 reagents have been employed to convert the

dl-compound into cis- and trans-stilbenes, the yield of
the former varying from 0 to 96%. In fact, apart from
a few two-electron reductants, the eliminating agents
generate mainly trans-stilbene. The selectivity of
the dl-stilbene dibromide toward debromination appears
to offer a useful guide to reaction mechanism. As the
trans-alkene is much more stable than the cis-form over
the temperature range usually employed, any intermediate, which allows free rotational processes to establish an equilibrium between precursors leading to the
two alkenes, will strongly favor formation of the
trans-alkene. Additionally, Miller and his group have
also considered the reverse reaction (electrophilic
addition to alkenes), and competing substitution and
rearrangement reactions in proposing their all-embrac-
ing reaction scheme for dehalogenation. In this section, we concentrate on the iodide debromination.

In both methanol and DMF,[126],[156] meso-stilbene
dibromide is converted quantitatively into trans-stilbene by sodium iodide. The anti-stereospecificity and
the first-order dependence of the rate of reaction in
both substrate and iodide ion are consistent with a
one-step elimination proceeding via transition states
53 or 54. The other nucleophiles present, the solvents
and bromide ion, are insufficiently reactive to compete
with the excess of iodide ion for the substrate.

In proposing a more complex mechanism for debromination, Miller and his co-workers have considered a
number of observations concerning the electrophilic
addition reaction. In hydroxylic solvents, the rate
of bromine addition to an alkene often takes the form
of Eq. 5.[151],[157-159] The initial bracket could refer
to a concerted bimolecular or trimolecular reaction.
However, it is generally accepted that the addition
reaction occurs in a stepwise fashion.[151],[157] The
addition usually occurs with anti-stereospecificity
and, in the presence of another nucleophile, products
of mixed addition (Br-alkene-X) are formed. In carbon
disulfide, bromine addition to cis-stilbene gives

$$-d(Br)/dt = [k_1 + k_2(\overline{Br})][(Alkene)(Br_2)] \quad (5)$$

dl-stilbene dibromide. In more polar solvents, such as
nitrobenzene and ethanol, some meso-stilbene dibromide
is also produced, but in the presence of added bromide
ion, a complete reversion to apparent anti-addition is
observed. Polar solvents by stabilizing the intermediate "onium" species, allow partial or complete equilibration before their capture by bromide ion or another

nucleophile.[160-165] Invoking the principle of microscopic reversibility, Miller and his co-workers[126]
suggest that the same intermediates could be involved
in both addition to and elimination forming alkenes
(Scheme 31). Their scheme, which is far more detailed
than the simple concerted mechanisms discussed in the
previous section, embraces dehalogenation, addition,
and solvolysis-substitution. For meso-stilbene dibromide, they propose attack of iodide ion on the substrate or 60 in the rate-determining step, the transition state being similar to 58, which is essentially
54. Back reactions (e.g., 59→58) seem unimportant, as
(in acetone) added bromide ion does not cause a rate
retardation.[143] The highly favorable orbital overlap
and the lack of steric interactions between the two
aryl rings make the elimination much more facile than
other alternative processes.

In methanol, dl-stilbene dibromide is much less
reactive than the meso-isomer toward iodide ion.
Eclipsing interactions between the phenyl groups raise
the energy of activation for anti elimination, leading
to cis-stilbene and displacement, and solvolytic reactions become competitive. Of several reaction products
(see Scheme 32), cis-stilbene (30%), trans-stilbene
(15%), and meso-α,α'-dimethoxybibenzyl (30%) are the
most prominent. Interconversion of the reaction products under the reaction conditions is not significant.
Independent studies showed that erythro-α-bromo-α'-
methoxybibenzyl was converted mainly into meso-α,α'-
dimethoxybibenzyl (75%) and trans-stilbene (23%) by
iodide ion in methanol, but at a much slower rate than
the dehalogenation of dl-stilbene dibromide. From
these observations, Miller proposed the operation of
the entire Scheme 31 for the reaction of dl-stilbene
dibromide with iodide in methanol. However, their data
also accord with concurrent E2, S_N2-E2, and S_N1-E1
mechanisms.

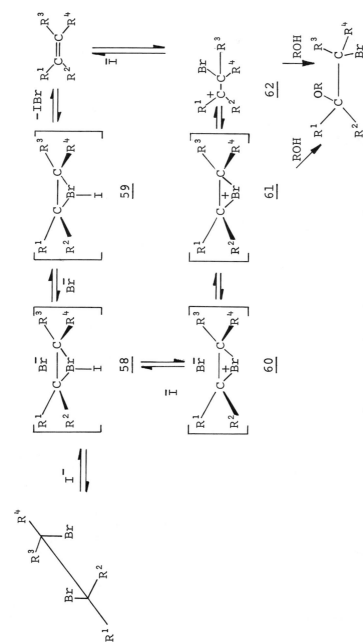

Scheme 31

for meso-stilbene dibromide, $R^1 = H = R^4$, $R^2 = C_6H_5 = R^3$

for dl-stilbene dibromide, $R^1 = C_6H_5 = R^3$, $R^2 = H = R^4$

Scheme 32

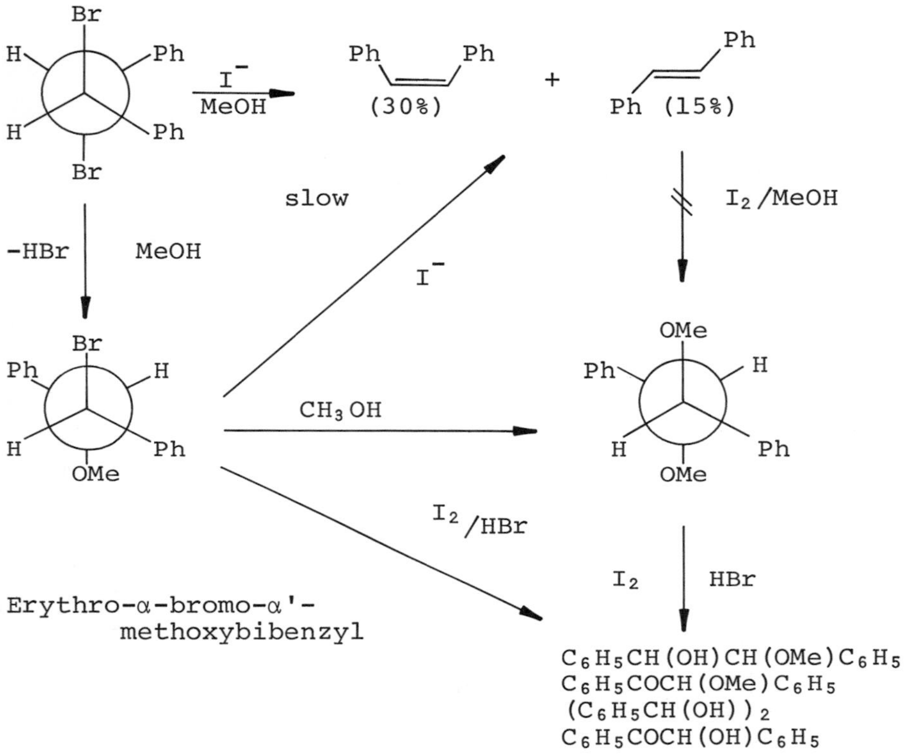
Br
H
Ph
H
Ph
Br
I⁻
MeOH
Ph
Ph
(30%)
+
Ph
Ph (15%)
slow
I₂/MeOH
-HBr
MeOH
I⁻
Br
Ph
H
H
Ph
OMe
CH₃OH
OMe
Ph
H
H
Ph
OMe
I₂/HBr
I₂
HBr
Erythro-α-bromo-α'-methoxybibenzyl
C₆H₅CH(OH)CH(OMe)C₆H₅
C₆H₅COCH(OMe)C₆H₅
(C₆H₅CH(OH))₂
C₆H₅COCH(OH)C₆H₅

Solvolytic reactions do not complicate iodide-induced debromination in DMF, and dl-stilbene dibromide is converted into cis-stilbene (88%) and trans-stilbene (12%).[159] Despite the much reduced reactivity relative to the meso-substrate (k_{meso}/k_{dl} = 300 for 25 to 75° for anti elimination), the reaction is remarkably stereoselective. Miller and Mathai again prefer a stepwise mechanism for debromination (Scheme 31), but they do indicate that the trans-stilbene could arise via a S_N2-E2 sequence as the observed rate coefficient is very similar to the value predicted for iodide displacement from calculations using data from model compounds.

The essential difference between Miller's approach and the simple mechanisms discussed in the previous section lies in the explanation of the origin of the multitude of reaction products. Miller prefers a common rate-determining step followed by fast simultaneous modes of decomposition of the intermediate to the reaction products. Within the framework of the elimination transition states 53 and 54, the source of substitution and elimination products not of direct anti elimination origin, are explained by simultaneous rate-determining mechanisms. Within our present know-ledge of kinetics, clear-cut distinction between these schemes cannot be made, and the choice of mechanism remains one of personal conjecture rather than proved scientific fact.

Miller has proposed an approach for comparing the energetics of the anti eliminations of meso- and dl-stilbene dibromides, using Eq. 6.[167] The terms G, $G^{\ddagger}$,

$$(G^{\ddagger}_{dl} - G^{\ddagger}_{meso}) = (\Delta G^{\ddagger}_{dl} - \Delta G^{\ddagger}_{meso}) + (G_{dl} - G_{meso}) \qquad (6)$$

and $\Delta G^{\ddagger}$ refer to the ground state, transition state, and activation free energies, respectively. For ben-zene at 80°, $(G_{dl} - G_{meso}) = 0.78$ kcal/mole, a value that is assumed to be appropriate for the other sol-vents. From the kinetic data at 80°, $(\Delta G^{\ddagger}_{dl} - \Delta G^{\ddagger}_{meso})$ ≃ 2.6 kcal/mole in methanol and 3.3 kcal/mole in DMF. Therefore, the values of $(G^{\ddagger}_{dl} - G^{\ddagger}_{meso})$ are 3.4 kcal/mole and 4 kcal/mole in methanol and DMF, respective-ly.[126,156] For the product stilbenes, $(G_{cis} - G_{trans})$ = 3.7 kcal/mole.[150] Consequently, these calculations lead to the prediction of product-like transition states in the dehalogenations, a conclusion that accords with the substituent effects discussed in the previous section. The rates of addition of bromine to cis- and trans-stilbenes are very similar, resulting in a prediction of transition states that are energet-ically-like the reactant stilbenes. Thus this dissec-tion of reaction energetics provides support for the assumption that addition and elimination reactions pass through common intermediates.

(b) Chloride and Bromide-Induced Dehalogenation. Un-like iodide ion, chloride and bromide ion are inef-fective reagents for debromination of vicinal dibro-mides in protic solvents.[126] A number of factors con-tribute to this change in reactivity. Bromine and chlorine being more electronegative than iodine, chloride and bromide are less easily oxidized than io-dide. In protic media, small ions are solvated by hydrogen-bonding interactions with solvent molecules

more strongly than larger ions. Thus the nucleophilicity of the halides in displacement reactions at saturated C, N and S decreases markedly along the series
$\bar{I}$ > $\bar{Br}$ > $\bar{Cl}$.[169] Finally, the halogen molecules liberated during the dehalogenation reaction can undergo
an electrophilic addition with the olefinic product,
thereby causing the product to revert to starting
material. This process is obviously more serious in
suppressing the extent of dehalogenation for bromide-
and chloride- than for iodide-induced eliminations.
We see subsequently that the addition reaction can complicate kinetic studies of dehalogenation and can cause
the reaction to follow thermodynamic rather than kinetic control.

Recently, the advantage of using dipolar aprotic
solvents as media for reactions involving anionic reagents has been realized.[170] These solvents usually
possess a molecular dipole in which the positive end
is sterically shielded, and anions are therefore poorly
solvated and possess highly localized and reactive
electron pairs. For bimolecular nucleophilic displacement reactions at saturated C, N, or S, the reactivity
order $\bar{I}$ < $\bar{Br}$ < $\bar{Cl}$ (the reversal of the reactivity sequence in protic media), is observed.[156,169] Success-
ful debrominations with bromide and chloride ion have
been reported in acetone and DMF.[142,143,166,168,171]
We confine our subsequent discussion to the stilbene
dibromides.

In DMF, meso-stilbene dibromide undergoes debromination with anti-stereospecificity to give trans-
stilbene quantitatively, with all three tetra(n-butyl)
halides. At 25°, the relative reactivities of the
halides are as follows: $\bar{I}$: $\bar{Br}$: $\bar{Cl}$ = 520:7.4:1.[142]
This order of reactivity is the opposite of that ob-
served for nucleophilic substitution at saturated C in
this solvent,[158] suggesting that halide ion-carbon atom
interactions are of minor importance in the transition
state for dehalogenation. The merged transition states
(55 and 56) and the S_N2-E2 mechanism seem unlikely.

The k_{Br}/k_{Cl} ratio (dehalogenation of meso-stilbene
dibromide compared to that of erythro-1-bromo-2-chloro-
1,2-diphenylethane) is similar for the reaction with
KI(260) and NaBr(285 at 50° in DMF).[142] For a more
certain comparison of the reactivities of the halides,
a common counterion, preferably an n-tetraalkylammonium
ion, which has little tendency toward ion association,
is really required. However, the counterion possibly
has little influence on the rupture of the C-Cl(Br)
bond in this case and, obviously, the bonding between
the carbon atom and the halogen, which departs as an

anion, is considerably weakened in the transition state for both bromide- and iodide-induced debrominations (Section IX.B.2.c).

As for the iodide debrominations, all substituents increase the reactivity of the meso-stilbene dibromide molecule in both bromide and chloride ion reactions in DMF.[142] The substituent effect is slightly greater for the latter two reactions, and the rate ratio, $k[(\text{meso-}p\text{-NO}_2\text{C}_6\text{H}_4\text{CHBr})_2]/k[\text{meso-stilbene dibromide}]$, varies quite markedly with the halide: $\bar{\text{I}}$, 4; $\bar{\text{Br}}$, 24; $\bar{\text{Cl}}$, 100 (at 25°). Thus the nucleophilic reactivity ratios $\bar{\text{I}}$: $\bar{\text{Br}}$: $\bar{\text{Cl}}$ at 25° decrease significantly as the substrate is changed from meso-stilbene dibromide (520: 7.4: 1) to its 4,4'-dinitro derivative (22: 1.5: 1). Baciocchi and Schiroli[142] suggest that halide ion-α-carbon interaction may be more important when the halide is $\bar{\text{Cl}}$, less so with $\bar{\text{Br}}$, and of little significance with $\bar{\text{I}}$, which from measurements in aqueous solution has the greatest bromine nucleophilicity.[172] This interaction is only important for the dinitro compound which possesses the most electrophilic α-carbon atom of the substrates studied.

The debromination of dl-stilbene dibromide by lithium bromide in DMF is complicated by competing dehydrobromination by the solvent and by addition of the liberated bromine to the olefinic products.[168] To remove the liberated bromine, the reaction was performed in the presence of stannous chloride. This reagent also depresses the rate of dehydrobromination and, by careful monitoring of the $(\text{LiBr})/(\text{SnCl}_2)$ ratio, it is possible to measure the debromination by either of these reagents.

dl-Stilbene dibromide is debrominated by lithium bromide in DMF to trans-stilbene (83%) and cis-stilbene (17%). (In the absence of the bromine scavenger, only the trans-alkene is isolated). These figures contrast markedly to the sodium iodide reaction which gives cis-stilbene (88%) and trans-stilbene (12%). The iodide reaction therefore shows a greater preference for anti-stereospecificity. Miller and his group[168] explain this change in terms of Scheme 33. It is not clear why 63 decomposes via 64 to a greater extent when bromide rather than iodide is the nucleophilic agent.

The products of the dl-stilbene dibromide dehalogenation could have arisen by various combinations of anti elimination, syn elimination, and S_N2-E2 reactions. The latter can reasonably be excluded as the meso-substrate reacts 50 times faster at 59° than the dl-substrate with lithium bromide, and $k_{\bar{\text{I}}}/k_{\bar{\text{Br}}}$ for the latter is about 10. For the temperature variation

Scheme 33

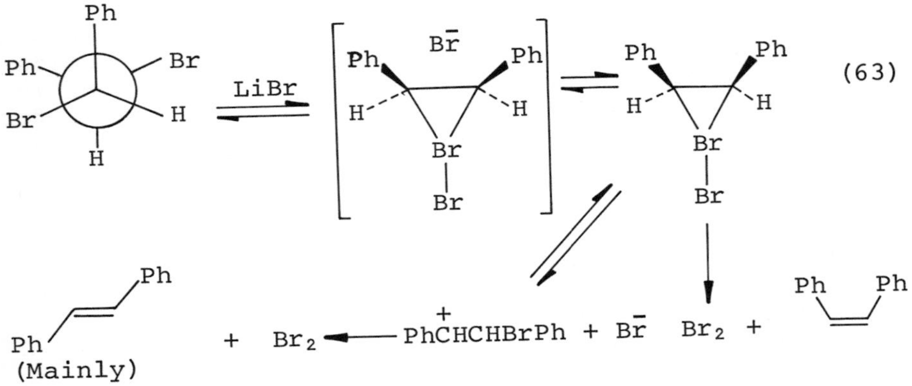

(63)

$$PhCHCHBrPh + Br^- + Br_2 +$$

PhCH=CHPh (Mainly) + Br$_2$ ← PhC̊HCHBrPh + Br̄   Br$_2$ +

64

60 to 100°, the proportion of trans-stilbene in the bromide-induced debromination varies from 84.6 to 79%, a change regarded as insignificant by Miller and his co-workers,[168] who regard this as evidence supporting a single rate-determining step, followed by the rapid partitioning of unstable intermediates. However, the change in product proportions is of a magnitude similar to that observed in the reaction of isopropyl bromide and sodium ethoxide in 80% aqueous ethanol, which occurs by competing S_N2 and E2 mechanisms [%S_N2; 42 (50°), 34 (100)].[173] Small variations in the product distribution with temperature need not necessarily be indicative of a single rate-determining step.

For the iodide-induced dehalogenations, the greater reactivity of the meso-substrate arises from its smaller activation energy, the entropies of activation being similar for both isomers (see Table 7), but in the bromide reaction, a significant difference in the entropy of activation is also observed.

On the evidence available, it is not possible to exclude syn elimination involving an ion pair (Li̇B̄r) and dl-stilbene dibromide as the origin of trans-stilbene. Certainly, the smaller Li̇ and B̄r ions are more likely to associate in DMF than the larger Nȧ and Ī ions. The difference in the yields of trans-stilbene between the iodide and bromide reactions could be a measure of the extent of association. Until the role of "free" ions and ion pairs in the dehalogenation is unfolded, it will not be possible to distinguish

TABLE 7 Activation Parameters for the Reaction of Stilbene Dibromides with Halide Ions at 59.6° (Elimination Components Only)

Substrate	Solvent	Halide Ion	$\Delta H^{\ddagger}$(kcal/mole)[a]	$\Delta S^{\ddagger}$(cal/deg)[b]	Reference
meso	MeOH	$\overset{+}{N}a\ \bar{I}$	19.6	-13	126
dl	MeOH	$\overset{+}{N}a\ \bar{I}$	22.5	-12	126
meso	DMF	$\overset{+}{N}a\ \bar{I}$	16.3	-15	156
dl	DMF	$\overset{+}{N}a\ \bar{I}$	22.3	- 7	156
meso	DMF	$\overset{+}{L}i\ \bar{B}r$	20.6	- 9	168
dl	DMF	$\overset{+}{L}i\ \bar{B}r$	28.9	+ 8	168

a ±1.0 kcal/mole.
b ±3 cal/deg.

between competitive syn and anti elimination and the
various contributions of "onium" ion intermediates proposed by the Miller group. Dehalogenation studies with
the tetraalkylammonium halides on dl-stilbene dibromide
may provide the necessary evidence.

(c) Various Nucleophilic Reagents Other Than Halide
Ions. The stereochemical consequences of the reductive
eliminations of dl-stilbene dibromide with a number of
nucleophilic reagents are shown in Table 8. Some of
the nucleophilic reagents (entries 3 to 6) show very
similar stereoselectivity to iodide ion, indicative of
the operation of a similar mechanism. The temperature
dependency of the yield of cis-stilbene in the LiAlH$_4$
reduction lies in the ability of this reagent to promote cis-trans isomerization. Interestingly, the
smaller ions $\bar{H}$ and $\bar{B}H_4$ simulate $\bar{B}r$ in debromination,
syn elimination or some stereoelectronically equivalent
mechanism being more facile than anti elimination.

From a synthetic standpoint, a number of reagents
have a rather limited but specific application. Simple
vicinal dibromides are converted by trialkylphosphites
into mono- and di-phosphonates but vicinal dihalides,
flanked by α-carbonyl functions, debrominate in high
yield.[175] The latter substrates, when treated with
sodium iodide in acetone, give tars, and zinc metal
also reduces the olefinic product. Care has to be
taken to use only a molar equivalent of the phosphite,
greater amounts causing formation of the phosphonates.

As a planar dienophile is required for the Diels-
Alder reaction, the facile reaction of cyclooctatetraene, 65, with maleic anhydride is explained as
occurring through bicyclo (4,2,0) octa-2,4,7-triene,
its valence tautomer, 66, which contributes about 0.01%
to the total structure at equilibrium at room temperature.[176] Bromination of cyclooctatetraene yields its
dibromide 67, but attempts to debrominate this substrate with sodium iodide in acetone at room temperature for 45 hr led only to the formation of cyclooctatetraene.[177] However, at -78° in dimethyl ether, the
highly reactive reductant, disodiumphenanthrene, induces debromination in 95% yield to a product that has
the spectral properties expected for 66.[178] The NMR
spectrum shows the presence of 4 vinylic protons in a
six-membered ring, and 2 in a four-membered ring, and
2 aliphatic doubly-allylic protons. Catalytic hydrogenation of the substrate gives bicyclo (4,2,0) oc-
tane. A rise in temperature causes the substrate to
convert into cyclooctatetraene and at 0°, the half-
life for this reaction is only 14 min. Not

TABLE 8 The Yields of cis-Stilbene in the Total Stilbene Fraction in the Debromination of dl-Stilbene Dibromide by a number of Nucleophilic Reagents[120]

	Nucleophile	Solvent[a]	cis-Stilbene %[b]	Reference
1.	NaI	DMF, MeCN, EtOH, DMSO MeOH, n-PrOH, DME, Acetone	89 to 96	120
2.	LiBr	DMF	16	168
3.	K_2PtCl_4	DMSO	85	120
4.	$C_6H_5SO_2Na$	DMSO, DMF	88	120
5.	$p-CH_3C_6H_4SNa$	EtOH, DMF, DMSO	> 75	120
6.	$LiAlH_4$	THF at -10°	95	144
		at 25°	50 to 65	144
7.	NaH	HMPA at 35°	27	174
8.	$NaBH_4$	Diglyme at 25° (30 min)	25	174

[a]The reactions were usually run for about 48 hr under reflux or at 60 to 70° for DMF and DMSO, unless specified.
[b]% cis-Stilbene in cis-trans mixture.

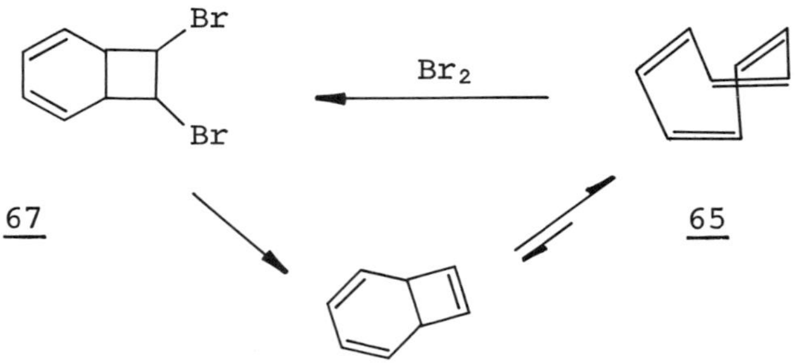

67

65

66

suprisingly, 66 was not isolated in the sodium iodide
reaction.

Sodium dihydronaphthylide and biphenylide have
been reported as very effective reagents for debromination to cyclobutenes.[179]   These reagents are conveniently prepared by dissolving sodium metal in an ethereal solution of the aryl compound under nitrogen. Fil-
tration removes the metal traces, and the dihalide is
then added. Excess reagent is destroyed by exposure
of the reaction mixture to the air. Sodium dihydronaphthylide is converted into polycyclic fused ring
systems by 1,3- and 1,4-dichloroalkanes.[180]

The products of reaction of the erythro-dibromides
68 and 69 with various nucleophiles illustrate the importance of coplanarity of the fragmenting bonds in the
transition state for anti elimination.[181]   These substrates can undergo either dehydrobromination or debromination, the most favorable reactant conformations
leading to the transition states for anti elimination
being 70 and 71, respectively. Steric interactions are
less severe for debromination and for the amide 68,
this is the sole mode of reaction for the following reagents: aqueous solutions of Na_2CO_3, KOH, and
CH_3COONa, Ph_3P in $CHCl_3$, and pyridine/thioacetic acid
in benzene. For the ester, 69, the latter two reaction
solutions produce the debromination product, but the
more basic reagents (greater nucleophilicity toward
hydrogen) cause mainly dehydrobromination. Clearly,
COOEt is more acid strengthening than $CONEt_2$, and the
increase in C-H acidity partially offsets the unfavorable steric factor. Perhaps this increase in acidity
causes a change to a transition state for dehydrobromination with greater carbanion character (Sections

II.A.3 and III.E.2), double-bond character being less developed in the transition state and eclipsing inter- actions between substituents destined to be cis-relat- ed in the product being thereby reduced.

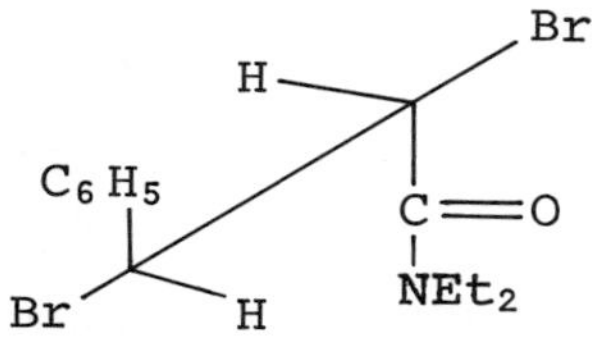

68

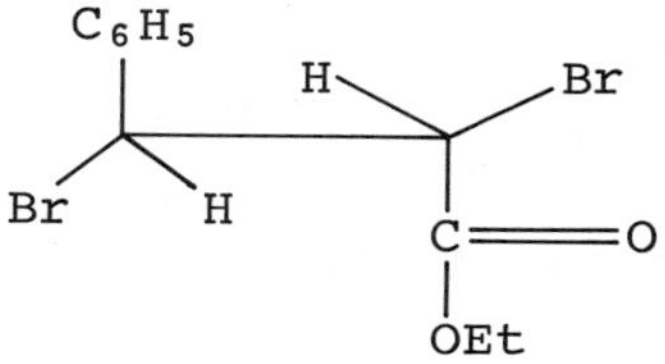

69

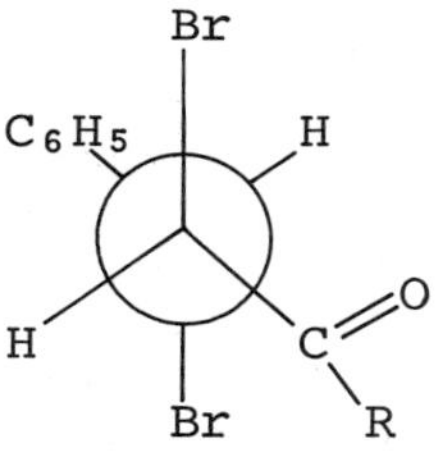

70

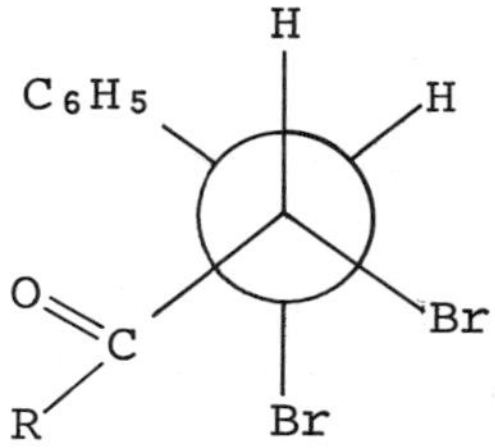

71

3 Metal-Promoted Dehalogenations

Many metals have been used in a variety of solvents as reductants for debromination of vicinal dibromides. The reduction is usually assumed to occur at the sur- face of the metal, and the reactivity of this surface is maintained by the presence of the solvent, which re- dissolves the products and acts as the transport medium for the substrate to the metal surface.

Of the metals, zinc has been used to effect de- halogenation most frequently. The products anticipated for anti elimination are formed in the reaction of zinc with meso-dideuterodibromoethane and the diastereoiso- meric-2,3-dibromobutanes.[110,112] However, departure from such stringent stereospecificty is found for the reactions of the dibromo-pentanes, -hexanes and -octanes,[182] but the purity of the starting materials was less certain in these cases. In the cyclodecyl series, vicinal dibromides show a marked preference for syn elimination.[183] These trends have a parallel in the base-catalyzed Hofmann eliminations of quaternary ammonium salts (Section III.E.6). For these reactions, the change from anti to syn elimination can be accounted for in terms of steric inhibition to the approach of the base by the bulky alkyl groups in the conformation leading to anti elimination. Similar steric interactions could cause preferential surface absorption of conformers leading to syn rather than anti elimination.

Products of substitution and dehydrohalogenation* are never found simultaneously with those of metal-promoted dehalogenation. This observation excludes the unlikely possibility that the role of the metal is of electrophilic nature, inducing a carbonium ion mechanism. Some authors have attributed the change in stereospecificity to the intervention of carbanion intermediates in which rotation about the C_α - C_β bond is more rapid than or is of a similar rate to the ejection of bromide ion.[109,182] However, any carbanion must have a very transient existence as products characteristic of carbanion capture by the solvent are not observed in metal-promoted dehalogenations. Consequently, if any carbanions are involved, they must be associated with the metal surface. (Similar restrictions apply to the role of radical intermediates, which are more probable for monovalent than divalent metals.) An extension of this view is to invoke organometallic intermediates (e.g., $BrZnCR^1R^2CR^3R^4$) which decompose by concerted processes.[109] Whatever type of intermediate is involved, the metal appears to exert a role right up to the product-forming step.[120]

Recent results on the debromination of dl-stilbene dibromide with a variety of metals are compiled in Table 9. The yields of cis-stilbene are controlled mainly by the choice of metal but, in some instances, a solvent sensitivity is also noted. The presence of significant proportions of cis-stilbene is inconsistent with the reaction occurring entirely via an intermediate in which rotation about the C_α -C_β bond is faster than product formation.

*Several reductants (Zn, Zn-Hg, Cd, Cd-Hg, Mg, Cu, Al, Pb, Sn) convert dl-stilbene dibromide into stilbenes contaminated by bromostilbene in DMF. However, the origin of the bromostilbene could be due to DMF rather than metal-induced dehydrohalogenation.[120]

TABLE 9 Metal Promoted Elimination from dl-Stilbene Dibromide[120]

Solvent	Metals*							
	Zn	Zn-Hg	Cd	Cd-Hg	Mg	Al	Pb	Cu
CH₃CN	3							
MeOH	18		26					
THF	11	21	25					
MeOCH₂CH₂OH	11		28		10[110]			
Acetone	5		24					
MEK	11	25	16					
n-Propanol	15		21					
EtOH	14	14	12	4				
CH₃COOEt	8							
DMSO	25		25		45			
95% aq.EtOH	13[161]					53	7	
H₂O	12[109,110]							18[161]

Reactions run at the boiling point of solvent or 60 to 70° for DMF, DMSO for 4 to 5 hr.

*% of cis-stilbene in total stilbene fraction listed.

355

Elimination has been achieved for a variety of halogen compounds possessing β-Y groups in which Y can be OH, OR, OCOR, and NH_2. For these substrates, metals (e.g., Zn,[184] Mg,[185] and Na[109,186]) and metal cations are found to be better eliminating reagents than iodide or phosphines. However, the poorer leaving groups retard elimination at the expense of protolytic reduction (see Eq. 7), leading to poorer yields of the alkene than are obtained from the corresponding dehalogenation. Additionally, more vigorous reaction conditions are required to effect elimination, and a lower stereoselectivity is usually observed than in the dehalogenation reaction. Concerted elimination for dehalogenation, but a carbanion mechanism for elimination involving the poorer leaving groups (Y = OH, NH_2, OR) are suggestions to account for this change in stereoselectivity.[109]

$$\underset{X \quad Y}{>C-C<} \;+\; 2e^- \;+\; \overset{+}{H} \longrightarrow \underset{H \quad Y}{>C-C<} \tag{7}$$

(X = Br, Cl.   Y = OH, NH_2, OR)

4   Metal-ion Promoted Dehalogenation

Metal cations used in preparative work as dehalogenating reagents for vicinal dihalides include vanadium II,[114] iron II,[113] titanium III,[114] cobalt II,[187] chromium II,[118,119] and copper I.[120] The majority of the reactions involving these reagents are homogeneous, and they offer a more realistic challenge to the kineticist than the heterogeneous metal reductions, yet few mechanistic studies have been reported.

With a number of these cationic reductants, meso-stilbene dibromide eliminates entirely to trans-stilbene, which is also the major product formed from the reactions of dl-stilbene dibromide (see Table 10).[120] Superficially, the anti-debrominations, e.g., meso $\longrightarrow$ trans, dl $\longrightarrow$ cis-stilbene, might seem straightforward E2 type reactions, the lesser reactivity of the dl-substrate residing in unfavorable eclipsing interactions in the transition state between the phenyl groups, which also enables side reactions to become more competitive than for the meso-isomer. However, the yields of cis-stilbene are greatly reduced compared to the reactions of dl-stilbene dibromide with nucleophilic reductants. With the exception of Sn^{II}, all the other cations undergo only a one-electron oxidation during the reaction, and the stoicheiometry of the reaction requires two molecules of reductant be consumed by each substrate molecule. Most likely, the difference in stereoselectivity arises from a change in the initial electron transfer to the substrate from two to one as the reductant goes from a nucleophilic reagent to a metal cation. For the dl-stilbene dibromide, loss of asymmetry can occur in the initially formed radical by rotation about the C_α - C_β bond, enabling the trans-stilbene to stem from equilibrium rather than kinetic control (see Eq. 8). The small yields of cis-stilbene may reflect the time scale for bond rotation in and subsequent attack on this radical by a second molecule of reductant which completes the elimination.

TABLE 10   Metal Ion Promoted Debromination of dl-Stilbene Dibromide[120]

(% cis-Stilbene in Total Stilbene Fraction Listed)

Reductant	Solvent	% cis-Stilbene
SnCl₂	DMF (50 to 75%)	6±3
CuCl	DMF	0-5
	DMSO	0-2
FeCl₂	EtOH, pyridine-N-oxide	2
	DMF	0
	DMSO	0
CrCl₂	EtOH	0-4
	DMF	0-2
	DMSO	0-3
TiCl₃	EtOH	4
	DMF	12

The reaction of stannous chloride with the stilbene dibromides in DMF has been studied in detail.[188] Both debrominations are first order in each reactant, and the difference in reactivity between the substrates is only a factor of 11.1 at 59.4°, much less than observed in the reaction with iodide or bromide ions (Section VII.2.a.iv, 2.b). This result suggests a minor role for concerted anti elimination. The oxidation of Sn^{II} to Sn^{IV} could occur in one step involving initial two-electron transfer to the substrate to give a carbanion, or in two steps via Sn^{III}, which arises

meso-Stilbene dibromide $\xrightarrow{M}$ (Newman projection: Br, H, Ph, Ph, H, M) $\rightleftharpoons$

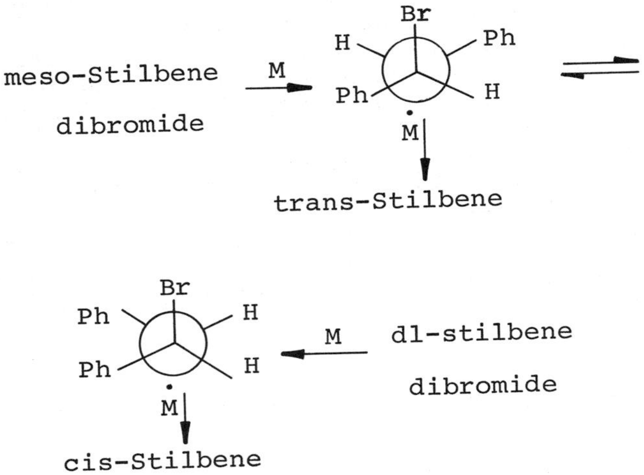

trans-Stilbene

(Newman projection: Br, Ph, Ph, H, H, M) $\xleftarrow{M}$ dl-stilbene dibromide (8)

cis-Stilbene

(M = metal-ion)

from an initial one-electron transfer to give an intermediate radical. Kwok and Miller prefer the latter scheme, and it should be possible to establish the intermediacy of Sn^{III}, a valence state that is capable of reducing trioxalatocobaltate^{III}.[189]

Of the metal cations, Cr^{II} has attracted most attention as an elimination reagent. Initially chromous salts were used to effect reductive elimination of dihalides and α-keto-epoxides.[190] A number of structural effects in the formation of alkenes from vicinal dihalides were subsequently reported by Kray and Castro.[191] Related eliminations with Cr^{II} to produce ethylene in low yield were only accomplished under vigorous conditions (e.g., elimination from $BrCH_2CH_2OH$ and $BrCH_2CH_2NH_2$), accompanying protolytic reduction being observed.[192] Chromous salts can be used to reduce simple alkyl halides, but the reaction is only practicable with the more reactive α-haloketones,[193] and with allylic and benzylic halides.[194] However, facile reduction occurs when the chromous ion is complexed with ethylenediamine (Cr^{II}en) or related ligands.[195] The enhanced reactivity of this reagent is attributed to a more exothermic step involving halogen transfer. The ethylenediamine exerts a larger field stabilization than the solvent molecules, which

are the ligands for the uncomplexed Cr^{II}.[119] Kochi
and his co-workers have reported detailed systematic
investigations of the influence of the following structural variations on the stereochemistry and products of
dehalogenation and related 1,2-elimination: (1) variation of Cr^{II} ligands, (2) solvent, and (3) substrate
structure and the eliminating groups.[118,119] We confine our discussion mainly to those reactions in which
the product is either ethylene[119] or cis- and trans-2-
butenes.[118]

     At room temperature, ethylene is formed in quanti-
tative yield in 15 min when a series of substrates of
general formula $BrCH_2CH_2Y$ are treated with Cr^{II}en in
10% v/v H_2O-DMF, Y being Br, Cl, OH, NH_2, $ClCOO$, PhO,
NMe_3^+, CH_3COO, C_2H_5O and OTs. For Y = PhS, the alkene
yield is greatly reduced and for Y = CN, phthalimido,
benzamido, and sulfonato, the only reaction product is
that of protolytic reduction. Reactions of the corresponding chloro series give similar product distributions, but the reaction times are increased, in some
cases to more than 24 hr. The relative rates of reductive elimination by Cr^{II}en and Cr^{II} are compiled in
Table 11. Rate coefficients are first order in both
chromous ion and substrate. As two moles of reductant
are consumed by the substrate during the reaction (see

TABLE 11 Relative Rates of Reduction of β-Substituted
Alkyl Halides by Cr^{II}en and Cr^{II}.[119]

Substituent Y	$BrCH_2CH_2Y$ ($+Cr^{II}$en)	$ClCH_2CH_2Y$ ($+Cr^{II}$en)	$CH_3CHBrCHYCH_3$ ($+ Cr^{II}$)	
			erythro	threo
Br	25	–	690	310
$N(Me)_3^+$	21	28		
Cl	5	8	26	12
OTs	2	4	4	3
CH_3COO	1	1	1	1
OH	–	–	0.3	0.3

Solvent: 99.9% v/v DMF.H_2O; room temperature; Cr^{II}
0.0052M, en 0.022M.

Eq. 9), the reductive elimination of vicinal dibromides by chromous ion must be a multistep process.[196] Since the chloro-compounds are much less reactive than the corresponding bromides, by analogy to the simple reductions of alkyl halides and peroxides, the most probable initial step is the slow ligand transfer of halide from the substrate to Cr^{II} to generate a β-haloalkyl radical (see Eq. 10). Supporting evidence for this view is provided by the much greater rate of elimination from trans-dibromo- than cis-dibromo- cyclohexane.[196] Only with an anti-orientation of the bromine atoms is orbital overlap at a maximum between the forming one-electron orbital and the lone pairs of the β-situated bromine atom (72). Of course, a similar orientation is required for facile concerted anti elimination by nucleophiles such as iodide ion. However, whereas elimination of Br-Br and Br-OTs occur at similar rates with iodide ion, the tosylate elimination is much slower with Cr^{II}. In fact, the lesser effectiveness of OH, acetoxy, and tosylate than Cl and Br as neighboring groups in homolytic reactions contrasts to their reactivity in intramolecular nucleophilic reactions.[197] Considering also that β-bromine is known to assist in homolytic reactions involving addition[198] and abstraction,[199] there seems little doubt that the initial rate-determining step involves a one-electron transfer to give a β-haloalkyl radical (73). We now consider the fast decompositions of this radical.

$$\begin{array}{c}>C{-}C< \\ \ \ | \ \ \ \ | \\ \ \ X \ \ \ \ Y \end{array} + 2Cr^{II} \longrightarrow >C{=}C< +$$

$$2Cr^{III}(\bar{X}\ ,\ \bar{Y}) \tag{9}$$

(X = Br, Cl; Y= various groups, see Table 11)

$$\begin{array}{c}>C{-}C< \\ \ \ | \ \ \ \ | \\ \ \ Br \ \ \ Br \end{array} + Cr^{II} \longrightarrow \left[\begin{array}{c} Br \cdots \\ >C{-}C< \\ \ \ \ \ \ \ \ Br{-}{-}{-}{-}Cr^{II} \end{array} \right] \longrightarrow$$

$$\underline{72}$$

$$>C\overset{Br}{-}C< + Cr^{III}Br \tag{10}$$

$$\underline{73}$$

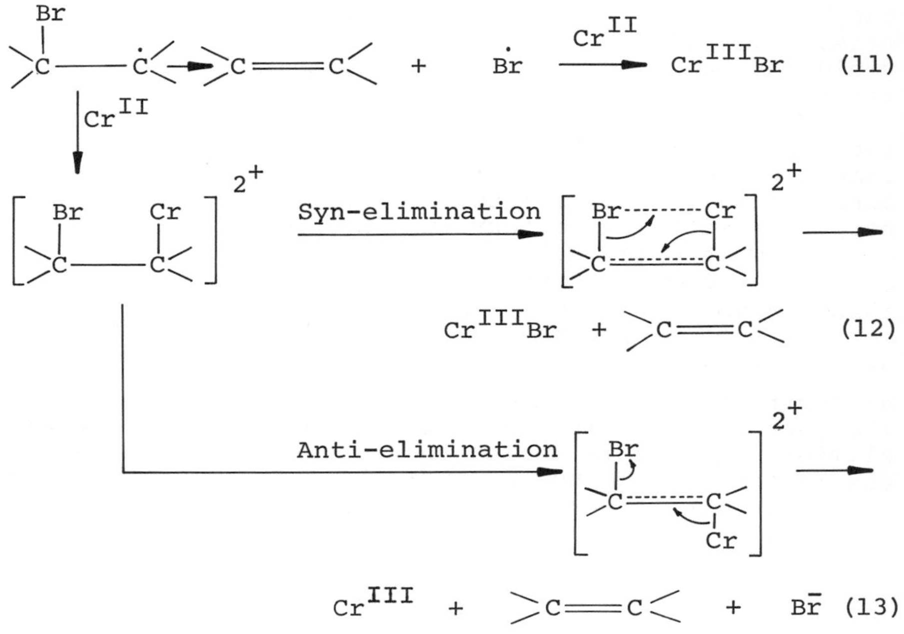

In chromous ion debromination, all the bromine initially present in the substrate can be accounted for as either "free" bromide ion or as $Cr^{III}Br$ at completion of the elimination.[196] $Cr^{III}Br$ and related complexes are relatively stable to hydrolysis and can be separated from Cr^{III} by ion-exchange chromatography.[118] As the concentration of chromous ion is increased, relative to the substrate, the free bromide ion concentration rises to a maximum, which is then maintained with further increases in reductant. The ratio of $Cr^{III}Br$ to $\overline{Br}$ is dependent on the alkyl moeity of the substrate and on the nature of the leaving group Y (see Table 11 and Eq. 9). To account for these variations, Kochi and his co-workers[118,119,196] have suggested the intermediate β-bromoalkyl radical decomposes by three simultaneous routes (see Eqs. 11 to 13). Bromide ion is formed only in the anti elimination from the intermediate chromium alkyl. When the reductant concentration is similar to that of the substrate, unimolecular decomposition of the β-bromoalkyl radical occurs more rapidly than the bimolecular reaction with a second molecule of the reductant. Facile elimination of $\overset{\circ}{Br}$ from radicals is well known, the activation ener-being only 7 kcal/mole for conversion of $[C_2H_4Br]^{\bullet}$ into ethylene and $\overset{\circ}{Br}$, and 5 kcal/mole for ejection of $Br^{\bullet}$ from the bromoisobutyl radical.[200] With increasing chromous ion concentration, the bimolecular process becomes dominant and the ratio of $Cr^{III}Br$ to $\bar{Br}$ is determined by the relative ease of anti and syn elimination from the chromium alkyl. For syn elimination, the leaving group must possess a reasonably available electron pair for coordination to chromium in the transition state. For debromination, all three modes of radical decomposition contribute to product formation. However, for the other substituents, Y = Cl, OH, acetoxy, alkoxy, and amino (see Eq. 9), C-Y homolytic dissociation is less facile, and the greater availability of the electron pair on Y leads to syn elimination only from the intermediate chromium alkyl. When Y = CN, phthalimido and benzamido, the bonding to chromium is too weak and the leaving groups are reluctant to depart as $\bar{Y}$ and protolytic reduction products only are formed.

The mechanism described in Eqs. 10, 11, 12, and 13 can be used to explain the stereochemistry of reductive elimination of 2-bromo-3-butyl derivatives.[118] The yields of the isomeric 2-butenes obtained under conditions of kinetic control are listed in Table 12.

With the exception of the vicinal dibromides, all the other epimeric 2-bromo-3-Y-butyl pairs gave the same mixture of cis- and trans-2-butenes under common conditions (in addition to the results in Table 12, experiments with Y = acetoxy and OH support this statement). This observation is accommodated by the involvement of a common intermediate for each erythro- and threo-3-substituted butyl bromide. Rate-determining formation of a free radical in which rapid rotation about the C_α—C_β bond allows loss of the original asymmetry of the substrate satisfies the requirements. The debromination is the most facile and yet the only elimination in this series that shows an obvious stereoselectivity. We now discuss the influence of conditions on this latter reaction in more detail.

TABLE 12 The Effect of Leaving Group Y on the Products of Reductive Elimination of 2-Bromo-3-butyl Derivatives by Chromous Ion[118]

2-Bromo-3-Y-butane	Concentration (M)	Cr^{II} Concentration (M)	Tempera-ture °C	Solvent[b]	Butene-2 trans	cis
$\underline{Y}$						
1. meso-Br	0.125	0.063^a	0	EtOH	73	27
dl-Br	0.124	0.063^a	0	EtOH	61	39
2. meso-Br	0.133	0.063^a	25	DMF	79	21
dl-Br	0.150	0.063^a	25	DMF	48	52
3. meso-Br	0.0123	0.095	25	DMSO	97	3
dl-Br	0.0122	0.095	25	DMSO	8	92
4. erythro-Cl	0.0131	0.0565^a	0	DMF	74	26
threo-Cl	0.0126	0.0565^a	0	DMF	72	28
5. erythro-Cl	0.0128	0.095	25	DMSO	72	28
threo-Cl	0.0123	0.095	25	DMSO	67	33
6. erythro-OTs	0.0514	0.0515^a	0	DMF	54	46
threo-OTs	0.0441	0.0515^a	0	DMF	54	46

TABLE 12 Continued

2-Bromo-3-Y-butane	Concentration (M)	Cr^{II} Concentration (M)	Temperature °C	Solvent[b]	Butene-2 trans	cis
Y						
7. erythro-OTs	0.005	0.095	25	DMSO	54	46
threo-OTs	0.004	0.095	25	DMSO	54	46

[a]Perchloric Acid 0.90 M added. [b]EtOH was diluted to 75% (v/v) ethanol–water, DMF to 90%–DMF–water, and DMSO to 89% (v/v) DMSO–water.

TABLE 13   Effect of Reaction Temperature and Reactant Ratio on the Stereochemistry of Reductive Debromination of dl-2,3-Dibromobutane by Cr^{II} in DMSO[118]

Cr^{II}/Substrate	Temperature °C	cis-/trans-2-Butene
1. 37	0	19.4
2. 37	20	15.7
3. 37	73	7.3
4. 6.2	20	14.2
5. 1.8	20	12.9

DMSO diluted to 89% vol. DMSO-water.

From Table 12, anti-debromination of both epimeric 2,3-dibromobutanes is optimized in DMSO. Within this solvent, variations in temperature and the ratio of Cr^{II} to substrate have a small but significant effect on the stereoselectivity (Table 13). Of a range of solvents, only the basic pyridine and 2,4,6-collidine promote stereoselectivity of a comparable magnitude to DMSO (less for HMPT, sulfolane, formamide, and methyl cyanide). Finally, the ligands markedly alter the stereoselectivity (see Table 14), and simply increasing the concentration of ethylenediamine ($Cr^{II} \rightarrow Cr^{II}$en) causes a rise in the yield of the anti-debromination product. All of these experimental observations reinforce the mechanism outlined in Eqs. 11 to 14 for reductive eliminations.

Of all the groups Y evaluated (see Eq. 9), only Br affords stabilization to the initially formed radical. The interaction shown in 73 imparts some restriction to rotation about the C_α—$\overline{C_\beta}$ bond in the radical (see Eq. 14). Greater concentrations of reductant relative to substrate reduce the lifetime of the radical and thereby diminish the chance for loss of asymmetry. However, stereospecific formation of the β-bromoalkylchromium intermediate 74 does not alone ensure complete stereospecificity of reaction as both syn and anti elimination can occur (see Eqs. 12 and 13). To explain the effects of solvent and ligands, a role for these in determining the stereochemistry of

TABLE 14   Effect of Ligand Concentration on the Stereochemistry of Reductive Elimination of dl-2,3-dibromo-
butane by Cr^{II} at 25°C[18]

Ligand: ethylene diamine, substrate 0.0166 M,
Cr^{II} 0.083 M

Ligand concentration (M)	1.63	3.6	7.2	10.8
trans-2-Butene	22	14	9	8
cis-2-Butene	78	86	91	92
DMF/ligand %	78	65	44	22

elimination from the alkyl chromium intermediate 74
must be formulated. Solvents, such as DMSO and pyridine, and the ligands, such as ethylenediamine, promote anti elimination. Kochi and Singleton[118] regard
these reagents as more effective electron donors than
water [e.g., $Cr^{II}(H_2O)_6$], and conclude a tightly coordinated chromium species such as 75 is less likely to
be involved in an intramolecular transfer of a β-bro-
mide to form $Cr^{III}L_5Br$, which is required for syn-
elimination. Steric size of the chromium leaving group
could also exert a controlling influence, the ligands
that promote anti-stereospecificity for the overall debromination being much larger than water molecules.

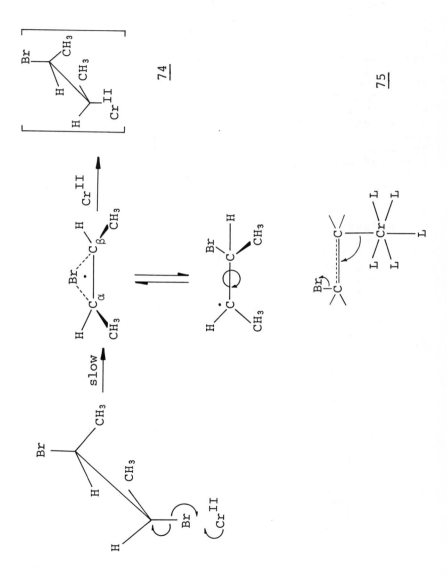

(14)

74

75

5 General Comments

The stereochemistry of dehalogenation varies widely with the type of reductant used. It is unsound to assume a mechanism from stereochemical studies alone, as both radical and heterolytic reactions can apparently give products of anti-stereospecificity, especially if the alkene formed is the thermodynamically more stable geometric isomer. For a reaction that only in recent years has been studied in more detail from a kinetic view, all the possible types of reaction intermediates have been suggested. These include radical, carbanion, and carbonium ion intermediates, and also a concerted single-step process. Miller's examination of the stereochemistry of the debromination of dl-stilbene dibromide is a classic approach. His results enable the generalization of reaction mechanism into classes determined primarily by the reductant and the probable extent of electron transfer to the substrate in the initial step.

REFERENCES

1.   R. Piria, Ann. Chem., 68, 348(1848).
2.   F. C. Whitmore and D. P. Langlois, J. Am. Chem. Soc., 54, 3441(1932).
3.   E. H. White, J. Am. Chem. Soc., 77, 6008(1955).
4.   E. H. White, J. Am. Chem. Soc., 77, 6011(1955).
5.   E. H. White and C. A. Aufdermarsh, Jr., J. Am. Chem. Soc., 83, 1179(1961).
6.   E. H. White and D. W. Grisley, Jr., J. Am. Chem. Soc., 83, 1191(1961).
7.   E. H. White and C. A. Aufdermarsh, Jr., J. Am. Chem. Soc., 83, 1174(1961).
8.   E. H. White and L. A. Dolak, J. Am. Chem. Soc., 88, 3790(1966); J. Org. Chem., 31, 3038(1966).
9.   T. Koenig and M. Deinzer, J. Am. Chem. Soc., 88, 4518(1965).
10. J. H. Cooley, P. T. Jacobs, M. A. Kahn, L. Heasley, and W. D. Goodman, J. Org. Chem., 30, 3062(1965).
11. E. H. White and H. Scherrer, Tetrahedron Lett., 758(1961).
12. E. H. White, H. Maskill, D. J. Woodcock, and M. A. Schroeder, Tetrahedron Lett., 1713(1969).
13. J. H. Ridd, Q. Rev., (London), 15, 418(1961).
14. A. Streitwieser, Jr., J. Org. Chem., 22, 861 (1957).
15. R. J. Baumgarten, J. Chem. Educ., 43, 398(1966).

16. D. V. Banthorpe, "Rearrangements involving Amino Groups" in "The Chemistry of the Amino Group," S. Patai, Ed., Interscience Publishers, London, 1968.

17. E. H. White and D. J. Woodcock, "Cleavage of the Carbon-Nitrogen Bond," in "The Chemistry of the Amino Group," S. Patai, Ed., Interscience Publishers, London, 1968.

18. E. Müller, H. Haiss, and W. Rundel, Chem. Ber., 93, 1541(1960).

19. (a) P. A. S. Smith and H. G. Pars, J. Org. Chem., 24, 1325(1959); (b) J. Glazer, E. D. Hughes, C. K. Ingold, A. T. James, G. T. Jones, and E. Roberts, J. Chem. Soc., 2671(1950).

20. P. Brewster, F. Hiron, E. D. Hughes, C. K. Ingold, and P. A. D. S. Rao, Nat., 166, 179(1950).

21. C. K. Ingold, "Structure and Mechanism in Organic Chemistry," Cornell University Press, Ithaca, 1953, p. 383.

22. V. Prelog, Angew. Chem., 70, 145(1958).

23. R. A. Raphael, Proc. Chem. Soc., 97(1962).

24. A. Streitwieser, Jr., and W. D. Schaeffer, J. Am. Chem. Soc., 79, 2888(1957).

25. J. D. Roberts, R. E. McMahon, and J. Hine, J. Am. Chem. Soc., 72, 4237(1950).

26. C. C. Lee, C. P. Slater, and J. W. T. Spinks, Can. J. Chem., 35, 1416(1957).

27. F. C. Whitmore and R. S. Thorpe, J. Am. Chem. Soc., 63, 1118(1941).

28. L. G. Cannell and R. W. Taft, Jr., J. Am. Chem. Soc., 78, 5812(1956).

29. D. W. Adamson and J. Kenner, J. Chem. Soc., 838 (1934).

30. H. C. Brown and M. Nakagawa, J. Am. Chem. Soc., 77, 3614(1955).

31. D. J. Cram and J. E. McCarty, J. Am. Chem. Soc., 79, 2866(1957).

32. J. G. Burr, Jr., and L. S. Ciereszko, J. Am. Chem. Soc., 74, 5426(1952).

33. L. S. Ciereszko and J. G. Burr, Jr., J. Am. Chem. Soc., 74, 5431(1952).

34. (a) W. E. Bachmann and J. W. Ferguson, J. Am. Chem. Soc., 56, 2081(1932). (b) W. E. Bachmann and F. H. Moser, J. Am. Chem. Soc., 54, 5431(1932).

35. D. Y. Curtin and M. C. Crew, J. Am. Chem. Soc., 76, 3719(1954).

36. (a) B. J. Benjamin, H. J. Schaeffer, and C. J. Collins, J. Am. Chem. Soc., 79, 6160(1957);

(b)   J. C. Martin and W. G. Bentrude, J. Org. Chem., 24, 1902(1959).

37.   J. D. Roberts and J. A. Yancey, J. Am. Chem. Soc., 74, 5943(1952).

38.   (a)   D. Semenow, C. Shih, and W. G. Young, J. Am. Chem. Soc., 80, 5472(1958);
(b)   E. J. Corey, J. Casanova, Jr., P. A. Vatakencherry, and R. Winter, J. Am. Chem. Soc., 85, 169(1963).

39.   (a)   O. E. Edwards and T. Sano, Can. J. Chem., 47, 3489(1969);
(b)   P. Yates and R. J. Crawford, J. Am. Chem. Soc., 88, 1561(1966);
(c)   M. Hanack and J. Dolde, Tetrahedron Lett., 321(1966).

40.   J. D. Roberts, C. C. Lee, and W. H. Saunders, Jr., J. Am. Chem. Soc., 76, 4501(1954).

41.   H. Zollinger, "Azo and Diazo Chemistry," Interscience, New York, 1961, pp. 96, 129.

42.   D. Bethell and V. Gold, "Carbonium Ions," Academic Press, London 1967, p. 41.

43.   J. A. Mills, J. Chem. Soc., 260(1953).

44.   C. W. Shoppee, R. E. Lack, and P. Ram, J. Chem. Soc., C, 1018(1966).

45.   W. Hückel and K. Heyder, Chem. Ber., 96, 220 (1963).

46.   G. Drefahl and S. Huneck, Chem. Ber., 93, 1961 (1960).

47.   M. Chérest, H. Felkin, J. Sicher, F. Šipǒs, and M. Tichý, J. Chem. Soc., 2513(1965).

48.   D. V. Stolow and W. Maya, J. Am. Chem. Soc., 81, 330(1959).

49.   D. V. Banthorpe, "Elimination Reactions," Elsevier, Amsterdam, 1963.

50.   Ref. 17, p. 466.

51.   C. A. Bunton, "Nucleophilic Substitution at a Saturated Carbon Atom," Elsevier, Amsterdam, 1963, p. 101.

52.   J. D. Roberts and C. M. Regan, J. Am. Chem. Soc., 75, 2069(1953).

53.   C. C. Lee, Can. J. Chem., 31, 761(1953).

54.   M. C. Caserio, R. D. Levin, and J. D. Roberts, J. Am. Chem. Soc., 87, 5651(1965).

55.   J. D. Roberts and A. W. Fort, J. Am. Chem. Soc., 78, 584(1956).

56.   M. S. Kharasch and A. L. Flenner, J. Am. Chem. Soc., 54, 674(1932).

57.   P. S. Skell and I. Starer, J. Am. Chem. Soc., 82, 2971(1960).

58.   G. J. Karabatsos and C. E. Orzech, Jr., J. Am.

Chem. Soc., 84, 2823(1962).

59. For a review on the role of protonated cyclopropanes as intermediates see: C. C. Lee, Prog. in Phys. Org. Chem., 7, 1970, p. 129. Wiley, Interscience, New York.

60. G. J. Karabatsos, C. E. Orzech, Jr., J. L. Fry, and S. Meyerson, J. Am. Chem. Soc., 92, 606(1970).

61. (a) C. C. Lee, J. E. Kruger, and E. W. C. Wong, J. Am. Chem. Soc., 87, 3985(1965); (b) C. C. Lee and J. E. Kruger, J. Am. Chem. Soc., 87, 3986(1965).

62. R. Huisgen and C. Rüchardt, Ann. Chem., 601, 1 (1956); 601, 21(1956).

63. E. H. White and J. E. Stuber, J. Am. Chem. Soc., 85, 2168(1963).

64. E. H. White and F. W. Bachelor, Tetrahedron Lett., 77, (1965).

65. J. A. Berson and A. Remanich, J. Am. Chem. Soc., 86, 1749(1964).

66. T. Cohen and A. R. Daniewski, J. Am. Chem. Soc., 91, 533(1969).

67. J. Levisalles and J. F. Mouchard, Bull. Soc. Chim. Fr., 678(1970).

68. D. J. Cram and M. R. V. Sahyun, J. Am. Chem. Soc., 85, 1257(1963).

69. M. Cocivera and S. Winstein, J. Am. Chem. Soc., 85, 1702(1963).

70. P. S. Skell and W. L. Hall, J. Am. Chem. Soc., 85, 2851(1963).

71. R. A. More O'Ferrall, Adv. Phys. Org. Chem., 5, 331(1967).

72. T. Cohen and E. Jankowski, J. Am. Chem. Soc., 86, 4217(1964).

73. E. H. White, H. P. Tiwari and M. J. Todd, J. Am. Chem. Soc., 90, 4734(1968).

74. E. H. White and C. A. Elliger, J. Am. Chem. Soc., 89, 165(1967).

75. T. Cohen and E. Jankowski, Private Communication quoted in Ref. 17, p. 469.

76. D. Semenow, C. Shih, and W. G. Young, J. Am. Chem. Soc., 80, 5472(1958).

77. R. Huisgen and H. Reimlinger, Ann. Chem., 599, 161(1956).

78. K. Heyns and W. Badenburg, Ann. Chem., 595, 55 (1955).

79. E. H. White, J. Am. Chem. Soc., 77, 6014(1955).

80. B. A. Gingras and W. A. Waters, J. Chem. Soc., 1920(1954).

81. R. Huisgen and H. Reimlinger, Ann. Chem., 599, 183(1956).

82.   E. H. White and R. J. Baumgarten, J. Org. Chem.,
       29, 2070(1964).
83.   R. A. Moss, J. Org. Chem., 31, 1082(1966).
84.   A. Streitwieser, Jr., and W. D. Schaeffer, J.
       Am. Chem. Soc., 79, 2893(1957).
85.   E. H. White and C. A. Elliger, unpublished re-
       sults quoted in Ref. 17, p. 445.
86.   D. B. Denney and D. G. Denny, J. Am. Chem. Soc.,
       79, 4806(1957).
87.   H. Maskill, R. M. Southam, and M. C. Whiting,
       Chem. Comm., 496(1966).
88.   J. Hine, J. Am. Chem. Soc., 72, 2438(1950).
89.   J. Hine and A. M. Dowell, Jr., J. Am. Chem.
       Soc., 76, 2688(1954).
90.   J. Hine, E. L. Pollitzer, and H. Wagner, J. Am.
       Chem. Soc., 75, 5607(1953).
91.   P. S. Skell and I. Starer, J. Am. Chem. Soc.,
       81, 4117(1959).
92.   M. Freund and F. Lenze, Ber., 24, 2150(1891).
93.   L. Friedman and H. Shechter, J. Am. Chem. Soc.,
       81, 5512(1959).
94.   P. S. Skell and I. Starer, J. Am. Chem. Soc.,
       84, 3962(1962).
95.   O. A. Reutov and T. N. Shatkina, Tetrahedron,
       18, 237(1962).
96.   P. S. Skell and R. J. Maxwell, J. Am. Chem. Soc.,
       84, 3963(1962).
97.   P. S. Skell, I. Starer and A. P. Krapcho, J. Am.
       Chem. Soc., 82, 5257(1960).
98.   W. A. Sanderson and H. S. Mosher, J. Am. Chem.
       Soc., 83, 5033(1961); 88, 4185(1966).
99.   F. C. Whitmore, C. S. Rowland, S. N. Wrenn, and
       G. W. Kilmer, J. Am. Chem. Soc., 64, 2970(1942).
100.   I. Starer, Ph.D. thesis, The Pennsylvania State
       University, U.S.A., June 1960, quoted in Ref. 98.
101.   C. J. Collins, V. F. Raaen, and M. D. Eckart, J.
       Am. Chem. Soc., 92, 1787(1970).
102.   C. C. Lee and Kwok-Ming Van, J. Am. Chem. Soc.,
       91, 6416(1969).
103.   C. C. Lee and J. E. Kruger, Tetrahedron, 23,
       2539(1967).
104.   C. C. Lee and B. Hahn, Can. J. Chem., 45, 2129
       (1967).
105.   J. Hine, A. D. Ketley and K. Tanabe, J. Am. Chem.
       Soc., 82, 1398(1960).
106.   C. G. Swain and C. B. Scott, J. Am. Chem. Soc.,
       75, 246(1953).
107.   (a) Ref. 42, pp. 231-235;
       (b) J. L. Coke, F. E. McFarlane, M. C. Mourning,
       and M. G. Jones, J. Am. Chem. Soc., 91, 1154

(1969).

108. W. H. Perkin, J. Chem. Soc., 24, 37(1871).

109. H. O. House, and R. S. Ro, J. Am. Chem. Soc., 80, 182(1958).

110. W. M. Schubert, B. S. Rabinovitch, N. R. Larson, and V. A. Sims, J. Am. Chem. Soc., 74, 4590 (1952).

111. J. Sicher, M. Havel, and M. Svoboda, Tetrahedron Lett., 4269(1968).

112. W. M. Schubert, H. Steadly, and B. S. Rabinovitch, J. Am. Chem. Soc., 77, 5755(1955).

113. H. Bretschneider and M. Ajtai, Monatsh. Chem., 74, 57(1941).

114. L. H. Slaugh and J. H. Raley, Tetrahedron, 20, 1005(1964).

115. F. Weygand and H. G. Peine, Rev. Chim. (Bucharest), 7, 1379(1962).

116. A. J. Speziale and C. C. Tung, J. Org. Chem., 28, 1353, 1521(1963).

117. S. Dershavitz and S. Proskaner, J. Org. Chem., 26, 3595(1961).

118. J. K. Kochi and D. M. Singleton, J. Am. Chem. Soc., 90, 1582(1968).

119. J. K. Kochi, D. M. Singleton, and L. J. Andrews, Tetrahedron, 24, 3503(1968).

120. I. M. Mathai, K. Schug, and S. I. Miller, J. Org. Chem., 35, 1733(1970).

121. D. R. James, R. W. Rees, and C. W. Shoppee, J. Chem. Soc., 1370(1955).

122. E. Baciocchi and A. Schiroli, J. Chem. Soc., B, 554(1969).

123. J. Hine and W. H. Brader, J. Am. Chem. Soc., 77, 361(1955).

124. S. Winstein, D. Pressman, and W. G. Young, J. Am. Chem. Soc., 61, 1645(1939).

125. W. G. Young, D. Pressman, and C. D. Coryell, J. Am. Chem. Soc., 61, 1641(1939).

126. C. S. Tsai Lee, I. M. Mathai, and S. I. Miller, J. Am. Chem. Soc., 92, 4602(1970).

127. W. G. Lee and S. I. Miller, J. Phys. Chem., 66, 655(1962).

128. C. F. Van Duin, Rec. Trav. Chim., 43, 341 (1924); 45, 345(1926); 47, 715(1928).

129. E. Biilman, Rec. Trav. Chim., 36, 313(1917).

130. J. Hine, R. Weisboeck, and R. G. Chirardelli, J. Am. Chem. Soc., 83, 1219(1961).

131. S. I. Miller, J. Phys. Chem., 66, 978(1962).

132. D. H. R. Barton and W. J. Rosenfelder, J. Am. Chem. Soc., 72, 1066(1950).

133. D. H. R. Barton and W. J. Rosenfelder, J. Chem.

Soc., 1048(1951).

134. H. L. Goering and H. H. Espy, J. Am. Chem. Soc., 77, 5023(1955).

135. S. J. Cristol, J. Q. Weber, and M. C. Brindell, J. Am. Chem. Soc., 78, 598(1956).

136. C. L. Stevens and J. A. Valicenti, J. Am. Chem. Soc., 87, 838(1965).

137. S. Angyal and R. J. Young, Aust. J. Chem., 14, 8(1961).

138. A. B. Forster and W. G. Overend, J. Chem. Soc., 3452(1951).

139. P. Bladon and L. N. Owen, J. Chem. Soc., 598 (1950).

140. E. S. Gould, "Mechanism and Structure in Organic Chemistry," Holt, Rinehard and Winston, New York, 1959, p. 494.

141. Ref. 49, p. 140.

142. E. Baciocchi and A. Schiroli, J. Chem. Soc., B, 554(1969).

143. J. Mulders and J. Nasielski, Bull. Soc. Chim. Belg., 72, 322(1963).

144. J. F. King and R. G. Dews, Can. J. Chem., 42, 1294(1964).

145. J. Csapilla, Chim. (Switzerland), 18, 37(1964).

146. R. T. Dillon, W. G. Young, and H. J. Lucas, J. Am. Chem. Soc., 52, 1953(1930).

147. R. T. Dillon, J. Am. Chem. Soc., 54, 952(1932).

148. W. Hückel and H. Waiblinger, Ann., 666, 17(1963).

149. D. Pressman and W. G. Young, J. Am. Chem. Soc., 66, 705(1944).

150. G. Fischer, K. A. Muszkat, and E. Fischer, J. Chem. Soc., B, 1156(1968).

151. J. Hine, "Physical Organic Chemistry," McGraw-Hill, New York, 1962, pp. 209-220.

152. L. H. Schwartzman and B. B. Corson, J. Am. Chem. Soc., 78, 322(1956).

153. H. H. Jaffé, Chem. Rev., 53, 191(1953).

154. T. L. Davis and R. F. Heggie, J. Org. Chem., 2, 470(1938).

155. A. J. Parker, Adv. Phys. Org. Chem., 5, 173 (1967).

156. I. M. Mathai and S. I. Miller, J. Org. Chem., 35, 3416(1970).

157. P. B. D. de la Mare and R. Bolton, "Electrophilic Additions to Unsaturated Systems," Elsevier, Amsterdam, 1966, Chapters 1,7.

158. J. E. Dubois and E. Bienvenna-Goetz, Bull. Soc. Chim. Fr., 2086(1968).

159. R. P. Bell and M. Pring, J. Chem. Soc., B, 1119 (1966).

160. R. E. Buckles, J. L. Miller, and R. J. Thurmaier, J. Org. Chem., 32, 888(1967).

161. R. E. Buckles, J. M. Bader, and R. J. Thurmaier, J. Org. Chem., 27, 4523(1962).

162. G. Heublein, J. Prakt. Chem., 31, 84(1966).

163. D. R. Dalton, V. P. Dutta and D. C. Jones, J. Am. Chem. Soc., 90, 5498(1966).

164. P. D. Bartlett and D. S. Tarbell, J. Am. Chem. Soc., 58, 466(1936).

165. G. Heublein and P. Umbreit, Tetrahedron, 24, 4733(1968).

166. F. Badea, T. Constantinescu, A. Juvara, and C. D. Ninitzescu, Justus Liebigs Ann. Chem., 706, 20(1967).

167. S. I. Miller, Adv. Phys. Org. Chem., 6, 185 (1968).

168. W. K. Kwok, I. M. Mathai, and S. I. Miller, J. Org. Chem., 35, 3420(1970).

169. C. A. Bunton, "Nucleophilic Substitution at a Saturated Carbon Atom," Elsevier, Amsterdam, 1963, p. 134.

170. A. J. Parker, Q. Rev. (London), 16, 163(1962).

171. W. K. Kwok, Diss. Abstr., 27, B, 3040(1967).

172. R. L. Scott, J. Am. Chem. Soc., 75, 1550(1953).

173. C. K. Ingold, "Structure and Mechanism in Organic Chemistry," G. Bells, London, 1953, p. 461.

174. J. F. King, A. D. Allbutt, and R. G. Pews, Can. J. Chem., 46, 805(1968).

175. S. Dershowitz and S. Proskauer, J. Org. Chem., 26, 3595(1961).

176. R. Huisgen and F. Mietzsch, Angew. Chem. Int. Ed., 3, 83(1964).

177. R. E. Benson and T. L. Cairns, J. Am. Chem. Soc., 72, 5355(1950).

178. E. Vogel, H. Kiefer, and W. R. Roth, Angew. Chem. Int. Ed., 3, 442(1964).

179. C. G. Scouten, F. E. Barton, Jr., J. R. Burgess, P. R. Story, and J. F. Garst, Chem. Comm., 78 (1969).

180. D. Lipkin, F. R. Galiano, and R. W. Jordan, Chem. and Ind., 1657(1963).

181. A. J. Speziale and C. C. Tung, J. Org. Chem., 28, 1353, 1521(1963).

182. W. G. Young, S. J. Cristol, and T. Skei, J. Am. Chem. Soc., 65, 2099(1943).

183. J. Sicher, M. Havel, and M. Svoboda, Tetrahedron Lett., 4269(1968).

184. (a) L. Fieser and R. Ettore, J. Am. Chem. Soc., 75, 1700(1953);
      (b) L. Crombie and S. Harper, J. Chem. Soc.,

1705, 1715(1950); 136(1956);
(c) S. J. Cristol and L. Rademacher, J. Am.
Chem. Soc., 81, 1600(1959).

185. R. Fuson, "Advanced Organic Chemistry," Wiley,
New York, 1950, p. 143 ff.

186. R. Paul and S. Tchelitcheff, C. R. Acad. Sci.
Paris, 224, 475(1947); 238, 2089(1954).

187. J. Halpern and J. P. Maher, J. Am. Chem. Soc.,
87, 5361(1965).

188. W. K. Kwok and S. I. Miller, J. Am. Chem. Soc.,
92, 4599(1970).

189. E. A. M. Wetton and W. C. E. Higginson, J. Chem.
Soc., 5890(1965).

190. P. Julian, W. Cole, A. Magnani, and E. Meyer, J.
Am. Chem. Soc., 67, 1728(1945).

191. W. Kray and C. Castro, J. Am. Chem. Soc., 86,
4063 (1964).

192. (a) F. Anet and E. Isabelle, Can. J. Chem., 36,
589(1958);
(b) J. Cornforth, R. Cornforth and K. Mathew, J.
Chem. Soc., 112(1959).

193. W. Kray and C. Castro, J. Am. Chem. Soc., 85,
2768(1963); 86, 4603(1964).

194. (a) J. K. Kochi and D. Davis, J. Am. Chem. Soc.,
86, 5264(1964);
(b) J. K. Kochi and D. Buchanan, J. Am. Chem.
Soc., 87, 853(1965).

195. J. K. Kochi and P. Mocaldo, J. Am. Chem. Soc., 88,
4094(1966).

196. J. K. Kochi and D. M. Singleton, J. Am. Chem.
Soc., 89, 6547(1967).

197. J. Hine, "Physical Organic Chemistry," 2nd ed.,
McGraw-Hill, New York, 1962, p. 141.

198. H. Goering, P. Abell, and B. Aycock, J. Am.
Chem. Soc., 74, 3588(1952).

199. P. S. Skell, D. Tuleen, and P. Readio, J. Am.
Chem. Soc., 85, 2850(1963).

200. (a) R. Barker and A. Maccoll, J. Chem. Soc.,
2839(1963);
(b) P. Abell and R. Anderson, Tetrahedron Lett.,
3727(1964);
(c) W. Haag and E. Heiba, Tetrahedron Lett.,
3683(1965).

VIII   PYROLYTIC ELIMINATIONS

A   INTRODUCTORY COMMENTS

Many organic compounds undergo thermal decomposition to form alkenes at elevated temperatures in the absence of added reagents (Eq. 1). Most often X is a poly-atomic group, being monatomic only for the halides. Of

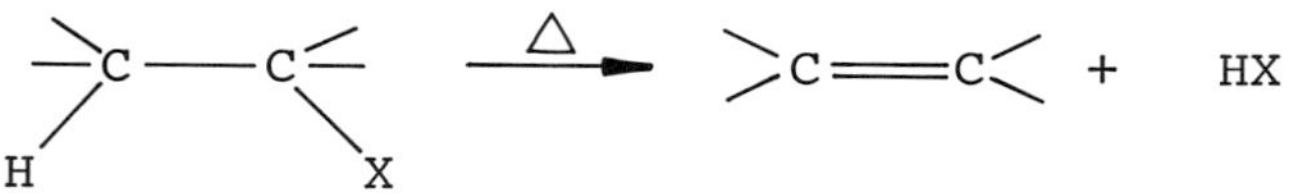

$$\underset{\overset{|}{H}}{\overset{|}{>}}C-\underset{\overset{|}{X}}{\overset{|}{C}}< \quad \xrightarrow{\Delta} \quad >C=C< \quad + \quad HX \qquad (1)$$

these, the alkyl fluorides are reluctant to eliminate to alkenes even at elevated reaction temperatures.[1] Alkyl iodides decompose thermally to give a mixture of alkenes and alkanes, the latter arising from a side reaction of the alkyl iodide with the liberated hydro-gen iodide (Eqs. 2 and 3).[2,3] The elimination of alkyl bromides and chlorides, which occur smoothly between 300 to 600°C in most cases, in the vapor phase, have

$$(CH_3)_2 \; CHI \xrightarrow[290-357°]{\triangle} CH_3 CH = CH_2 + HI \qquad (2)$$

$$HI + (CH_3)_2 \; CHI \longrightarrow CH_3 CH_2 CH_2 + I_2 \qquad (3)$$

attracted considerable interest. However, the results
are of more theoretical than practical value because of
the reversibility of the reaction. At high temperatures, the equilibrium lies in favor of the alkene, but
cooling to room temperature to isolate the product
shifts the equilibrium in favor of the alkyl halide.
From a preparative standpoint, it is necessary to carry
out the pyrolysis in the presence of a base such as
ammonia, which by consuming the liberated acid, effectively "freezes" the high temperature equilibrium. In
the absence of a surface-catalyzed reaction, the pyrolysis of an alkyl halide is often unimolecular. Initial C-X cleavage can occur either as a homolytic dissociation to give radical intermediates or by a heterolytic dissociation. In the latter case, a concerted
elimination, even if wholly constrained to a syn-
periplanar arrangement of the eliminating groups, would
involve considerable angle strain in attaining the
four-membered transition state. As we see subsequently,
C-X bond breaking runs ahead of C-H bond breaking, the
transition state possessing carbonium-ion character
and simulating those involved in El reactions in solution. As illustrated in 1, the cyclic syn elimination
is called the Ei (elimination-intramolecular) mechanism.

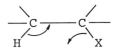

<u>1</u>

        Cases in which X (Eq. 1) is a polyatomic group
fall into two distinct categories. In one series the
unimolecular eliminations involve five-membered cyclic
transition states, typical examples being the Cope
elimination of amine oxides (Eq. 4),[4] the closely related pyrolysis of sulfoxides,[5] and the thermolysis of
the intermediate phosphine oxides in the reactions of
carbonyl compounds with the Wittig reagent.[6] In all
three of these reactions, the leaving groups must
attain a syn-periplanar conformation with respect to

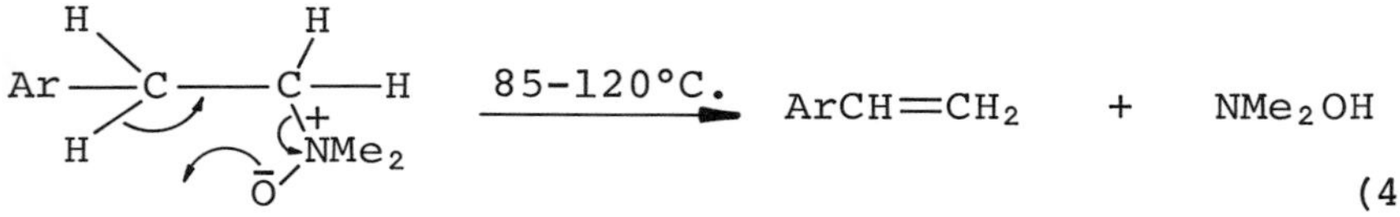

$$\underset{H}{\overset{H}{\diagdown}}C\underset{\overset{|}{\underset{\overset{+}{NMe_2}}{}}}{\overset{H}{\diagup}}C-H \xrightarrow{\;85-120°C\;} ArCH{=}CH_2 \quad + \quad NMe_2OH$$

$$(4)$$

each other, to enable the necessary proximity between them for effective bond formation in the transition state. This stereochemical requirement dominates the orientation of elimination in alicyclic compounds, particularly in the cyclohexyl series.

The other series of unimolecular eliminations involve six-membered cyclic transition states, which by analogy with the stable chair conformation of the cyclohexane ring, should be staggered. The leaving groups should, in these cases, be orientated syn-clinal with respect to each other, an arrangement that is present in 1-X, 2-H substituted cyclohexanes if either X or H is equatorial. Reactions falling into this category include the pyrolysis of carboxylic acid esters[7] (Eq. 5) and the Chugaev reaction of S-methyl xanthates (Eq. 6).[8]

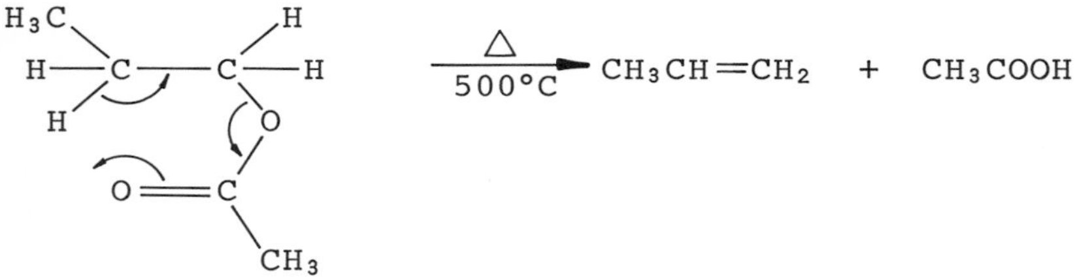

$$\xrightarrow[500°C]{\triangle} CH_3CH{=}CH_2 \quad + \quad CH_3COOH \quad (5)$$

$$\xrightarrow[180°-200°C]{\triangle} CH_3CH{=}CH_2 \quad + \quad COS \\ + \quad CH_3SH$$

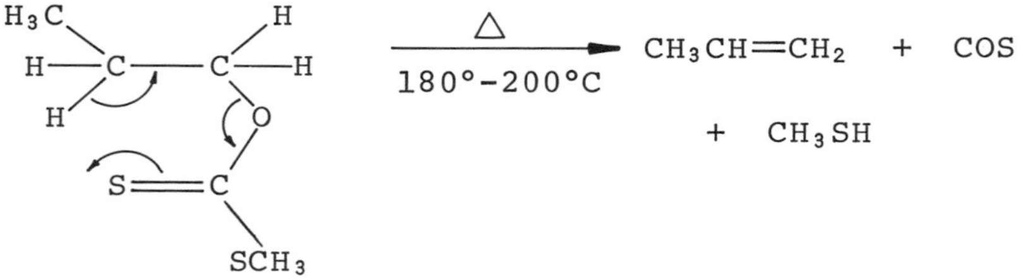

$$(6)$$

Of the reactions mentioned thus far, the Cope elimination and the Chugaev reaction are the most useful in synthesis, occurring at the lowest reaction temperatures in inert solvents in many cases. Of all the pyrolytic reactions, the Cope elimination exhibits the greatest syn-stereospecificity. The mild reaction conditions and the weakly acidic (excluding the alkyl halides) or basic nature of the reaction media makes pyrolytic eliminations ideal for the synthesis of labile alkenes. For example, the acid catalyzed dehydration of pinacolyl alcohol gives skeletally rearranged olefins, but the pyrolysis of the xanthates leads to the unrearranged alkene (Eq. 7).[9]

$$(CH_3)_3C\!-\!\overset{\overset{\displaystyle OH}{|}}{C}HCH_3 \quad \xrightarrow[\substack{2.\quad CS_2 \\ 3.\quad CH_3I}]{1.\quad K} \quad (CH_3)_3C\!-\!\overset{\overset{\displaystyle O-\overset{\overset{\displaystyle S}{\|}}{C}-SCH_3}{|}}{C}HCH_3 \quad \xrightarrow{\triangle}$$

$$(CH_3)_3C\!-\!CH\!=\!CH_2 \qquad (7)$$

In the succeeding sections we divide our discussion on the mechanism of pyrolysis into classes governed by the structure of the reactants (e.g., alkyl halides, acetates, N-oxides, etc). The mechanisms fall into three categories; heterogeneous or surface catalyzed, radical, and homogeneous Ei reactions. The most certain mechanistic criterion is kinetic analysis, and the alkyl halide pyrolyses illustrate the methodology used to distinguish the various reaction schemes. The kinetics of pyrolysis of the other reactants have been less thoroughly documented, and mechanistic deductions have frequently been made on stereochemical and orientational behavior only.

In the limited space available, it is only possible to discuss the mechanism of elimination of a few types of reactants, but they illustrate the general mechanisms involved and the experimental techniques and theoretical procedures used in their elucidation.

B   PYROLYSIS OF ALKYL HALIDES

1   Experimental Methods

The kinetics of the pyrolysis of alkyl halides, the subject of a number of reviews,[1,10-14] is usually measured in the temperature range 300 to 600°C in the gas phase. Not suprisingly, the experimental procedures differ quite markedly from those used for reactions in solution and, therefore, a brief outline of the methods follows. For more specific details the reader is referred to the recent review by Maccoll.[1]

In the static method,[15] the pyrolysis is carried

out at constant volume, and the extent of reaction is
determined from the change in pressure. As the liberated hydrogen halides attack mercury, an all-glass
system, including a glass diaphragm gauge is used.[15],
[16] To verify that the pressure change definitely reflects the rate of reaction, the rate coefficients obtained can be compared with those derived by monitoring
the liberated hydrogen halide using a sampling-titri-
metric procedure,[17] or those obtained from analysis of
the alkene product using a sampling-glc analysis
method.[18]

The flow method employs toluene vapor to sweep
the reactant through the reaction vessel.[19] The rate
coefficient is derived from a knowledge of the rate of
introduction of substrate, the expulsion of product,
and the flow of the carrier gas.[20] The toluene also
acts as a radical trap, and the presence of bibenzyl
among the reaction products indicates the involvement
of radical intermediates.

The single-pulse shock tube has been used by
Tsang[21] in a comparative study of the rates of pyrolysis of a number of alkyl halides. In this method a
wide range of temperature is available, the heating is
homogeneous, and as the reaction times are very short,
heterogenous reactions do not cause complications. A
fourth experimental method, termed chemical activation,
has been used in particular for alkyl fluoride reactions, and we discuss that approach subsequently (Section VIII.B.6). The activation parameters, of an Ei
reaction, derived by the different experimental methods
are usually in close agreement.[1]

2   Mechanisms of Pyrolysis of Alkyl Chlorides and Bromides

Heterogeneous reactions, which give rise to variable
kinetics, can complicate the study of dehydrohalogenation in clean Pyrex vessels. An increase in the rate
coefficient with the surface/volume ratio or a variation with surface material provide clear indications of
a surface-catalyzed reaction. For example, the dehydrohalogenation of t-butyl chloride occurred much
faster in a clean Pyrex vessel than in an "aged" reac-
tor.[22] Packing of the clean vessel with Pyrex glass
(surface/volume ratio changed from 3 to 10) also increased the rate of elimination threefold. The initial
runs exhibited zero-order kinetics, becoming first
order as the reaction vessel aged. During continual
usage, carbonaceous deposits collect on the vessel
walls, a process that is termed "seasoning" or "aging".

The carbon coating is believed to eliminate surface reactions by covering the catalytically active sites.[23]
This process of seasoning is of great importance to the
gas kineticist, for only when the vessel is aged are
reproducible data obtained. Aging can be accelerated
by leaving the reaction products in contact with the
vessel[24] or by decomposing allyl bromide, which during
pyrolysis deposits a carbonaceous coating, in the
vessel.[25]

Shapiro, Swinbourne, and Young have reported an
interesting comparison of the relative rates of dehydrohalogenation of ethyl, n-propyl, i-butyl, and neopentyl chlorides in clean (heterogeneous reaction) and
coated (homogeneous reaction) vessels.[23] For ethyl
chloride, the activation energy is 32 kcal/mole less
for the heterogeneous reaction. For both· series, the
reactivity increases from ethyl to n-propyl to i-butyl,
but neopentyl chloride is the most reactive in the
heterogeneous reaction and the least reactive in the
homogeneous series.

Three mechanisms have been recognized in the
thermolysis of alkyl halides under homolytic conditions
in seasoned vessels; (1) the Ei mechanism involving
synchronous elimination of HX in a cyclic unimolecular
transition state as shown in $\underline{1}$, (2) a radical-chain
mechanism (Scheme 1), and (3) a radical nonchain mechanism (Scheme 2). All three mechanisms usually follow

Scheme 1   Radical Chain Mechanism

$$RCH_2CH_2X \xrightarrow{\text{slow}} R\overset{\bullet}{C}H_2\overset{\bullet}{C}H_2 + \overset{\bullet}{X} \qquad \text{Initiation}$$

$$\overset{\bullet}{X} + RCH_2CH_2X \longrightarrow HX + R\overset{\bullet}{C}HCH_2X \;\Big\} \quad \text{Propagation}$$

$$R\overset{\bullet}{C}HCH_2X \longrightarrow RCH{=}CH_2 + \overset{\bullet}{X}$$

$$\overset{\bullet}{X} + \overset{\bullet}{X} \longrightarrow X_2 \qquad \text{Termination}$$

Scheme 2   Radical Nonchain Mechanism

$$RCH_2CH_2X \xrightarrow{\text{slow}} R\overset{\bullet}{C}H_2\overset{\bullet}{C}H_2 + \overset{\bullet}{X} \qquad \text{Initiation}$$

$$\overset{\bullet}{X} + RCH_2CH_2X \longrightarrow HX + R\overset{\bullet}{C}HCH_2X \qquad \text{Propagation}$$

$$R\overset{\bullet}{C}HCH_2X + R\overset{\bullet}{C}H_2\overset{\bullet}{C}H_2 \longrightarrow RCH{=}CH_2 + RCH_2CH_2X \qquad \text{Termination}$$

first-order kinetics, but the Ei and radical non-chain
reactions should not exhibit an induction period, and
the reaction rates are not affected by radical inhibitors such as nitric oxide, propylene, cyclohexene, or
toluene. The activation energy for the nonchain mechanism should approximate to the homolytic bond dissociation energy of the carbon-halogen bond. A lower
activation energy is anticipated for the radical chain
mechanism, the observed rate coefficient being dependent on the rate coefficients of the individual steps.
In most instances, the alkyl halides decompose by a
mixture of Ei and radical chain pathways, but the
latter can be suppressed by increasing the concentration of an inhibitor until a constant rate coefficient
is observed.

3   Radical Mechanisms for Thermolysis of Alkyl Chlorides and Bromides

(a)   The Radical Nonchain Mechanism. This mechanism is
exemplified by the thermolysis of allyl bromide in a
seasoned vessel.[25],[26]   The pressure increase of 50%
during the reaction is consistent with Eq. 8.[25]   The
reaction products observed are hydrogen bromide,

$$C_3H_5Br \longrightarrow \tfrac{1}{2}(C_6H_8) + HBr \qquad (8)$$

benzene, and a carbonaceous coating. The reaction does
not exhibit an induction period, and the rate is unaffected by low pressures of oxygen and propene, observations ruling against a chain mechanism. The activation energy [45,400 kcal/mole (static),[25] 47,500
kcal/mole (flow method),[26]] is in close agreement with
the homolytic bond dissociation of the C-Br bond in
allyl bromide (45,500 kcal/mole).[27]
     The mechanism of pyrolysis of allyl chloride is
more complicated. In the temperature range 540 to
710°C, Shilov has explained his results using a flow
method in terms of a radical nonchain mechanism.[28]
The gaseous products were propylene (90%) and allene
(10%). The thermolysis showed first-order kinetics,
but the rate coefficient was reduced by half when
toluene, rather than benzene (or no carrier), was
employed as the carrier gas. The bibenzyl/HCl ratio,
which varied from 0.2 to 0.25, was explained in terms
of Scheme 3. [For a radical mechanism, involving alkyl
halides that do not possess a β-hydrogen, a bibenzyl/
HCl ratio of unity is predicted when toluene is employed as the carrier gas in the flow method (Scheme

4),[1] if the alkyl radical is less stable than the benzyl radical.] The activation energy of 59.3 kcal/mole
should reflect the homolytic bond dissociation energy
of the C-Cl bond.

Scheme 3

$$CH_2=CH-CH_2Cl \longrightarrow CH_2=CH-\overset{\bullet}{C}H_2 + \overset{\bullet}{Cl}$$

$$\overset{\bullet}{Cl} + Ph-CH_3 \longrightarrow HCl + Ph\overset{\bullet}{C}H_2$$

$$2Ph\overset{\bullet}{C}H_2 \longrightarrow PhCH_2CH_2Ph$$

$$CH_2=CH-\overset{\bullet}{C}H_2 + Ph\overset{\bullet}{C}H_2 \longrightarrow CH_2=CH-CH_2CH_2Ph$$

$$2CH_2=CH-\overset{\bullet}{C}H_2 \longrightarrow CH_2=CH-CH_2CH_2CH=CH_2$$

Scheme 4

$$RX \longrightarrow \overset{\bullet}{R} + \overset{\bullet}{X}$$

$$\overset{\bullet}{R} + PhCH_3 \longrightarrow RH + Ph\overset{\bullet}{C}H_2$$

$$\overset{\bullet}{X} + PhCH_3 \longrightarrow HX + Ph\overset{\bullet}{C}H_2$$

$$2Ph\overset{\bullet}{C}H_2 \longrightarrow PhCH_2CH_2Ph$$

At a lower reaction temperature range of 370 to
475°C, Goodall and Howlett concluded the pyrolysis of
allyl chloride occurs by simultaneous chain, nonchain,
and heterogeneous mechanisms.[29] Even in a seasoned
vessel, the rate was dependent on the surface/volume
ratio, and homogeneous rate coefficients were calculated by extrapolation to zero surface/volume ratio. For
the homogeneous component, the first-order rate coefficients give an activation energy of 46,000 kcal/mole.
This figure is close to that observed for allyl bromide,
and suggests that the mechanism is dissimilar as the
C-Cl bond should have a greater homolytic bond dissociation energy than the C-Br bond. Propylene causes a

small rate reduction, indicating that about 10% of the reaction follows a chain mechanism and the remainder presumably follows the nonchain mechanism. Under the reaction conditions, the initially formed allene polymerized more rapidly than the allyl chloride decomposed.

(b) The Chain Mechanism. Radical chain mechanisms have been demonstrated for the pyrolysis of a variety of polychloroalkanes.[22,30-35] The pyrolysis of 1,2-dichloroethane was the first example to be reported in detail.[32] This reaction exhibits an induction period, the length of which depends on the reaction temperature, followed by a first-order decomposition. The reaction can be inhibited by propene and is very sensitive to oxygen,[30,31] acetaldehyde,[36] and mercury vapor,[35] when a glass diaphragm gauge is not employed.[37] The rate depression by radical inhibitors demonstrates the intermediacy of free radicals but does not prove what extent the chain reaction constitutes of the total decomposition. However, by continually increasing the concentration of inhibitor, a position should be reached after which further addition causes no rate change, and at this stage the unimolecular component can be gauged. Scheme 5 depicts the chain mechanism proposed by Howlett for the pyrolysis of 1,2-dichloroethane and accommodates the first-order kinetics and inhibition by propene.[33]

Scheme 5

$ClH_2CCH_2Cl \longrightarrow H_2\overset{\cdot}{C}CH_2Cl + \overset{\cdot}{C}l$ Initiation

$\overset{\cdot}{C}l + ClH_2CCH_2Cl \longrightarrow HCl + ClH_2C\overset{\cdot}{C}HCl$ $\left.\begin{array}{c} \\ \\ \end{array}\right\}$ Propagation

$ClH_2C\overset{\cdot}{C}HCl \longrightarrow \overset{\cdot}{C}l + H_2C=CHCl$

$\overset{\cdot}{C}l + H_2\overset{\cdot}{C}CH_2Cl \longrightarrow H_2C=CHCl + HCl$ Termination

$C_3H_6 + ClH_2C\overset{\cdot}{C}HCl \longrightarrow C_3\overset{\cdot}{H}_5 + ClH_2CCH_2Cl$ $\left.\begin{array}{c} \\ \\ \end{array}\right\}$ Inhibition

$C_3H_6 + H_2\overset{\cdot}{C}CH_2Cl \longrightarrow C_3\overset{\cdot}{H}_5 + CH_3CH_2Cl$

A radical chain mechanism will only occur as long as the propagation steps occur more rapidly than the

combined termination and inhibition steps. Thus an
alkyl chloride is unlikely to decompose by a chain
mechanism if either the reaction products or reactant
is an inhibitor for the chains.[22],[33] As propene and
the higher alkenes are inhibitors, radical chain mechanisms are not expected, and indeed are not observed,
for the pyrolysis of the monochloroalkanes containing
three or more carbon atoms.[22],[38],[39] The pyrolysis of
ethyl chloride is an example of a self-inhibited reaction. Radical chain chlorination of ethyl chloride
gives mainly 1,1-dichloroethane.[42] Therefore attack
of a chlorine on ethyl chloride is expected to give
the 1-chloroethyl (2) rather than the 2-chloroethyl
radical (3). As 2 cannot eject a chlorine atom without
a hydrogen transfer, it acts as a stopping radical. By
the same argument, 1,1-dichloroethane also acts as its
own inhibitor,[40] but 1,2-di-, and tri-substituted
chloroalkanes decompose by chain mechanisms,[34],[38] as
they can give rise to propagating radicals of type 3,
which can easily lose a chlorine atom to form a stable
alkene.

$$\underline{2} \quad H_3C\overset{\bullet}{C}HCl \qquad\qquad \underline{3} \quad H_2\overset{\bullet}{C}CH_2Cl$$

Alkyl bromides are more prone than alkyl chlorides to decompose by a chain mechanism. For instance,
whereas ethyl chloride and 1,1-dichloroethane pyrolyze
by molecular mechanisms, the corresponding bromides decompose mainly by radical-chain mechanisms.[43],[44],[45]
The pyrolysis of n-propyl bromide has been reported by
a number of workers.[15],[46-48] In a coated reaction
vessel, the schools of both Semenov[46] and Maccoll[47]
have reported that the rate of reaction follows an order of 1.5 in the n-propyl bromide, is inhibited by
propene, and is independent of the surface/volume ratio. To accommodate these facts, Maccoll proposed a
chain mechanism (Scheme 6) in which an initially formed
stopping radical is converted into a propagating radical (Eq. 13). A similar approach was adopted by
Semenov, but he preferred Eq. 15 as the termination
step. More recently, Cross and Stimson have reported

Scheme 6

$$n-C_3H_7Br \xrightarrow{\quad k_1 \quad} CH_3CH_2\overset{\bullet}{C}H_2 + \overset{\bullet}{B}r \qquad (9)$$

$$\overset{\bullet}{B}r + CH_3CH_2CH_2Br \xrightarrow{\quad k_2 \quad} CH_3\overset{\bullet}{C}HCH_2Br + HBr \qquad (10)$$

$$\xrightarrow{k_2^!} \overset{\bullet}{C}H_2CH_2CH_2Br + HBr \qquad (11)$$

$$(\text{or } CH_3CH_2\overset{\bullet}{C}HBr)$$

$$CH_3\overset{\bullet}{C}HCH_2Br \xrightarrow{k_3} CH_3CH=\!\!=\!\!CH_2 + \overset{\bullet}{B}r \qquad (12)$$

$$\overset{\bullet}{C}H_2CH_2CH_2Br + CH_3CH_2CH_2Br \xrightarrow{k_4} CH_3\overset{\bullet}{C}HCH_2Br + CH_3CH_2CH_2Br$$

$$(\text{or } CH_3CH_2\overset{\bullet}{C}HBr) \qquad (13)$$

$$\overset{\bullet}{C}H_2CH_2CH_2Br + \overset{\bullet}{B}r \xrightarrow{k_5} BrCH_2CH_2CH_2Br \qquad (14)$$

$$(\text{or } CH_3CH_2\overset{\bullet}{C}HBr)$$

$$2CH_3CH_2\overset{\bullet}{C}HBr \longrightarrow \text{end of chain} \qquad (15)$$

that the chain mechanism for pyrolysis of n-propyl bromide is catalyzed by both bromine and hydrogen bromide. Consequently, here, the unimolecular decomposition can stimulate the chain mechanism. The inhibition by propene is attributed to its ability to intercept bromine atoms partially, thereby reducing the probability of reactions described by Eqs. 10 and 11.

Despite the fact that an olefinic inhibitor is one of the products, the pyrolysis of all primary alkyl bromides, reported thus far, occur by simultaneous unimolecular and radical-chain mechanisms.[49] The relative contribution of the chain mechanism, which can be suppressed by effective inhibitors such as cyclohexene, decreases with increasing size of the alkyl moeity. This is not surprising, as the probability that bromine attack on the alkyl bromide will give rise to a stopping radical instead of propagating radical increases with the alkyl chain length.[50]

Secondary and tertiary alkyl bromides pyrolyze almost exclusively by the unimolecular mechanism in seasoned vessels.[51] The reason for the change in mechanism is clearly understood by comparing the steps required for the chain mechanism for isopropyl bromide to those outlined in Scheme 6 for n-propyl bromide.[52] Initial bromine attack on n-propyl bromide to give a propagating radical (Eq. 10) is on a secondary hydrogen, but to give a stopping radical (Eq. 11), is on a primary hydrogen. Therefore, k_2 should be greater than k_2'. However, for isopropyl bromide, a propagating radical only forms if the bromine attacks a primary hydrogen, attack on the tertiary hydrogen giving rise to a stopping radical. In this case, k_2' is expected to be greater than k_2 on statistical grounds. In addition, k_4 will be much smaller for isopropyl bromide, as this involves conversion of a secondary radical into a primary radical (Eq. 13).

4 Pyrolysis of Alkyl Iodides

The general picture of alkyl iodide pyrolysis is considerably more complicated than those of alkyl chlorides and bromides, but both radical and unimolecular mechanisms can be recognized. Benson[53] and his co-workers[54] have provided much of the available information. He regards the rate-determining step in alkyl iodide pyrolysis as the formation of hydrogen iodide.[53] In the pyrolysis of n-propyl iodide, this involves secondary-hydrogen abstraction by an iodine atom (Eq. 16), the reaction exhibiting an order of 1.5 in the substrate (Eq. 17).[55] With isopropyl iodide, iodine

$$\overset{.}{I} + CH_3CH_2CH_2I \longrightarrow HI + CH_3\overset{.}{C}HCH_2I \longrightarrow CH_3CH = CH_2 + \overset{.}{I}$$

$$(16)$$

$$\frac{-d[n\text{-}PrI]}{dt} = k_{1.5}[n\text{-}PrI][I_2]^{\frac{1}{2}}$$

attack on a primary hydrogen is a slower process than unimolecular decomposition (Eq. 18), and a first-order law in substrate is followed (Eq. 19).[56] The pyrolysis of s-butyl iodide involves simultaneous decomposition

$$CH_3CHICH_3 \longrightarrow HI + CH_3CH = CH_2 \qquad (18)$$

$$\frac{-d[CH_3CHICH_3]}{dt} = k_1[CH_3CHICH_3] \qquad (19)$$

by both mechanisms,[57] whereas ethyl iodide[58] and t-butyl iodide[21] decompose predominantly by the unimolecular mechanism.

Secondary reactions involving liberated hydrogen

iodide can complicate the kinetics of pyrolysis of alkyl iodides. Alkyl iodides rapidly convert to an alkane and an iodine molecule in the presence of hydrogen
iodide (see Eq. 3), a reaction that has been studied in
detail to aid understanding of the pyrolysis mechanism.[54] Hydrogen iodide addition to the alkene product, the reverse of the unimolecular elimination (Eq.
18) also occurs, and Benson and his co-workers[54] have
studied this reverse process to gain insight into the
mechanism of elimination. By applying the principle of
microscopic reversibility, they have obtained similar
activation parameters by using this approach to those
obtained in the direct pyrolysis of t-butyl iodide.[21]
We discuss the significance of the activation parameters for the unimolecular eliminations in the next section.

5   Details of the Ei Mechanism

A mass of reliable kinetic data concerning the Ei reaction of alkyl halides has been accumulated by using
seasoned vessels and employing the systematic addition
of an inhibitor until the rate coefficient ceases to
change. The necessary conditions for this unimolecular
reaction are: (a) first-order kinetics at high pressures, (b) Lindemann falloff at low pressures, (c) the
absence of induction periods and the lack of effect of
inhibitors, and (d) the absence of stimulation of the
reaction by atoms or radicals.[1,10-14] In the limited
space available in this chapter, it is only possible
to list a fraction of the results, and the reader is
referred elsewhere for more detailed coverage.[1,10-13]

(a)   Homolytic versus Heterolytic C-X Bond Dissociation. The activation parameters for the pyrolysis of
a number of simple alkyl halides are listed in Table 1.
The variations in the log A factors of the Arrhenius
equation with reactant structure are, in many cases,
within experimental error. For a series of about 100
results, embracing a wider range of structures, Maccoll
has calculated a mean log A factor of 13.40.[1] Changes
in the rate with reactant structure, therefore, reflect
variations in the activation energy, and as this is
considerably less than the unimolecular bond dissociation energy (see Table 2), the first-order-kinetic behavior does not accord with the radical nonchain

TABLE 1 Activation Parameters for Alkyl Halide Pyrolysis

(A in sec^{-1}, E_A in kcal/mole)

Alkyl Group	Cl		Br		I		References
	Log A	E_A	Log A	E_A	Log A	E_A	
C_2H_5	14.60	60.8	12.86	52.3	14.10	52.8	59, 68, 74
	13.16	56.4	13.45	53.9	13.66	50.0	21, 68, 75
	13.46	56.6	12.95	52.2	13.02	48.3	43, 67, 76
	14.03	58.4	13.19	53.7			60, 21
	13.51	56.6					61
$n-C_3H_7$	13.45	55.0	13.00	50.7			38, 69
	13.50	55.1	12.90	50.7			61, 70
			13.18	51.9			48
$i-C_3H_7$	13.40	50.5	13.60	47.7	14.46	48.2	59, 69, 2
	13.64	51.1	13.62	47.8	14.79	48.0	21, 52, 3
			12.74	47.0	12.96	43.5	46, 77
					13.67	45.1	21
$s-C_4H_9$	13.62	49.6	12.63	43.8	15.20	47.9	18, 51, 3
	14.00	50.6	13.05	45.5			62, 71
	14.07	50.8	13.53	46.5			63, 51
$t-C_4H_9$	12.40	41.4	13.30	40.5	13.75	38.0	22, 73, 21
	13.77	45.0	14.00	42.00	13.53	38.3[a]	64, 72, 1
	13.70	44.9	13.50	41.50	12.52[a]	36.4[a]	65, 21, 78[a]

TABLE 1 Continued

Alkyl Group	Cl		Br		I		References
	Log A	E_A	Log A	E_A	Log A	E_A	
t-C_4H_9	13.74	44.7	13.23	41.00			21, 71
	13.90	46.2					21
	14.41	46.6					66

^{a}Calculated by a study of the reverse reaction (HI + $\rangle\!=\;\longrightarrow$ t-C_4H_9I).

TABLE 2 Bond Dissociation Energies (kcal/mole) for the C-X Bond in Alkyl Halides.[79]

R-X	Cl		Br		I	
	Homolytic	Heterolytic	Homolytic	Heterolytic	Homolytic	Heterolytic
C_2H_5	79	200	65	191	52	183
$i-C_3H_7$	77	181	63	172	50	164
$t-C_4H_9$	75	155	61	144	48	138

mechanism. For a cyclic unimolecular mechanism, dissociation of the C-X bond can occur either homolytically, 4, as is the case for the radical mechanisms,
or in a heterolytic manner, 5, thereby leading to
charge separation in the transition state. Correlations involving the activation energies and bond

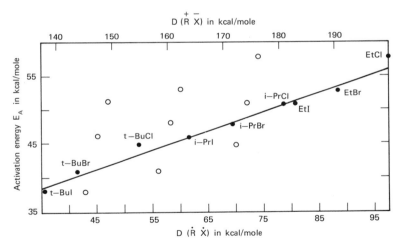

dissociation energies, measured by mass spectrometric
methods,[79] clearly support the idea of heterolytic
dissociation. There is little correlation when homolytic bond dissociation energies are employed, but
there is a very good correlation using the heterolytic
values (see Tables 1 and 2 and Figure 1).[1,80] This
observation led Maccoll and Thomas in 1955 to propose
a "quasi-heterolytic" transition state (6) for pyrolytic elimination.[80] Since that time considerable

Figure 1 The relationship between the Arrhenius activation energies for the pyrolyses of the corresponding
acetates and alkyl halides (data taken from compila-
tions in Refs 1, 13, and 14).

experimental evidence has been amassed to support this view.   There are many analogies concerning the influence of substituents on the rates of pyrolysis and unimolecular solvolysis in solution, but none with substituent effects on known homolytic decompositions. For example, α-methylation leads to enhanced rates of pyrolytic elimination of alkyl chlorides [k(C$_2$H$_5$Cl): k(i-PrCl): k(t-BuCl) = 1 : 150 : 25,000 at 360°C],[22], [32,41] but it has little effect on the rate of homolytic thermolysis of cyclobutane [k(☐)/k(☐) = 1.3].[81] Maccoll[10] and Benson and Bose[79a] have shown that the activation energies for pyrolysis can be calculated assuming an ion-pair model in which the charge separation of the carbon-halogen bond in the transition state is less developed than in the intimate ion-pair.

(b)   Charge Distribution in the Transition State.
Within the bounds of heterolytic C-X bond dissociation, a range of transition states is available for the Ei reaction of alkyl halides.   These vary from those with considerable carbonium ion character, as depicted in 6, in which C-X bond breaking is well advanced compared to other bond changes, to those possessing greater double-bond character in which the C-H bond is also extensively weakened (7, 8).

In the preceding section, the kinetic evidence presented suggests that C-X bond breaking is well advanced in the transition state.   Further support for this view is provided by the additional substituent effects listed in Table 3.   Substituents, α to the halogen, which possess an electron-releasing mesomeric effect, markedly enhance the rate of pyrolysis.   These substituents, of which methoxyl is the most effective, are capable of stabilizing carbonium-ion character at C$_\alpha$ by direct resonance interaction (9).   On the other hand, substitution at C$_\alpha$ by the strongly electron-withdrawing carbonyl group causes destabilization of carbonium-ion character in the transition state and opposes C-X bond breaking, giving a marked reduction in reaction rate.

TABLE 3   The Effect of α-Substituents on Rate of Pyrolysis of Alkyl Halides

RX X=	$k_{Relative}$		References
	Br (T°C)	Cl (T°C)	
α-Methylation			
CH_3CH_2X	1 400	1 400	52, 67 to 69
CH_3CHXCH_3	170 400	130 400	72 22,32,41
$(CH_3)_3CX$	32,000 400	11,800 400	
α-Halogenation			
CH_3CHX_2	9.6 416	8.2 437	45 32
CH_3CX_3	– –	32 437	22

α-Methoxylation

CH$_3$OCH$_2$CH$_2$X   10^9 200   82

α-Carbonyl Substitution

CH$_3$CHXCH$_3$ 1   390   22,32,41
CH$_3$CHXCOCH$_3$ 0.064   390   83

α-Phenylation

PhCHXCH$_3$ 366   280   28   280   85   84

9 

The effect of β-substitution on the rate of pyrolysis is much smaller than that of α-substitution (Table 4). From the β-position, the influence of a

TABLE 4 The Effect of β-Substituents on Rate of Pyrolysis of Alkyl Halides

RX X=	$k_{Relative}$ Br	(T°C)	Cl	(T°C)	References	
β-Methylation						
CH_3CH_2X	1	400	1	400	69,86	38,39
$CH_3CH_2CH_2X$	3	400	5	400	69,86	38,39
$(CH_3)_2CHCH_2X$	5	400	5	400	69,86	38,39
β-Halogenation						
XCH_2CH_2X	0.25	416	0.55	437	1	1
β-Phenylation						
$CH_3CH_2CH_2X$	1	385	1	425	69,86	38,39
$PhCH_2CH_2X$	0.7	385	1.8	425	1	1

substituent on C_α-X bond breaking is restricted to a polar (field or inductive) effect. The absence of activation by the electron-withdrawing β-halogen and β-phenyl, both substituents that enhance E2 reactions in basic solution (Section II.C.2) demonstrates that β-hydrogen acidity is of minor importance in pyrolytic elimination. Double-bond character is also little developed in the transition state, as the phenyl substituent is much less activating in the β- than α-position. Clearly all these results accord with a transition state of type 6 rather than 7 or 8.

Blades and his co-workers have measured the intra-molecular isotope effects [k($C_2D_4HX \rightarrow C_2D_4$)/k($C_2D_4HX \rightarrow C_2D_3H$)] in the pyrolysis of ethyl chloride[87] and ethyl bromide.[88] The values of 2.20 and 2.10, respectively, correspond to the maximum predicted for complete loss of the C-H stretching vibration in the transition state at the elevated reaction temperature of 500°C. However, the rate ratios measured comprise not only a primary isotope effect, but also a secondary isotope effect, which being almost temperature invariant, could be the dominant factor at the elevated reaction temperature.[14] The temperature invariance of the rate ratio, k(isopropyl bromide)/k (d_6-isopropyl bromide), accords with a secondary rather than a primary isotope effect. Clearly, the interpretation of isotope effects for alkyl halide pyrolysis at elevated reaction temperatures is not unambiguous, and isotope effects should only be used as supporting evidence for deductions made using more certain kinetic methods.

(c) Analogy between Unimolecular Pyrolysis and Solvolysis of Alkyl Halides. A method that is often used in the interpretation of substituent effects involves comparison of their effect on the reaction under investigation with their effect on a reaction of known mechanism. Linear correlations imply that similarly charged transition states or intermediates are involved in the two reactions. This approach is the basis on which the Hammett equation[89] and its modifications[90] and the acidity function technique,[91] both examples of linear free energy relationships, have been developed and used so successfully in the elucidation of mechanisms of organic reactions. Hoffmann and Maccoll have shown that the rate constants for gas phase elimination of a series of alkyl chlorides, RX, (R=α-phenylethyl, t-butyl, s-butyl, cyclopentyl, cyclohexyl) are linearly related in a logarithmic manner to their rate constants for unimolecular solvolysis in acetonitrile.[92] They regard the β-H as fulfilling the role in the pyrolysis of the weakly nucleophilic solvent in the solvolysis. The correlation is less successful if solvolytic data for hydroxylic solvents are employed, in which a bimolecular component or solvent assisted unimolecular C-X bond dissociation is probable.

Wagner-Meerwein rearrangements, characteristic of carbonium ion reactions in solution,[93] have been observed in the gas phase pyrolysis of some organic halides. Neopentyl chloride, which lacks a β-hydrogen, decomposes by concurrent radical chain and unimolecular mechanisms (Scheme 7).[94],[95] About 40% of the total reaction leading to 2-methyl-1-butene and somewhat less of that leading to 2-methyl-2-butene are unimolecular, a 1,2-methyl shift being involved (Scheme 8). Despite

Scheme 7

$$(CH_3)_3CCH_2Cl \xrightarrow[-HCl]{75\%} (CH_3)_2C=CHCH_3 + (CH_3)_2CHCH=CH_2$$
$$+ H_2C=C(CH_3)CH_2CH_3$$

$$\begin{cases} \longrightarrow (CH_3)_2C=CH_2 + CH_3Cl \\ \xrightarrow[-CH_4]{25\%} (CH_3)_2C=CHCl + H_2C=C(CH_3)CH_2Cl \end{cases}$$

Scheme 8

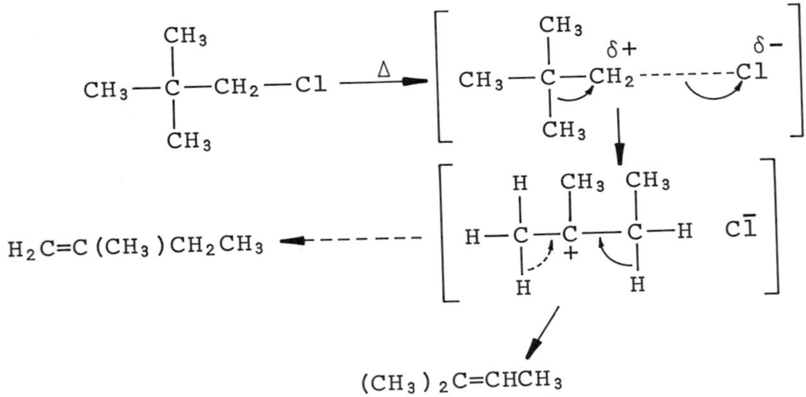

$$(CH_3)_2C=CHCH_3$$

possessing the required coplanarity for a four-membered cyclic transition state, both bornyl and isobornyl chlorides pyrolyze to give mainly the Wagner-Meerwein rearrangement products, camphene and tricyclene (Scheme 9). Bornylene, the β-elimination product, is formed in comparatively low yield and is not actually isolated as it undergoes a retro-Diels-Alder reaction to tri-methylcyclopentadiene.[96] The rate enhancement, iso-bornyl: bornyl of 20 at 400°C is very small compared to the value of 10^5 at 25°C, observed in solvolysis,[97] indicating that neighboring group assistance is much less in gas phase pyrolysis, presumably carbonium-ion character being less developed at the transition state. Additional support for this view is provided by the observation that racemization of optically-active D-(+)-2-chlorooctane does not occur during its pyrol-ysis.[92] On the other hand, it is well established

that ion-pair return, leading to racemization, is a
much faster process than product formation in solvolysis reactions.[98]   Thus the transition state for pyrolysis is best regarded as possessing an elongated carbon-halogen bond rather than as a carbonium ion-halide
ion pair.[1]

Scheme 9

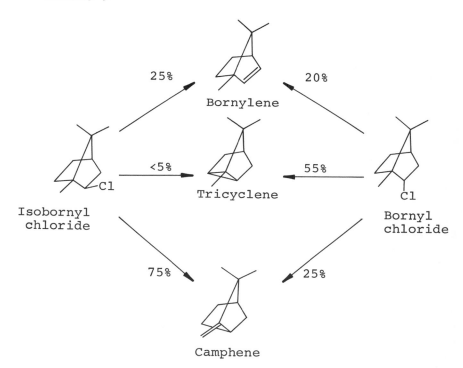

The rates of pyrolysis of a series of substituted
α-phenylethyl chlorides correlate with the Hammett
equation if the substituent constants σ^+ are used.   The
need to use the σ^+ values (Section VI.B.6) and the
magnitude of the reaction constant $\rho^+ = -1.36$ at 335°C
imply a fair degree of carbonium-ion character in the
transition state.[84]   It is difficult to compare the
magnitude of this reaction constant with that of -4.95
observed in the solvolysis of the chlorides at 45°C
because of the temperature differential.[99]   Assuming
that the reaction constant varies inversely with temperature,[89] a value of -2.6 is predicted for hypothetical pyrolysis at 45°C.[1]   Even if additional allowance
is made for the difference in media between the two

reactions, the pyrolytic value is still much smaller than that recorded for the solvolysis. Again this result accords with considerable, but less bond breaking in the transition state for pyrolysis than solvolysis.

(d) <u>Stereochemistry and Orientation</u>. Stereochemical and orientational features of alkyl halide pyrolysis have received little attention. This situation may have arisen because of the complications arising from isomerization of the olefinic products by the liberated acidic gases. The presence of added ammonia gas during the pyrolysis can reduce this complication,[18] but even so, it is wise to extrapolate product proportions back to zero reaction time.

It is generally accepted that unimolecular pyrolyses of alkyl halides occur via 4-membered cyclic transition states in which the leaving groups are coplanar.[100] This view is supported by the pyrolysis of menthyl and neomenthyl chloride (Table 5). Menthyl chloride pyrolyzes to give a 3:1 ratio of the more stable 3-menthene: 2 menthene, an orientation in accord with the Saytzev rule. In the most stable conformation, syn elimination can occur in either direction and, although statistical considerations favor 2-menthene formation, the direction of elimination is controlled by thermodynamic considerations. The product ratio is very similar to that observed in solvolysis. However, neomenthyl chloride eliminates to give mainly the less stable alkene. In the most stable conformation, syn elimination can only lead to 2-menthene, and the orientation is controlled by stereochemical considerations. The solvolysis result is markedly contrasting in this case, the almost exclusive 3-menthene formation illustrating control by thermodynamic influences, which can play a greater role in solvolysis than pyrolysis due to the greater carbonium ion character of the transition state (Section V.C).

TABLE 5 Products of Elimination of Menthyl and Neomenthyl Chlorides

Reaction	Product	Menthyl Chloride	Neomenthyl Chloride	Reference
Unimolecular Gas phase Pyrolysis	% 3-menthene	75	15	101,102
	% 2-menthene	25	85	101,102
Unimolecular Solvolysis in ethanol	% 3-menthene	68	99	103
	% 2-menthene	32	1	103

403

6   Pyrolysis of Alkyl Fluorides

The pyrolyses of alkyl fluorides cannot be studied using the static method, as the liberated hydrogen fluoride attacks the carbonaceous coating and, subsequently, the glass walls of the reaction vessel. However, this limitation does not apply to the more recently developed chemical activation and shock-tube methods, and these have been used successfully in the study of alkyl fluoride reactions.

Tschuikow-Roux and his co-workers[105] have reported the activation parameters for the pyrolysis of a series of polyfluorinated alkanes. The dehydrofluorinations were measured in a single-pulse shock tube in an argon atmosphere in a stainless steel vessel. The product-reactant ratio was determined by GLC analysis and first-order rate coefficients were measured in the temperature range 1200 to 1400°K. The activation energy increases with the degree of fluorination (Table 6), in contrast with the results available for the chloride series.

The chemical activation technique requires the synthesis of molecules of high energy content by exothermic reactions[104,107] and the interpretation of their rates of elimination using the RRKM (Rice-Ramsperger-Kassel-Marcus) theory of unimolecular reactions.[108] Pritchard and his co-workers have generated activated fluoroalkanes by the photolysis of fluorinated acetone. The Trotman-Dickinson school have prepared activated alkyl fluorides by the reaction of fluorine with an alkyl radical (Eq. 20), by combination of radicals (Eq. 21), and by methylene insertion (Eq. 22).[104] The activated molecules can undergo deactivation by molecular collision, a process that decreases with decreasing reaction pressure, or by unimolecular elimina-

$$E^* \text{ (kcal/mole)}$$

$$\overset{\bullet}{C_2H_5} + F_2 = C_2H_5F^* + \overset{\bullet}{F} \qquad 66 \qquad (20)$$

$$\overset{\bullet}{C}H_3 + \overset{\bullet}{C}H_2F = C_2H_5F^* \qquad 90 \qquad (21)$$

$$\overset{\bullet}{\cdot}CH_2 + CH_3F = C_2H_5F^* \qquad 105 \qquad (22)$$

tion of HF with critical activation energy E_0. If n is the number of effective oscillators, then from a knowledge of E^*, E_0 can be estimated for any value of n using Eq. 23. Activation energies derived by the

TABLE 6 Activation Parameters for Alkyl Fluorides and Chlorides

Reactant X=	F		Cl		References
	E_A	log A	E_A	log A	
H_3CCH_2X	59.9	13.4	57.7	13.7	104, Table 1, this chapter
H_3CCHX_2	61.8	13.9			105, 1
H_3CCX_3	68.7	14.0	50.4	12.4	105, 106
X_2HCCHX_2	70.1	13.4	54.0	14.0	105
X_2HCCX_3	72.3	13.7			105

chemical activation method are in good agreement with

$$k = A \left(1 - \frac{E_O}{E^*}\right)^{n-1} \tag{23}$$

those found by the shock-tube approach, but are generally about 2 kcal/mole lower.[104] By photolyzing a
mixture of d_6-acetone and 1,1,3,3-tetrafluoroacetone,
Perona, Bryant, and Pritchard have prepared activated
1,1,1-trideutero-2,2-difluoroethane.[107] Product analysis reveals formation of both HF and DF, consistent
with competitive elimination by both β- and α- pathways (Scheme 10). The activation energies were reported as 64 kcal/mole for α-elimination, and 53 kcal/
mole for β-elimination. The authors comment that the

Scheme 10

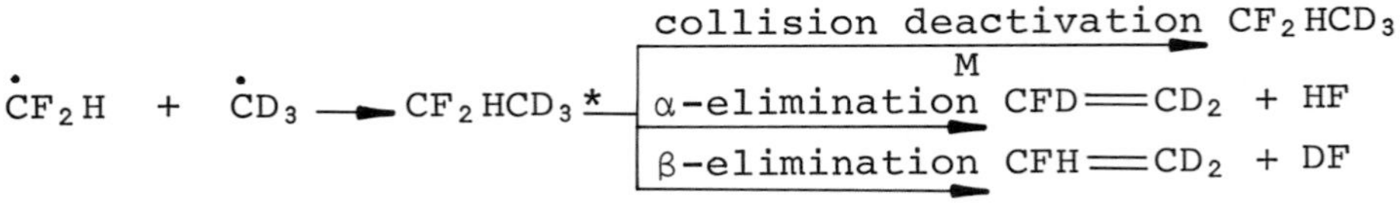

$$\overset{\bullet}{C}F_2H \;+\; \overset{\bullet}{C}D_3 \longrightarrow CF_2HCD_3{}^* \begin{array}{l} \xrightarrow{\text{collision deactivation}} CF_2HCD_3 \\[2pt] \xrightarrow{\alpha\text{-elimination}} CFD{=}CD_2 + HF \\[2pt] \xrightarrow{\beta\text{-elimination}} CFH{=}CD_2 + DF \end{array}$$

presence of two fluorine atoms on one carbon appears
necessary for incursion of the α-elimination. The
value of labeled reactants in revealing the duality of
mechanism is apparent.
    Pritchard and Perona have recently outlined the
limitations of the RRKM theory of unimolecular decompositions.[107b]

C   PYROLYSIS OF ALKYL ACETATES

1   Introductory Comments

Carboxylic acid esters, which possess a β-hydrogen
atom, undergo thermal elimination to give an alkene
and the carboxylic acid when heated in the vapor phase.
As esters can easily be prepared from alcohols, the
pyrolysis constitutes a useful method for converting
alcohols into alkenes. Product isomerization is much
less than for the alkyl halide pyrolysis as the carboxylic acids are weaker acids than the hydrogen halides. However, the reaction temperatures usually range
between 300 to 500°C, rather drastic conditions for
thermally sensitive alkenes.
    The thermal stability of the ester relates to the

molecular weight of the carboxylic acid.[109] High molecular weight esters such as stearates eliminate at
their boiling point, but higher reaction temperatures
are required to cause elimination from propionates and
acetates. Most of the investigations have been concerned with acetate pyrolysis, benzoate and methylcarbonate esters being the most popular alternatives.

Acetate pyrolyses usually follow first-order kinetics, are homogeneous, and exhibit a negative entropy
of activation, the log A factor averaging about 12.4,
a value that is slightly smaller than that of 13.4 recorded for the pyrolysis of alkyl halides.[1] The salient features of mechanism have been discussed previously by DePuy and King[13] and Banthorpe.[14]

2   Stereochemistry

It is generally accepted that the pyrolysis of esters
involves a six-membered cyclic transition state, in
which the eliminating groups are orientated syn-periplanar or syn-clinal with respect to each other. Syn
stereospecificity is clearly demonstrated by comparison
of the reactions of diastereoisomeric pairs in the
acyclic series. For example, trans-stilbene is formed
in the pyrolysis of both erythro-10 and threo-11, but
in the former case is labeled with deuterium (Eqs. 24
and 25).[110] In a base-catalyzed E2 elimination, which
occurs with anti-stereospecificity, the labeled trans-
stilbene is formed from the threo-isomer, 11. The
preference for trans- over cis-stilbene formation may
reflect a combination of steric and thermodynamic
effects.

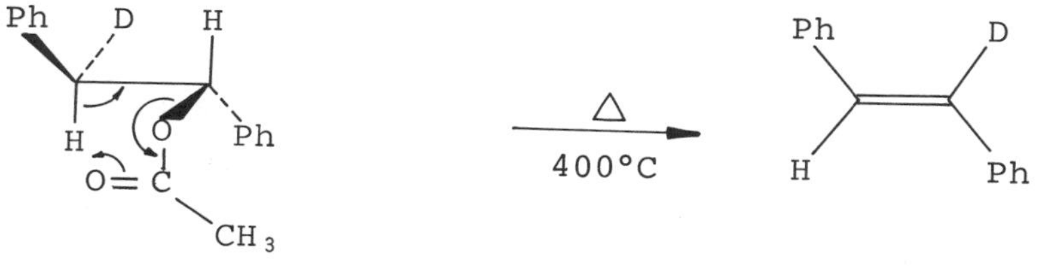

$$\xrightarrow[400°C]{\triangle}$$

(24)

10

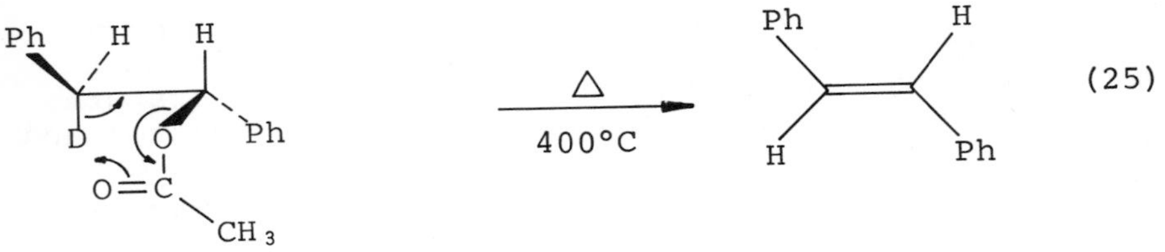

$$\text{(25)}$$

11

In the rigid indanyl system, Alexander and Mudrak have shown that the isomer **12**, which can undergo syn-elimination, decomposes at a much lower reaction temperature than **13**, which is confined to anti-clinal elimination (Eq. 26).[111]   Barton has used the syn-stereospecificity of acetate pyrolysis to interpret the structure of a number of natural products possessing five- and six-membered alicyclic rings.[112]

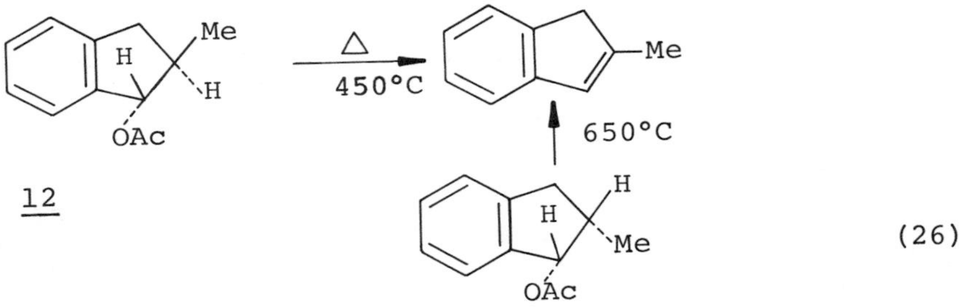

12

13

$$\text{(26)}$$

Five- and six-membered lactones (cyclic esters) are unable to attain the required syn-orientation of the leaving groups and are, therefore, thermally stable. Larger lactones are more flexible and undergo elimination (Eq. 27).[113]

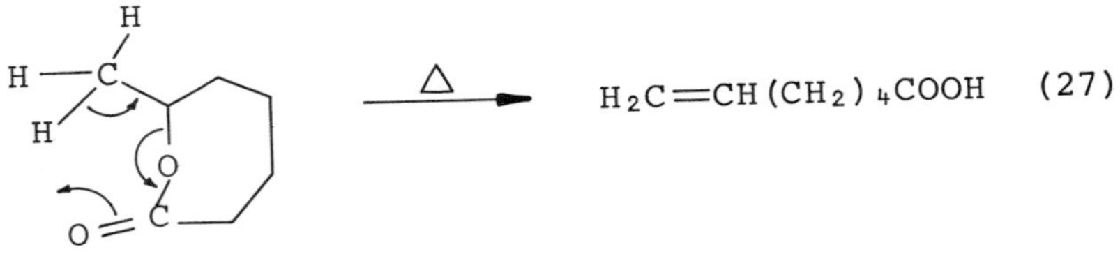

$$H_2C=CH(CH_2)_4COOH \qquad \text{(27)}$$

3 Orientation

(a) <u>Controlling Features</u>. DePuy and King recognized
that three principal factors controlled the orientation
of acetate pyrolysis in cases where more than one alkene is formed.[13] Of these, the statistical factor,
determined by the number of hydrogen atoms on each
β-carbon, is the dominant influence. Steric and thermodynamic effects were also noted. For alicyclic substrates, the syn-stereochemical requirement for elimination poses an additional complicating factor.

(b) <u>Acyclic Systems</u>. At the elevated reaction temperatures, there is little resistance to conformational
rotation in the simple aliphatic esters, and the orientation of elimination is determined by the statistical
factor.[114,115] Thus the 2-ene/1-ene ratio approximates
to 2:3 and 1:3 for the elimination of 2-butyl acetate
and 2-methyl-2-butyl acetate, respectively (Eqs. 28 and
29). The products of pyrolysis of <u>14</u> clearly reflect
the operation of a steric effect, imposed by the bulky

$$CH_3CH_2CH(OAc)CH_3 \xrightarrow{\triangle} CH_3CH=CHCH_3 + CH_3CH_2CH=CH_2 \qquad (28)$$

$$43\% \qquad\qquad 57\%$$

$$CH_3CH_2C(CH_3)_2OAc \xrightarrow{\triangle} CH_3CH_2C(CH_3)=CH_2 + CH_3CH=C(CH_3)_2$$

$$76\% \qquad\qquad 24\% \qquad (29)$$

t-butyl substituent.[116] For both alkene fractions, the
trans-isomer comprises the major fraction (Eq. 30).

$$CH_3CH_2CH(OAc)CH_2C(CH_3)_3 \xrightarrow{\triangle} CH_3CH_2CH=CHC(CH_3)_3 +$$

$$\underline{14} \qquad\qquad\qquad 70\%$$

$$5\% \text{ cis; } 65\% \text{ trans}$$

$$CH_3CH=CHCH_2C(CH_3)_3 \qquad (30)$$

$$30\%$$

$$9\% \text{ cis; } 21\% \text{ trans}$$

The pyrolysis of 3-phenyl-2-acetoxypropane, 15, accords with a combination of statistical and thermodynamic control (Eq. 31).[117] The activating effect of the aryl substituent is, however, small compared to its directing effect in base-catalyzed E2 reactions (Section III.E.2), implying little double-bond character in the transition state. Thermodynamic effects are more important than β-H acid-strengthening influences, in this case, as elimination occurs preferentially toward a p-anisyl substituent in the pyrolysis of 16 (Eq. 32).[118]

$$PhCH_2CH(OAc)CH_3 \xrightarrow{\triangle} PhCH{=}CHCH_3 + PhCH_2CH{=}CH_2 \qquad (31)$$

15   75%   25%

$$PhCH_2CH(OAc)CH_2\text{—}\langle\text{—}\rangle\text{—}OCH_3 \xrightarrow{\triangle}$$

16

$$PhCH{=}CHCH_2\text{—}\langle\text{—}\rangle\text{—}OCH_3 + PhCH_2CH{=}CH\text{—}\langle\text{—}\rangle\text{—}OCH_3 \qquad (32)$$

25%   75%

(c) **Alicyclic Systems.** Most of the orientational studies have been concerned with the cyclohexyl ring system. A more rigid molecular framework would have been desirable (eg., steroidal systems) as the ease of chair-chair inversion in many of the substrates complicates interpretation of the results. At the elevated reaction temperatures, boat conformations may contribute a larger proportion to the total structure than at room temperature, but to simplify the discussion we consider only chair conformations for the transition states.

1-Methylcyclohexyl acetate undergoes pyrolysis to give an endo/exo-alkene ratio of 3:1 (Eq. 33).[119] On statistical grounds, if conformations 17 and 18 make an equal contribution to the transition state, the alkenes should be formed in equal yield. Thermodynamically, the endo isomer is more stable. It has also been suggested that entropy considerations favor endo-alkene formation as the freedom of rotation of the methyl group is lost in the transition state for exo-alkene formation.[14] The corresponding figures for the cyclopentyl and cycloheptyl analogues are 15:85 and 24(exo):76(endo), respectively.[119] Thus acetate

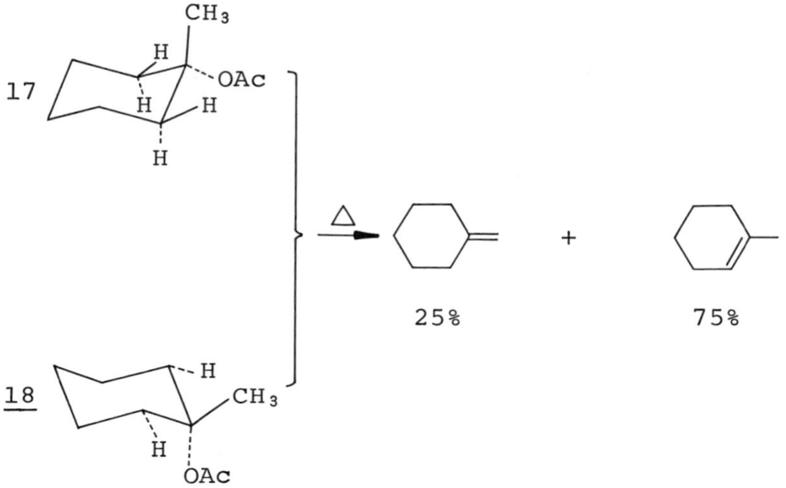

$$17 \quad \cdots \quad \overset{\triangle}{\longrightarrow} \quad \text{(methylenecyclohexane)} \quad + \quad \text{(1-methylcyclohexene)} \tag{33}$$

25%   75%

18

pyrolysis does not constitute a useful synthesis of exo alkenes, a result which contrasts sharply with the orientation in the Cope and other eliminations that utilize a five-membered cyclic transition state (Section VIII.E,F). Sufficient flexibility is present in the six-membered transition states to allow the nucleophilic atom to come within bonding distance of the ring hydrogens without having to attain coplanarity.

If conformation 18 is utilized in endo-alkene formation, elimination must involve an equatorial hydrogen (ae-elimination). In conformation 17, the equatorial acetate can abstract either an axial (ea) or an equatorial hydrogen (ee-elimination). It is only possible to decide definitely between these two possibilities by specific isotopic-labeling experiments (Section VII.A.1.f), but evidence in favor of ae-elimination being more facile than ee-elimination is provided by pyrolysis studies on 2-substituted cyclohexyl acetates.

Pyrolysis of trans-2-methylcyclohexyl acetate gives a mixture of 3- and 1-methylcyclohexene but the cis-isomer eliminates almost entirely to give 3-methylcyclohexene.[120] For the trans-isomer, 19, both chair forms can eliminate with syn-clinal stereospecificity to the 1-ene with an ae-conformation of the leaving groups. (Conformation 19b should be much less stable than 19a because of 1,3-diaxial interactions involving the acetoxy and methyl substituents.) The cis-isomer can only eliminate with syn-clinal

(a)   19   (b)   20

stereospecificity to the 1-ene from conformation 20, in
which both leaving groups are situated equatorial. At
the elevated reaction temperature, a significant proportion of the cis-substrate should exist in conformation 20 as the acetoxy and methyl substituents have
similar spatial requirements. Thus the low yield of
1-ene from the cis-isomer suggests that the ee-elimination is much less favorable than ae-elimination from
the trans-isomer.

Banthorpe[13] has proposed that buckling of the
carbon skeleton, which occurs as the six-membered ring
develops a certain amount of planarity in the transition state, will cause increased steric interaction between axial substituents for diequatorial eliminations,
but will result in a reduction in 1,3-diaxial interactions for an ae-elimination. DePuy and King[13] have
reported that ae-elimination from cis-4-t-butylcyclo-
hexyl acetate (21), in which the bulky t-butyl group
should confine the acetoxy group mainly to an axial
position, occurs 1-5 times faster than diequatorial
elimination from trans-4-t-butylcyclohexyl acetate (22).

21   22

β-Phenyl substituents exert a much weaker activating effect on acetate pyrolysis than on E2 reactions of
alkyl halides and onium salts in solution (Section II.
C.2 and III.E.2), suggestive of lesser double-bond and/
or carbanion character in the transition state (Eq.
34). For cis-2-phenylcyclohexyl acetate, 23, the conjugated alkene is the minor product, an ee-elimination
being required to effect its formation.[111] For the
trans-reactant, 24, despite the activating potential

(34)

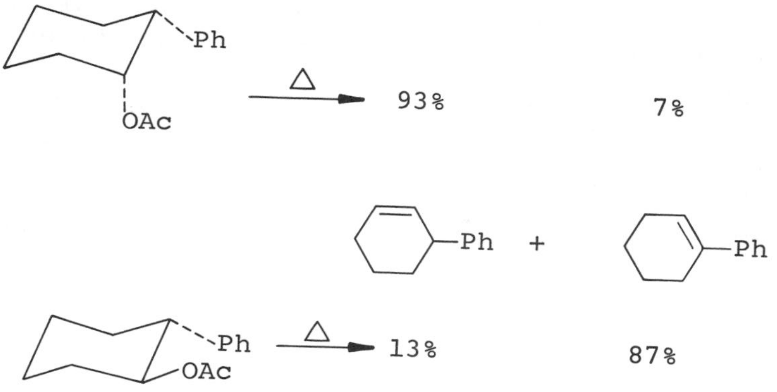

of the phenyl group, and the ae orientation of eliminating fragments, the yield of 3-phenylcyclohexene is
not negligible. Elimination from trans-2-carbome-
thoxycyclohexyl acetate gives a 97% yield of the conjugated ester,[121] a proportion that is greater than
predicted from the difference in thermodynamic stabilities of the conjugated and nonconjugated alkene. Thus
the activating effect of the carbomethoxy group on the
acidity of the β-hydrogen must in this case be important.

4   Details of the Cyclic Transition State

Some primary esters and those lacking β-hydrogens eliminate via radical intermediates,[122] and these may in-
trude into many of the product studies for which cyclic
mechanisms have been proposed. However, in the presence of inhibitors, a collection of reliable unimolecular homogeneous kinetic data have been compiled and,
fortunately, as many similar substituent studies have
been employed in both acetate and alkyl halide pyrolysis, it is possible to compare the transition states
for the two eliminations.

     The syn requirement of the leaving groups in ace-
tate pyrolysis accords with a cyclic six-membered
transition state. The small influence of conjugating
substituents on the orientation of elimination suggests
that little double-bond character is developed in the
transition state. The log A factors of the Arrhenius
equation vary slightly more widely than in alkyl halide
pyrolysis, and average about 12.3,[216] a figure consistent with a cyclic unimolecular transition state.
Changes in rate with substituent therefore reflect
variations in the activation energy and, for simple

alkyl substrates, activation energies for acetate pyrolysis parallel those for pyrolysis of the corresponding alkyl halides (Figure 2), suggesting that similar
transition states are involved in the two reactions.
    The cyclic transition state for the Ei reaction
of an alkyl acetate comprises two bond-breaking and one
bond-making process and three changes in bond multiplicity (25). The effect of α-methyl substitution on the
rate of acetate pyrolysis is less marked than on alkyl

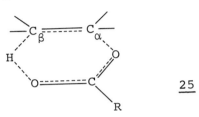

25

halide pyrolysis, but it is still significant (Table
7).[124],[125] This observation suggests that a certain

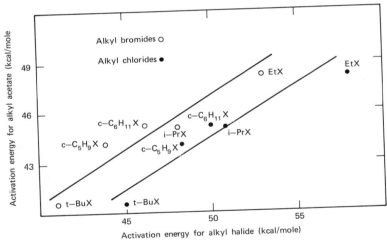

Figure 2   The relationship between the Arrhenius activation energies for the pyrolyses of the corresponding
acetates and alkyl halides (data taken from compila-
tions in Refs 1,13, and 14).

amount of C_α-O heterolysis occurs in the transition
state, but this is less than C_α-X heterolysis in the
alkyl halide pyrolysis. β-Methyl substitution retards
the rate of pyrolysis of primary acetates (Table 7)
but causes a slight rate enhancement for secondary and

tertiary acetates.[124],[126] As primary carbonium ions are less stable than secondary and tertiary carbonium ions, C_α-O heterolysis is probably less advanced in the transition state for primary acetates, but C_β-H heterolysis is more developed as a compensating factor. Thus

TABLE 7   Relative Rates of Pyrolysis of some Alkyl Acetates and Alkyl Halides

RX	Acetate		Chloride	Bromide
CH_3CH_2X	1	1	1	1
$(CH_3)_2CHX$	26	24	130	170
$(CH_3)_3CX$	1660	2000	11,800	32,000
T°C	400	400	400	400
Reference	124	125	22,32,41	52,67-9,72
CH_3CH_2X	1		1	1
$CH_3CH_2CH_2X$	0.60		3	5
$(CH_3)_2CHCH_2X$	0.48		5	5
T°C	489		400	400
Reference	125		38,39	69,86

the rate reduction by a β-methyl substituent may originate in its electron-releasing effect reducing the acidity of the β-hydrogen. Alternatively, the rate reduction could arise from statistical considerations. For the secondary and tertiary acetates, the rate increase is attributable to either an increase in alkene stability (thermodynamic effect) or to a second-order effect on C_α-O heterolysis. In the alkyl halide series, Maccoll and Wang have preferred the latter interpretation as the rate of elimination increases continually from 26 to 29, whereas alkene stability increases from the products of 26 to 28 but then decreases to that of 29.[64]

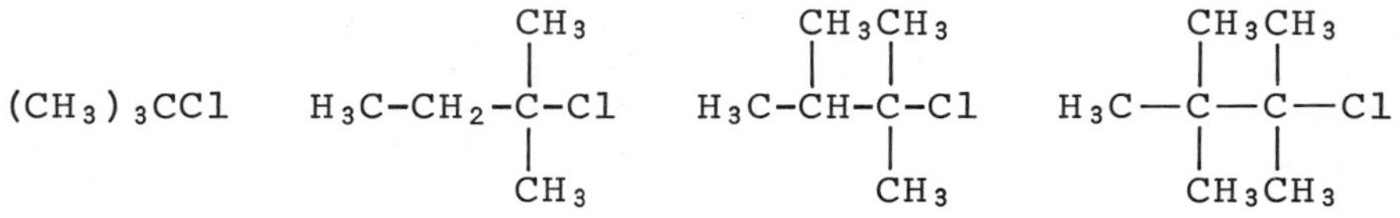

$$(CH_3)_3CCl \qquad H_3C-CH_2-\underset{\underset{CH_3}{|}}{\overset{\overset{CH_3}{|}}{C}}-Cl \qquad H_3C-\underset{\underset{CH_3}{|}}{CH}-\underset{\underset{CH_3}{|}}{\overset{\overset{CH_3}{|}}{C}}-Cl \qquad H_3C-\underset{\underset{CH_3}{|}}{\overset{\overset{CH_3}{|}}{C}}-\underset{\underset{CH_3}{|}}{\overset{\overset{CH_3}{|}}{C}}-Cl$$

26   27   28   29

Bailey and Hewitt noted the rate of pyrolysis of twelve esters of 4-methyl-2-pentanol increased with the acidity of the liberated acid.[127]   Smith and Wetzel found a similar correlation for the pyrolysis of some cyclohexyl esters,[128] and Emovon has reported that the rates of pyrolysis of the dichloroacetate, chloroacetate, and acetate of t-butyl alcohol at 250°C are 18.6: 4.4:1.[129]   These results imply a minor role for the nucleophilicity of the carbonyl group in acetate pyrolysis.   Smith and his co-workers have demonstrated that an α-phenyl substituent markedly enhances the rate of acetate pyrolysis, but a β-phenyl substituent causes only a slight increase in rate.[130-132]   Thus C_α-O bond breaking is of greater importance than C_β-H bond breaking and C_α-C_β double bond formation.

The substituent effects discussed so far support a transition state of type 30 for the pyrolysis of acetates.   Of the bond changes, C_α-O bond breaking has

30

progressed to the greatest extent.   This pictorial representation is supported by the systematic Hammett equation studies of Smith and his co-workers (Table 8).[130-132]   The σ^+ correlations and the sign of the reaction constant for the α-phenyl substituents imply carbonium ion character at C_α.   [Rates of pyrolysis of acetate derivatives have been measured to yield σ^+ values of heterocyclic rings.   The values agree well with those obtained from typical carbonium ion reactions in solution such as electrophilic substitution reactions.[133]]   The magnitude of the reaction constant is much smaller than that of -1.36 observed for the pyrolysis of substituted 1-arylethyl chlorides,[84] inferring greater carbonium-ion character in the alkyl halide elimination.   The reaction constant for β-phenyl substitution is smaller than that for α-substitution but positive in sign.   This accords with carbanion character at C_β being much less than carbonium ion character at C_α.   The small positive reaction constant for substitution in the benzoate ring implies anion

TABLE 8 Hammett Reaction Constants for the Pyrolysis of Arylethyl Acetates

Reaction Series	Temperature °C	Reaction Constant	Reference
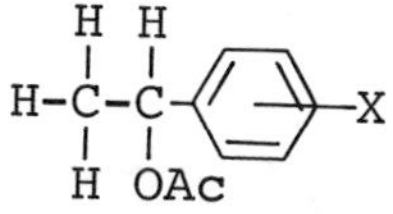 H H H–C–C–X H OAc (para-X phenyl)	337	$\rho^+ = -0.66$	132
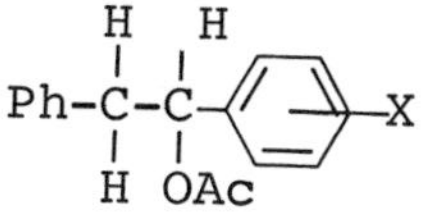 H Ph–C–C–X H OAc (para-X phenyl)	337	$\rho^+ = -0.62$	132
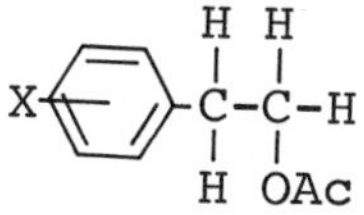 H H H–C–C–H H OAc (X-phenyl)	387	$\rho = 0.30$	132
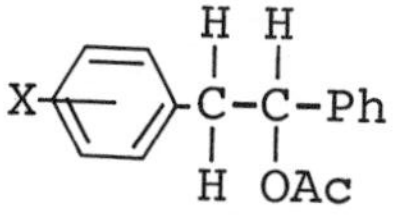 H H H–C–C–Ph H OAc (X-phenyl)	337	$\rho = 0.08$	130
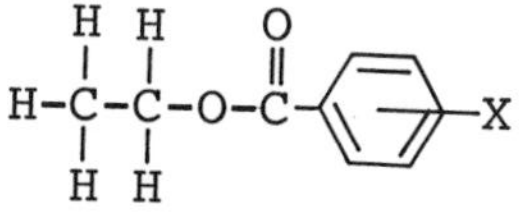 O=C–O–C–H H H (X-phenyl, ethyl)	515	$\rho = 0.20$	131

stability is of greater importance than carbonyl nucleophilicity.

5 Pyrolysis with Rearrangement

Rearrangement seldom accompanies β-elimination in acetate pyrolysis. However, Kwart and Hoster have shown that neophyl derivatives (acetate, methylcarbonate, xanthate, and halides), which lack a β-hydrogen, undergo rearrangement to give mainly β,β-dimethyl styrene (31) and 2-benzyl propene (32) (Eq. 35).[134] In the pyrolysis of neophyl acetate, as the reaction temperature is increased from 550 to 600°C, the ratio of 32/31 increases from 1.5 to 3.3. The preferential formation of the less conjugated alkene and the almost exclusive migration of the phenyl group do not accord with the intermediacy of a gas phase ion pair. Any carbonium ion intermediate is expected to rearrange preferentially to the more stable conjugated alkene, the circumstance found in the solvolysis of neophyl acetate.[135] To explain the formation of 32, a seven-membered cyclic transition state, 33, was proposed.[134]

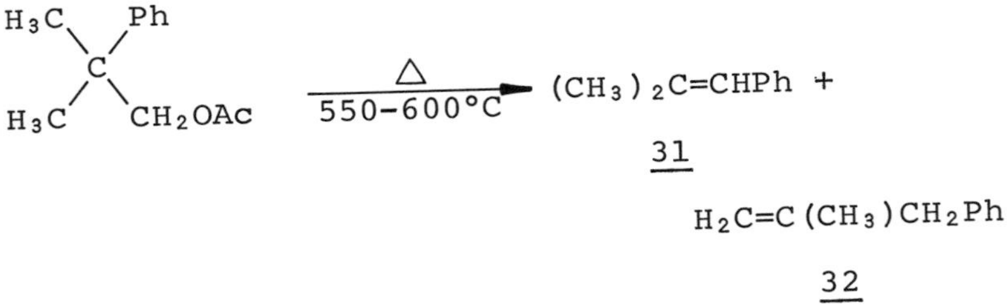
H3C
Ph
C
H3C
CH2OAc
Δ
550-600°C
(CH3)2C=CHPh +
31
H2C=C(CH3)CH2Ph
32

(35)

Subsequent work on the pyrolysis of deuterated neophyl methyl carbonate 34,[136] revealed that a minor fraction of the β,β-dimethyl styrene (which is dideuterated) arises by a 1,3-sigmatropic hydrogen shift[137] in the decomposition of the initially formed 2-benzylpropene

33

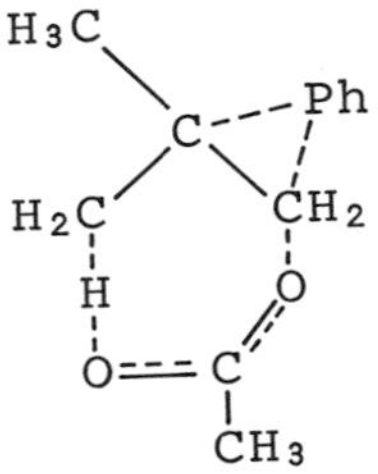
H3C
C
Ph
H2C
CH2
H
O
O
C
CH3

34

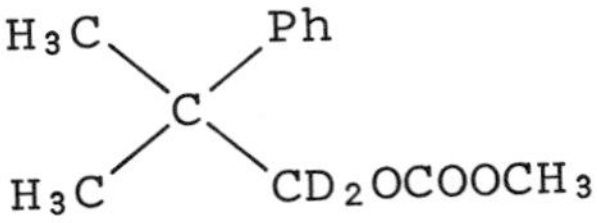
H3C
Ph
C
H3C
CD2OCOOCH3

(Eq. 36). Most of the β,β-dimethyl styrene is only monodeuterated, arising directly from an α-elimination with rearrangement of the neophyl ester (Eq. 37).

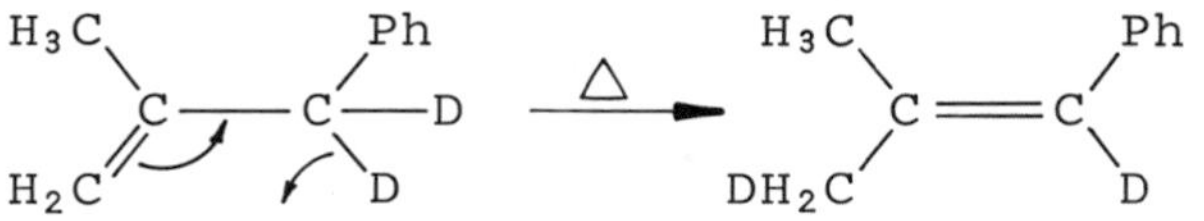

$$
\begin{array}{c}
\underset{H_2C}{\overset{H_3C}{>}}C\text{—}C\underset{D}{\overset{Ph}{<}}\!\!\text{—D} \quad\xrightarrow{\Delta}\quad \underset{DH_2C}{\overset{H_3C}{>}}C=C\underset{D}{\overset{Ph}{<}}
\end{array} \qquad (36)
$$

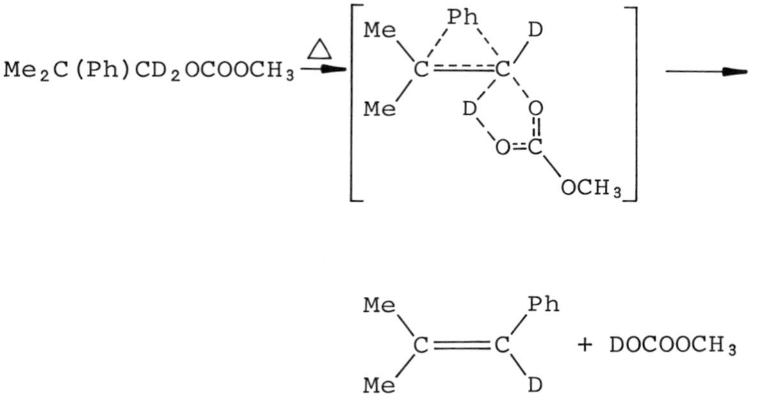

$$
Me_2C(Ph)CD_2OCOOCH_3 \xrightarrow{\Delta}
\left[
\begin{array}{c}
\underset{Me}{\overset{Me}{>}}C\!\!\cdots\!\!C\underset{\underset{O=C}{D}}{\overset{Ph\,\cdots\,D}{}}\\
\quad\quad\quad\quad O\overset{\|}{\underset{}{}}C\\
\quad\quad\quad\quad\quad OCH_3
\end{array}
\right]
\longrightarrow
$$

$$
\underset{Me}{\overset{Me}{>}}C=C\underset{D}{\overset{Ph}{<}} \quad + \quad DOCOOCH_3 \qquad (37)
$$

Lewis and Newman have recently shown that the rearrangements of norbornyl and cyclopropylcarbinyl trifluoroacetates are heterogeneous reactions, even in aged reaction vessels.[138] They suggest that the rearrangements observed in alkyl halide pyrolysis could be of a similar origin.

trans-1-Hydroxy-2-acetoxycyclohexanes decompose thermally to give a mixture of the anticipated allylic alcohol and a ring-contracted ketone (Eq. 38).[139] Neighboring group participation was invoked to account for the ring-contracted product. Participation

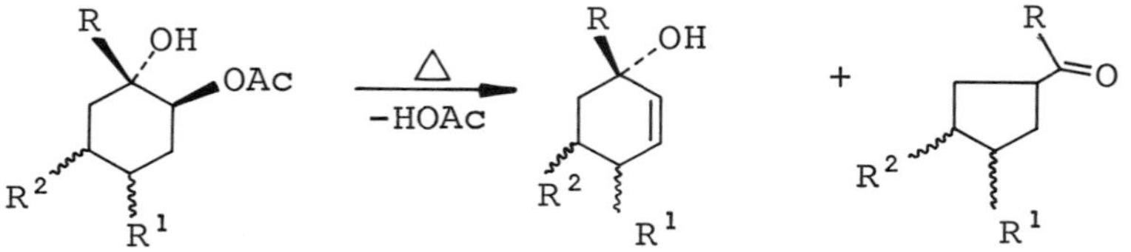

$$(38)$$

requires that the hydroxyl and acetoxyl groups be orientated syn-clinal with respect to each other and,
not suprisingly, the yield of ring-contracted product
is greater from 35a, in which all the bulky groups are
equatorial, than from 35b, in which the isopropyl group
constrains the hydroxyl and acetoxyl groups into a
diaxial conformation in a greater proportion of the
molecules (Scheme 11).

Scheme 11

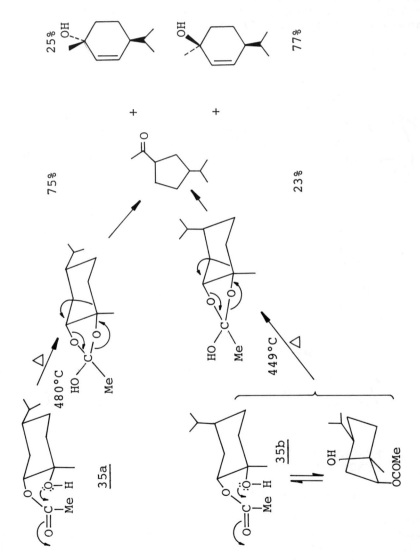

421

6   Assisted Anti Elimination

A departure from the usual syn stereospecificity has been observed in the pyrolysis of acetates which possess a cis-related carbomethoxy group in the β-position.[140]   The apparent anti elimination has been ascribed to a neighboring group effect (Scheme 12).   A similar observation has been made in the more rigid trans-decalin system (Eq. 39).[141]

Scheme 12

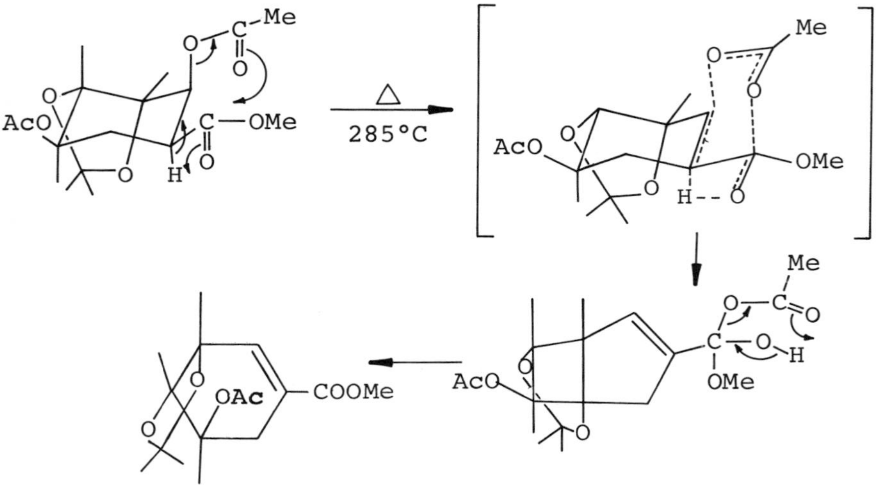

D   PYROLYSIS OF XANTHATE ESTERS: THE CHUGAEV
    REACTION

1   General Comments

Chugaev discovered that xanthate esters of alcohols, which possess at least one β-hydrogen, undergo facile thermal elimination to an alkene, carbon oxysulfide and a mercaptan (see Eq. 6),[142] while investigating their optical properties in 1899.   Since that time, the

Chugaev reaction has become one of the methods of choice for the synthesis of alkenes from secondary and to a lesser extent tertiary alcohols. Few examples of the decomposition of xanthate esters of primary alcohols have been reported and, in general, only low yields of terminal alkenes are obtained. Xanthate esters usually decompose at a temperature about 200°C less than the corresponding acetate esters, but preparation of the xanthate ester is not as facile as acetylation of the alcohol. The sodium xanthate is normally prepared by treating the alcohol with sodium hydride and carbon disulfide in ether, and is alkylated most often with methyl iodide (see Eq. 7). The pyrolysis of the S-methyl xanthate is usually carried out under reduced pressure to minimize sulfur contamination of the alkene product, as at elevated temperatures (200-300°C) thiols add to alkenes.[143] Nace has given full experimental details in his review on the Chugaev reaction.[144]

2   Stereochemistry

Syn stereospecificity has been clearly demonstrated for the elimination of S-methyl xanthates of both acyclic and alicyclic alcohols. Cram showed that the threo- and erythro-isomers of 2-butyl-3-phenyl-S-methyl xanthate eliminate mainly to cis- and trans-2-phenyl-2-butene, respectively, and a lesser yield of the terminal alkene (Eqs. 40,41).[145]

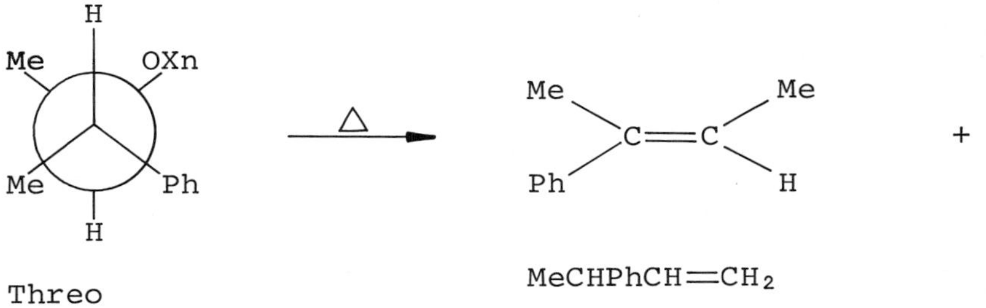

Threo

MeCHPhCH=CH$_2$ (40)

Erythro

MeCHPhCH=CH$_2$ (41)

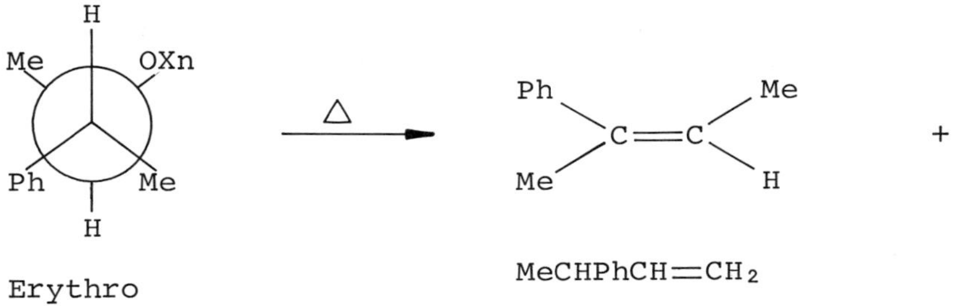

By using a bulky t-butyl substituent to effectively "lock" the conformation of cis-2-t-butylcyclohexyl-
S-methyl xanthate, Bordwell and Landis showed that only
3-t-butylcyclohexene, the product of syn-clinal elimination, was formed (Eq. 42).[146] As for corresponding

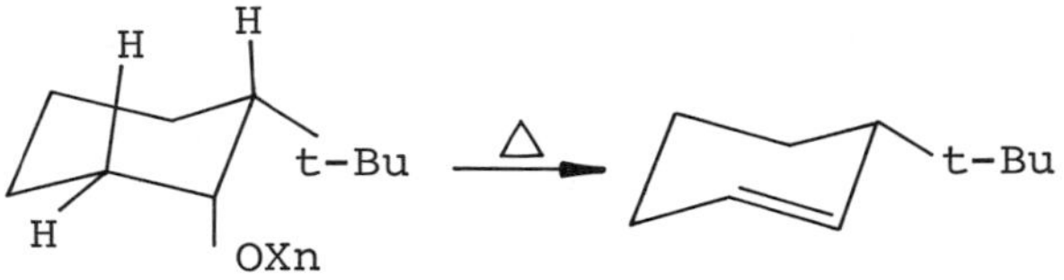

(42)

acetates, Alexander and Mudrak found that the cis-2-
methyl ester, 36, was stable at temperatures considerably greater than that required to effect elimination
from the trans-isomer, 37.[111,147]

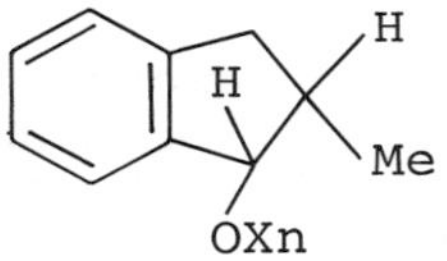

3   Mechanism of the Ei Reaction

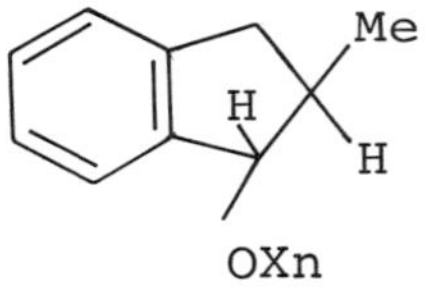

The observation of first-order kinetics[148] and the syn
stereospecificity are consistent with an Ei mechanism,
but three reaction schemes fit these requirements. Although neopentyl-S-methyl xanthate, which lacks a β-
hydrogen, rearranges to the more stable dithiocarbonate,[149] dithiocarbonate intermediates[150] can be excluded in the Chugaev reaction, as they are more stable
than the corresponding xanthates.[151] Two cyclic eliminations involving the xanthate ester are possible,
the β-hydrogen being abstracted by the thiol sulfur
(Eq. 43)[152,153] to give all three reaction products
from one cyclic process, or by the thion sulfur (Eq.
44)[145,147,154] to yield an intermediate dithiocarbonate
that rapidly decomposes.

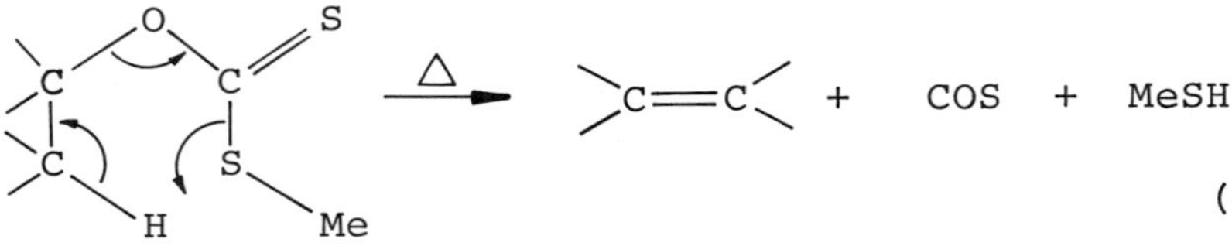

(43)

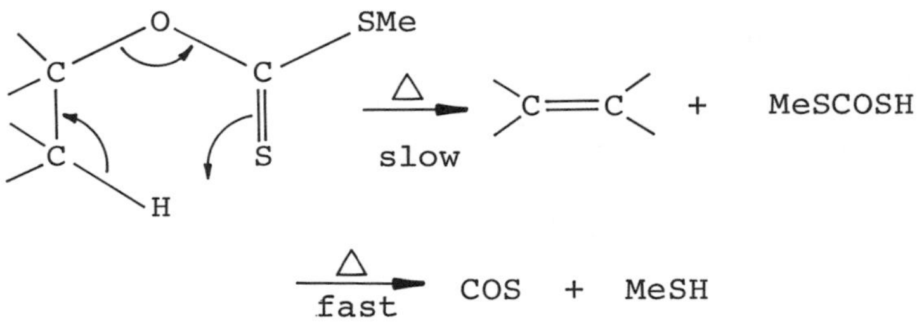

$$\xrightarrow[\text{slow}]{\Delta} \quad \text{>C=C<} \quad + \quad \text{MeSCOSH}$$

$$\xrightarrow[\text{fast}]{\Delta} \quad \text{COS} \quad + \quad \text{MeSH} \tag{44}$$

Bader and Bourns have shown by an elegant combination of sulfur and carbon isotope effects (Table 9) that the "thion-abstraction" mechanism (Eq. 44) operates in the Chugaev reaction.[155]

The thion-mechanism is, of course, analogous to the cyclic scheme accepted for acetate pyrolysis, and not suprisingly the activation entropies for the two reactions are very similar. However, the activation energies for the Chugaev reaction run 8 to 10 kcal/mole less than those observed in pyrolysis of the corresponding acetates.[148] The extra driving force for the Chugaev reaction comes from the conversion O-C=S $\longrightarrow$ O=C-S, for which summation of the bond energies[157] suggests an exothermic reaction to the extent of 20 kcal/mole.[13,144,148] Thus only part of this energetic gain is realized in the transition state.

The work of Salomaa illustrates the importance of thiocarbonyl to carbonyl conversion in Ei reactions.[151] Compounds of structure 38 and 39 are unstable to distillation, but those of type 40, 41, and 42 can be purified by distillation, the cyclic transition state for elimination involving the unfavorable weakening of C=O and the making of C=S or little change (e.g., O-C=O $\longrightarrow$ O=C-O). In the latter cases, higher reaction temperatures are required to effect elimination.

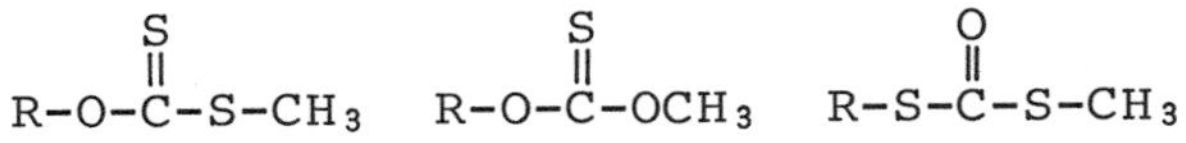

$$\overset{S}{\overset{\|}{R-O-C-S-CH_3}} \qquad \overset{S}{\overset{\|}{R-O-C-OCH_3}} \qquad \overset{O}{\overset{\|}{R-S-C-S-CH_3}}$$

<u>38</u>   <u>39</u>   <u>40</u>

$$\overset{O}{\overset{\|}{R-O-C-O-CH_3}} \qquad \overset{O}{\overset{\|}{R-O-C-S-CH_3}}$$

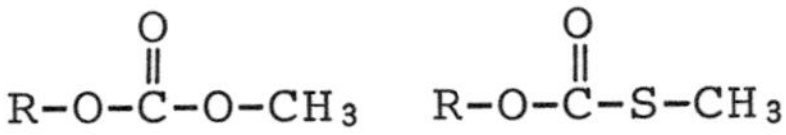

<u>41</u>   <u>42</u>

TABLE 9 Carbon and Sulfur Isotope Effects in the Pyrolysis of S-Methyl-2-trans-methylindanyl Xanthate at 78°C[155]

	Thiol Sulfur (k_{32}/k_{34})	Thion Sulfur (k_{32}/k_{34})	Carbonyl Carbon (k_{12}/k_{13})
Experimental results	1.0021 ± 0.0007	1.0086 ± 0.0016	1.0004 ± 0.0006
Theoretical Predictions[a]			
for Mechanism(43)	1.012	Unity	1.03 to 1.04
Mechanism(44)	Unity	1.007 to 1.010	Unity

[a]Calculations made using simple models and the Bigeleisen equation.[156]

O'Connor and Nace found the rate of pyrolysis of a series of S-substituted cholesteryl xanthates related in a linear manner to the pKa of the corresponding carboxylic acids.[148] Electron-withdrawing substituents promote elimination, a direct analogy with acetate pyrolysis. This observation implies that stabilization of the developing xanthate ion is more important than the nucleophilicity of the thion sulfur in the transition state. Further substituent studies are required before the extent of C_β-H and C_α-O bond breaking can be assessed with more certainty.

4   Orientation

(a) Acyclic Compounds. The direction of elimination of xanthates of secondary and tertiary alcohols in the simple acyclic series reflects statistical control.[13] In more complex molecules the effect of thermodynamic and steric factors on orientation can be recognized.[144] In general, the product proportions for the Chugaev reaction and pyrolysis of the corresponding acetate are very similar, despite the temperature differential of about 200°C between the two reactions. The results listed in Table 10 illustrate this similarity, and accord with the operation of a steric effect, this being more marked on the trans/cis ratio than the 3-ene/2-ene ratio.[158] The impact of thermodynamic influences on the statistical factor is evident in the pyrolysis of the xanthate esters of 3-phenyl-2-butanol[145] and 3-p-tolylthio-2-butanol[159] which both eliminate to give mainly the 2-butene.

TABLE 10 Elimination Products from Pyrolysis of S-Methylxanthates and Acetates of 1-R-2-butanols[158]

$$RCH_2CHXCH_2CH_3 \xrightarrow{\Delta} \underset{\text{3-Ene (cis and trans)}}{RCH{=}CHCH_2CH_3} + \underset{\text{2-ene (cis and trans)}}{RCH_2CH{=}CHCH_3}$$

X = S-Methylxanthate (OXn) or Acetate (OAc)

R in- $RCH_2CHXCH_2CH_3$	X in-	Temperature (°C)	% 3-ene trans	% 3-ene cis	% 2-ene trans	% 2-ene cis
CH_3	OXn	250			63	37
	OAc	450			60	40
C_2H_5	OXn	250	33	17	34	16
	OAc	450	33	15	35	17
$(CH_3)_2CH$	OXn	200	49	6	34	11
	OAc	450	50	5	33	12

$(CH_3)_3C$					
OXn	150	67	2	25	6
OAc	425	66	4	21	9

(b)   Alicyclic Compounds

(i)   2-Substituted cyclohexyl derivatives.   The pyrolysis products of a number of 2-substituted cyclo-hexyl xanthates are listed in Table 11.[146]   For the trans-series, the most stable conformation is 43, in which an ae-syn-clinal elimination can lead to both alkenes.   Clearly, the predominant formation of the

TABLE 11   Products of Pyrolysis of S-Methyl cis- and trans-2-substituted Cyclohexyl Xanthates[146]

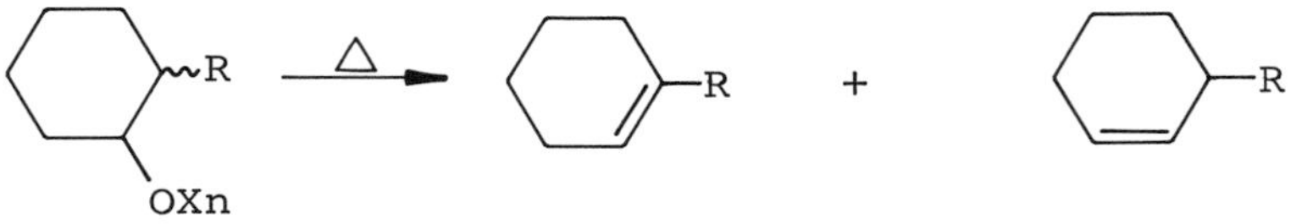

R		%	%	Reference
cis	i-Pr[a]	30	70	153
cis	t-Bu	0	100	146
cis	Ph	4	96	111
cis	SC_7H_7	10	90	160
trans	Ph	88	12	111
trans	SC_7H_7	90	10	160

[a]Menthyl not cyclohexyl derivative.

1-ene (Saytzev product) reflects thermodynamic control. For the cis-series, from conformation 44, in which the 2-R substituent is equatorial, syn-clinal elimination

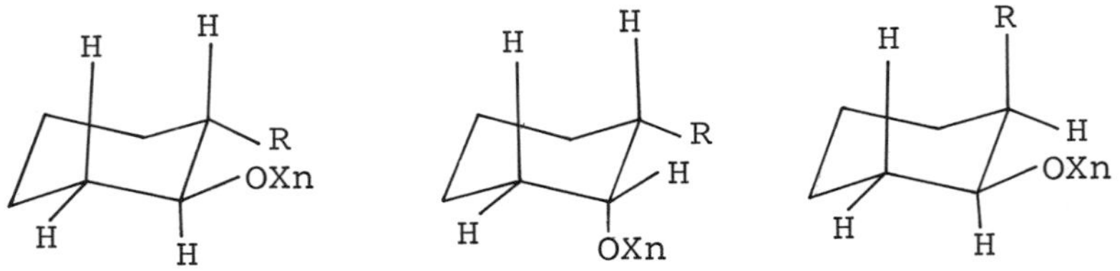

43   44   45

can only lead to the Hofmann product, the 3-R-cyclo-hexene.   With the bulky t-butyl substituent, which con-strains the smaller xanthate group to an axial

position, all the reaction appears to occur via conformation 44. For the other smaller R-groups, the contribution of conformation 45 to the total structure becomes more significant. Either ae or ee syn-clinal
elimination can give the Hofmann olefin, but the
Saytzev product must arise via an ee syn-clinal elimination. As for acetate pyrolysis, ae elimination
appears more favorable since, even in presence of the
conjugative phenyl substituent, the reaction gives
mainly the Hofmann product. A more certain interpretation could be made if the contributions of 44 and 45
to the transition state were known, and studies using
the more rigid trans-decalin system would be worthwhile.

(ii)   Endo- versus exo-alkene formation. The
pyrolysis of xanthates is not a good synthetic method
for the synthesis of an olefinic bond exocyclic to a
cyclohexyl ring. Even when the statistical factor
favors exo-alkene formation, 1-alkylcyclohexyl xanthates
eliminate to give mainly the thermodynamically more
stable endo-alkene (Table 12).[161] The two other compounds listed in Table 12 decompose to give mainly the
terminal olefin rather than that exocyclic to the cyclohexyl ring. In these cases, the terminal alkene is
the more stable product and statistical and thermodynamic effects reinforce each other. However, even so the
yield of the terminal alkene in the last example is not
as great as the 6:1 statistical factor alone should
determine, and this may reflect an unfavorable entropy
effect. In the transition state for terminal alkene
formation, the rotational entropy of one of the methyl
groups is lost.

TABLE 12 Olefinic Products in the Pyrolysis of Some S-Methyl Cyclohexylxanthates[161]

Compound	T°C	% Olefin Yield	% exo-Alkene	% endo-(Terminal) alkene
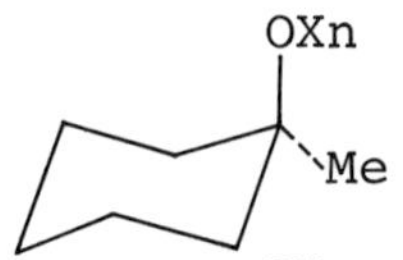 cyclohexyl with OXn, Me	200	49	21	79
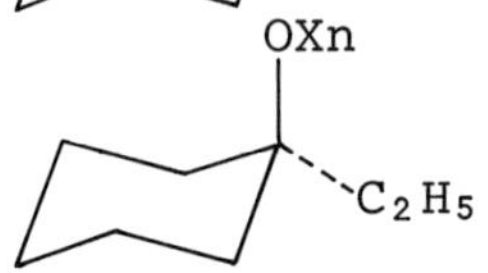 cyclohexyl with OXn, C_2H_5	200	52	12	88
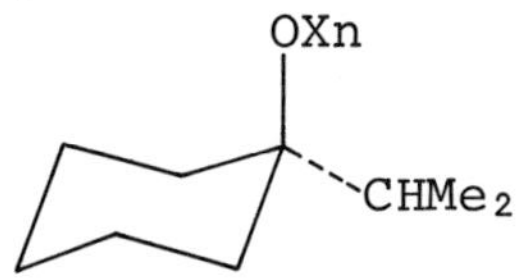 cyclohexyl with OXn, CHMe₂	100	46	22	78
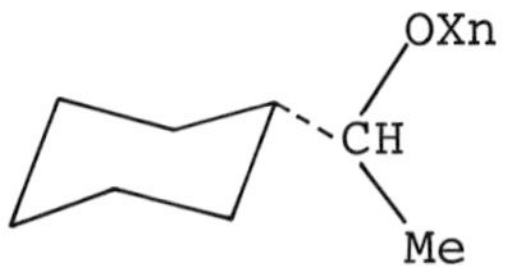 cyclohexyl with CH(OXn)Me	250	52	38	(62)

150   51   21   (79)

433

   (iii)   Alicyclic compounds excluding cyclohexyl
derivatives. Pyrolysis of cyclobutyl xanthate leads to
ring fragmentation (Eq. 45),[162] as is also observed in
the pyrolysis of the diesters 46 and 47.[163] In the

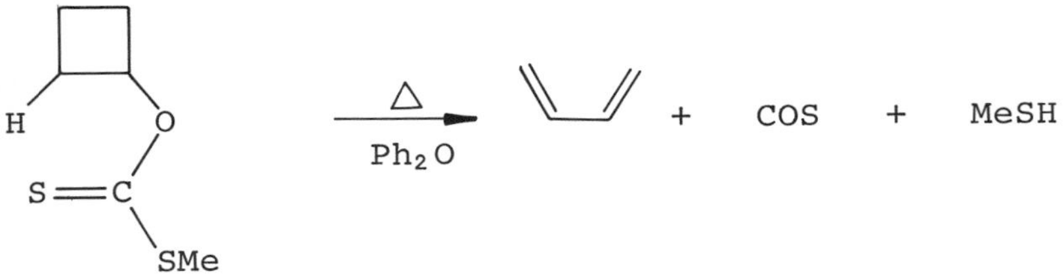

$$\xrightarrow[\text{Ph}_2\text{O}]{\Delta} \quad \diagup\!\!\!\diagup\diagdown\!\!\!\diagdown \quad + \quad COS \quad + \quad MeSH \quad (45)$$

less strained cyclopentyl cases, cyclopentene is isolated.

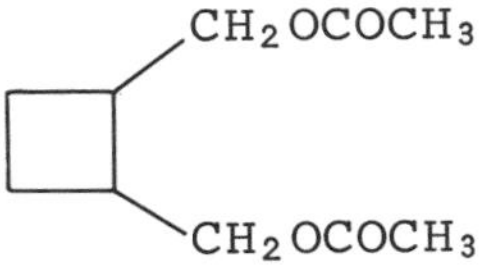

46

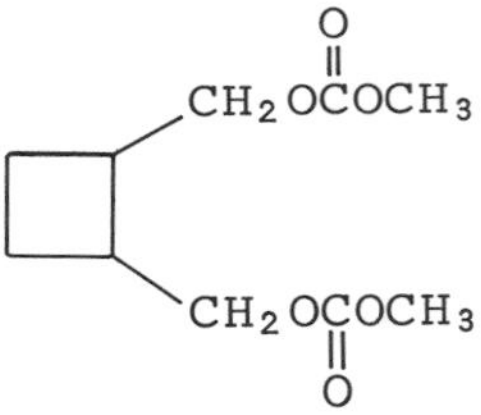

47

   The pyrolysis of the eight, nine, and ten-membered
alicyclic S-methyl xanthates are listed in Table 13.
At elevated reaction temperatures, a significant amount
of 1,8-nonadiene is formed in the Chugaev reaction of
the cyclononyl derivative. This diene arises mainly
from the facile decomposition of the initially formed
trans-cyclononene and to a lesser extent from cis-
cyclononene. The conformation of trans-cyclonene is
particularly favorable for a transannular hydride shift
of the quasi-axial hydrogen at C-5 to occur with ring
opening (Eq. 46). In the cis-cyclononene, conformation

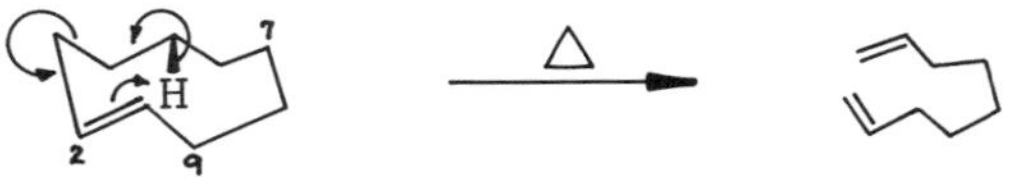

$$\xrightarrow{\Delta} \qquad\qquad (46)$$

Trans-cyclononene

TABLE 13 Olefinic Reaction Products in the Pyrolysis of Cyclooctyl, Cyclononyl and Cyclodecyl S-Methyl Xanthates

Compound	T°C	% Diene	% cis-Cycloalkene	% trans-Cycloalkene	Reference
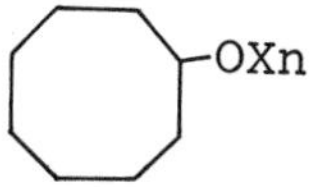 Cyclooctyl—OXn	130 to 290	–	100	<0.05	164
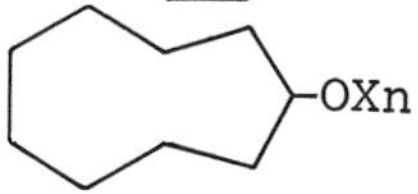 Cyclononyl—OXn	120 to 130	–	60	40	165
	360 ± 10	–	55	45	
	400 ± 10	Trace	78	19	
	450 ± 10	10	72	18	
	500 ± 10	50[a]	50	–	
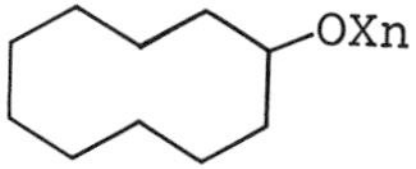 Cyclodecyl—OXn	150 to 220	–	18	82	164
	130	–	14	86	166

[a] Product: 1,8-nonadiene.

restrictions prevent the hydrogen at C-5 approaching effective bonding distance from the double bond.[165]

Equilibration experiments using acidic catalysts reveal that the cis-cycloalkenes (particularly the eight- and nine, and to a lesser extent the ten-membered ring) are thermodynamically more stable than the trans-cycloalkenes, yet the latter are the major reaction products under kinetically controlled conditions for the Chugaev reactions of the nine- and ten-membered systems.[164-166] Examination of molecular models shows that steric interactions are less for the syn-clinal elimination leading to trans-cycloalkene than that leading to cis-cycloalkene, and for the larger ring systems this interaction is the dominating factor. This position is analogous to that of E2 reactions of cycloalkyltrimethylammonium salts in which trans-cycloalkenes arise via a syn elimination (Section III.E.5).

The corresponding acetate and benzoate pyrolyses only occur above 450°C, and the products are those of thermodynamic rather than kinetic control.[165,166]

5   Nonstereospecific Chugaev Reactions

A number of xanthate pyrolyses are not consistent with the operation of a concerted cyclic Ei mechanism. Menthyl xanthate usually decomposes to menthenes on distillation,[167] but if it is washed with ferrous salts to remove peroxide impurities, higher temperatures are required to effect elimination, although the product proportions remain the same.[168] This result suggests a possible role for radical intermediates in the Chugaev reaction, although the function of the peroxides is at present not understood. Nace and O'Connor demonstrated the insensitivity of rates of pyrolysis of some cholesteryl xanthates to radical inhibitors and glass wool.[148]

Under conditions of kinetic control, both threo-48 and erythro-49 eliminate to give mainly the cis-2-butene (Scheme 13).[159] To explain the unexpected anti- elimination, Bordwell and Landis proposed a dipolar ion intermediate 50, in which internal rotation leads to loss of stereochemical control and formation of the more stable product. However, a sole common intermediate is not involved as the two reactants give a different product ratio. A similar intermediate ion has been suggested to explain the formation of 1-p-tolylsulfonylcyclohexene as the sole elimination product in the pyrolysis of cis-2-p-tolylsulfonylcyclohexyl-S-methyl xanthate.[160] Thus it appears that strongly electron-withdrawing β-substituents can promote β-hydrogen acidity sufficiently to overcome the normal syn-clinal stereochemistry of the Chugaev reaction.

A knowledge of the optical rotations of the products has been used by Bunton and his co-workers to show that esters of optically active borneol and isoborneol decompose simultaneously by concerted and carbonium ion mechanisms (Table 14).[169] Bornyl xanthate decomposes to bornylene with retention of configuration, presumably via an Ei mechanism, but isobornyl xanthate gives partially racemized alkene, suggesting a competing carbonium ion pathway. On the other hand, camphene isolated from bornyl xanthate or either benzoate has lost most of its optical activity, implying carbonium ion intermediates. However, camphene is formed with complete retention of configuration from isobornyl xanthate, and a concerted rearrangement was proposed (51). Of the two xanthates, only the bornyl derivative, in which the ester is favorably placed to participate in a cyclic γ-elimination 52, gives a significant yield of tricyclene.

438

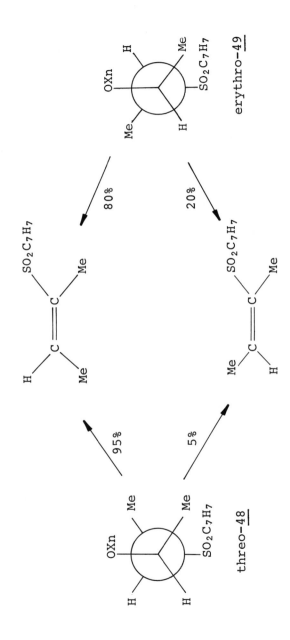

Scheme 13

erythro-49

threo-48

80%

20%

95%

5%

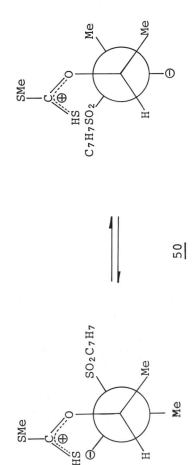

$\underline{50}$

439

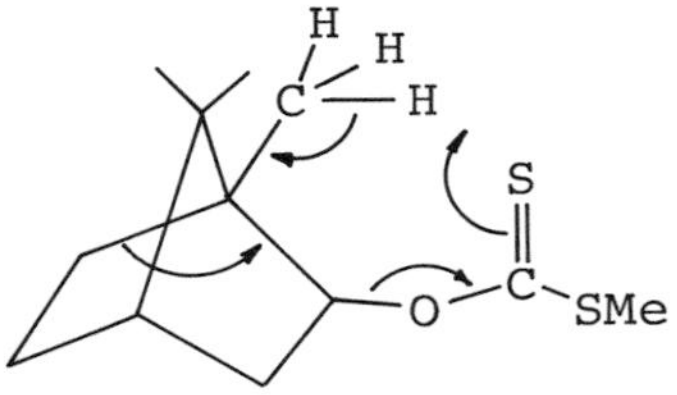

51

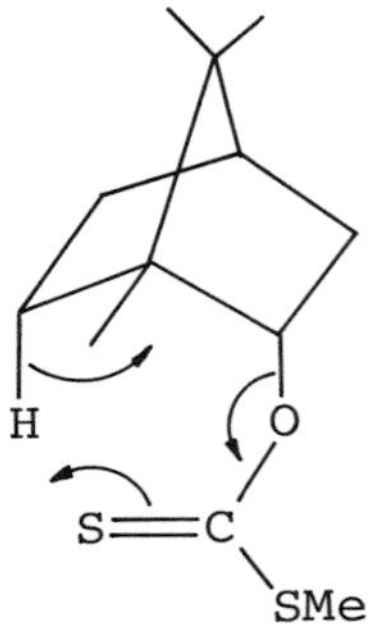

52

TABLE 14   Products of Pyrolysis of Bornyl and Isobornyl Esters[169]

Compound	% 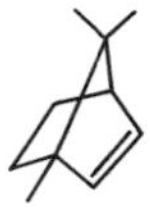 Bornylene	%  Tricyclene	% 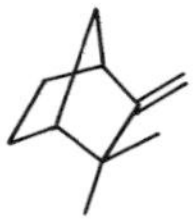 Camphene
Bornyl xanthate	70(100) [a]	13.5	16.5(≈30)
Isobornyl xanthate	38.5(92)	-	61.5(100)
Bornyl benzoate	24.5(97)	21.5	54(≈30)
Isobornyl benzoate	-	13	87(≈30)

[a]Figures in parentheses represent % retention of configuration.

To explain the interesting array of products formed in the pyrolysis of the S-p-bromophenacyl xanthates of some primary alcohols, competing radical pathways and an Ei mechanism have been suggested (Scheme 14).[170]

Scheme 14

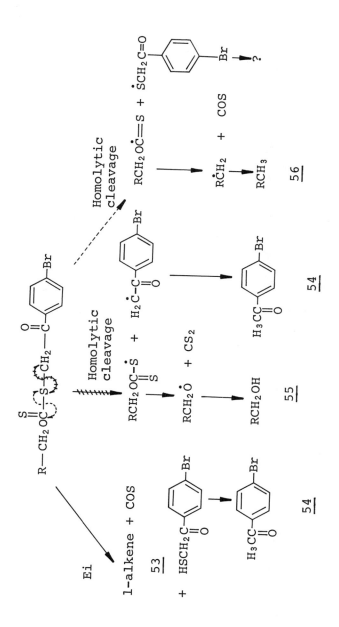

441

Scheme 14 Continued

R	%53	%54	%55	%Other Products
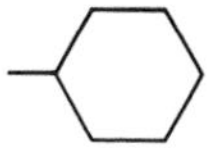 (methylcyclohexyl structure)	20	33	25	13, toluene; 15, 1-bromomethyl-cyclohexene; 18, 1-methylcyclohexene
$-(CH_2)_4CH_3$	29	22	35	1, $CH_3(CH_2)_2CH=CHCH_3$; 24, $CH_3(CH_2)_4CH_2Br$
$-CH_2Ph$	26	22	25	16, 2-phenylethyl bromide; 4, acetophenone; 13, ethylbenzene

442

Table 15 lists the products of pyrolysis of cis-
and trans-2-methyl-1-cyclohexyl-S-methyl xanthates and
their 2-d_1 analogues at 250°C in a seasoned vessel.[171]
From the 1-ene/3-ene proportions, the kinetic isotope
effect, k_H/k_D, for 1-ene formation from the trans-isomer is about 1.6, a value that is consistent with the
operation of an Ei mechanism. As expected, the 1-ene
fraction is almost nondeuterated, and the 3-ene fraction is monodeuterated. On the other hand, the alkene
proportions give an isotope effect of unity for 1-ene
formation from the cis-isomer and the 1-ene fraction is
almost half monodeuterated. These facts caused
Djerassi and Briggs to suggest initial slow formation
of an ion-pair, in which a 1,2-deuteride shift leads to
a more stable tertiary carbonium ion, from which either
hydrogen or deuterium can be lost to give the 1-ene
(Scheme 15). The 11% nondeuterated product in the
3-ene fraction is more difficult to explain and, at
this stage, a surface-catalyzed reaction (e.g., on the
carbonaceous support) cannot be excluded.

Scheme 15

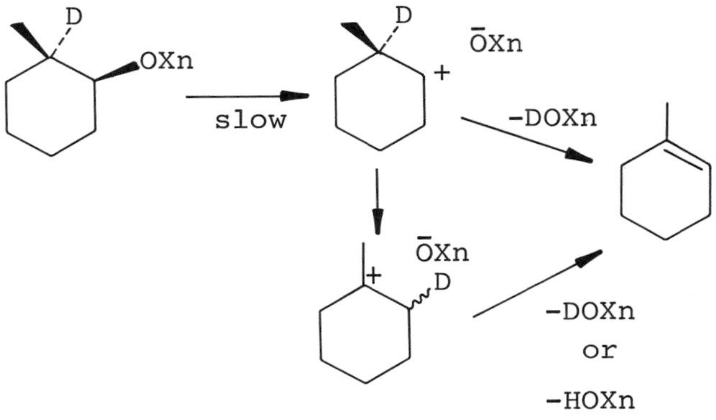

TABLE 15 Products of Pyrolysis of cis- and trans-2-Methylcyclohexyl-S-Methyl Xanthates in a Seasoned Vessel at 250°C[171]

Xanthate	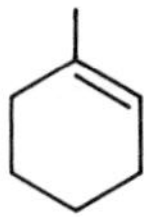 % Yield[a]	% D[b]	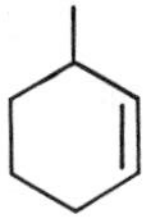 % Yield[a]	% D[b]	k_H/k_D
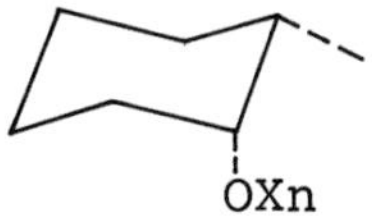	29	–	71	–	
	65	–	35	–	
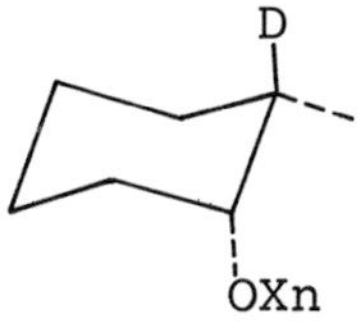	28	56 d_o 44 d_1	72	11 d_o 89 d_1	~1.0

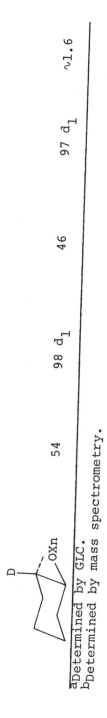

54 98 d_1 46 97 d_1 ~1.6

aDetermined by GLC.
bDetermined by mass spectrometry.

E PYROLYSIS OF AMINE OXIDES: THE COPE ELIMINATION

1 Experimental Details

Tertiary amine oxides decompose thermally to give alkenes and hydroxylamines in high yield. The synthetic
utility of this reaction was first recognized in 1949
by Cope, Foster and Towle,[172] and Cope and Trumbull
reviewed the field in 1960.[173] Most tertiary amines
can be converted into the amine oxide by 35% aqueous
hydrogen peroxide, although for sterically hindered
compounds, the more reactive 40% peroxyacetic[174] or
monoperoxyphthalic acids[175] are usually employed. For
safety reasons, peroxide traces are destroyed by plati-
num black or catalase[176] before commencing the pyrolysis, which usually occurs in the temperature range of
80 to 160°C in a high boiling solvent. In dipolar
aprotic solvents, facile elimination even occurs at
room temperature.[177]

2 Stereochemistry

Of all the pyrolytic eliminations, the Cope elimination
exhibits the greatest syn stereoselectivity, in accord
with a five-membered Ei transition state. In the
acyclic series, the products of elimination of the isomeric pair of 1,2-diphenylpropyl-[178] and 3-phenyl-2-
butyl-dimethylamine oxides[179] exemplify this stereochemical adherence.

1-Methylcyclohexyldimethylamine oxide, 57, decomposes to give almost entirely the less stable exo-
alkene (Eq. 47).[180] This result contrasts markedly to
the pyrolyses of the corresponding esters, which give
mainly the endo-alkene, and demonstrates the greater
need for coplanarity of the leaving groups in the
transition state in the Cope elimination. For endo-
alkene formation, a planar transition state requires
that the elimination occurs from a boat conformation
rather than from the more stable chair conformation of
the cyclohexane ring.

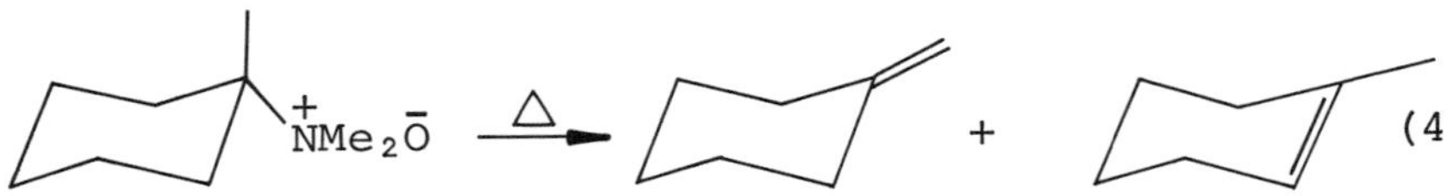

$$\overset{+}{N}Me_2\bar{O} \quad \xrightarrow{\Delta} \qquad + \qquad \qquad (47)$$

57   97.2%   2.8%

Both the l-methylcyclopentyl and l-methylcyclo-
heptyl dimethylamine oxides eliminate to give mainly
the thermodynamically more stable endo-alkenes as the
major reaction products.[173]   In the cyclopentyl case,
ring geometry confines the N-oxide group and the two
cis-β-hydrogens to the required coplanarity for elimination to the endo-alkene.   In the seven-membered alicyclic ring, sufficient flexibility is present to
enable the requisite coplanarity to be achieved.

3   Orientation

(a)   Acyclic Compounds.   The direction of elimination
in secondary and tertiary dimethylamine oxides is controlled principally by the statistical factor for simple acyclic compounds, a direct analogy with the pyrolyses of alkyl halides and esters (Table 16).[181]   The
impact of a steric factor is evident for the t-butyl
substituent, and the significant activating effect of
a β-phenyl substituent has been attributed to its influence on β-hydrogen acidity.   The magnitude of this
latter activation is considerably greater than that
observed in ester pyrolyses (cf. Eq. 31, Section VIII,
C.3.b), but smaller than noted for E2 reactions in
solution [eg., k(2-phenylethyl)/k(ethyl), corrected for
number of hydrogens = 530 (at 55°C for bromides with
ethoxide in ethanol); = 650 (at 64°C for dimethylsulfonium salts with ethoxide in ethanol).[182]]

TABLE 16 Orientation of Elimination in the Pyrolysis of $\overset{+}{R}NMe(C_2H_5)\bar{O}$[181]

R	Et	i-Pr	t-Bu	n-Pr	n-Bu	i-Am	n-Decyl
Elimination into R relative to C_2H_5, corrected for the number of β-hydrogens	1.00	1.32	2.02	0.90	1.20	1.14	1.32

	i-Bu	PhCH$_2$CH$_2$
	1.33	100

(b) <u>Cyclohexyl Compounds</u>. Menthyldimethylamine oxide
eliminates to give 2-menthene (64%) and 3-menthene
(36%) (Eq. 48).[183] There is one hydrogen cis-related
to the leaving group at both C-2 and C-4, so on statistical grounds both alkenes should be formed in 50%
yield. Thermodynamic considerations favor 3-menthene,
which is the main product from pyrolysis of the esters
and alkyl halides. Obviously neither of these factors
is dominating, as the Hofmann alkene, 2-menthene is
the major product, and Cope and Acton[183] suggested a
steric factor. 1,3-Interactions (◄─►) are greater in

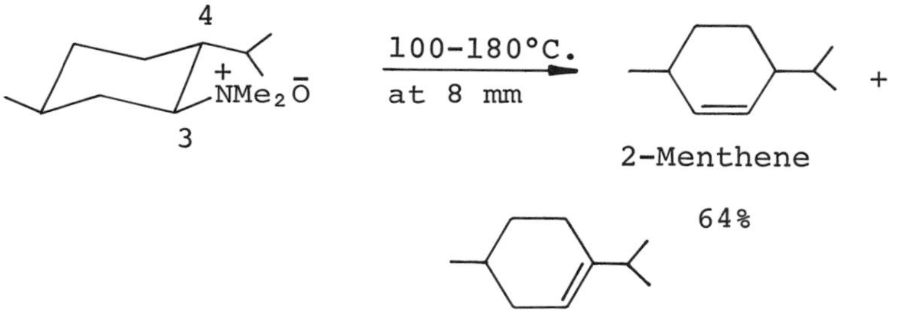

2-Menthene

64%

(48)

3-Menthene

36%

the boat conformations <u>58</u> and <u>59</u>, which lead to 3-men-
thene than in <u>60</u>, which gives 2-menthene.

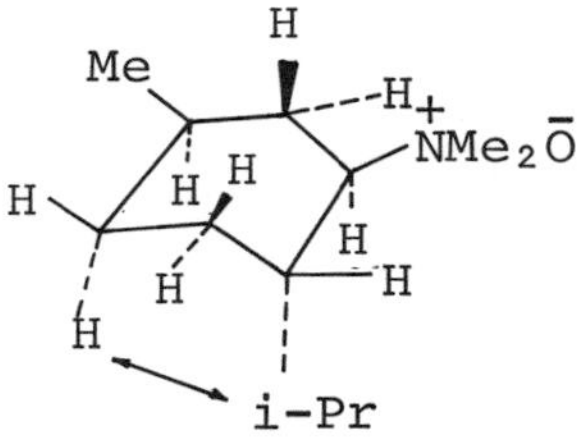

<u>58</u>

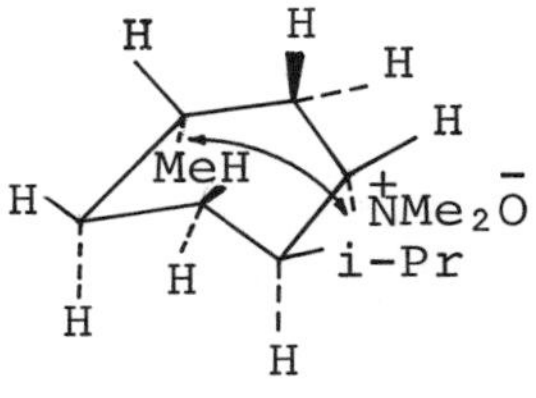

<u>59</u>

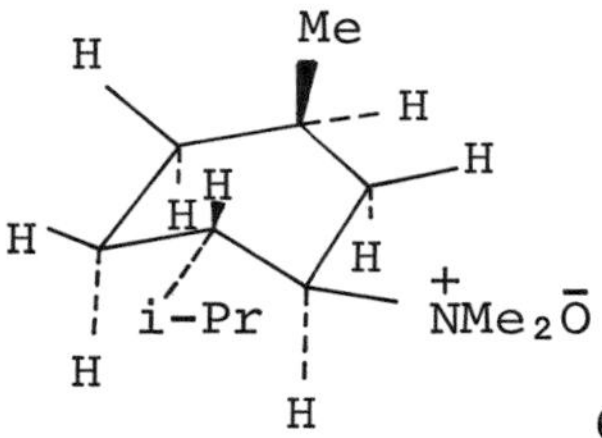

<u>60</u>

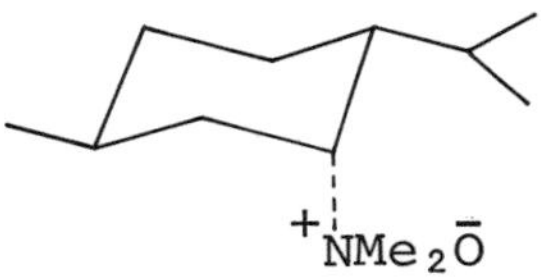

<u>61</u>

A syn-periplanar transition state is not possible for neomenthyldimethylamine oxide (61) involving the C-4 hydrogen and only 2-menthene is formed on pyrolysis.[183]   Likewise cis-2-phenylcyclohexyldimethylamine oxide eliminates almost entirely to 3-phenylcyclo- (Eq. 49), but the trans-isomer yields 1-phenylcyclo- hexene as the major product (Eq. 50).[184]   However, the activating effect of the phenyl is much less than noted in the acyclic series and is possibly reduced by steric interactions.   In the boat conformations leading to the 1-ene, (62,63) an additional 1,3-interaction involving a large group is present compared to conformation 64, which gives the 3-ene.   By comparison, both cis- and trans-2-phenylcyclohexyltrimethylammonium

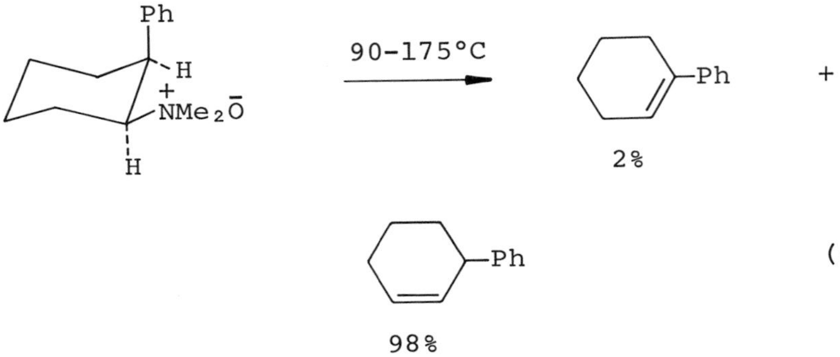

$$(49)$$

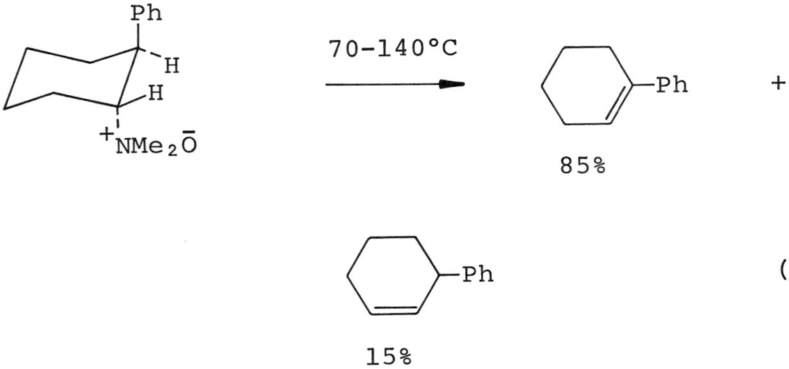

$$(50)$$

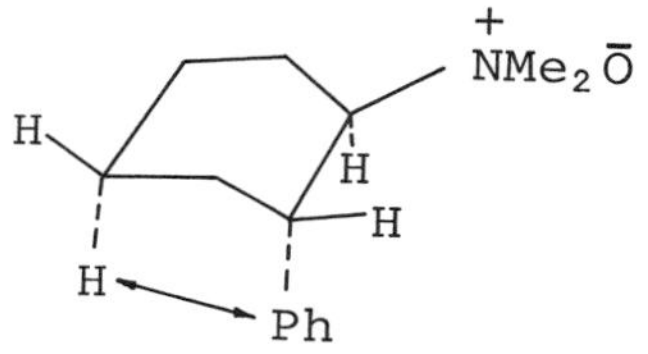

62

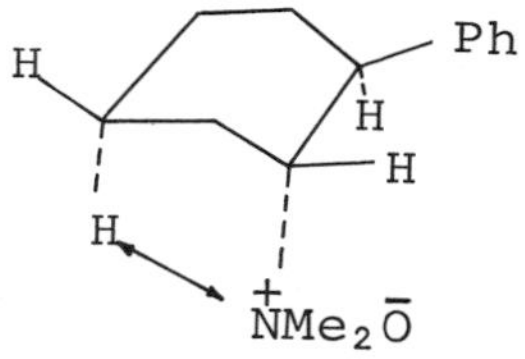

63

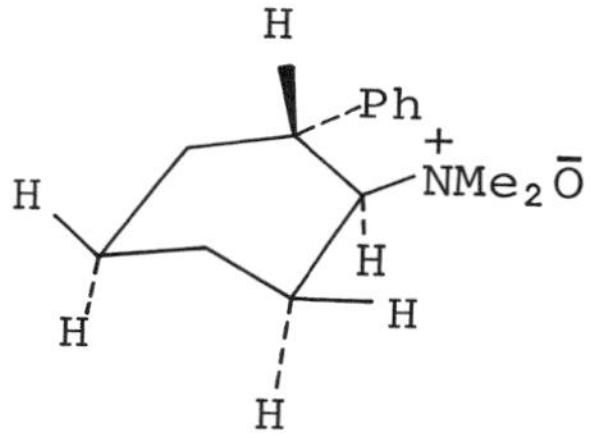

64

hydroxides eliminate to give only 1-phenylcyclohex-
ene.[185] Thus the stereochemical integrity of the Cope
elimination is maintained in the presence of activating
substituents, whereas the Hofmann elimination can
change from anti- to syn-clinal stereospecificity.

(c) Various Alicyclic Compounds. Dramatic variations
in the rate of the Cope elimination with changes in the
size of the alicyclic ring have been reported (see
Figure 1, Section III.E.5).[186] A rate differential of
2.5×10^5 exists between the most reactive (cyclodecyl-
dimethylamine oxide) and the cyclohexyl compound which
is the least reactive. Rate variations with ring size
arise from two sources: (a) the energy needed to
achieve the syn-periplanar transition state for elimination, and (b) the energy required to achieve bond
angle enlargement. These adjustments are highly energy
demanding in the cyclohexyl system,[187] but are achieved
with relative ease, because of greater flexibility, in
the cyclodecyl case.[188]
     Foote[189] and Schleyer[190] found that a linear re-
lationship exists between the rates of solvolysis of
cycloalkyl tosylates and the carbonyl stretching frequency of the corresponding cycloalkanone. A similar
expression is obeyed for the rates of formation of
both cis- and trans-cycloalkenes by the Cope

elimination.[186]   The rationale behind these relationships is that the rate processes proceed more rapidly
as bond angle enlargement becomes more facile.

The relative yields of cis- and trans-cycloalkenes
formed in the Cope elimination are listed in Table 17.
Angle strain allows only the cis-alkene for the seven-
and smaller membered rings.   The larger rings possess
sufficient flexibility to enable both geometric isomers
to exist, but the cis-alkene is thermodynamically more
stable up to C-10.   However, as we note in the Chugaev
reaction (Section VIII.D.4.b.iii) and in the Hofmann
elimination (Section III.E.5), the trans-alkenes are
the major reaction products for all rings greater than
C-8.   As mentioned previously, for syn-periplanar eliminations, steric interactions involving the hydrogens
of the ring system are less severe in the transition
state leading to the trans-alkene.

Sicher and his co-workers have pioneered the use
of variations in rate of a reaction with ring size as
a criterion of stereochemistry of a reaction.   The Cope
elimination is used as the reference reaction for a
bona fide syn process, and its use to imply the stereochemistry of E2 reactions in solution is illustrated
earlier in this book (see Section III.E.5).

TABLE 17 Cycloalkene Composition in the Cope Elimination in t-Butyl Alcohol at 70.6°C[186]

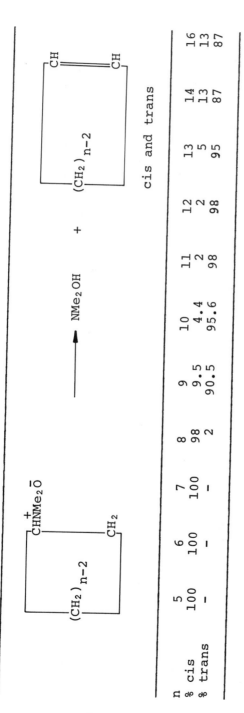

n	5	6	7	8	9	10	11	12	13	14	16
% cis	100	100	100	98	9.5	4.4	2	2	5	13	13
% trans	–	–	–	2	90.5	95.6	98	98	95	87	87

4   Kinetics of the Ei Mechanism

α-Methyl substitution has a much smaller effect on amine oxide pyrolysis than on thermal elimination of acetates. At 100°C, along the series ethyl, isopropyl, t-butyl, the rate of the Cope elimination increases from 1 to 2.64 to 6.06,[173] while extrapolated data for the acetate pyrolysis increase from 1 to 10^2 to 10^5.[13] Thus carbonium ion character at C_α is considerably less in the transition state in the Cope elimination.   On the other hand, the greater activation in the Cope elimination by a β-phenyl substituent (Section VIII.E.3) implies carbanion character at C_β is greater in the Cope elimination.

The elimination of the diastereoisomeric N,N-dimethyl-3-phenyl-2-butylamine oxides, reported by Sahyun and Cram, constitutes the most detailed kinetic study of the Cope elimination.[191]   In a variety of solvents the reactions are first order and completely syn-stereospecific (Scheme 16, erythro-65 → 67 + 69; threo-66 → 68 + 69).

Scheme 16

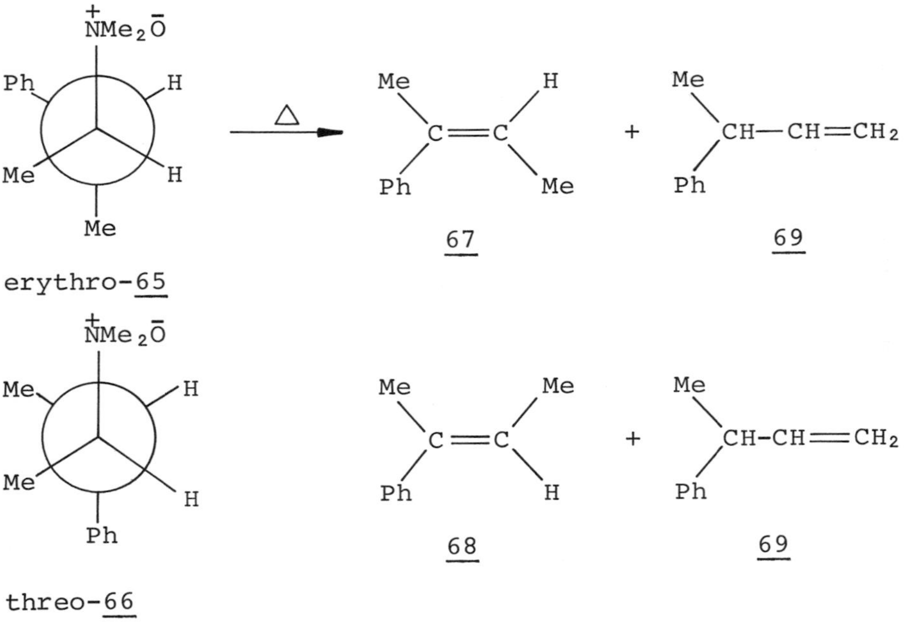

erythro-65

threo-66

The rates of elimination are much greater in aprotic than protic media, the latter reducing the proton-abstracting ability of the oxide function by hydrogen-bonding interactions. In the aprotic media, the reaction is several times faster in tetrahydrofuran than in the more polar dimethyl sulfoxide, as is expected for a reaction in which charge dispersal occurs in passing into the transition state. Traces of water dramatically reduce the rate in tetrahydrofuran, but they have less impact on the dimethyl sulfoxide that acts as a more effective "desiccant." β-Hydroxy groups are known to reduce the rate of the Cope elimination in cycloalkyl systems due to H-bonding interactions.[192]

To a certain extent, the variations in the ratios, $k_{threo}/k_{erythro}$, k_{67}/k_{69}, k_{68}/k_{69}, k_{68}/k_{67}, afford a measure of the type of transition state. At equilibrium, by analogy with other open-chain diastereoisomers,[193] the threo and erythro isomers have similar thermodynamic stability, but the alkene ratio 68/67 is about five.[191] Thus the rate ratio $k_{threo}/k_{erythro}$

of 4.5 in tetrahydrofuran (at 25°C) implies a product-like transition state, but that of only 1.2 on addition of 20 mole % of water suggests a reactantlike transition state. The ratios of conjugated (67 or 68) to nonconjugated alkene (69) vary from 6 to 50, being greater for the threo-isomer which gives the more stable alkene, 68. After statistical correction, the conjugated alkene always dominates the Hofmann product by a factor in excess of ten, emphasizing the activating effect of the β-phenyl substituent.

5 Pyrolysis of Azacycloalkanes

Cleavage of N-oxides of N-methylazacycloalkanes on heating does not occur when the ring is six-membered but becomes progressively easier as the ring size is increased to eight atoms (Eq. 51).[194] Wittig and Polster have shown that steric limitations prevent the operation of the ylide mechanism in the reaction of N,N-dimethylpiperidinium iodide with lithium phenyl as the $\overset{+}{N}-\overset{-}{C}H_2$ and β-C-H bond cannot attain coplanarity.[195]

$$\xrightarrow[165°C]{\triangle}\quad H_2C{=}CH(CH_2)_4N(CH_3)OH \quad (51)$$

Of the two forms of N-methyl-α-picoline oxide, the trans-isomer 70 undergoes elimination to

N-methyl-N-5-hexenylhydroxylamine, but the cis-isomer
71 is inert, the O⁻ and CH₃ groups being too distant
for a cyclic Ei mechanism.[194]

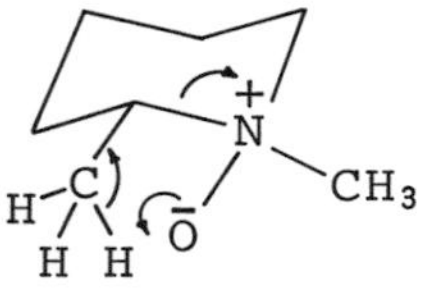

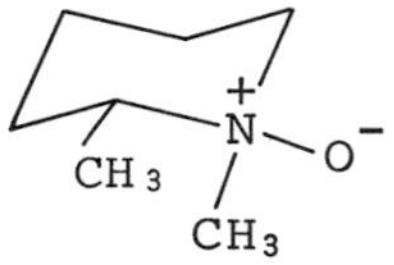

| **70** | **71** |

6   Thermal Decomposition of Amine-Imides

The thermal decomposition of amine imides is closely
related to the Cope elimination. Under reduced pressure at 125°C, the cyclooctyl derivative, **72**, gives
cyclooctene (82% yield), of which 3.5% is the trans-
alkene (Eq. 52).[196] This compares with a 0 to 2%
yield of trans-isomer for the corresponding Cope elimination,[188,197] and compares with a 15% yield in the
Hofmann ylide mechanism using phenyllithium as the
base.[198] Thus the yield of cis-cyclooctene appears to
increase with the electronegativity of the proton ab-
stracting atom (e.g., $\overset{+}{N}-\overset{-}{C}H_2$, $\overset{+}{N}-\overset{-}{N}-Ac$, $\overset{+}{N}-\overset{-}{O}$). The amine-

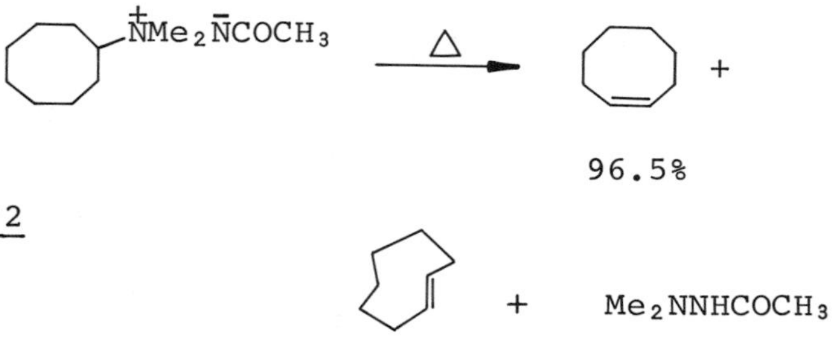

$$Me_2NNHCOCH_3 \qquad (52)$$

imide elimination is probably another example of a
five-membered cyclic Ei mechanism, **73**, although at
this stage an alternative seven-membered transition
state cannot be excluded (**74**).[196]

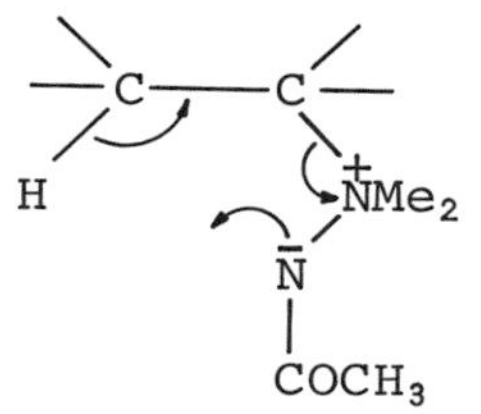

F   SULFOXIDE PYROLYSIS

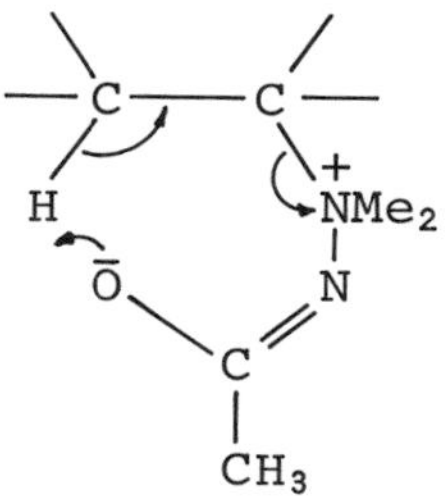

1   Experimental Details

Only in recent years has sulfoxide pyrolysis attracted attention as a synthetic method, compounds possessing a β-hydrogen giving alkenes in high yield. Cram and Kingsbury[199] found that erythro- and threo-1,2-diphenyl-1-propylphenyl sulfoxides eliminate in a range of solvents with greater syn stereospecificity and at a lower reaction temperature (80°C) than with all but the corresponding Cope elimination. The reaction temperatures for other derivatives are as follows: Chugaev reaction (180°C), acetates (400°C), sulfites and carbonates (120 to 260°). Inert solvents are the ideal reaction medium for a kinetic study, but sulfoxide eliminations have been accomplished in the melt or gaseous phase under nitrogen[200] and in the injection port of a gas chromatograph.[201]

Sulfoxides can be easily prepared by the oxidation of sulfides,[202] which can, in turn, be obtained by the alkylation of sodium thiolates with alkyl halides or esters. Johnson and Entwistle prepared alkylmethyl sulfoxides directly from the alkylating agent by treatment with an equivalent of sodium dimsylate in dimethyl sulfoxide.[203] Elimination was effected by raising the reaction temperature to 50 to 80°C. Under these conditions, however, the elimination could occur by concurrent pyrolytic and E2 pathways.

2   Duality of Mechanism and Stereochemistry

Sulfoxides lacking β-hydrogen can rearrange to aldehydes on heating (Eq. 53),[204] but much higher reaction temperatures are required than for β-eliminations, and the rearrangement mode can be reasonably excluded for substrates possessing a β-hydrogen.

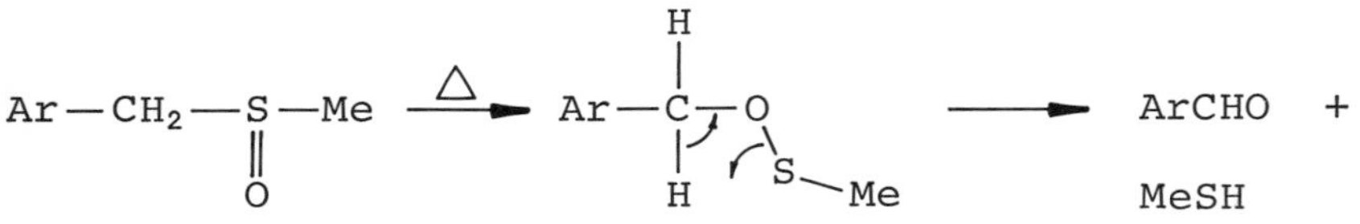

$$Ar-CH_2-\overset{\overset{\displaystyle \|}{O}}{S}-Me \quad \overset{\triangle}{\longrightarrow} \quad Ar-\overset{\overset{\displaystyle H}{|}}{\underset{\underset{\displaystyle H}{|}}{C}}-O \quad \longrightarrow \quad ArCHO \; + \; MeSH$$

(53)

1,2-Diphenyl-1-propyl phenyl sulfoxide possesses three asymmetric centers and, therefore, exists as a mixture of four diastereomeric racemates (75, a,b,c,d). Kingsbury and Cram separated the racemates and decomposed them independently in a range of solvents.[199]

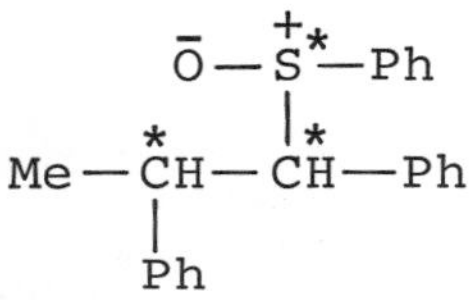

75

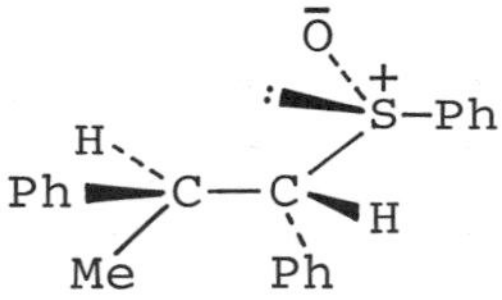

a

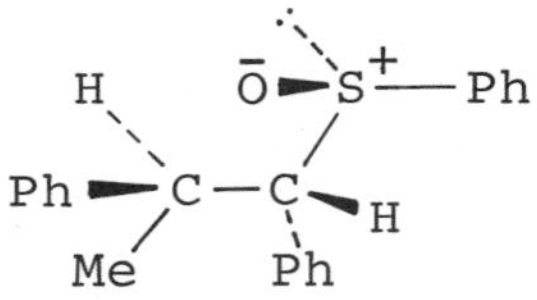

b

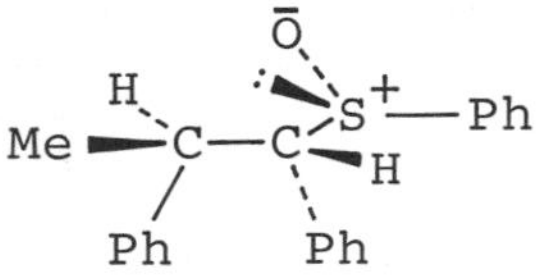

c

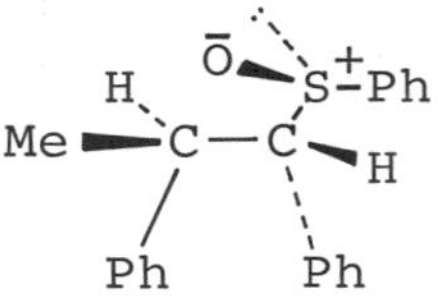

d

*Asymmetric centers.

At 80°C, all the reactions are highly stereoselective, the product of syn elimination dominating over that of anti elimination by a factor of between 3 and 16. The first-order rate coefficients vary only slightly with quite a marked variation in medium polarity (suggesting that neither carbonium ion or carbanion intermediates are involved) and between 70 and 80°C lead to slightly negative activation entropies. These facts were interpreted in terms of a cyclic Ei mechanism, analogous to the Cope elimination. The threo isomers, which eliminate mainly to the less stable cis-α-methyl

stilbene, react slightly faster than the erythro compounds and as the stereoselectivity is similar, these
facts suggest that the transition states do not reflect
the products but are reactantlike in character.

At higher temperatures, the reactions are less
stereoselective. Isomer 75d gives 84% cis-α-methyl
stilbene at 80°C in dioxane but gives only 37% at 120°C.
Values of $\Delta H^{\ddagger}$ vary continually with temperature, and at
higher temperatures, ΔS becomes positive. Clearly more
than one mechanism is involved, and a radical-pair
mechanism was invoked to explain the higher temperature
observations (Scheme 17). The stereoselectivity is
determined by the relative rates of hydrogen abstraction and configurational rotation of the radical pair
within the solvent cage.

Although higher reaction temperatures are required
for the pyrolysis of alkyl sulfoxides than for the
benzylic sulfoxides of Cram and Kingsbury, the activation parameters are consistent with the operation of
the unimolecular Ei mechanism.[205,206] It appears that
the radical-pair mechanism only becomes competive with
the Ei mechanism if the potential alkyl radical is
stabilized by adjacent substituents such as phenyl.
This view is supported by studies using optically
active sulfoxides, the alkyl derivatives racemizing at
elevated temperatures by a pyramidal inversion, process,[207] while the benzyl sulfoxides racemize via an
achiral radical-pair and partly decompose to give bibenzyl among the products.[208]

The Hammett reaction constant for the pyrolysis
of substituted phenyl n-propyl sulfoxides in diphenyl
ether at 175°C is 0.51 (r = 0.995).[205] The activation
entropies range from -11.5 to -16 cal/(deg)(mole) and
the activation energies from 25 to 28 kcal/mole. These
results suggest a highly ordered cyclic transition
state in which charge separation develops across the
C_α-S bond (76). The reaction constant resembles that

76

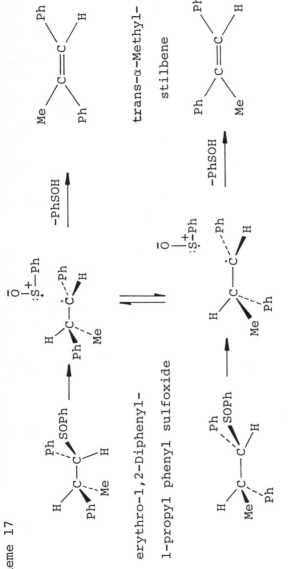

Scheme 17

erythro-1,2-Diphenyl-
1-propyl phenyl sulfoxide

trans-α-Methyl-
stilbene

threo-1,2-Diphenyl-1-
propyl phenyl sulfoxide

cis-α-Methyl-
stilbene

460

observed in the pyrolysis of substituted ethyl benzoates (see Section IX.C.4).

3   The Fate of the Sulfoxide Group

We have paid no attention so far to the fate of the
sulfur atom in the sulfoxide pyrolysis, although in
Scheme 17 we implied the formation of a sulfenic acid.
Such acids are extremely unstable and have only been
isolated as anthraquinone derivatives.[209]   Shelton and
Davis[210] followed the pyrolysis of di-t-butyl sulfoxide
(Eqs. 54 and 55) at 80°C in various solvents by nmr
spectroscopy and observed a t-butyl signal, in addition
to those of the reactant and thiolsulfinate product, in
partially decomposed solutions.   This signal was attributed to t-butyl sulfenic acid which was trapped as an
adduct when the pyrolysis was carried out in the presence of electrophilic olefins or alkynes (Scheme 18).

$$Me_3C-\overset{\overset{O}{\uparrow}}{S}-CMe_3 \xrightarrow{\triangle} Me_3CSOH + Me_2C{=}CH_2 \qquad (54)$$

$$2\ Me_3CSOH \longrightarrow Me_3C-\overset{\overset{O}{\uparrow}}{S}-S-CMe_3 + H_2O \qquad (55)$$

Scheme 18

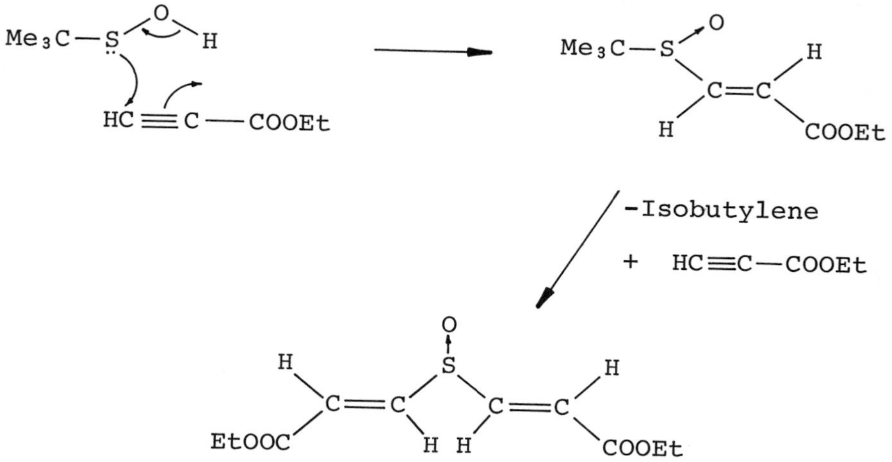

4   Orientation

(a) Acyclic Compounds. Thermal elimination of ethyl
n-butyl sulfoxide at 190°C gives ethylene (62.5%) and
1-butene (36.8%), a ratio of products close to that expected on statistical grounds.[200] However, secondary
alkyl groups are cleaved more readily than primary
groups, the olefin fraction from the pyrolysis of ethyl
s-butyl sulfoxide at 180°C being as follows: ethylene
15%, 1-butene 59%, trans-2-butene 17%, and cis-2-butene
8%. Surprisingly, although elimination occurs preferentially into the secondary group, the major product is
the less substituted butene, the yield of which is
greater than predicted on statistical grounds. The
presence of a secondary alkyl substituent also increases the rate of elimination of ethylene from the ethyl
group. These unexpected results could arise from a
steric effect.

(b) Alicyclic Compounds. The products of the Cope,
Hofmann, and sulfoxide eliminations of 1-substituted-1-
ethylcycloalkyl derivatives are listed in Table 18.[211],
[212] For all three reactions, the relative yield of the
endo-alkene is smallest for the cyclohexyl series. The
changes in the exo/endo ratios with ring size reflect
ease of attainment of the transition state for anti
elimination for the Hofmann elimination and syn elimination for the Cope elimination. The yield of

TABLE 18   Comparison of the Products of the Cope,
Sulfoxide, and Hofmann Eliminations of 1-Ethylcyclo-
alkyl Derivatives

X	% exo			% endo		
	n = 0	1	2	n = 0	1	2
PhSO	5	75	28	95	25	72
Me_2NO	58	95	37	42	5	63
Me_3NOH	80	95	71	20	5	29

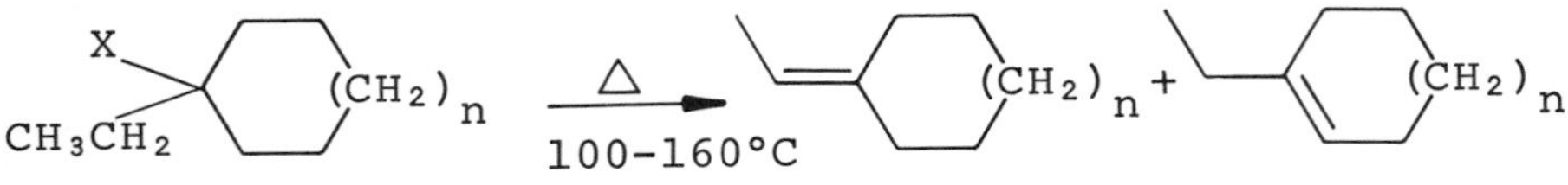

endo-alkene is noticeably greater in the sulfoxide pyrolysis than the Cope elimination, suggesting a less
stringent need for coplanarity in the transition state.
Perhaps, the tertiary nature of the alkyl substituent
may confer sufficient stability to allow elimination to
occur in part by the radical-pair mechanism.

Kice and Campbell found that the rates of pyrolytic elimination of cyclopentyl, cyclohexyl, and cycloheptyl phenyl sulfoxides at 130°C vary in the ratio
25:1:120.[213] Thus both the elimination rate and the
endo/exo ratio attain a minimum in the cyclohexyl system.

Jones and Green have attributed the incursion of
the radical-pair mechanism to steric interactions
assisting unimolecular dissociation.[202b] They analyzed
the alkene fraction obtained in the pyrolysis of the
eight isomeric structures described by the general
name 3- and 4-phenylsulfinyl-5α-cholestane (Table 19).
To achieve the required syn-periplanar transition state
for the Ei mechanism, the A ring of the steroid must
distort into a half-chair conformation, as shown in the
Table.

The 4β-isomers, 77 and 78, possess only one hydrogen cis-related to the sulfinyl group. As expected for
the Ei mechanism, 78 eliminates only to the 3-ene, but
77, in which phenyl-10-methyl steric interactions are
severe in the conformation leading to 3-ene, gives the
3- and 4-enes in equal proportions, and a radical-pair
mechanism was suggested. The products of elimination
of the 4α-isomers, 79 and 80, accord with transition
states with the weaker phenyl-framework steric interactions being preferred.

In the 3-phenylsulfinyl series, 81-84, even when
steric interactions are unfavorable, the yield of the
thermodynamically more stable 2-ene is quite appreciable. Jones and Green suggest this result indicates
that transition states with considerable double-bond
character are involved, a proposal that contrasts with
the reactantlike transition states favored by Kingsbury
and Cram in the pyrolysis of 1,2-diphenyl-1-propylphen-
yl sulfoxides.[199] There is an obvious need to determine the mechanism and not just the stereochemistry of
elimination.

A classic example of the sulfoxide pyrolysis is
the conversion of phenoxymethylpenicillin sulfoxide
methyl ester into the corresponding cephalosporin.[219]
In this instance the intermediate sulfenic acid adds to
the terminal alkene, thereby completing the ring ex-
pansion process (Eq. 56).

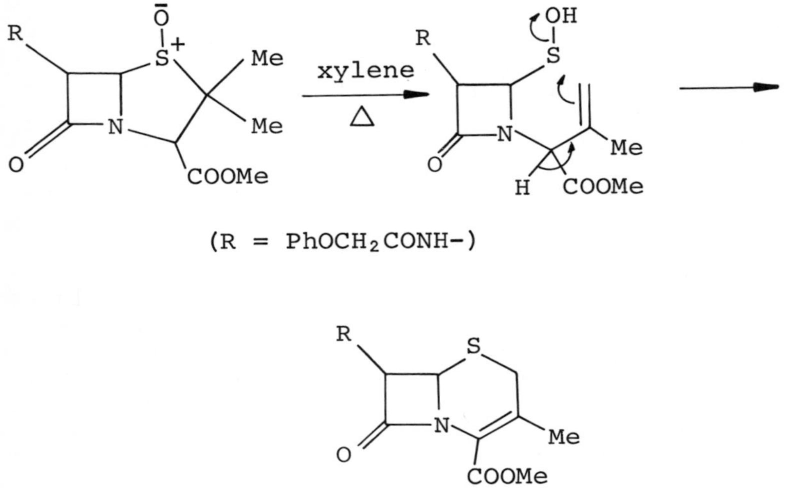

(R = PhOCH$_2$CONH-)

(56)

TABLE 19 Olefin Yields in the Pyrolysis of Some Steroidal Sulfoxides[202b] (Solvent: Benzene or Toluene)

Sulfoxide		Reaction Temperature (°C)	Reaction Time (hr)	% Yield Recovered Starting Material	% Yield Olefins[a]	% Olefin Composition 2-	3-	4-ene
77 S-4β	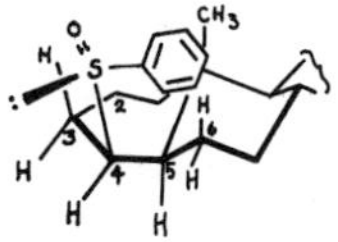	100	72	26	70	0	50	50
78 R-4β	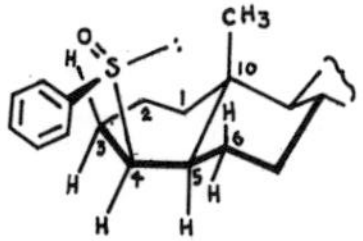	80	12	0	85	0	100	0
79 R-4α	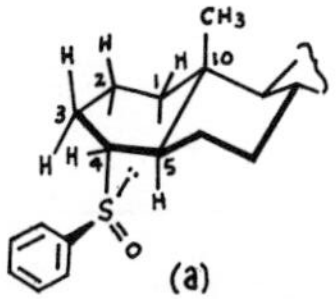 (a) 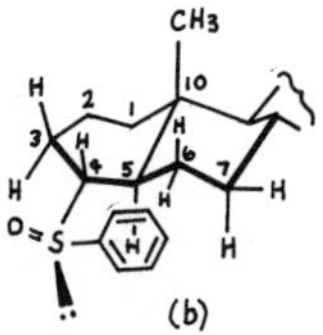 (b)	100	70	40	85	0	46	54

TABLE 19 Continued

Sulfoxide	Reaction		% Yield Recovered Starting Material	% Yield Olefins[a]	% Olefin Composition		
	Temperature (°C)	Time (hr)			2-	3-	4-ene
80 S-4α	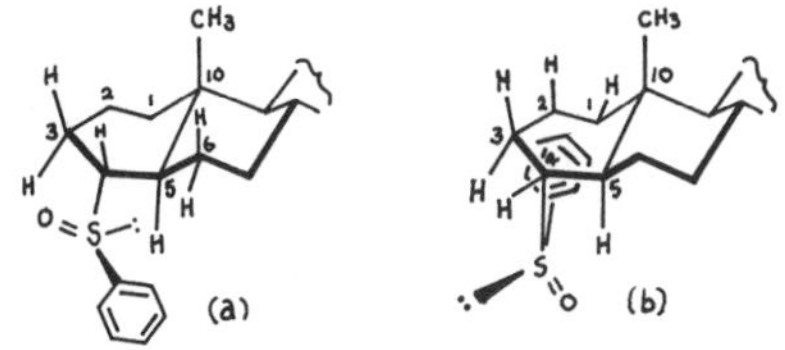 100	70	40	80	0	71	29
81 R-3α	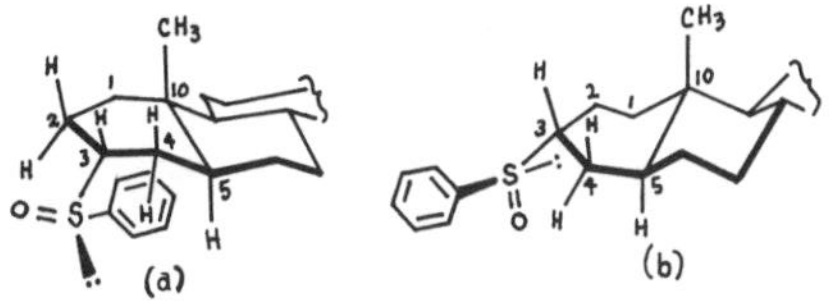 100	20	50	78	62	38	0
82 S-3α	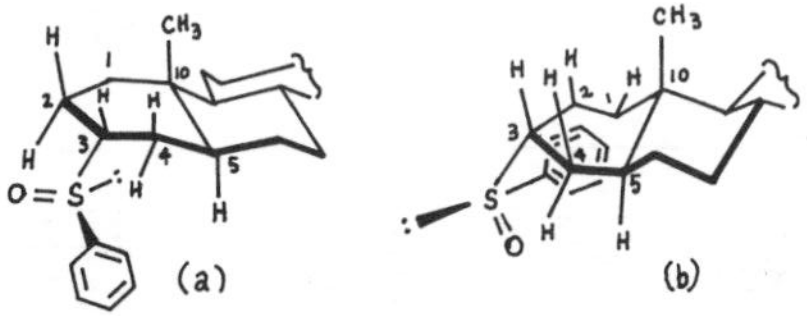 100	20	50	95	83	17	0

466

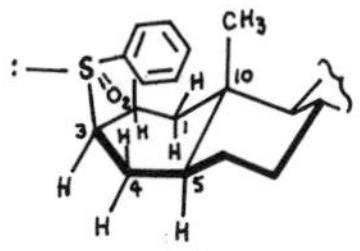 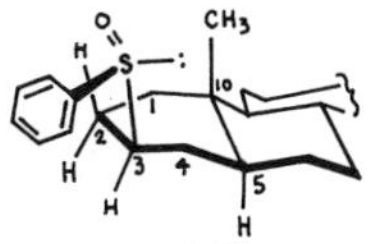 83 R-3β (a) (b)	110	24	85	80	82	18	0
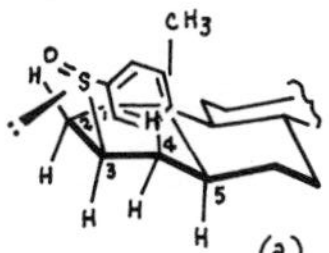 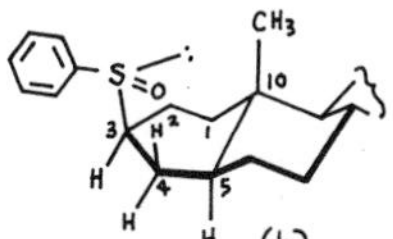 84 S-3β (a) (b)	110	160	67	76	48	52	0

aBased on starting material consumed.

G   PYROLYSIS OF N-SULFONYLSULFILIMINES

The thermal decomposition of N-sulfonylsulfilimines is closely related to the sulfoxide pyrolysis and the Cope elimination.   Spectroscopic studies indicate that the S-N bond in these compounds is more dipolar than S-O bond in sulfoxides,[214] and Oae, Tsujihare, and Furukawa have found that sulfonylsulfilimines decompose under milder conditions.[215]   The products of pyrolysis of simple alkyl derivatives are controlled by statistical factors.
      The rates of elimination of substituted ethyl aryl-N-sulfonylsulfilimines (Eq. 57) follow first-order kinetics in both benzene and dimethyl sulfoxide, the rate being six times greater in the former solvent at 80.3°C.   The Hammett reaction constant $\rho_X = 0.88$ (Y=Me, X=Me,H,OMe,NO$_2$,Br) accords with an increase in electron density at the S atom in the transition state. The ρ_Y value of -0.60 (X-H,Y=Me,Br,H) suggests that the nucleophilicity of the N atom is a significant factor.   Comparison of the rates of elimination of ethyl phenyl-N-p-tolylsulfonylsulfilimine and its 1,1,1-trideuteroethyl derivative affords an isotope effect of 3.03 in benzene at 80.3°C.   This isotope effect is too large to be mainly of secondary origin and implies that C$_\beta$-H bond breaking has progressed significantly in the transition state (cf.85).

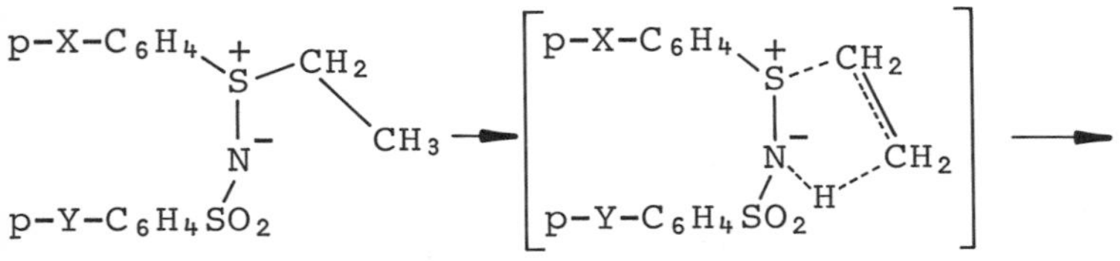

$$85$$

$$H_2C{=}CH_2$$

$$+ \qquad\qquad\qquad\qquad (57)$$

$$p\text{-}X\text{-}C_6H_4\text{-}S\text{-}NHSO_2\text{-}C_6H_4\text{-}p\text{-}Y$$

H   PYROLYSIS OF THIO- AND ISOTHIO-CYANATES

The activation parameters for the pyrolysis of some

alkyl thio- and isothiocyanates are listed in Table 20.[216-218] The stoicheiometry of the isothiocyanate pyrolyses is represented by the reversible equilibrium 58 which lies mainly to the product side.[217,218] In the pyrolysis of s-butyl thiocyanate, the back reaction leads to the formation of a small amount of s-butyl isothiocyanate.[216] The rates of elimination of all the compounds follow first-order kinetics and are unaffected by the surface/volume ratio and the addition of radical inhibitors. These observations and the log A factors of about 12 to 13 suggest the eliminations are unimolecular homogeneous reactions involving a cyclic Ei transition state.

$$R_2 CHCH(NCS)R \rightleftharpoons R_2 C = CHR + HNCS \qquad (58)$$

The β-hydrogen in the elimination of s-butyl thiocyanate could be abstracted by either the sulfur (e.g., 86) or the nitrogen atom (e.g., 87), but the authors preferred the six-membered transition state 87 as the log A factor of 12.30 is closer to that of a series of esters (12.3) than for a series of alkyl halides (13.5).[1] However, for the isothiocyanate pyrolysis, a four-membered cyclic transition state was preferred

86   87   88

(88) as the linear nature of the -N=C=S group prevents the sulfur atom approaching effective bonding distance to the β-hydrogen. The activation energies for the pyrolysis of the alkyl isothiocyanates are linearly related to the heterolytic bond dissociation energy of the R-X (X=NCS) bond but not to the homolytic bond dissociation energy (cf. Section VIII.B.5.a), and of the pyrolytic eliminations, the influence of α-alkyl substituents on the rate of elimination relates most closely to the effects observed in acetate pyrolysis (see Figure 3). These results show the isothiocyanates eliminate through a transition state that has a definite carbonium ion character at C_α, but this is comparable to that involved in ester pyrolysis and much

TABLE 20 Activation Parameters for the Pyrolysis of Some Thio- and Isothiocyanates

Compound	Log A (sec^{-1})	E_A (cal/mole)	Temperature Range (°C)	Reference
s-BuSCN	12.30	38,820	245 to 306	216
EtNCS	12.40	45,350	327 to 394	217
i-PrNCS	12.98	42,870	276 to 329	217
t-BuNCS	13.00	39,460	225 to 278	217
s-BuNCS	12.46	41,330	273 to 323	218

less (polar) than that for alkyl halides.[218]

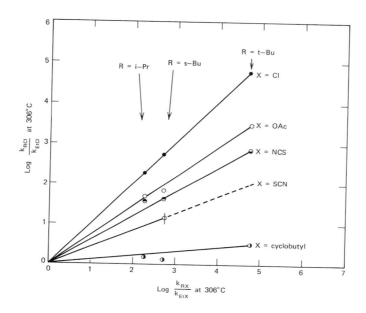

Figure 3   Comparison of the effect of alkyl substituents on the rates of a number of pyrolytic eliminations at 306°C (data taken from Ref. 217).

I   CONCLUDING REMARKS

We discuss a spectrum of transition states for concerted heterolytic thermal 1,2-elimination in this
chapter. The sulfoxide, xanthate and acetate pyrolyses
are highly concerted reactions, which involve transition states with mild carbonium ion character, while
the N-oxides decompose through less ionic, concerted
transition states. All four of these reactions are
syn stereoselective. At higher reaction temperatures,
if the reactants possess radical-stabilizing groups
such as α-phenyl substituents, homolytic elimination
becomes competitive in the sulfoxide pyrolyses, and the
reactions decrease in stereoselectivity. However, of
all the pyrolytic eliminations, the alkyl halide pyrolyses exhibit the greatest carbonium ion character. In
addition, homolytic elimination must be suppressed in
many cases to allow the measurement of the heterolytic
rates. Unlike the other pyrolyses, the alkyl halide
eliminations involve the movement of four rather than

six bonding electrons. The significance of this difference is now apparent with the advent of the
Woodward-Hoffmann rules concerning the conservation of
orbital symmetry.[220] Concerted thermal elimination is
expected for transition states involving a shift of
4n + 2 electrons(n = 1,2, etc.) as the reaction is
symmetry allowed. However, for transition states involving a 4n-electron shift, the concerted reaction is
symmetry forbidden, and thermal elimination must occur
homolytically or in a stepwise heterolytic manner.
These orbital considerations reinforce the stereochemical features, the five- and six-membered cyclic transition states allowing the necessary proximity of the
leaving groups for concerted reaction, which is precluded in the highly strained four-membered cyclic
transition states involved in the alkyl halide
thermolyses.

In the light of the orbital symmetry rules, it
is unlikely that cyclobutyl xanthate (see Eq. 45) decomposes in a concerted manner directly to butadiene,
as this would require an eight-electron shift. More
probably, cyclobutene is formed in a normal syn elimination, and then undergoes rapid ring opening.

In the limited space available, we confine our
discussion to pyrolytic 1,2-elimination. However, it
is worth noting that these reactions are closely related to a series of concerted, symmetry-conserved,
[3,2]-sigmatropic rearrangements, which have been
recognized as a class of reactions by Baldwin.[221] Equation 59 depicts the general case with the obvious
analogy to pyrolytic elimination.

$$\tag{59}$$

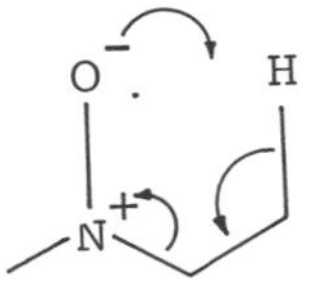

$$X = CR_2,^{222} \ \overset{+}{N}R_2,^{223} \ O,^{224} \ \overset{+}{S}R,^{225} \ \overset{+}{C}l,^{226} \ -\overset{+}{P}\!\!<,^{227}$$

REFERENCES

1. A. Maccoll, Chem. Rev., 69, 33 (1969).
2. J. L. Holmes and A. Maccoll, Proc. Chem. Soc., 175 (1957).
3. J. L. Holmes and A. Maccoll, J. Chem. Soc., 5919 (1963).
4. A. C. Cope and E. R. Trumbull, Org. React. 11, 361 (1960).
5. C. A. Kingsbury and D. J. Cram, J. Am. Chem. Soc., 82, 1810 (1960).
6. S. Trippett, Q. Rev., 17, 406 (1963).
7. C. D. Hurd and F. H. Blunck, J. Am. Chem. Soc., 60, 2419 (1938).
8. H. R. Nace, Org. React., 12, 57 (1962).
9. W. Formin and N. Sochanski, Ber., 46, 246 (1913).
10. A. Maccoll, in "Theoretical Organic Chemistry," papers presented to the Kekulé Symposium, London, 1958, p. 230, Butterworth's Scientific Publications, London, 1959.
11. A. Maccoll, Chem. Soc. Spec. Publ., No. 16, 158 (1962).
12. A. Maccoll, Adv. Phys. Org. Chem., 3, 91 (1965).
13. C. H. DePuy and R. W. King, Chem. Rev., 60, 431 (1960).
14. D. V. Banthorpe, "Elimination Reactions," Vol. 2, Elsevier, Amsterdam, 1963, Chap. 7.
15. E. T. Lessig, J. Phys. Chem., 36, 2325 (1932).
16. E. S. Swinbourne, Aust. J. Chem., 11, 314 (1958).
17. J. H. S. Green and A. Maccoll, J. Chem. Soc., 2449 (1955).
18. A. Maccoll and R. H. Stone, J. Chem. Soc., 2756 (1961).
19. (a) M. Swarzc, J. Chem. Phys., 16, 128 (1948); (b) M. Szwarc and B. N. Ghosh, J. Chem. Phys., 17, 744 (1949).
20. M. F. R. Mulcahy and D. J. Williams, Aust. J. Chem., 14, 534 (1961).
21. W. Tsang, J. Chem. Phys., 40, 1171 (1964); 40, 1498 (1964); 41, 2487 (1964).
22. D. H. R. Barton and F. P. Onyon, Trans. Faraday Soc., 45, 725 (1949).
23. J. S. Shapiro, E. S. Swinbourne, and B. C. Young, Aust. J. Chem., 17, 1217 (1964).
24. D. Brearley, G. B. Kistiakowsky, and C. H. Stauffer, J. Am. Chem. Soc., 58, 43 (1936).
25. A. Maccoll, J. Chem. Soc., 965 (1955).
26. M. Szwarc, B. N. Ghosh, and A. H. Sehon, J. Chem. Phys., 18, 1142 (1950).
27. O. H. Gellner and H. A. Skinner, J. Chem. Soc.,

1145 (1949).

28. A. Shilov, Dokl. Akad. Nauk. SSSR, 98, 601 (1954).
29. A. M. Goodall and K. E. Howlett, J. Chem. Soc.,
    2596 (1954).
30. D. H. R. Barton, J. Chem. Soc., 148 (1949).
31. D. H. R. Barton, Nat., 157, 626 (1946).
32. D. H. R. Barton and K. E. Howlett, J. Chem. Soc.,
    155 (1949).
33. K. E. Howlett, Trans. Faraday Soc., 48, 25 (1952).
34. D. H. R. Barton and K. E. Howlett, J. Chem. Soc.,
    2053 (1951).
35. K. E. Howlett, Nat., 165, 860 (1950).
36. D. H. R. Barton and K. E. Howlett, Trans. Faraday
    Soc., 45, 735 (1949).
37. R. Baldt and E. Cremer, Monatsh, 80, 153 (1949).
38. D. H. R. Barton, A. J. Head, and R. J. Williams,
    J. Chem. Soc., 2039 (1951).
39. K. E. Howlett, J. Chem. Soc., 4487 (1952).
40. D. H. R. Barton and K. E. Howlett, J. Chem. Soc.,
    165 (1949).
41. D. H. R. Barton and A. J. Head, Trans. Faraday
    Soc., 46, 114 (1950).
42. W. E. Vaughan and F. F. Rust, J. Org. Chem., 5,
    449 (1940).
43. N. Capon and R. A. Ross, Trans. Faraday Soc., 62,
    1560 (1966).
44. A. E. Goldberg and F. Daniels, J. Am. Chem. Soc.,
    79, 1314 (1957).
45. P. T. Good, Ph.D. thesis, University of London,
    1956.
46. N. N. Semenov, G. B. Sergeev, and G. A. Kaprilova,
    Dokl. Akad. Nauk. SSSR. 105, 301 (1955).
47. P. J. Agius and A. Maccoll, J. Chem. Soc., 973
    (1955).
48. J. T. D. Cross and V. R. Stimson, Aust. J. Chem.,
    21, 973 (1968).
49. A. Maccoll et al., J. Chem. Soc., 973 (1955);
    5033 (1957); 1192, 1197 (1959).
50. J. H. S. Green, A. Maccoll, and P. J. Thomas,
    J. Chem. Soc., 184 (1960).
51. A. Maccoll et al., J. Chem. Soc., 979, 2445, 2449,
    2454 (1955); 5020, 5024, 5028 (1957); 3016 (1958).
52. A. Maccoll and P. J. Thomas, J. Chem. Soc., 979
    (1955).
53. S. W. Benson, J. Chem. Phys., 38, 1945 (1963).
54. S. W. Benson et al., J. Chem. Phys., 34, 514
    (1961); 37, 540, 1081 (1962); 38, 878, 882 (1963);
    39, 132 (1963); 41, 530 (1964); J. Am. Chem. Soc.,
    86, 2773 (1964); 88, 3196 (1966).
55. J. L. Jones and R. A. Ogg, Jr., J. Am. Chem. Soc.,

59, 1931 (1937).
56. (a) J. L. Holmes and A. Maccoll, J. Chem. Soc.,
5919 (1963);
(b) H. Teranishi and S. W. Benson, J. Chem.
Phys., 40, 2946 (1964).
57. R. A. Ogg and M. Polanyi, Trans. Faraday Soc., 31,
482 (1935).
58. A. N. Bose and S. W. Benson, J. Chem. Phys., 37,
2935 (1962).
59. D. H. R. Barton and A. J. Head, Trans. Faraday
Soc., 46, 114 (1950).
60. K. A. Holbrook and A. R. W. Marsh, Trans. Faraday
Soc., 64, 643 (1967).
61. H. Hartmann, H. G. Bosche, and H. Heydtmann, Z.
Phys. Chem., (Frankfurt am Main), 42, 329 (1964).
62. H. Heydtmann and G. Rinck, Z. Phys. Chem.,
(Frankfurt am Main) 30, 250 (1961).
63. H. Heydtmann and G. Rinck, Z. Phys. Chem.,
(Frankfurt am Main), 36, 75 (1963).
64. A. Maccoll and S. C. Wong, J. Chem. Soc., 1492
(1968).
65. R. L. Failes and V. R. Stimson, Aust. J. Chem.,
15, 437 (1962).
66. B. Roberts, Ph.D. thesis, University of London,
1961; quoted in Ref. 1.
67. A. T. Blades, Can. J. Chem., 36, 1043 (1958).
68. P. J. Thomas, J. Chem. Soc., 1192 (1959).
69. A. T. Blades and G. W. Murphy, J. Am. Chem. Soc.,
74, 6219 (1952).
70. A. Maccoll, and P. J. Thomas, J. Chem. Soc., 5033
(1957).
71. G. B. Sergeev, Dokl. Akad. Nauk. SSSR, 106, 299
(1955).
72. G. D. Harden and A. Maccoll, J. Chem. Soc., 2454
(1955).
73. G. B. Kistiakowsky and C. H. Stauffer, J. Am.
Chem. Soc., 59, 165 (1937).
74. J. H. Yang and D. C. Conway, J. Chem. Phys., 43,
1296 (1965).
75. A. N. Bose and S. W. Benson, J. Chem. Phys., 37,
2935 (1962).
76. R. A. Lee, M.Sc. thesis, University of London;
quoted in Ref. 1.
77. H. Teranishi and S. W. Benson, J. Chem. Phys., 40,
2946 (1964).
78. A. N. Bose and S. W. Benson, J. Chem. Phys., 37,
1081 (1962).
79. (a) S. G. Szabo and T. Berces, Acta Chim. Acad.
Sci. Hung., 22, 461 (1960);
(b) S. W. Benson and A. N. Bose, J. Chem. Phys.,

$\underline{39}$, 3463 (1963).

80. A. Maccoll and P. J. Thomas, Nat., $\underline{176}$, 392 (1955).

81. G. T. Genaux, F. Kern, and W. D. Walters, Acta Chim. Acad. Sci. Hung., $\underline{75}$, 6191 (1963).

82. (a) P. J. Thomas, J. Chem. Soc., 136 (1961); (b) R. L. Failes and V. R. Stimson, Aust. J. Chem., $\underline{15}$, 437 (1967).

83. M. Dakubu, Ph.D. thesis, University of London; quoted in Ref. 1.

84. M. R. Bridge, D. H. Davies, A. Maccoll, R. A. Ross, and O. Banjoko, J. Chem. Soc., $\underline{B}$, 805 (1968).

85. B. Stephenson, Ph.D. thesis, University of London; quoted in Ref. 1.

86. G. D. Harden and A. Maccoll, J. Chem. Soc., 1197 (1959).

87. A. T. Blades P. W. Gilderson, and M. G. H. Wallbridge, Can. J. Chem., $\underline{40}$, 1527 (1962).

88. A. T. Blades, P. W. Gilderson, and M. G. H. Wallbridge, Can. J. Chem., $\underline{40}$, 1533 (1962).

89. H. H. Jaffé, Chem. Rev., $\underline{53}$, 191 (1953).

90. C. D. Ritchie and W. F. Sager, Prog. Phys. Org. Chem., $\underline{3}$, 323 (1964).

91. M. A. Paul and F. A. Long, Chem. Rev., $\underline{57}$, 1 (1957).

92. H. M. R. Hoffman and A. Maccoll, J. Am. Chem. Soc., $\underline{87}$, 3774 (1965).

93. C. K. Ingold, "Structure and Mechanism in Organic Chemistry," G. Bells and Sons, Cornell University, Ithaca, Ithaca, 1953, p. 482.

94. A. Maccoll, and E. S. Swinbourne, J. Chem. Soc., 149 (1964).

95. J. S. Shapiro and E. S. Swinbourne, Can. J. Chem., $\underline{46}$, 1341 (1968).

96. (a) R. C. L. Bicknell, Ph.D. thesis, University of London, 1962; (b) R. C. L. Bicknell and A. Maccoll, Chem. Ind., London, 1912 (1961).

97. P. Beltrame, C. A. Bunton, A. Dunlop, and D. Whitaker, J. Chem. Soc., 658 (1964).

98. S. Winstein, B. Appel, R. Baker, and A. Diaz in "Organic Reaction Mechanisms," Chem. Soc., Spec. Publ., No. 19, London (1965).

99. (a) Y. Yukawa, M. Sawada, and Y. Tsuno, Bull. Chem. Soc. Jap., $\underline{39}$, 2274 (1966); (b) C. Mechelynck-David and P. J. C. Fierens, Tetrahedron, $\underline{6}$, 232 (1959).

100. D. H. R. Barton, J. Chem. Soc., 2174 (1949).

101. D. H. R. Barton, A. J. Head, and R. J. Williams, J. Chem. Soc., 453 (1952).

102. T. Bamkole, Ph.D. thesis, University of London, 1964.

103. E. D. Hughes, C. K. Ingold, and J. B. Rose, J. Chem. Soc., 3839 (1953).

104. P. Cadman, M. Day, A. W. Kirk, and A. F. Trotman-Dickinson, Chem. Comm., 203 (1970).

105. G. E. Millward, R. Hartig, and E. Tschuikow-Roux, Chem. Comm., 465 (1971).

106. D. H. R. Barton and F. P. Onyon, J. Am. Chem. Soc., 72, 988 (1950).

107. (a) M. J. Perona, J. T. Bryant, and G. O. Pritchard, J. Am. Chem. Soc., 90, 4782 (1968); (b) G. O. Pritchard and M. J. Perona, Int. J. Chem. Kinet., 2, 281 (1970).

108. (a) S. W. Benson and G. Haugen, J. Phys. Chem., 69, 3898 (1965); (b) J. T. Bryant and G. O. Pritchard, J. Phys. Chem., 71, 3439 (1967).

109. (a) F. Krafft, Ber. Deut. Chem. Ges., 16, 3018 (1883); (b) G. G. Smith and W. H. Wetzel, J. Am. Chem. Soc., 79, 875 (1957).

110. D. Y. Curtin and D. B. Kellom, J. Am. Chem. Soc., 75, 6011 (1953).

111. E. R. Alexander and A. Mudrak, J. Am. Chem. Soc., 72, 3194 (1950); 73, 59 (1951).

112. D. H. R. Barton, J. Chem. Soc., 2174 (1949).

113. (a) W. J. Bailey and C. N. Bird, A. C. S. Meeting, Miami, April, 1957, Abstr. pp 44-0; (b) A. C. Cope and N. A. LeBel, J. Am. Chem. Soc., 82, 4656 (1960).

114. D. H. Froemsdorf, C. H. Collins, G. S. Hammond, C. H. DePuy, J. Am. Chem. Soc., 81, 643 (1959).

115. W. O. Haag and H. Pines, J. Org. Chem., 24, 877 (1959).

116. R. A. Benkeser, J. J. Hazdra, and M. L. Burrous, J. Am. Chem. Soc., 81, 5374 (1959).

117. W. J. Bailey and C. King, J. Org. Chem., 21, 858 (1956).

118. C. H. DePuy and R. E. Leary, J. Am. Chem. Soc., 79, 3705 (1957).

119. W. J. Bailey, and W. F. Hale, J. Am. Chem. Soc., 81, 651 (1959).

120. (a) R. T. Arnold, G. G. Smith and R. M. Dodson, J. Org. Chem., 15, 1256 (1950); (b) W. J. Bailey and L. Nicholas, J. Org. Chem., 21, 854 (1956).

121. W. J. Bailey and R. A. Baylouny, J. Am. Chem. Soc., 81, 2126 (1959).

122. E. M. Bilger and H. Hibbert, J. Am. Chem. Soc.,

58, 823 (1936).

123. T. O. Bamkole and E. U. Emovon, J. Chem. Soc.,
      (B), 187 (1969).

124. A. Maccoll, J. Chem. Soc., 3398 (1958).

125. J. C. Sheer, E. C. Kooyman, and F. L. J. Sixma,
      Rec. Trav. Chim., 82, 1123 (1963).

126. E. U. Emovon and A. Maccoll, J. Chem. Soc., 335
      (1962).

127. W. J. Bailey and J. J. Hewitt, J. Org. Chem.,
      21, 543 (1956).

128. G. G. Smith and W. H. Wetzel, J. Am. Chem. Soc.,
      7, 875 (1957).

129. E. U. Emovon, J. Chem. Soc., 1246 (1963).

130. G. G. Smith, F. D. Bagley, and R. Taylor, J. Am.
      Chem. Soc., 83, 3647 (1961).

131. G. G. Smith, D. A. K. Jones, and D. F. Brown,
      J. Org. Chem., 28, 403 (1963).

132. R. Taylor, G. G. Smith, and W. H. Wetzel, J. Am.
      Chem. Soc., 84, 4817 (1962).

133. R. Taylor, J. Chem. Soc., (B), 1559 (1968).

134. H. Kwart and D. P. Hoster, Chem. Comm. 1155
      (1967).

135. W. H. Saunders Jr., and R. H. Paine, J. Am. Chem.
      Soc., 83, 882 (1961).

136. H. Kwart and H. G. Ling, Chem. Comm., 302 (1969).

137. R. Hoffmann and R. B. Woodward, Acc. Chem. Res.,
      1, 17 (1968).

138. E. S. Lewis and E. R. Newman, J. Am. Chem. Soc.,
      91, 7455 (1969).

139. J. C. Leffingwell and R. E. Shackelford,
      Tetrahedron Lett., 2003 (1970).

140. E. E. Smissmann, J. T. Suh, M. Oxman, and R.
      Daniels, J. Am. Chem. Soc., 81, 2909 (1959); 84,
      1040 (1962).

141. E. E. Smissman, J. P. Li, and M. W. Creese, J.
      Org. Chem., 35, 1352 (1970).

142. L. Chugaev, Ber. 32, 3332 (1899).

143. W. A. Pryor, "Mechanism of Sulfur Reactions,"
      McGraw-Hill, New York, 1962, p. 71.

144. H. R. Nace, Org. React., 12, 57 (1962).

145. D. J. Cram, J. Am. Chem. Soc., 71, 3883 (1949).

146. F. G. Bordwell and P. S. Landis, J. Am. Chem.
      Soc., 80, 6379 (1958).

147. E. R. Alexander and A. Mudrak, J. Am. Chem. Soc.,
      72, 1810 (1950).

148. G. L. O'Connor and H. R. Nace, J. Am. Chem. Soc.,
      74, 5454 (1952); 75, 2118 (1953).

149. P. V. Laakso, Chem. Abstr., 34, 5059 (1940).

150. D. S. Tarbell and D. P. Harnish, Chem. Rev., 49,
      60 (1951).

151. E. Salomaa, Ann. Acad. Sci. Fenn. Ser., $\underline{A}$, II, No. 94, 1 (1959).

152. P. D. Bartlett, "Organic Chemistry, An Advanced Treatise," Vol. 3, H. Gilman, Ed., J. Wiley, New York, 1953, p. 53.

153. W. Hückel, W. Tappe, and G. Legutke, Ann., $\underline{543}$, 191 (1940).

154. P. G. Stevens and J. H. Richmond, J. Am. Chem. Soc., $\underline{63}$, 3132 (1941).

155. R. F. W. Bader and A. N. Bourns, Can. J. Chem., $\underline{39}$, 348 (1961).

156. (a) J. Bigeleisen, J. Chem. Phys., $\underline{56}$, 823 (1952); J. Phys. Chem., $\underline{17}$, 675 (1949); (b) J. Bigeleisen and M. G. Mayer, J. Phys. Chem., $\underline{15}$, 261 (1947).

157. L. Pauling, "The Nature of the Chemical Bond," Cornell University Press, Ithaca, 1940, pp. 52, 130.

158. R. A. Benkeser and J. J. Hazdra, J. Am. Chem. Soc., $\underline{81}$, 5374 (1959).

159. F. G. Bordwell and P. S. Landis, J. Am. Chem. Soc., $\underline{80}$, 6383 (1958).

160. F. G. Bordwell and P. S. Landis, J. Am. Chem. Soc., $\underline{80}$, 2450 (1958).

161. R. A. Benkeser and J. J. Hazdra, J. Am. Chem. Soc., $\underline{81}$, 228 (1959).

162. J. D. Roberts and C. W. Sauer, J. Am. Chem. Soc., $\underline{71}$, 3925 (1949).

163. W. J. Bailey, C. H. Cunov, and L. Nicholas, J. Am. Chem. Soc., $\underline{77}$, 2789 (1955).

164. A. C. Cope and M. J. Youngquist, J. Am. Chem. Soc., $\underline{84}$, 2411 (1962).

165. A. T. Blomquist and A. Goldstein, J. Am. Chem. Soc., $\underline{79}$, 3505 (1957).

166. A. T. Blomquist and A. Goldstein, J. Am. Chem. Soc., $\underline{77}$, 1001 (1955).

167. I. M. McAlpine, J. Chem. Soc., 1114 (1931); 906 (1932).

168. H. R. Nace, D. Manly, and S. Fusco, J. Org. Chem., $\underline{23}$, 687 (1958).

169. C. A. Bunton, K. Khaleeluddin, and D. Whittaker, Nat., $\underline{190}$, 715 (1961).

170. R. E. Gilman et al., Can. J. Chem., $\underline{48}$, 970 (1970).

171. C. Djerassi and W. S. Brigøs, J. Org. Chem., $\underline{33}$, 1625 (1968).

172. A. C. Cope, T. T. Foster, and P. H. Towle, J. Am. Chem. Soc., $\underline{71}$, 3929 (1949).

173. A. C. Cope and E. R. Trumbull, Org. React., $\underline{11}$, 361 (1960).

174. A. C. Cope and H. H. Lee, J. Am. Chem. Soc., 79, 964 (1957).
175. M. A. T. Rogers, J. Chem. Soc., 769 (1955).
176. F. F. Caserio, Jr., S. H. Parker, R. Piccolini, and J. D. Roberts, J. Am. Chem. Soc., 80, 5507 (1958).
177. D. J. Cram, M. R. V. Sahyun, and G. R. Knox, J. Am. Chem. Soc., 84, 1734 (1962).
178. M. Cocivera and S. Winstein, J. Am. Chem. Soc., 85, 1702 (1963).
179. D. J. Cram and J. E. McCarty, J. Am. Chem. Soc., 76, 5740 (1954).
180. A. C. Cope, C. L. Bumgardner, and E. E. Schweizer, J. Am. Chem. Soc., 79, 4729 (1957).
181. A. C. Cope, N. A. LeBel, H. H. Lee, and W. R. Moore, J. Am. Chem. Soc., 79, 4720 (1957).
182. C. K. Ingold, "Structure and Mechanism in Organic Chemistry," Cornell University Press, Ithaca, 1953, Chap. 8.
183. A. C. Cope and E. M. Acton, J. Am. Chem. Soc., 80, 355 (1958).
184. A. C. Cope and C. L. Bumgardner, J. Am. Chem. Soc., 79, 960 (1957).
185. R. T. Arnold and P. N. Richardson, J. Am. Chem. Soc., 76, 3649 (1954).
186. J. Zavada, J. Krupička, and J. Sicher, Coll. Czech. Chem. Comm., 31, 4273 (1966).
187. E. L. Eliel, N. L. Allinger, S. J. Angyal, and G. A. Morrison, "Conformational Analysis," Interscience, New York, 1965, p. 126.
188. J. Sicher, J. Jonaš, M. Svoboda, and D. Knessl, Coll. Czech. Chem. Comm., 22, 2141 (1958).
189. C. S. Foote, J. Am. Chem. Soc., 86, 1853 (1964).
190. P. von R. Schleyer, J. Am. Chem. Soc., 86, 1854 (1964).
191. M. R. V. Sahyun and D. J. Cram, J. Am. Chem. Soc., 85, 1263 (1963).
192. A. C. Cope, E. Ciganek, and J. Lazar, J. Am. Chem. Soc., 84, 2591 (1962).
193. D. J. Cram and F. A. Abd Elhafez, J. Am. Chem. Soc., 75, 739 (1953).
194. A. C. Cope and N. A. LeBel, J. Am. Chem. Soc., 82, 4656 (1960).
195. G. Wittig and R. Polster, Ann., 599, 13 (1956).
196. D. G. Morris, B. W. Smith and R. J. Wood, Chem. Comm., 1134 (1968).
197. A. C. Cope, R. A. Pike, and C. F. Spencer, J. Am. Chem. Soc., 75, 3212 (1953).
198. G. Wittig and T. F. Burger, Ann., 632, 85 (1960).
199. C. A. Kingsbury and D. J. Cram, J. Am. Chem.

Soc., 82, 1810 (1960).

200. D. W. Emerson, A. P. Craig, and I. W. Potts, Jr., J. Org. Chem., 32, 102 (1967).

201. S. I. Goldberg and M. S. Sahli, Tetrahedron Lett., 4441 (1965).

202. (a) C. R. Johnson and D. McCants, J. Am. Chem. Soc., 87, 1109 (1965);
(b) D. N. Jones and M. J. Green, J. Chem. Soc., (C), 532 (1967).

203. I. D. Entwistle and R. A. W. Johnstone, Chem. Comm., 29 (1965).

204. I. D. Entwistle, R. A. W. Johnstone, and B. J. Millard, J. Chem. Soc., C, 302 (1967).

205. D. W. Emerson and T. J. Korniski, J. Org. Chem., 34, 4115 (1969).

206. C. Walling and L. Bollyky, J. Org. Chem., 29, 2699 (1964).

207. D. R. Raynor, A. J. Gordon, and K. Mislow, J. Am. Chem. Soc., 90, 4854 (1968).

208. E. G. Miller, D. R. Rayner, H. T. Thomas, and K. Mislow, J. Am. Chem. Soc., 90, 4861 (1968).

209. (a) T. C. Bruice and P. T. Markiw, J. Am. Chem. Soc., 79, 3150 (1957);
(b) W. Jenny, Helv. Chim. Acta., 41, 317, 326 (1958).

210. J. R. Shelton and K. E. Davis, J. Am. Chem. Soc., 89, 718 (1967).

211. E. Marongin, Rend. Semin. Fac. Sci. Univ. Cagliari, 36, 290 (1966); Chem. Abstr., 68, 77823 (1968).

212. G. Gelli and V. Solinas, Rend. Semin. Fac. Sci. Univ. Cagliari, 36, 279 (1966); Chem. Abstr., 68, 77827m (1968).

213. J. L. Kice and J. D. Campbell, J. Org. Chem., 39, 1631 (1967).

214. K. Tsujihara, N. Furukawa, and S. Oae, Bull. Chem. Soc. Jap., 18, 793 (1970).

215. S. Oae, K. Tsujihara, and N. Furukawa, Tetrahedron Lett., 2663 (1970).

216. N. Barroeta, A. Maccoll, M. Cavazza, L. Congiu, and A. Fava, J. Chem. Soc., (B), 1264 (1971).

217. N. Barroeta, A. Maccoll, M. Cavazza, L. Congiu, and A. Fava, J. Chem. Soc., (B), 1267 (1971).

218. N. Barroeta, A. Maccoll, and A. Fava, J. Chem. Soc., B, 347 (1969).

219. R. B. Morin, B. G. Jackson, R. A. Mueller, E. R. Lavagnino, W. B. Scanlon, and S. L. Andrews, J. Am. Chem. Soc., 85, 1896 (1963).

220. R. B. Woodward and R. Hoffmann, Angew. Chem. Int. Ed., 11, 781 (1969).

221.   (a)   J. E. Baldwin, W. F. Erickson, R. E.
Hackler, and R. M. Scott, Chem. Comm., 576
(1970);
(b)   J. E. Baldwin, J. E. Brown, and R. W.
Cordell, Chem. Comm., 31 (1970).

222.   J. E. Baldwin and F. J. Urban, Chem. Comm., 165
(1970).

223.   (a)   S. W. Kantor and C. R. Hauser, J. Am. Chem.
Soc., 73, 4122 (1951);
(b)   R. W. Jemison and W. D. Ollis, Chem. Comm.,
294 (1969).

224.   J. E. Baldwin, J. DeBernardis, and J. E. Patrick,
Tetrahedron Lett., 353 (1970).

225.   J. E. Baldwin R. E. Heckler, and D. P. Kelly,
Chem. Comm., 537, 538, 539, 899, 1083 (1968).

226.   W. Kirmse, M. Kapps, and R. B. Hager, Chem.
Ber., 99, 2855 (1966).

227.   J. E. Baldwin and C. H. Armstrong, Chem. Comm.,
631 (1970).

A   β-Eliminations Leading to Multiple Bonds Between Carbon and Other Elements

1   General Comments

Many chemists associate β-elimination and even the more general term "elimination reaction" with olefin-forming processes. Understandably, this situation has arisen as most of the mechanistic studies have been concerned with alkene formation and all the reviews on β-elimination have not suprisingly focused attention on the same area. However, β-eliminations yielding multiple bonds between carbon and a heteroatom are very common reactions, occurring with particular facility and exhibit-many of the characteristics of the alkene-forming processes.

   Most often one of the eliminating fragments is a hydrogen atom, and if this is situated on the hetero-atom, the substrate ionizes rapidly to an anion that decomposes subsequently in a slow step. This mechanism, which is analogous to the ElcB mechanism of olefin-formation (Section I.B.1) is exemplified by the elimination of HCN from cyanohydrins (Eq. 1). Of somewhat greater complexity is the aldol condensation (Scheme 1). In this case the successive equilibria comprise a

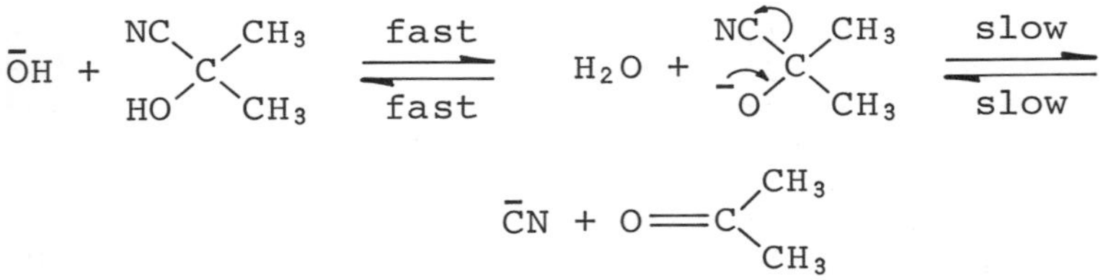

$$\overline{OH} + \underset{HO}{\overset{NC}{\diagdown}}\underset{CH_3}{\overset{CH_3}{\diagup}}C \underset{fast}{\overset{fast}{\rightleftharpoons}} H_2O + \underset{\overline{O}}{\overset{NC}{\diagdown}}\underset{CH_3}{\overset{CH_3}{\diagup}}C \underset{slow}{\overset{slow}{\longrightarrow}}$$

$$\overline{CN} + O{=}C\underset{CH_3}{\overset{CH_3}{\diagdown}} \qquad (1)$$

dehydration (k_4) and a two-step carbonyl-forming elimination (k_{-3} and k_{-2}). Depending on the reaction conditions, intermediate carbinolamine $\underline{1}$ can undergo either an imine or carbonyl-forming elimination

   Scheme 1

$$\overline{OH} + CH_3COPh \underset{k_{-1}}{\overset{k_1}{\rightleftharpoons}} H_2O + H_2\overline{C}{-}COPh$$

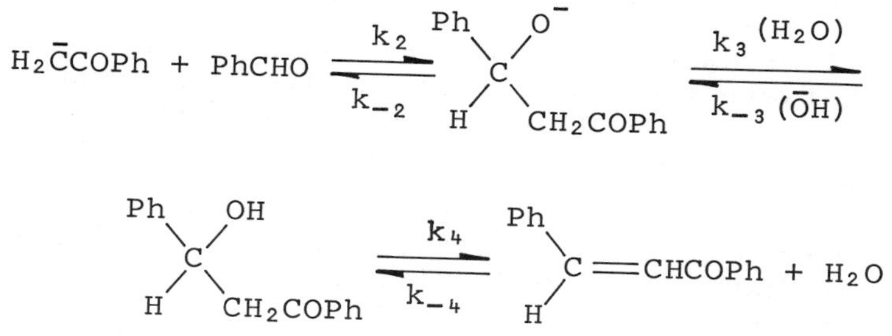

$$H_2\bar{C}COPh + PhCHO \underset{k_{-2}}{\overset{k_2}{\rightleftharpoons}} \underset{H}{\overset{Ph}{>}}C\underset{CH_2COPh}{\overset{O^-}{<}} \underset{k_{-3}(\bar{O}H)}{\overset{k_3(H_2O)}{\rightrightarrows}}$$

$$\underset{H}{\overset{Ph}{>}}C\underset{CH_2COPh}{\overset{OH}{<}} \underset{k_{-4}}{\overset{k_4}{\rightleftharpoons}} \underset{H}{\overset{Ph}{>}}C=CHCOPh + H_2O$$

(Eq. 2).[1] In these reactions, the greater acidity of

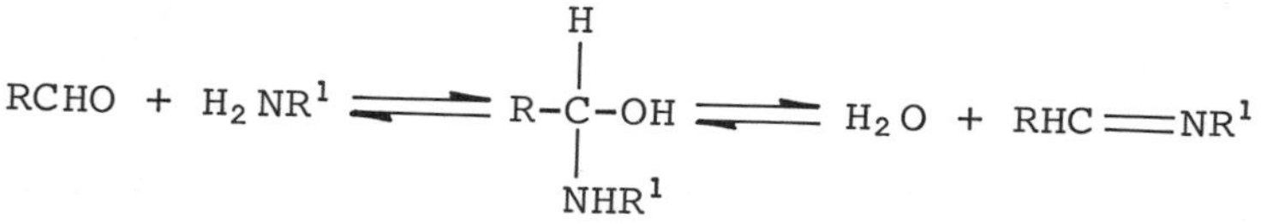

$$RCHO + H_2NR^1 \rightleftharpoons R-\overset{\overset{\displaystyle H}{|}}{\underset{\underset{\displaystyle NHR^1}{|}}{C}}-OH \rightleftharpoons H_2O + RHC=NR^1 \qquad (2)$$

$$\underline{1}$$

O-H and N-H bonds, than C-H bonds increases the probability of anionic two-step eliminations relative to the
situation in olefin formation. However, these reactions are more realistically regarded as the reversal
of nucleophilic additions to double bonds, and are thus
outside the scope of this book.
    When hydrogen is removed from a carbon atom ($\underline{2}$ -
$\underline{5}$), the β-elimination of HX to form a multiple bond between carbon and a heteroatom occurs much more easily
than the formation of the corresponding carbon-carbon
multiple bond ($\underline{6}$, $\underline{7}$). This difference in reactivity
can be attributed to a number of factors.[2] In most

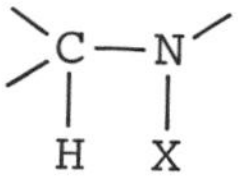

$$\underset{\underset{H \quad X}{|\quad|}}{>C-N<} \qquad \underset{\underset{H}{|}}{C=N-X} \qquad \underset{\underset{H}{|}}{>C-O-X} \qquad \underset{\underset{H}{|}}{>C-S-X} \qquad \underset{\underset{H \quad X}{|\quad|}}{>C-C<}$$

$$\underline{2} \qquad\qquad \underline{3} \qquad\qquad \underline{4} \qquad\qquad \underline{5} \qquad\qquad \underline{6}$$

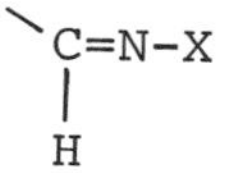

$$\underset{\underset{H \qquad X}{}}{>C=C<}$$

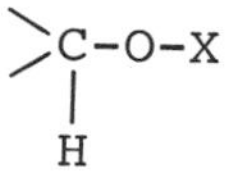

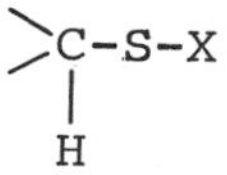

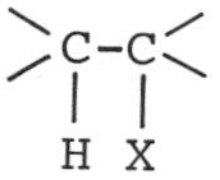

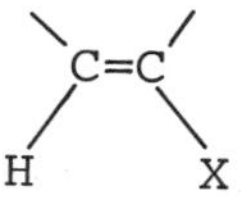

$$\underline{7}$$

instances X is an electronegative atom, or a group with
strongly electron-attracting properties, or a group
attached by an electronegative atom. The strength of
the bond to X increases with the difference in electro-
negativities between the atoms constituting that bond.
Consequently X is more strongly bound (i.e., less prone
to dissociate as X$^-$) in substrates 6 and 7 than substrates 2 to 5. Usually in β-elimination the bond
changes are well advanced in the transition state,
which reflects to a certain extent the stability of the
products. With increasing bond multiplicity (C-C → C=C →
C≡C; C-N → C=N → C≡N), the bond strength rises more
rapidly when carbon is bonded to a heteroatom than to
another carbon atom. Activation energies for elimination leading to carbon-heteroatom multiple bonds should
be smaller than those for formation of the corresponding carbon-carbon multiple bond. A third property concerns the acidity of the hydrogen atom, which should be
greater in the presence of a β-heteroatom than a β-carbon atom.

The general mechanisms that have been demonstrated
for olefin-forming reactions provide an adequate framework for our discussion of mechanisms of reactions
leading to multiple-bond formation between carbon and
a heteroatom. In our discussion, we use the general
terms E2, ElcB, El, and Ei (Sections I.A.1,2 and I.B.
1; VIII) to indicate the mechanism with reference to
olefin formation, adding the subscripts CO, CN, CS, and
C=N to indicate the atoms in the substrate on which the
leaving groups are situated. Although in line with
previous nomenclature, only the term E_{CO}, for carbonyl
formation, has been used with regularity.[3,4] Within
this terminology, $E_{C=C}$ is used in the latter half of
this chapter to indicate alkyne formation. Equation 3
is an example of a carbonyl-forming elimination and
Eq. 4 depicts a nitrile-forming elimination.

$$\overset{H}{\underset{}{\underset{}{>}}}C-O-X \quad \xrightarrow{\quad E_{CO}\quad} \quad >C=O \;+\; \overset{+}{H} \;+\; \overset{-}{X} \tag{3}$$

$$\underset{H}{\overset{}{>}}C=N-X \quad \xrightarrow{\quad E_{C=N}\quad} \quad -C≡N \;+\; \overset{+}{H} \;+\; \overset{-}{X} \tag{4}$$

2 Thiocarbonyl-Forming Eliminations

Disulfide linkages are cleaved by anions such as $C\bar{N}$,
$SC\bar{N}$, and RS^-, by a bimolecular displacement mechanism

at sulfur.[5,7] However, for more basic nucleophiles, such as alkoxide and hydroxide, additional cleavage routes include E1cB[8,9] and E_{CS}1cB reactions.[10]

By choosing substrates lacking hydrogens beta to the disulfide, and substituting the α-carbon with electron-withdrawing groups, thus promoting C_α-H acidity, it should be possible to encourage the E_{CS} reaction at the expense of alkene formation and bimolecular displacement. This approach has been adopted by a group of Italian workers who have studied the reaction of diphenylmethyl disulfides with isopropoxide in 95% isopropanol-benzene under nitrogen.[11,12] The yield of substituted thiobenzophenone 9, as determined by visible spectroscopy, varied from 88 to 52%·at 20°C, increasing with the electron-withdrawing power of both Z and Y (see Eq. 5). The authors assumed that the main competing reaction; which followed first-order kinetics in each reactant, was a bimolecular nucleophilic displacement at sulfur leading to product 10. This deduction accords with the greater yield of thiol than thiobenzophenone for each reaction. Surprisingly the yield of thiobenzophenone decreased as the reaction temperature was increased (e.g., Z = Ph, Y = CH3, E_{CS}2 % 67 at 20°C, 58 at 30°C). Normally for substrates capable of undergoing competing E2 and SN2 reactions, the elimination component is increased by a rise in reaction temperature (Section I.C.1). For the α-deutero substrate (Z = Ph, Y = CH3), the E_{CS} reaction comprises only 31% of the total substrate decomposition at 30°C.

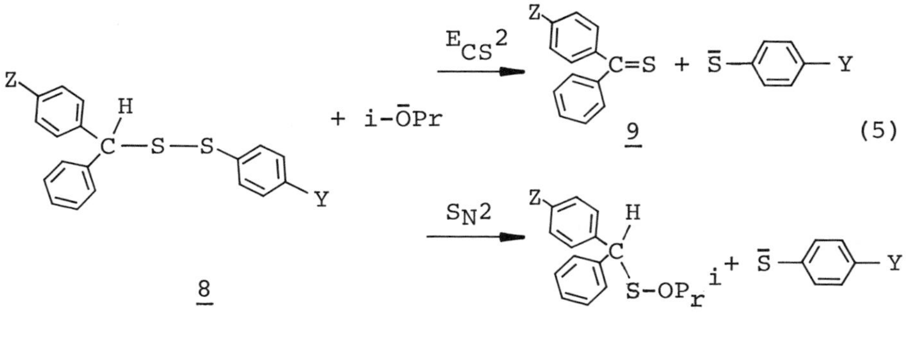

Comparison of the rates of conversion of substrate 8 (Z = Ph, Y = CH3) and its α-deutero analogue into

4-phenyl thiobenzophenone affords an isotope effect of 6.1 at 30°C. This value, which is near the maximum of 6.6 predicted for a simple proton transfer from carbon to oxygen[13] (Section II.C.3.a,b), suggests the proton is about half-transferred to the base in the transition state. The magnitude of this effect is inconsistent with a reversible E_{CS}lcB reaction, and not suprisingly tritium is not incorporated into unreacted substrate isolated in the early stages of reaction in tritiated 2-propanol [$(CH_3)_2$CHOT].

The rate coefficients for the E_{CS} reactions of the substituted diphenylmethyl aryl disulfides correlate with the Hammett equation, but in view of the few substrates examined, a degree of caution is necessary when interpretating the results, which are listed in Table 1. The ρ_Z values of 4.30 (Y = p-Cl) and 4.72 (Y = H) at 20°C are of a similar magnitude to the largest reaction constants observed for base-induced E2 reactions of substituted 2-phenylethyltrimethylammonium ions (ρ = 3.77 in $NaOC_2H_5/C_2H_5OH$ at 30°C)[14] and methyl sul- (ρ = 4.4 in $KOBu^t$ / t-BuOH/DMSO at 52.2°).[15] The large

TABLE 1   Hammett Reaction Constants for E_{CS} Reactions of Substituted Diphenylmethyl Phenyl Sulfides in 95% Isopropanol-benzene at 20°

Z^a	Y	$\rho_Z{}^c$	$\rho_Y{}^c$	r^b
Cl,Ph,H	Cl	4.30		0.998
Cl,Ph,H	H	4.72		0.999
H	Cl,H,Me		2.02	0.980
Ph	Cl,H,Me		2.05	0.999
Cl	Cl,H		2.12	-

[a]For definition of Z and Y, see structure 8.
[b]Correlation coefficient.
[c]σ^-(Ph) = 0.084, obtained from interpolation of rate of elimination from substituted diphenylmethyl thiocyanates.[18]

reaction constants suggest that the transition state for elimination from diphenylmethyl phenyldisulfides has considerable carbanionic character, a large negative charge residing on the central carbon atom. The decrease in ρ_Z with an increase in the electron-withdrawing power of the subtituent Y parallels the trend observed in elimination of substituted 2-phenyl-ethyl arenesulfonates (Section II.C.2.a, Table 1).[16]

The ρ_Y values of about 2 for all the substituents Z studied are of similar magnitude to that of 2.85 observed in the ionization of substituted benzenethiols in 95% aqueous ethanol at 20°C,[17] suggesting the S-S bond is relatively extended in the transition state, considerable negative charge accumulating on the sulfur atom of the leaving group.

The large reaction constants ρ_Z and ρ_Y and the magnitude of the hydrogen isotope effect k_H/k_D have been quoted by the Italian authors as evidence according with a transition state, 11, which possesses little double-bond character.[12] The barrier to delocalization of the charge accumulating on the central carbon may be attributable to the low effectiveness of π-electron overlap involving 3p orbitals of sulfur.

$$Pr^iO\text{------}H\text{------}\underset{\overset{|}{Ar}}{\overset{\overset{Ar}{|}}{C}}\overset{\delta^-}{=\!=\!=}\overset{\delta^+}{S}\text{------}\overset{\delta^-}{S}\text{---}Ar^1$$

11

The same group of Italian workers have also reported the thiocarbonyl elimination of diphenylmethyl thiocyanates (Eq. 6).[18] The Hammett reaction constant for this reaction (Z = Me, 4,4'-Me$_2$, 3-Me, H, 3-OMe,

$$\underset{C_6H_5}{\overset{Z-C_6H_4}{>}}\!\!\overset{H}{\underset{}{C}}\!\!-S-CN \quad \xrightarrow[Pr^iOH]{Pr^iONa} \quad \underset{C_6H_5}{\overset{Z-C_6H_4}{>}}C\!=\!S \; + \; HCN \qquad (6)$$

4-Cl) is 3.5 at 20°C, and the primary kinetic hydrogen isotope effect (k_H/k_D) is 3.0 (Z = p-Ph). As in the case of styrene formation in the E2 reactions of 2-phenylethyl compounds with bases, the combination of small isotope effect and large reaction constant was interpretated as indicating a transition state with considerable carbanion character in which the proton is more than half-transferred to the base[19] (Section II. C.3.b). The unsubstituted thiocyanate eliminates about 300 times faster than the corresponding disulfide.

This difference could reside in greater ionizing tendency of the leaving group in the thiocyanate case, but as cyanide and mercaptide ions possess similar "thionucleophilicity", an alternative explanation seems probable.[20]   The acidity of the C-H bond should be greater in the thiocyanate series as CN is more strongly electron withdrawing than SC_6H_5,[12] and Miotti and Ceccon regard the HCN elimination as involving a transition state with greater carbanion character.

3   Imine- and Nitrile-Forming Eliminations

There are comparatively few kinetic studies of β-eliminations giving imines or nitriles. Nitriles can be prepared by the acid-catalyzed dehydration of amides or aldoximes or by base catalyzed elimination from aldochlorimines or esters of aldoximes. Imines often arise in the attempted preparation of N-chloroamines, dehydrohalogenation occurring in the weakly alkaline hypochlorite solution, which is used as the chlorinating reagent (Eq. 7).[21] The mild reaction conditions required to effect this dehydrohalogenation contrast to

$$
\underset{\underset{NH_2}{|}}{\overset{\overset{H}{|}}{Ar-C-CN}} \xrightarrow{\text{"NaOCl"}} \underset{\underset{NCl_2}{|}}{\overset{\overset{H}{|}}{Ar-C-CN}} \xrightarrow{-HCl} \underset{\underset{N}{\|}\diagdown_{Cl}}{\overset{\overset{}{}}{Ar-C-CN}} \qquad (7)
$$

the more basic conditions needed to cause the corresponding elimination that produces an alkene. Even more contrasting is the facile dehydrofluorination of difluoroamines (Eq. 8).[22] Analogous alkyl fluorides are

$$
\underset{\underset{NF_2}{|}\;\underset{NF_2}{|}}{\overset{\overset{H}{|}\;\overset{H}{|}}{Ph-C-C-Ph}} \xrightarrow{Et_3N} \underset{\underset{}{}}{\overset{\overset{NF}{\|}\;\overset{NF}{\|}}{Ph-C-C-Ph}} \qquad (8)
$$

extremely reluctant to undergo β-elimination, requiring strongly basic conditions and elevated reaction temperatures to produce convenient rates of reaction.[23] Activation energies run 5-10 kcal/mole less for the imine-forming reactions than for E2 dehydrofluorinations.[26]

    The first reported kinetic study of dehydrohalogenation concerned the reaction of substituted aldochlorimines (12) with alkoxide ion in 92.5% aqueous ethanol (Eq. 9).[24] The Hammett reaction constant of 2.24 led Hauser and his co-workers to propose the operation of an $E_{CN}1cB$ reaction, although the experimental evidence accords equally well with an $E_{CN}2$ reaction in which C_β-H bond breaking runs ahead of N-Cl bond breaking. Subsequent studies by Jordan, Dyas, and Hill support the $E_{CN}2$ route.[25] The rate of elimination with a variety of bases afford a Bronsted co-efficient of 0.54, indicating the proton is about half-transferred to the base in the transition state in the rate-determining step of the reaction (Section II.C.4. b).

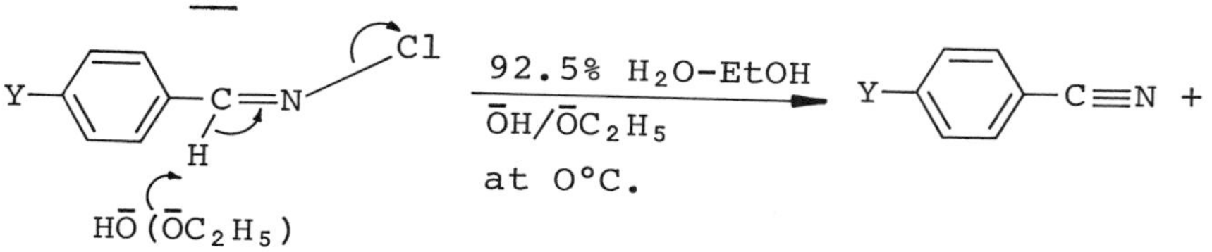

$$H_2O(C_2H_5OH) + \bar{C}l \qquad\qquad (9)$$

In recent years a number of authors have reported synthesis and subsequent elimination of alkyl difluoro-amines with a variety of mild bases.[27-32] An interesting synthetic observation concerns the relative ease of dehydrofluorination of stereoisomers 13 and 14. The trans-bis(difluoramine) 13 eliminates when heated under reflux with sodium fluoride in acetone, but the cis-isomer reacts only with the more basic sodium carbonate in methyl cyanide under reflux. Both reactions give rise to a mixture of syn- and anti-isomers as evidenced by two resonances in the F^{19}-NMR spectrum of the reaction product.[29] Brauman and Hill have reported the kinetics of dehydrofluorination of several mono-, bis-and tris-(N,N-difluoramino)alkanes in diglyme-water.[26,32,33]

Aldoximes can be converted into nitriles by a number of common acidic dehydrating reagents. Ben-zaldoxime, in fact, even dehydrates during gas chromatography on celite at 200°C.[34] Photodehydration of oximes to nitriles has also been reported.[35] However,

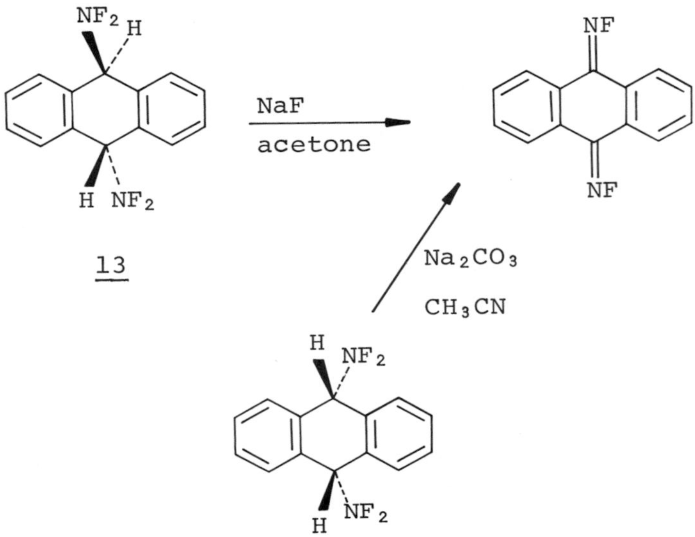

kinetic studies of $E_{C=N}$ reactions of oximes have been
limited to reactions of their N-acyl derivatives.[36,37]
The hydrolysis of N-acyl derivatives of substituted β-
benzaldoximes (15) in dilute alkaline solution gives
substituted nitriles and benzaldoximes (Eq. 10). The
elimination fraction is increased by electron-withdrawing substituents Y, which enhance the methine

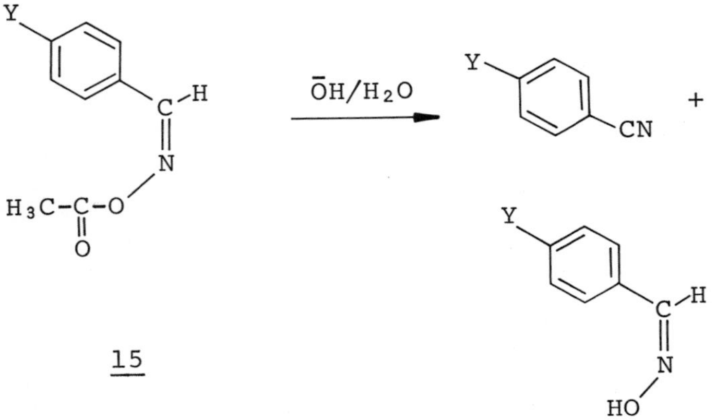

proton acidity but exert less influence on the more distant site of substitution.

The N-acyl derivatives of the α-benzaldoximes undergo only the substitution reaction to produce the α-benzaldoxime (Eq. 11). Two factors may contribute to the reduced competitiveness of the elimination route

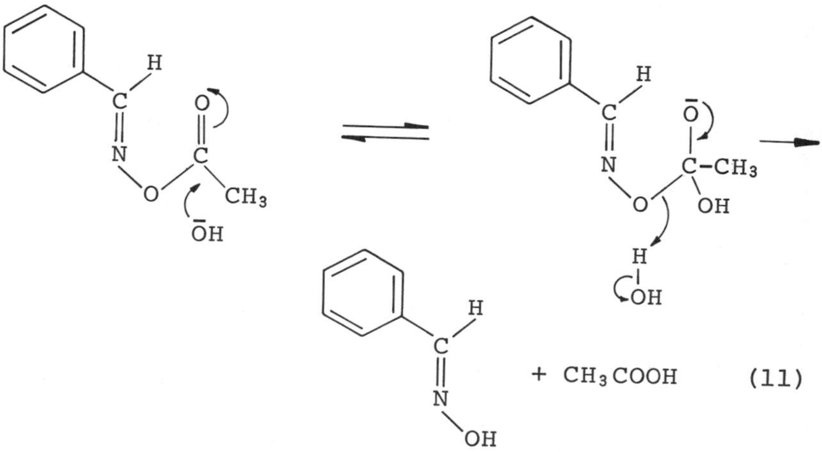

$$+ \; CH_3 COOH \qquad (11)$$

relative to the position for the β-series. The acyl group and the methine hydrogen are correctly situated in the α-isomer for effective internal hydrogen-bonding interaction (see 16). The proximity of the acyl group

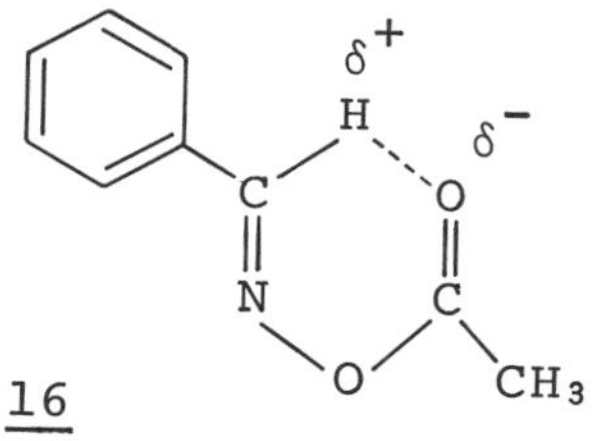

16

could hinder the approach of base to the methine hydrogen, but it is unlikely to affect the ease of the substitution reaction. Alternatively, the α-substrates must eliminate with syn stereospecificity, and overlap of developing orbitals is less energetically favorable than for the anti elimination from the β-series. If the stereochemical factor is the dominant cause for the product change, then these nitrile-forming reactions most probably occur by the $E_{C=N}2$ mechanism.

Oximes seldom dehydrate to nitriles in the presence of basic reagents, being converted usually only
into their corresponding anions. However, elimination
does occur in dilute alkaline solution if the methine
hydrogen is activated by acid-strengthening substituents, as is the case for α-o-nitrobenzaldoxime.[38]
    Ambrose and Brady[37] have measured the rates of
pyrolytic elimination of a series of N-acyl derivatives
of α- and β-benzaldoximes in xylene. Both geometrical
isomers eliminate at a convenient rate between 100 to
140°C, a much lower reaction temperature than required
to effect the Ei reactions of acetates to alkenes
(Section VIII.C). Not surprisingly the α-isomers,
which possess the required syn orientation of the leaving groups for a cyclic elimination, are more reactive
than β-derivatives. In fact, the reaction course
followed by the latter isomers is not clear, since in
the presence of traces of acid (e.g., carboxylic acids
liberated during pyrolysis), they are rapidly isomerized to the α-form. Consequently, mechanistic investigations were confined to the α-series.
    Electron-withdrawing substituents in the acyl
group R (17) enhance the rate of elimination. This result parallels the situation in acetate pyrolysis and
accords with N-O bond breaking being more advanced
than O-H bond formation in the transition state.
However, electron-releasing substituents X enhance the
rate of pyrolysis. This observation contrasts markedly
with the substituent effect in the pyrolysis of substituted 2-phenylethyl acetates in which a small positive reaction constant is found (Section VIII.C.4).
Obviously C-H bond breaking is of much less

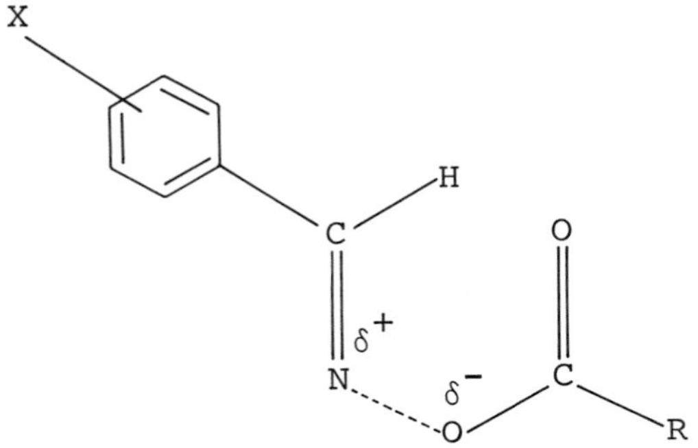

17

significance in the $E_{C=N}i$ reaction than in the Ei reaction of the acetates.

4 Carbonyl-Forming Eliminations

(a) <u>General Features</u>. Within the framework of heterolytic eliminations, the carbonyl-forming reactions can be separated into two categories: (1) those involving proton transfer (Eq. 12), and (2) others in which the β-hydrogen is eliminated as hydride (Eq. 13). Of course, the Hβ atom and group X may fragment in

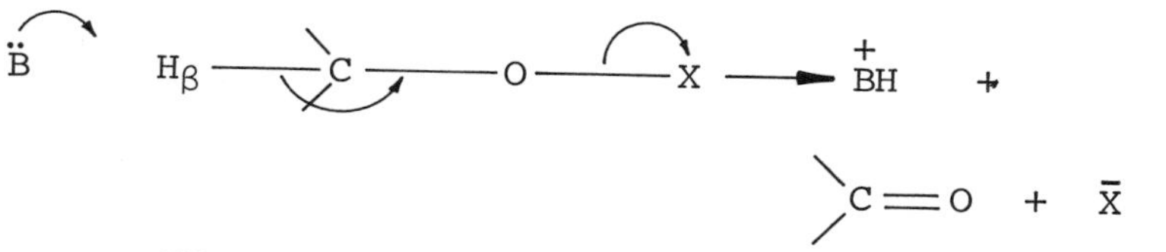

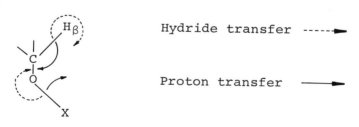

$$\diagup\!\!\!\!\diagdown C = O \;+\; \bar{X} \qquad (12)$$

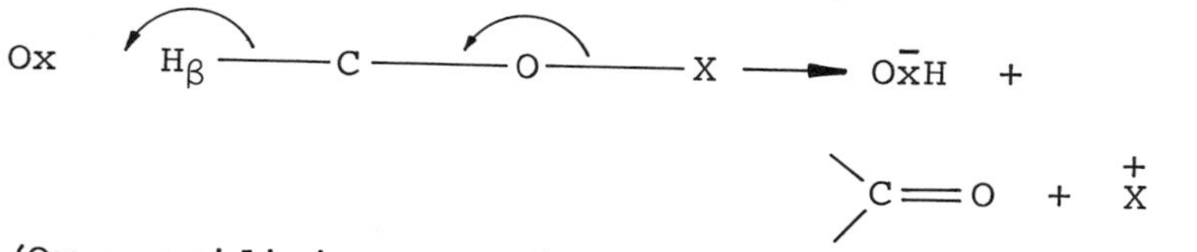

$$\diagup\!\!\!\!\diagdown C = O \;+\; \overset{+}{X} \qquad (13)$$

(Ox = oxidizing agent)

discrete steps, but in the absence of kinetic evidence we show a concerted reaction path. Both mechanisms may also follow a unimolecular cyclic course in which group X acts as a base or an oxidizing agent (see 18).

<u>18</u>

Hydride transfer ----→

Proton transfer ──→

Typical of the proton transfer category are the E_{CO} reactions of peroxides,[39] nitrate,[40] and sulfonate esters[41] and hypochlorites[39] with basic reagents.

Heterolytic oxidation reactions of alcohols, by reagents that cannot react via initial ester formation, are generally regarded as hydride-transfer processes.[42] The disproportionation of diaryl carbinols in perchloric acid[42] and the oxidation of alcohols by triarylcarbonium ion in dilute acidic solution (Eq. 14),[43] a reaction that is first order in the triaryl-carbinol, are typical examples.

$$(C_6H_5)_3 \overset{+}{C} \quad H-\overset{\overset{\displaystyle CH_3}{|}}{\underset{\underset{\displaystyle CH_3}{|}}{C}}-O-H \quad \longrightarrow \quad (C_6H_5)_3CH + \quad \overset{\displaystyle CH_3}{\underset{\displaystyle CH_3}{>}}C=O + \overset{+}{H}$$

$$(14)$$

The cyclic hydride transfer is exemplified by the reversible oxidation-reduction process termed the Meerwein-Pondorf-Verley-Oppenauer reaction, and cyclic proton transfer is typified by the thermal decomposition of nitronate esters (Eq. 15).[44]

$$(CH_3)_2 \quad C=\overset{+}{N}\overset{\overline{O}}{\underset{O}{\bigg\langle}} \overset{H}{\underset{H}{\diagdown}} C-Ar \quad \overset{\triangle}{\longrightarrow} \quad (CH_3)_2C=N-OH +$$

$$ArCHO \qquad (15)$$

Most detailed mechanistic studies are available for the base catalyzed eliminations of nitrates and the oxidation of alcohols by a variety of reagents, and we now briefly discuss these reactions.

(b) $\underline{E_{CO} \text{ Reactions of Nitrate Esters}}$. Baker and his co-workers[40],[45-48] have studied the effect of structural changes on the relative rates of E_{CO}, substitution, and olefin-forming eliminations of alkyl nitrates, while Bourns and his co-workers,[49],[50] have measured the nitrogen and hydrogen kinetic isotope effects for the E_{CO} reactions of benzyl and 9-fluorenyl nitrates. For the elimination reactions with alkoxide ion, the nitrogen and hydrogen kinetic isotope effects and the Hammett reaction constant (3.4 for the substituted benzyl nitrates) accords with the operation of $E_{CO}2$ mechanisms. The same mechanism is favored for the elimination of 9-fluorenyl tosylate.[41]

(c) <u>Elimination Mechanisms for Alcohol Oxidations.</u>

(i) <u>Oxidation by chromic acid.</u> The mechanism of oxidation of alcohols by chromic acid has been reviewed on many occasions.[42;51-55] Most chromic acid oxidations are carried out in either aqueous sulfuric or acetic acid solution, and the chromium valence state changes from Cr^{VI} to Cr^{III}, involving the movement to an inner shell of three valence electrons. Quantum restrictions prevent more than a pair of electrons, associated with any one atom, from behaving in an identical manner. Thus, the oxidation must be a stepwise process. Much of the kinetic work has concerned the oxidation of isopropyl alcohol and some simple secondary alcohols. Oxidation of 2-deutero-2-propanol affords an isotope effect of 6,[56] and the order in acid varies with medium acidity[54,57] Di-isopropyl chromate has been isolated from a partly reacted solution,[58] and electron-releasing substituents in the alkyl residue[59] or in 1-phenylethanols[60] increase the rate of oxidation. These facts are consistent with the mechanism of Scheme 2 in which the decomposition of an intermediate-chromate ester involves a cyclic hydride transfer in the slow step.

Scheme 2

(1) $HCrO_4^- + \overset{+}{H} \rightleftharpoons H_2CrO_4$ fast

(2) $H_2CrO_4 + R_2CHOH \rightleftharpoons R_2CHOCrO_3H + H_2O$ fast

(3) $R_2CHOCrO_3H + \overset{+}{H} \rightleftharpoons R_2CHOCr\overset{+}{O}_3H_2$ fast

(4) $R_2CHOCrO_3\overset{+}{H}_2$

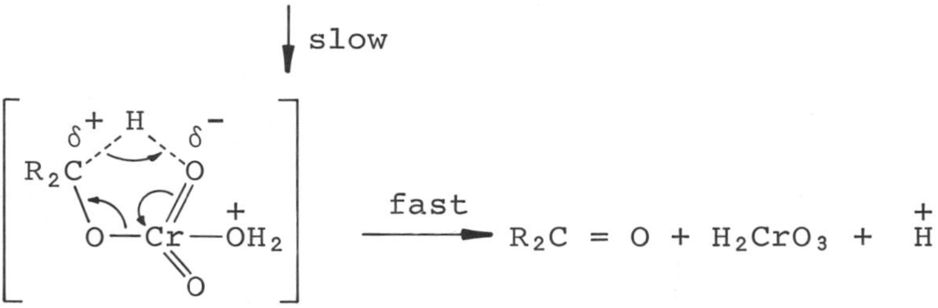

$\xrightarrow{\text{fast}} R_2C = O + H_2CrO_3 + \overset{+}{H}$

Kinetic isotope studies on the oxidation of the sterically hindered alcohol, 3β-28-diacetoxy-6β-hydroxy-18β-olean-12-ene, provide support for this ester-hydride mechanism.[61] In mineral acids, no isotope effect is observed, the rate-determining step being chromate-ester formation. In 60% aqueous acetic acid, containing sodium acetate, an isotope effect (k_H/k_D) of 2 is observed, ester formation and its subsequent decomposition contributing to the rate-determining step.

(ii) Oxidation by bromine-containing reagents. Primary and secondary alcohols can be oxidized to al- dehydes and ketones by a number of bromine-containing reagents, but subsequent bromination of the carbonyl products can complicate mechanistic studies and limit synthetic utility. Oxidation by bromine involves a cyclic transition state, 19,[62-66] whereas N-bromo- succinimide oxidations involve both the parent reagent and molecular bromine.[67-72] Rate-determining bromate ester formation has been proposed for oxidation of al- cohols by potassium bromate as the reactions are in- sensitive to substituent effects.[73]

19

B ALKYNE FORMING β-ELIMINATIONS

1 General Features

The most common alkyne-forming elimination is the dehy- drohalogenation of vinyl halides with basic reagents.[74] Vicinial dihalides can be dehalogenated by zinc, mag- nesium, or sodium in suitable solvents.[75,76] Vinyl fluorides do not react, and this method can be used to prepare fluorinated alkynes. Vinyl cations have for many years been accepted as intermediates in the reac- tion of alkynes with electrophiles but, only in recent years, have unimolecular solvolytic eliminations of vinyl halides to alkynes been proposed.[77]

A greater electronegativity is associated with sp^2- than sp^3- hybridized carbon and, consequently, olefinic carbon atoms accept positive charge less readily but negative charge more readily than the corresponding saturated carbons.[78] In support of this concept the electrophilic addition of bromine to phenylpropiolic acid occurs much more slowly than to either cis- or trans- cinnamic acids.[79] Thus base- induced eliminations of vinyl halides are more likely to occur via carbanion intermediates than are elimina- tions of alkyl halides. On the other hand, carbonium ion intermediates are less probable. Concerted eliminations should also show greater leaning toward transition states with extensive carbanion character.

Unlike the situation for saturated substrates, the stereochemical arrangement of the eliminating fragments is not complicated by conformational considerations. The rigidity of the reactant alkene constrains the leaving groups to planar syn- or anti-orientations. Which of these orientations is actually more favorable is determined principally by the reaction conditions, such as choice of solvent and base. In protic solvents, with alkoxide bases, anti elimination is preferred, but metal alkyls in ethereal solvents encourage syn elimination.

The majority of the mechanistic studies have been concerned with the reactions of vinyl halides, and we restrict our subsequent discussion to this subject.

2   Nucleophilic Reactivity of Activated Vinyl Halides

(a) <u>Introductory Remarks</u>. Simple vinyl halides are much more resistant to bimolecular nucleophilic substitution and base-catalyzed dehydrohalogenation than are the corresponding alkyl halides. This inertness has been attributed to the mesomeric electron-donating influence of the halogen atoms, <u>20</u>, which imparts double-bond character on the C-X bond, making it more difficult to break.[80] X-ray studies confirm that the

$$H_2C = CH - \ddot{C}l: \longleftrightarrow H_2C - CH = \overset{+}{C}l$$

<u>20</u>

C-X bond is shorter in vinyl than alkyl halides, and the ionization energy, as measured by gas phase dissociation to radicals which then ionize, is less for ethyl chloride than vinyl chloride by 0.7eV.[81] The effect of this partial double-bond character increases with the extent of bond cleavage in the transition state. Thus the reactivity ratio, $k_{alkyl\ halide}/k_{vinyl\ halide}$, will be large for unimolecular S_N1 (or E1) solvolyses, and smaller, but still significant, in displacement and elimination reactions in which C-X bond breaking is less progressed in the transition state.

The reactivity of vinyl halides toward nucleophiles can be greatly enhanced by the presence of strongly electron-attracting groups, attached directly

to the alkene double bond. Unfortunately, these activating groups not only enhance β-elimination but also
alternative eliminations and substitutions. In addition, the β-elimination products, the activated alkynes,
are often unstable under the reaction conditions,
undergoing addition reactions to give a vinylic derivative of apparent substitution. The products of elimination-addition and direct substitution in this case
are the same, and kinetic methods are required to elucidate the mechanism. The mechanisms of reactions of
vinyl halides with nucleophiles have attracted considerable interest in recent years, and an excellent
account of the numerous reaction paths has been compiled by Rappoport.[82] Initially we discuss the various
reaction schemes, the methods of distinguishing them
from each other, and their range of applicability.
Finally we discuss the β-elimination mechanisms, paying
attention to the stereochemistry of reaction and the
nature of intermediates.

(b) <u>Reaction Mechanisms</u>. The established mechanisms
for the reactions of vinyl halides with nucleophilic
reagents are listed in Eqs. 16 to 25. The terminology
is essentially that employed by Rappoport except for
the use of abbreviated symbols and $E_{C=C}$ for α,β-elimina-
tion, α-elimination (α-E) for β,β-elimination and E for
β,γ-elimination.[82] Only Eq. 16 contains an alkyne-
forming β-elimination ($E_{C=C}$). Equation 17 comprises an
α-elimination, a subject considered in greater detail
in Chapter X. Equations 18, 19, and 20 contain alkene-
forming β-eliminations (E), most probably of the E2
category. For substrates possessing an appropriately
situated vinylic hydrogen, Eq. 21 can also be regarded
as an $E_{C=C}1$ reaction. The less commonly encountered
routes, Eqs. 22 to 24, are really outside the scope of
this book, and the reader is referred elsewhere for a
discussion of them.[82] Equation 25 describes the reaction path for halogen-metal exchange followed by reaction with an electrophilic species.[83,84] We concentrate our discussion on the means of distinguishing
between the most commonly encountered mechanisms, outlined in Eqs. 16 and 19.

Reaction   Reaction Type

$$R^1CX = CYH \xrightarrow[Nu^-]{-HX} R^1C \equiv CY \xrightarrow{HNu}$$

$$\rightarrow R^1CNu = CHY \quad E_{C=C}\text{-Ad} \quad (16)$$

$$\rightarrow R^1CH = CNuY \quad E_{C=C}AdwR$$

$$HXC = CYR^1 \xrightarrow[Nu^-]{-HX} :C = CYR^1 \xrightarrow{NuH}$$

$$HNuC = CYR^1 \quad \alpha\text{-E} - \text{Ad} \quad (17)$$

$$R^1R^2CHCX = CYR^3 \xrightarrow[Nu^-]{-HX} R^1R^2C=C=CYR^3 \xrightarrow{NuH}$$

(18)

$$\rightarrow R^1R^2CHNu=CYR^3 \quad E\text{-Ad}$$

$$\rightarrow R^1R^2C=CNuCHYR^3 \quad E\text{-AdwR}$$

$$\rightarrow R^1R^2CNuCH=CYR^3 \quad E\text{-AdwR}$$

$$R^1CX=CYR^2 \xrightarrow{Nu^-} R^1CNuX-\bar{C}YR^2 \xrightarrow{-\bar{X}} R^1CNu=CYR^2 \quad (19)$$

Ad via
direct substitution

$$\xrightarrow{H^+} R^1CNuXCHYR^2 \xrightarrow{-HX}$$

$$R^1CNu=CYR^2 \quad Ad\text{-E}$$

$$R^1CX=CYH \xrightarrow{NuH} R^1CHXCHNuY \xrightarrow{-HX} R^1CH=CNuY \quad Ad\text{-EwR} \quad (20)$$

$$R^1CX=CR^2R^3 \xrightarrow{-\bar{X}} R^1\overset{+}{C} = CR^2R^3 \xrightarrow{Nu^-} R^1CNu = CR^2R^3$$

$$S_N1(E_{C=C}1) \quad (21)$$

Reaction   Reaction Type

$$R^1 R^2 CHCX = CYR^3 \longrightarrow R^1 R^2 C=CXCHYR^3 \xrightarrow{Nu^-}$$

$$R^1 R^2 C=CNuCHYR^3 \quad \text{P-S} \quad (22)$$

$$\downarrow$$

$$R^1 R^2 CHCNu=CYR^3 \quad \text{P-S-P}$$

$$R^1 CX=CYCHR^2 R^3 \longrightarrow R^1 CHXCY=CR^2 R^3 \xrightarrow{Nu^-}$$

$$R^1 CHNuCY=CR^2 R^3 \quad \text{P-S-} \quad (23)$$

$$\downarrow \qquad\qquad \text{allylic}$$

$$R^1 CNu=CYCHR^2 R^3 \quad \text{P-S-}$$

allylic-P

$$R^1 CX=CR^2 CR^3 R^4 Y \xrightarrow[-Y^-]{Nu^-} R^1 CNuXCR^2 =CR^3 R^4 \xrightarrow[-X^-]{+\overline{Y}}$$

$$RCNu=CR^2 CR^3 R^4 Y \quad S_N 2' - S_N 2'$$

(24)

$$R^1 CX=CR^2 R^3 \longrightarrow R^1 \overline{C}=CR^2 R^3 \xrightarrow{\overset{+}{E}} R^1 CE=CR^2 R^3$$

S via vinylic
carbanion   (25)

The symbols used in Eqs. 16 to 25 are as follows:

Nu^-	= nucleophile or base
X	= halogen atom
Y	= electron-withdrawing activating group
E	= olefin forming β-elimination
$E_{C=C}$	= alkyne forming β-elimination
α-E	= α-elimination
Ad	= addition to either alkene or alkyne
S	= substitution
P	= prototropic rearrangement
wR	= with rearrangement - refers to change in the relative positions of X and Nu in the reactant and final product, respectively

(c) <u>Kinetic Differences between the Ad-E and $E_{C=C}$-Ad</u>
<u>Routes</u>. Maioli and Modena have proposed three extremes
for the Ad-E mechanism (see Scheme 3).[85] In 21, bond
breaking and making occur simultaneously, and the reaction is analogous to the S_N2 mechanism for saturated

Scheme 3

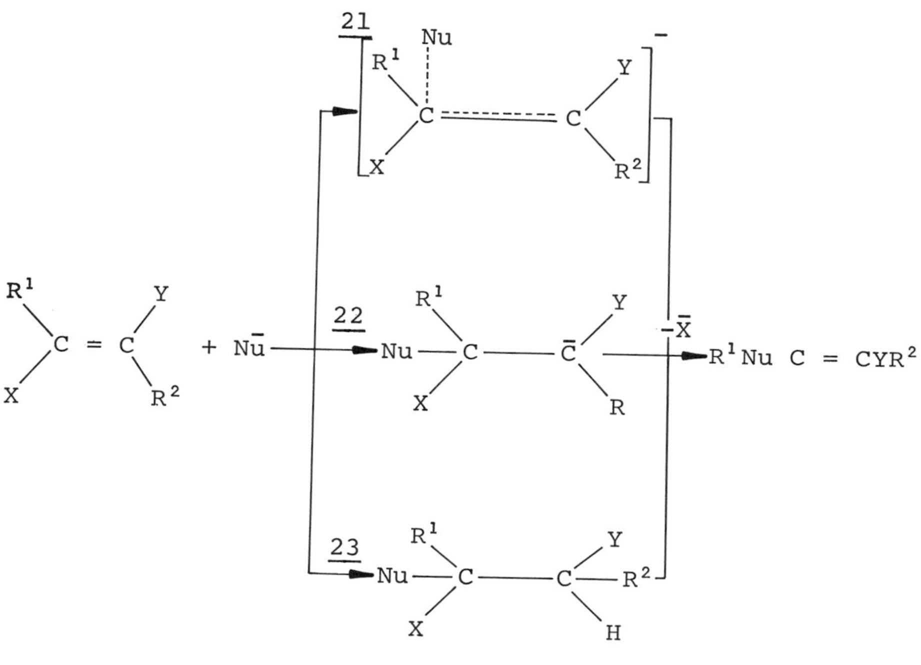

Y = activating group.

substrates. In 22, bond formation is considerably
ahead of bond breaking, and a tetrahedral carbanion intermediate in involved. In the third, 23 a real addition intermediate is formed that undergoes an elimination. This latter route has been characterized as
operable in the substitution-with-rearrangement route
illustrated in Eq. 20, but with the activating and
leaving groups on the same vinylic carbon in the substrate.[82] Of the three routes, the experimental observations usually accord with 22.
      Much of the experimental evidence for the Ad-E
mechanism has been obtained with substrates lacking

vinylic β-hydrogen, which cannot undergo the alternative $E_{C=C}$-Ad sequence. The so-called "element
effect"   has been used to provide useful information
on the role of the leaving group in the rate-determining step. The rates of reaction of compounds differing
only in the leaving group are compared. Along the
series C-F, C-Cl, C-Br, bond strength decreases markedly. Thus, if considerable C-X bond breaking occurs in
the slow step, large values of k_{Cl}/k_F and k_{Br}/k_{Cl} are
anticipated. On the other hand, if an addition intermediate is formed in the rate-determining step, the
substrate possessing the most polarized and least
sterically hindered double bond will be the most reactive. Both steric size, increasing in the order F <
Cl < Br and electronegativity decreasing in the order
F >> Cl ≈ Br, act in the same direction, namely F > Cl
≈ Br in activating properties. Therefore, for rate-
determining formation of the addition intermediate,
k_F/k_{Cl} should be large and k_{Cl}/k_{Br} should be about
unity.
      Clearly the carbanion intermediate is formed in
the slow step in the reactions shown in Eqs. 26 and 27.
In both cases k_F/k_{Cl} values are large, and an additional value of k_{Br}/k_{Cl} approximates to unity. In the
latter reaction, diphenylacetylene is also formed when

$$PhCOCMe = CHX \xrightarrow[C_2H_5OH,DMF]{} PhCOCMe = CHNC_5H_{10} + \bar{X}$$

$$\tag{26}$$

$$k_F/k_{Cl} = 204 (C_2H_5OH); 63 (DMF) \text{ at } 30° [86]$$

$$Ph_2C = CHX + \bar{O}C_2H_5 \longrightarrow Ph_2C=CHOEt + \bar{X} \tag{27}$$

$$k_F/k_{Cl} = 290 (100°) [87,88]; k_{Br}/k_{Cl} = 1.4 (120°) [88]$$

X = Cl(9%) and X=Br (60%), by an α-elimination with
rearrangement. (See the Fritsch-Buttenberg-Wiechell
rearrangement in Section X.B.3.b.iii.) In this reaction, an initial rapid equilibrium ionization of the
α-hydrogen is followed by a rate-determining fragmentation of the C-X bond from the vinyl carbanion. Not

surprisingly a k_{Br}/k_{Cl} ratio of 20 (120°) is observed.[89]

The isolation of significant quantities of the intermediate acetylene can be used to demonstrate the $E_{C\equiv C}$-Ad route. However, often the alkynes decompose more rapidly than they are formed. Two extremes of mechanism can be envisaged which, like all those of Scheme 3, should exhibit second-order kinetics (see Scheme 4). Of the five routes in Schemes 3 and 4, only <u>24</u> requires that isotopic exchange of β-vinylic hydrogen be a rapid process compared with halide ion loss from the starting material. However, observation of isotopic exchange with the solvent does not confirm the operation of an $E_{C\equiv C}$1cB reaction. It could

Scheme 4

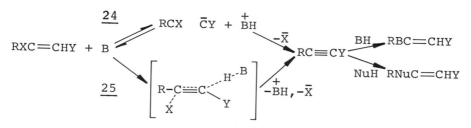

represent an irrelevant side reaction to an Ad-E sequence. In both the $E_{C\equiv C}$-Ad routes the β-hydrogen is removed to yield the acetylene, and hydrogen is then reintroduced from the solvent. Thus, for reactions in deuterated solvents, isotope incorporation into the vinylic β-position of the apparent substitution product is expected. Only route <u>25</u> should exhibit a primary kinetic hydrogen isotope effect. Finally the $E_{C\equiv C}$-Ad route should be characterized by large values of k_{Br}/k_{Cl}.

In addition to kinetic methods, stereochemical studies can provide a useful guide to mechanism. For a cis-trans alkene pair, the stereochemical outcome of the apparent substitution can be described as occurring with retention, (<u>26</u>→<u>27</u>, <u>28</u>→<u>29</u>), inversion (<u>26</u>→<u>29</u>, <u>28</u>→<u>27</u>), or with a mixture of the two (<u>26</u>→<u>27</u> + <u>29</u>, <u>28</u>→<u>27</u> + <u>29</u>).

In the $E_{C\equiv C}$-Ad route, the same alkyne intermediate is formed from both isomeric alkenes. Thus, as the stereochemistry of the product is determined in the subsequent addition reaction of the alkyne, both

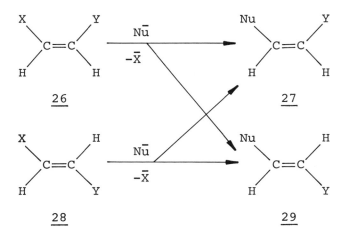

26   27

28   29

reactants will give rise to the same product mixture.
Normally nucleophilic additions of thioanions and, in
most instances, alkoxides to alkynes occur with anti
stereospecificity.[90]   Consequently, cis-alkene 26
should yield the substitution product 27 with retention
of configuration, but for the trans-alkene 28, this
product is one of inversion.   The addition of amines to
alkynes is more complex, and generalizations cannot be
made with any degree of certainty.[91]
        Substrates reacting by the Ad-E mechanism usually
give products with retention of configuration (see
Ref. 82 for an extensive compilation).   This stereochemical consequence is usually explained in terms of
a reaction via a short-lived carbanion intermediate
(cf. 22), in which the elimination of X̄ is more rapid
than internal rotation.   The lifetime of these carbanions is presumed long enough to allow bond rotation
to proceed through 60°, thereby attaining the stereochemical arrangement for expulsion of X̄ to give the
retention product.   For the more stable carbanion intermediates, bond rotation can attain 120°, resulting
in elimination of X̄ to give the reaction product with
inversion of configuration (see Scheme 5).[82,92-94]
        The work of Modena and his co-workers[95-97] pro-
vides a clear demonstration of the usefulness of kinetic and stereochemical methods to differentiate the
$E_{C=C}$-Ad and Ad-E mechanisms.   It also illustrates some
of the pitfalls and limitations of individual techniques.
        The element effects for the reactions of some sub-
stituted α-arylsulfonyl-β-halogenoethylenes with

Scheme 5

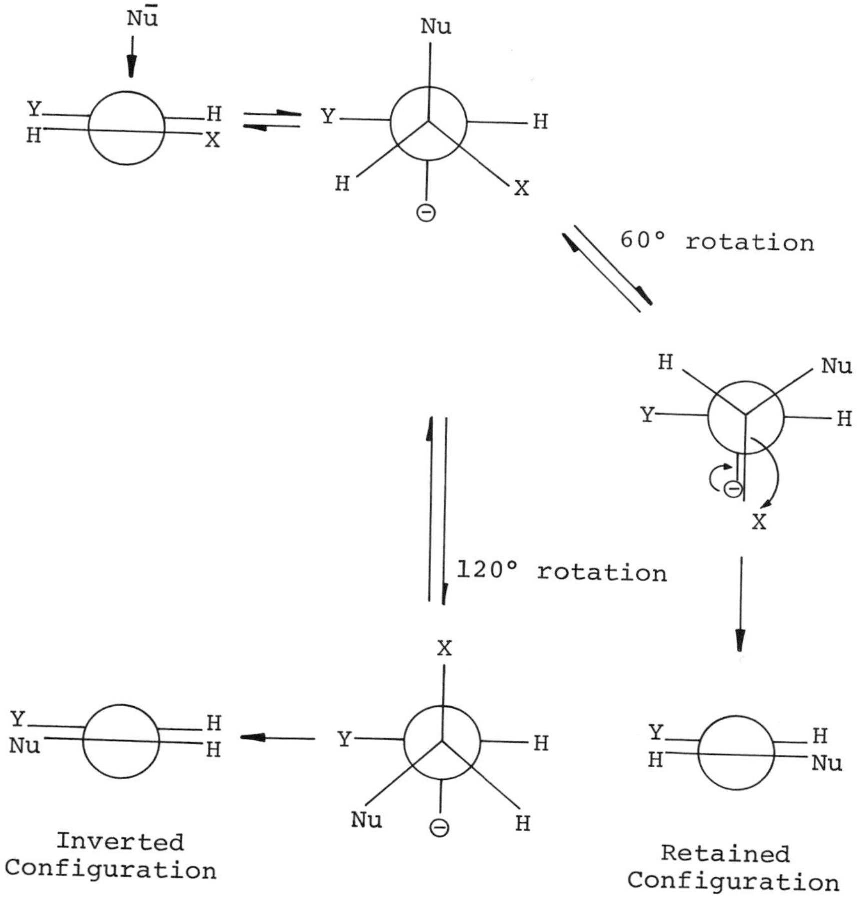

Inverted
Configuration

Retained
Configuration

methoxide and thiophenoxide ion in methanol are listed in Table 2. Compound 32 can only react by the Ad-E mechanism, and the k_{Br}/k_{Cl} ratios serve as a calibration of the magnitudes expected for this reaction. For both anions, the values are small, indicating slow formation of the intermediate carbanion. Similar small ratios are obtained for the reactions of both 30 and 31 with thiophenoxide, suggesting an Ad-E mechanism operates for these reactions also. The lack of isotopic exchange when the reaction is carried in CH_3OD,

TABLE 2 Element Effect (k_{Br}/k_{Cl}) for the Reaction of cis- and trans-
p-$R^1C_6H_4SO_2R^2C = CR^3X$ with Various Nucleophiles at 0° in Methanol

Nucleophile	R^1	R^2	R^3	Substrate Type	k_{Br}/k_{Cl} cis	k_{Br}/k_{Cl} trans	Reference
MeO⁻	p-NO$_2$	Me	H	32	0.93	0.85[a]	95
	p-Me	H	H	$\overline{30}$	185	0.84	96
	H	H	H	$\overline{30}$	144	0.84	96
	p-NO$_2$	H	Me	$\overline{31}$	109[a]	–	97
PhS⁻	p-NO$_2$	Me	H	32	2.6[a]	2.3[a]	95
	p-Me	H	H	$\overline{30}$	2.3	2.15	96
	H	H	H	$\overline{30}$	2.4	2.2	96
	p-NO$_2$	H	Me	$\overline{31}$	2.4	–	97

[a]At 25°.

$ArSO_2 CH = CHX$ $ArSO_2 CH = CMeX$ $ArSO_2 C(Me)=CHX$

<u>30</u>   <u>31</u>   <u>32</u>

and the formation of products with retention support this assignment.

The reaction with the more basic methoxide ion is more complex. For the trans-isomer of <u>30</u>, the k_{Br}/k_{Cl} ratio is similar to that for <u>32</u>, and the retention product is formed, both observations supporting the Ad-E mechanism. However, the exchange of β-vinylic hydrogen with labeled solvent occurs 75 times faster than chloride ion production. For the cis-isomers of <u>30</u> and <u>31</u>, the large values of k_{Br}/k_{Cl} for the methoxide ion reaction indicate considerable C-X bond breaking in the rate-determining step. Not surprisingly, both of these isomers exchange β-vinylic hydrogen with labeled solvent.[98] In both cases, the substitution products are those with retention of configuration, but here both the $E_{C=C}$-Ad and Ad-E routes predict the same stereochemical result. The absence of an isotope effect and the observation of rapid exchange [k(exchange/k(subs) = 50 (<u>30</u> - Cl)] suggest the choice of mechanism rests between an $E_{C=C}$lcB-Ad sequence and an Ad-E sequence competing alongside an irrelevant "parasitic" exchange reaction. The only method of solving this problem is to search for the intermediate alkyne.

Under suitable conditions, the alkyne is the main reaction product for the reaction of the cis-bromide <u>30</u>. However, for the chloro substrate the alkyne can only be detected in low concentration during the course of the reaction by infrared spectroscopy. By quenching the reaction at various times and monitoring the change in the alkyne concentration, Modena and his co-workers[99] were able to dissect the total rate of reaction into the $E_{C=C}$lcB-Ad and Ad-E components. At 0°C these routes make similar contributions, but at higher temperatures the $E_{C=C}$lcB-Ad route gains in importance [k(elim)/k(sub) = 3.0 at 25° for cis-Cl-<u>30</u>].

The conclusions from this work are as follows: (a) all the thiophenoxide reactions are of the Ad-E type; (b) the trans-<u>30</u>, <u>31</u> react with methoxide ion via the Ad-E route, with an accompanying inconsequential exchange reaction; and (c) the cis-<u>30</u>, <u>31</u> react by both the Ad-E and $E_{C=C}$lcB-Ad mechanisms with methoxide ion. The lack of elimination from the carbanion of

trans-30 and 31 presumably reflects unfavorable syn-
stereochemical arrangement of the leaving groups in the
reactant. The cis-substrates, 30 and 31 can undergo
an anti elimination.

Two clear lessons emerge from this work: (a) it
is unsound to draw mechanistic conclusions from only
one source of kinetic or stereochemical information;
and (b) rapid protium exchange alone should not be
taken as evidence for the $E_{C=C}$lcB mechanism.

Unactivated vinyl halides generally react by the
elimination (-addition) route. For the activated vinyl
halides, strong bases (affinity for hydrogen) such as
alkoxides usually induce the $E_{C=C}$-Ad sequence whereas
more polarizable nucleophiles (affinity for carbon)
such as thioanions, azides, amines, and halides prefer
to attack carbon rather than hydrogen, causing the
Ad-E sequence. With constant basicity, increased
steric size should favor attack at the more exposed
vinylic hydrogen rather than the more shielded carbon.

3 Metal Alkoxide Induced Dehydrohalogenations

Compared to the wealth of information available on
mechanistic aspects of alkene-forming eliminations,
alkyne-forming reactions have received scant attention.
Miller and Noyes[100] measured the rates of elimination
of 1,2- cis- and trans-dihalogenoalkenes with methoxide
ion in methanol. The rate of the reaction increased
markedly along the series Cl→Br→I and in all cases
the cis-isomers, which can eliminate with anti-stereospecificity, were converted into the halogenoacetylene
much more rapidly than were the corresponding trans-
isomers (see Table 3). The greater reactivity of the
cis-dichloro substrate resides in the entropy of
activation terms, whereas the enhanced reactivity of
the cis-diodo substrate, relative to its trans-isomer,
arises mainly from a lower activation energy. In fact,
for elimination from the cis-substrates the activation
energy decreases markedly along the series Cl→Br →I,
whereas for the trans-compounds the reverse sequence
is observed. Being aware of the preference for anti-
stereospecificity in the base-induced dehydrohalogenation of hexachlorocyclohexanes,[101] Miller and Noyes
proposed that the dramatic differences in activation
parameters reflected the operation of two mechanisms.
They suggested that the cis-substrates eliminate by a
concerted $E_{C=C}2$ mechanism, possessing the required
anti orientation of the leaving groups. However, they
preferred the alternative $E_{C=C}$lcB mechanism to explain

TABLE 3 Activation Parameters for Elimination from 1,2-Dihalogenoalkenes with Sodium Methoxide in Methanol[100]

Substrate		E_A (kcal/mole)	$\Delta S^{\ddagger a}$ [cal/(mole)(deg)]	k_{cis}/k_{trans}	k_{cis} [l./(mole)(sec)]
$C_2H_2Cl_2$	Cis	35.1 ± 0.3	22 ± 1	3.3 x 10^3	1.4 x 10^{-5}
	Trans	29.0 ± 1.0	-12 ± 3		
$C_2H_2Br_2$	Cis	28.1 ± 0.3	16 ± 1	5.4 x 10^5	1.8 x 10^{-2}
	Trans	33.4 ± 0.5	5 ± 2		
$C_2H_2I_2$	Cis	24.7 ± 0.3	14 ± 1	1.3 x 10^5	1.6

^{a}$\Delta S^{\ddagger}$, and the rate coefficients refer to 60°C.

the apparent syn elimination from the trans-compounds. The slow step could be either unimolecular decomposition of the carbanion or its inversion into the epimeric carbanion in which the configuration of the electron pair and leaving groups are correctly positioned for a rapid anti elimination.

Subsequently Miller and Lee found that vinylic protium exchange with the solvent occurs 25 times more rapidly than anti elimination at 26°, and the kinetic isotope effect $[k(C_2H_2Br_2/k(C_2D_2Br_2)]$ is only 1.03 at 35° for the reaction of cis- 1,2-dibromoethylene with sodium methoxide in methanol.[102] Thus even the cis-substrate appears to react by the $E_{C=C}$lcB route. Cis-dichloro- and dibromoethylenes were found to exchange vinylic protium at a similar rate, pK_a values of about 34 being estimated for these compounds. Exchange is slower for the trans-derivatives and decreases along the series, ClCH = CHCl > BrCH=CHBr > ICH=CHI. Increased electronegativity of the α-halogen and a β-halogen, which is trans- rather than cis-related to the ionizing C-H bond, therefore, enhance vinylic-hydrogen acidity. For the chloro substrates, exchange is not accompanied by loss of configuration, and the small configurational change observed for the bromides was attributed to the incursion of radical processes. This result indicates that vinylic carbanions have much greater configurational stability than saturated carbanions. The barrier to inversion in the related diimide system (HN = NH) has been estimated as 33 kcal/mole.[103] The greater influence of trans- rather than cis-related halogens on the acidity of a vinylic proton indicates greater interaction between the halogen and the electron pair in this configuration. Not surprisingly, this increased interaction leads to faster anti than syn elimination. Thus the exchange and stereochemical studies reinforce each other.

Trichloroethylene undergoes hydrogen exchange, in the absence of elimination, with calcium deuteroxide. In this instance, the second β-halogen enhances C-H acidity but retards unimolecular ejection of the other β-halogen from the carbanion.[104,105]

Unlike the haloalkenes, di- and triarylethenes do not exchange protium with sodium methoxide in methanol.[102] Thus carbanions derived by the ionization of sp^2-hybridized C-H bonds appear to be stabilized more effectively by α-halogen than α-aryl substituents.

Modena and his co-workers have studied the reactions of sulfonyl,[95-99] keto,[106] nitrophenyl,[108,109] and phenyl[108,109] halogenoethylenes with sodium methoxide in methanol and, in addition, potassium t-butoxide in t-butyl alcohol for the latter two series.[107] From their results, they suggest a spectrum of transition states is available for elimination, along the lines discussed in Chapter II for olefin-forming elimination.

As listed above, the activating groups are in or-der of decreasing electron-withdrawing power. The mechanisms of reaction of the sulfonyl derivatives 30 are discussed in the preceeding section. The trans-isomer reacts by the Ad-E route, but the cis-form fol-lows mainly on $E_{C=C}$lcB-Ad mechanism. A similar de-pendency of mechanism on stereochemical arrangement of the potential eliminating fragments is found for the keto-derivatives. Both substrates are converted into the dimethoxy product 33 (Eq. 26, see Table 4). For the trans-substrates, which cannot eliminate with anti-stereospecificity, no primary kinetic isotope effect or exchange of hydrogen with the solvent is observed. The small ratio of k_{Br}/k_{Cl} is consistent with the Ad-E route. For both the cis-compounds, a small primary kinetic isotope effect is observed, and the k_{Br}/k_{Cl}

$$ArCOCH=CHX + 2Me\bar{O} \xrightarrow{MeOH} ArCOCH\equiv CH(OMe) + \bar{X} \xrightarrow{MeOH}$$

$$ArCOCH_2-CH(OMe)_2 \quad (26)$$

33

ratio indicates a rate-determining breakage of the C-X bond. Infrared analysis of the reaction mixture revealed an alkyne intermediate. Thus, when anti elimination is possible, the $E_{C=C}$-Ad mechanism operates in preference to the Ad-E sequence. The lack of hydro-gen exchange with the solvent for the bromide suggests an $E_{C=C}2$ mechanism. However, for the chloride, the exchange process occurs to the extent of 10% at 48% de-composition, and the low isotope effect of 1.4 most likely reflects two simultaneous reactions; an $E_{C=C}2$ mechanism in which the proton is extensively trans-ferred to the base and an $E_{C=C}$lcB mechanism.

The reactions of cis-p-nitrophenyl and phenyl halogenoalkenes (Table 4) in both solvent systems ex-hibit measurable primary isotope effects, and signifi-cant positive element effects. In the absence of

TABLE 4 Hydrogen Isotope Effects and Element Effects in the Reaction of Some Activated Vinyl Halides with Alkoxide Ions

Compound	Base	k_H/k_D	k_{Br}/k_{Cl}*	Reference	$10^3 k_H^c$ [1/(mole)(sec)]
cis-4-OMeC$_6$H$_4$,COCD=CHBr	$\bar{O}$Me/MeOH	2.0±0.1	47[a]	106	—
cis-4-OMeC$_6$H$_4$,COCD=CHCl	$\bar{O}$Me/MeOH	1.4±0.1		106	—
trans-4-OMeC$_6$H$_4$COCD=CHBr	$\bar{O}$Me/MeOH	1.0	0.65[a]	106	—
trans-4-OMeC$_6$H$_4$COCD=CHCl	$\bar{O}$Me/MeOH	1.0		106	—
cis-C$_6$H$_5$CD=CHBr	$\bar{O}$Me/MeOH	2.9[b]	310[c]	107	0.0072
	$\bar{O}$Bu^t/Bu^{t}OH	4.6[c]	121[c]	108, 109,	23.0
cis-C$_6$H$_5$CD=CHCl	$\bar{O}$Me/MeOH	2.2[d]		107	0.000023
	$\bar{O}$Bu^t/Bu^{t}OH	3.2[c]		108, 109,	0.19
cis-4-NO$_2$C$_6$H$_4$CD=CHBr	$\bar{O}$Me/MeOH	2.2[e]	185[c]	107,	1.4
	$\bar{O}$Bu^t/Bu^{t}OH	4.3	38[c]	108, 109,	8300

cis-4-					
$NO_2C_6H_4CD=CHCl$	$\bar{O}Me/MeOH$	1.6		107	
	$\bar{O}Bu^t/Bu^tOH$	3.5		107, 109,	220
					0.0075
trans-4-					
$NO_2C_6H_4CD=CHBr$	$\bar{O}Bu^t/Bu^tOH$	1.8	9.0[c]	109	2.75
trans-4-					
$NO_2C_6H_4CD=CHCl$	$\bar{O}Bu^t/Bu^tOH$	1.6		109	0.31

*k_{Br}/k_{Cl} ratios refer to rates of the hydrogen substituted compounds.

a 0°C;
b 50°C;
c 30°C;
d 78.2°C;
e 45°C.

hydrogen exchange, these reactions appear to be of the $E_{C=C}2$ category. The kinetic ratios, however, do vary quite markedly and in a consistent manner. The following generalizations can be made: (a) the reactions occur much more rapidly in the stronger basic system; (b) the element effects are larger in methanol than in t-butyl alcohol and greater for the phenyl than p-nitrophenyl series; (c) the isotope effects are always smaller for the chlorides than the bromides under common conditions, and smaller in methanol than in t-butyl alcohol; and (d) anti elimination occurs much more rapidly than syn elimination and is accompanied by a larger isotope effect.

All the isotope effects are much less than the theoretically predicted maximum of between 6 and 8 for the temperature range studied (Section II.C.3). The small values could reflect either little or considerable proton transfer to the base in the transition state, but more probably the proton transfer is past the midpoint. In the corresponding eliminations from 2-phenylethyl halides, kinetic studies reveal proton transfer is about halfway for the bromide (in ethanol, $\bar{O}C_2H_5$) and is more extensive for the chloride. As vinyl C-X bonds are stronger than alkyl C-X bonds, greater assistance from the developing electron pair on the C_β-atom is required to cause fragmentation. Thus more complete proton transfer to the base is anticipated in the vinyl halide reactions. In addition, the keto-halogenoethylenes react partly by an $E_{C=C}1cB$ mechanism.

Accepting that the proton transfer to the base is past the midpoint, let us now consider the influence of basic strength. The higher isotope effect is in accord with the proton transfer having proceeded to a lesser degree in t-butyl alcohol than in methanol (Section II. C.3). Comparison of the element effect in the two solvents suggests that C-X bond breaking also is less progressed in t-butyl alcohol.

As both C-H and C-X bond breaking are reduced in t-butyl alcohol relative to methanol, an increase in carbanion character in the more basic system is only possible if C-X bond breaking is reduced to a greater extent. The substituent ratios (k_{nitro}/k_H) confirm this possibility. Greater values indicate more extensive carbanion character, the strongly electron-withdrawing nitro-group stabilizing the transition state more effectively than p-H in the unsubstituted phenyl group. For the chloride, k_{nitro}/k_H values are 1160 (t-butyl alcohol) and 325 (methanol) and for the

bromide, they are 360 (t-butyl alcohol) and 195 (methanol) at 30°C. These changes are much larger than
would be expected for a simple solvent effect on the
substituent influence.[110] The lower isotope effects
for the chlorides than bromides suggest that greater
carbanion character is associated with the transition
for elimination of the poorer leaving groups. The
higher k_{nitro}/k_H values for the chlorides than bromides
in both solvents accord with this conclusion.

Finally let us consider elimination from the
trans-p-nitrophenyl-β-halogenoethylenes in t-butyl alcohol. The primary kinetic isotope effects for both of
these halogen derivatives are very much smaller than
the value observed for the corresponding cis-isomers.
Likewise the k_{Br}/k_{Cl} values of only 9 is much smaller
than that of 38 observed for the corresponding cis-
isomers. These figures suggest that C-H bond breaking is more extensive and C-X bond breaking is less
extensive in these apparent syn eliminations than in
the equivalent anti eliminations. This result is not
surprising as weaker orbital interaction for syn-relat-
ed than anti-related groups necessitates greater charge
development at the β-carbon to exert the same electronic influence in aiding C_α-X bond breaking. The apparent syn eliminations are much slower than the reactions
of the cis-isomers, and almost total exchange of the
hydrogen alpha to the halogen atom was observed at only
40% completion of reaction. Consequently, the role of
an α-elimination with aryl migration cannot be excluded at this stage (Section X.B.3.b). In the shorter reaction times required for elimination from the cis-substrates, no exchange of either vinylic hydrogen with
the solvent was observed.

Marchase, Naso, and Modena[111] have reported the
reactions of cis- and trans- 34 with both thiophenoxide
and methoxide. Unlike the chloro and bromo derivatives, both of these fluorides are converted into the
thermodynamically more stable substitution product.
No hydrogen exchange or hydrogen isotope effect is
observed, and these observations accord with an Ad-E
mechanism in which the intermediate carbanion possesses
sufficient stability to allow internal rotation to
proceed to 180°, thereby permitting thermodynamic
rather than kinetic control of its decomposition to the
substitution derivative.

$$p-NO_2.C_6H_4.CH=CHF$$

34

Cristol and his co-workers have compared the rates of conversion of trans-1,2-diphenyl-1-chloroethylene, and 1,1-diphenyl-2-chloroethylene into diphenylacetylene by a variety of alkoxide-alcoholic media.[112-114] Their results clearly show that anti elimination occurs more easily than syn elimination, although the ratio k_{anti}/k_{syn} varies quite markedly with the reaction conditions. α-Elimination is much less favorable than either form of β-elimination in protic solvents. However, these generalizations do not hold true for aprotic solvents of low dielectric constant, such as ether, as we see in the next section.

4   Metal Alkyl-Induced Dehydrohalogenations

(a)   Metalated Halogenoalkenes as Intermediates. The carbanions, which have been proposed as intermediates in dehydrohalogenations of vinyl halides in protic solvents, have actually been isolated as their lithium salts in ethereal solution.[74,115] Most frequently the metalating reagent is either n-butyllithium or the slightly less reactive phenyllithium, and the reactions are accomplished in diethyl ether or tetrahydrofuran at reduced temperatures (-50 to -100°C). Metal-hydrogen exchange occurs most easily with hydrogen atoms situated alpha to a halogen, and all the simple chloroethylenes form lithiated derivatives, the stability of these increasing along the series 35 → 38 in tetrahydrofuran.[116] The order of stability of these derivatives depends on the stereochemical relationship between a β-halogen and the lithium atom in the carbanion, and as in the dehalogenations in protic solvents, reflects the fact that anti β-elimination occurs more easily than syn β-elimination or α-elimination. An interesting observation is the greater stability of

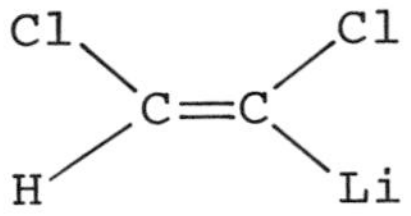

35 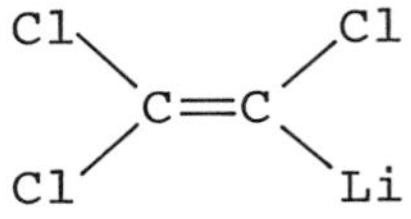   <   36

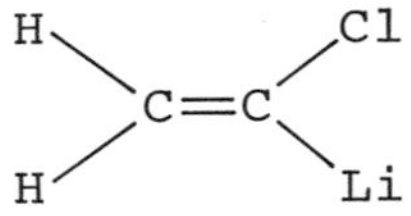

37 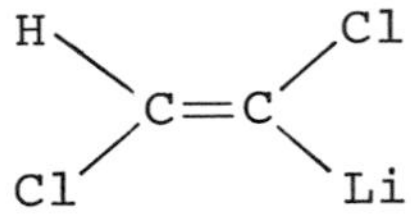   <   38

<u>38</u> than <u>37</u>, indicating apparent α-elimination is easier than syn elimination. We return to this point in the next section.

Several examples have been reported in which syn elimination appears to occur more rapidly than anti elimination.[74] The word "appears" is used as, in some cases, the reactions do not occur via the simple β-elimination route. In contrast to the reactions in protic solvents, 1,1-diphenyl-2-chloroethylene, is converted into diphenylacetylene much more rapidly than are the stilbene chlorides by phenyllithium in ether.[112] Of the stilbenes, the trans-isomer (syn elimination) is the more reactive. The greater facility of α-elimination is attributed to the greater acidity of vinylic hydrogen alpha to a halogen than to an aryl substituent. The involvement of a cyclic ion-pair mechanism is the most obvious explanation for the greater reactivity of trans- than cis-chlorostilbene (<u>39</u>).

<u>39</u>

The reaction of the isomeric β-halogenostyrenes with lithium phenyl or lithium alkyls in ethereal solvents provides another example of apparent syn elimination being preferred to anti elimination.[117] However, as these substrates possess a hydrogen alpha to the halogen, alternative mechanisms to β-elimination must also be considered. In Scheme 6 some of the possibilities are listed. The cis-isomer cannot decompose via α-elimination with rearrangement as the aryl group is incorrectly situated relative to the halogen, to exert a displacement by an internal backside attack. The incursion of α-elimination or the operation of an ion-pair mechanism could explain the greater reactivity of the trans-bromide. Certainly the hydrogen alpha to the halogen plays a major role in the reaction, for β-methyl-β-chlorostyrene reacts only partially with phenyllithium in ether during several days, whereas the β-chlorostyrenes[118] under the same conditions are consumed rapidly.[118]

Scheme 6

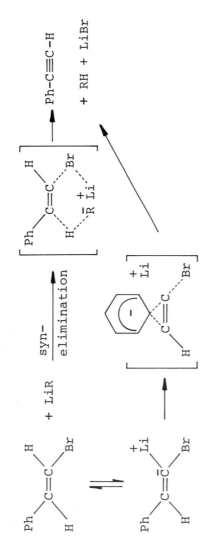

(b) The $E_{C=C}2cB$ Mechanism. The work of Schlosser and

Ladenberger[119,120] excludes the simple β-elimination
mechanisms and the α-elimination with rearrangement of
β-chlorostyrene into phenylacetylene by phenyl and
alkyl lithiums in ethereal solvents. These authors
propose an alternative route, which is called the
$E_{C=C}2cB$ mechanism (Eq. 27). The lithium salt of phenylacetylene is formed directly in an elimination step
and not via phenyl acetylene. At low temperatures in

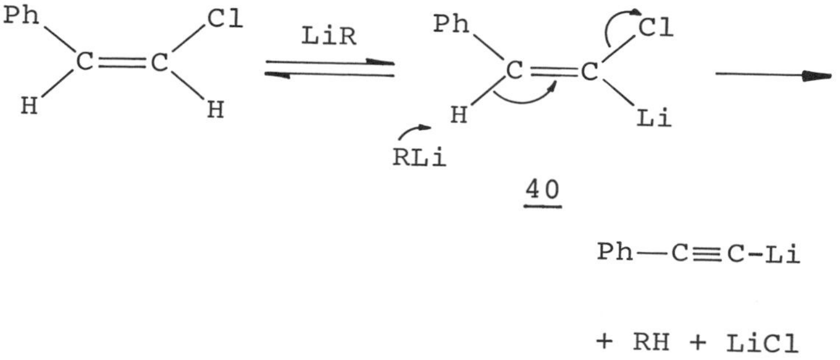

$$Ph-C{\equiv}C-Li$$

$$(27)$$

$$+ RH + LiCl$$

tetrahydrofuran the lithiated intermediate 40 can be
isolated. Thermal decomposition follows second-order
kinetics in the intermediate but, in the presence of
stronger bases, a first-order dependence, in both the
intermediate and the added base, is observed. Thus the
thermal decomposition involves one intermediate acting
as the base on a second molecule. The thermal decomposition of 40 occurs readily at -110°C and is accompanied by an isotope effect when the hydrogen is replaced by deuterium. The lithiated intermediate from
trans-β-chlorostyrene, which must eliminate with syn-
stereospecificity, is more stable than 40, decomposing
at-60°C.
     At 0°C, the lithiated intermediates decompose too
rapidly to be detected, and abstraction of the hydrogen, which is alpha to the halogen, becomes rate determining, as is clearly demonstrated by the observation
of primary isotope effects in ether, using phenyl lithium as the base (see 41 - 44). A tunneling phenomenon
was suggested to explain the large isotope effect in
the reaction of 42. This phenomenon was attributed to
steric inhibition to approach of the base by the cis-
positioned phenyl group. In 41, the phenyl substituent

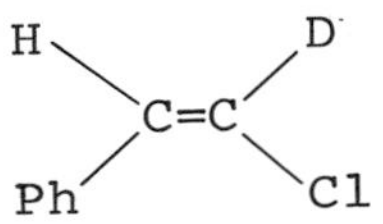

H—C=C—D (Ph, Cl) **41**   k$_H$/k$_D$ 8.1

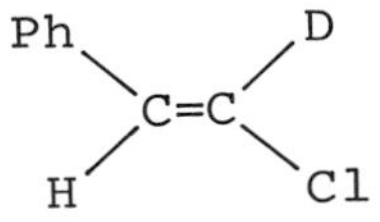

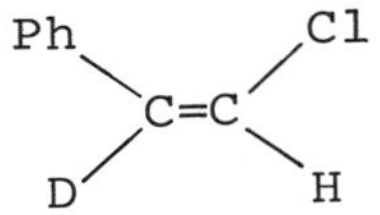

Ph—C=C—D (H, Cl) **42**   15.3

Ph—C=C—Cl (D, H) **43**   1.00

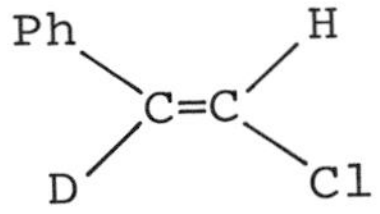

Ph—C=C—H (D, Cl) **44**   1.04

does not hinder the approach of the base, and the isotope effect, which is near the maximum predicted, on
theoretical grounds suggests the proton is half-transferred to the base in the transition state (see Section
II.C.3 for further discussion of the role of tunneling
in proton transfers).[21]

The E$_{C=C}$2cB mechanism affords an explanation of
the greater instability of 37 than 38.[115] Intermediate
37 can eliminate with anti stereospecificity (see 45),
whereas 38 is confined to syn elimination. Thus, despite trans-1,2-dichloroethylene being more acidic than
vinyl chloride, it eliminates to the alkyne more slowly.

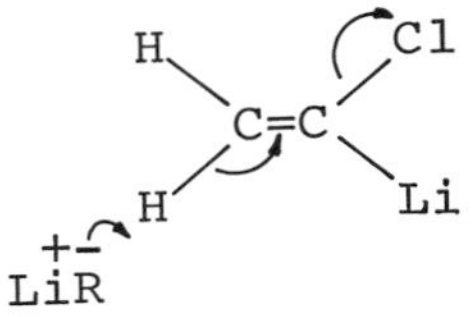

H—C=C—(Cl, Li), H, + LiR   **45**

The lesson to be learned from Schlosser and
Ladenbergers' work is that interpretation of the
stereochemistry of a reaction from a knowledge of the
configuration of the reactants and the structure of the
products should be made with considerable caution.

5 Amine-Induced Dehydrohalogenations

In a recent report, Kwok, Lee, and Miller have

described the kinetics of the reaction of cis-dibromo-
ethylene and the uncharged base, triethylamine in dimethyl formamide.[122]   Neither protium exchange with
labeled solvent nor a hydrogen isotope effect is observed.   An $E_{C=C}lcB$ - ion pair mechanism was proposed
in which an intimate ion pair collapses to reactants
or products without equilibration with the solvent
molecules.

6 Cyclic Alkyne Formation

Cycloalkynes can be prepared by dehydrohalogenation of
1-halocycloalkenes with bases,[123,124] dehalogenation of
1,2-cycloalkenyl dihalides with sodium,[125] or with
magnesium in ether,[126] oxidation of dihydrazones of
cyclodiones,[127] and photolytic decomposition of tria-
zole derivatives.[128]   Ring strain imposes considerable
instability on the cycloalkynes with less than nine
carbon atoms,[129] and indirect methods must be used to
prove their intermediacy.[74]   Labeling techniques have
been used to demonstrate the involvement of a symmetrical intermediate in the reaction of 1-chlorocyclo-
hexene (or 1-chlorocyclopentene) with phenyl lithium
in ether (Eq. 28).[130]   Alternatively, the trapping of
an intermediate alkyne as a Diels- Alder adduct can be
used to imply its existence (Eq. 29).[131]   As would be
predicted for an $E_{C=C}2$ reaction, the rate of dehydrobromination of substrates of structure 41 is quite

(28)

markedly dependent on the electron-withdrawing properties of the substituent in the cycloheptenyl ring.
The ketone 41 is much less active than the corresponding sulphone, but much more reactive than the weakly
electron-withdrawing dimethylacetal derivative.

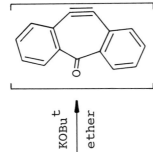

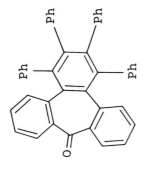

KOBu^t

ether

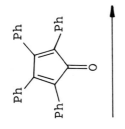

(29)

7   Solvolytic Eliminations of Vinyl Halides and Esters

Vinyl cations are accepted as intermediates in the electrophilic addition reactions of allenes[132] and alkynes.[133]   A typical example is the hydration of sub-stituted phenylacetylenes for which the reaction con-stant, $\rho = -3.84$ (Eq. 30).   The heterolytic cleavage of vinyl halides offers a third source of vinyl cations,

$$X-\langle C_6H_4\rangle-C\equiv C-H \; + \; \overset{+}{H} \longrightarrow X-\langle C_6H_4\rangle-\overset{+}{C}=CH_2 \; \xrightarrow{H_2O}$$

$$X-\langle C_6H_4\rangle-\underset{\underset{OH}{|}}{C}=CH_2 \qquad (30)$$

but only in more recent years have examples been dis-covered.

Grob and Cesh were the first authors to provide kinetic evidence in favor of vinyl cation formation during solvolysis of vinyl bromides.[134]   The solvolysis rate of α-bromostyrene is not altered by a fourfold variation in added triethylamine concentration.   Elec-tron-releasing substituents X (Eq. 31) markedly in-crease the rate of solvolysis, the reaction constant $\overset{+}{\rho}$ being -4.5 at 100°C.   The solvolysis, as measured by the appearance of bromide ion, increases tenfold as the solvent polarity is increased (80 → 50% ethanol / water).   All these results accord with rate-determining vinyl cation formation, and additional support is pro-vided by the observation of silver-ion catalysis.

β-Bromo-cis-cinnamic acid undergoes facile frag-mentation in alkaline solution to phenyl acetylene, but the trans-compound reacts much more slowly and also yields acetophenone (Scheme 7).[135]   Aromatic electron-releasing substituents enhance the fragmentation of the trans-isomer, and an initial zwitterion formation was proposed to accomodate the facts.

Rappoport and Apeloig[136] have analyzed the reac-tion products formed in the solvolysis of cis- and trans-1,2-dianisyl-2-phenylvinyl halides in 80% etha-nol-water (with added benzylthiolates), acetic acid (added $\overline{Cl}$, $O\overline{Ac}$), and in dimethylformamide (added $\overline{Cl}$). In all cases, both the bromide and chloride yield a mixture of stereoisomeric products in a 1:1- ratio. Thus the stereochemistry is independent of cation, anion, and solvent.   These unimolecular reactions are

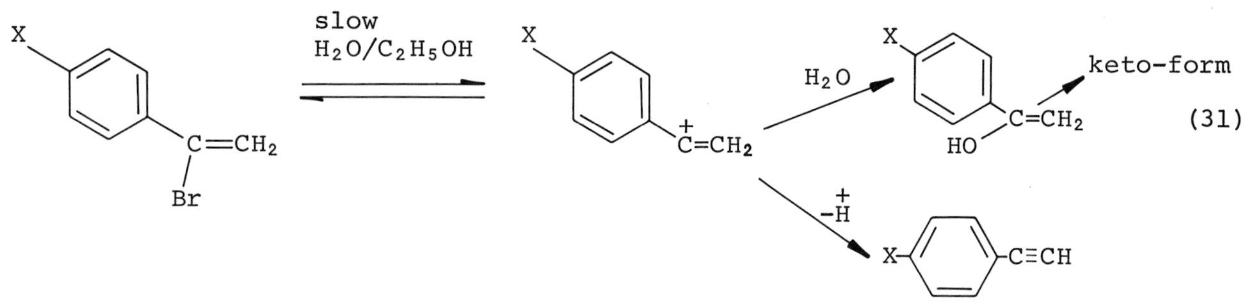

keto-form (31)

slow
H_2O/C_2H_5OH

H_2O

$\overset{+}{-H}$

$-\overset{-}{Br}$

H_2O

$Ph-C\equiv C-H$

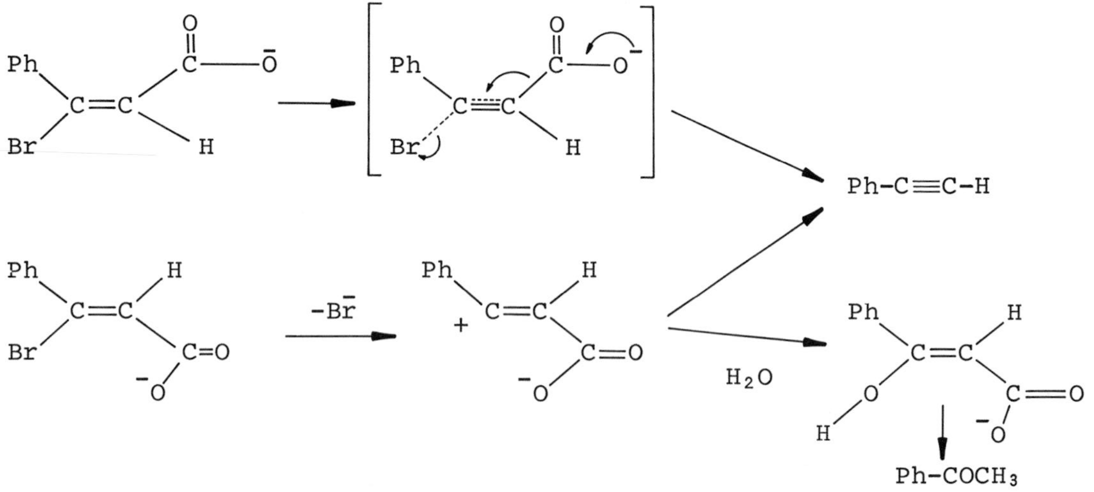

$Ph-COCH_3$

Scheme 7

therefore consistent with the intermediacy of a linear
vinyl cation 42 rather than a bent ion 43. For the
linear ion, nucleophilic addition should occur with
equal facility from both sides. On theoretical grounds,
Fahey[137] had predicted that a linear ion would be more
stable by about 75 kcal/mole. Other factors are enhanced stabilization due to shorter bond length with
higher bond energies and the overlap between the vacant
p-oribtal and aromatic π-orbitals. Steric, 1,2-inter-
actions are also less in the linear ion.[136]

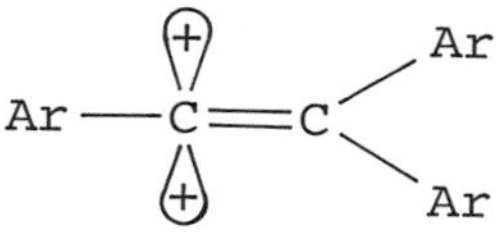

42 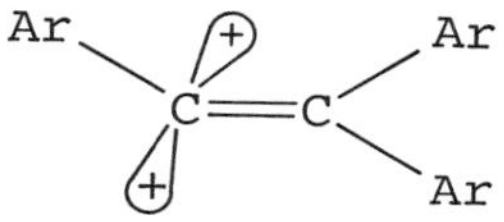   43

     The solvolysis of trans-2-buten-2-yl triflate, 44,
in 80% aqueous ethanol at 76°C yields dimethylacety-
lene. The cis-isomer 45 reacts 40 times more slowly
and also gives methyl allene and 2-butanone.[138] The
greater reactivity of the trans-isomer could reside in

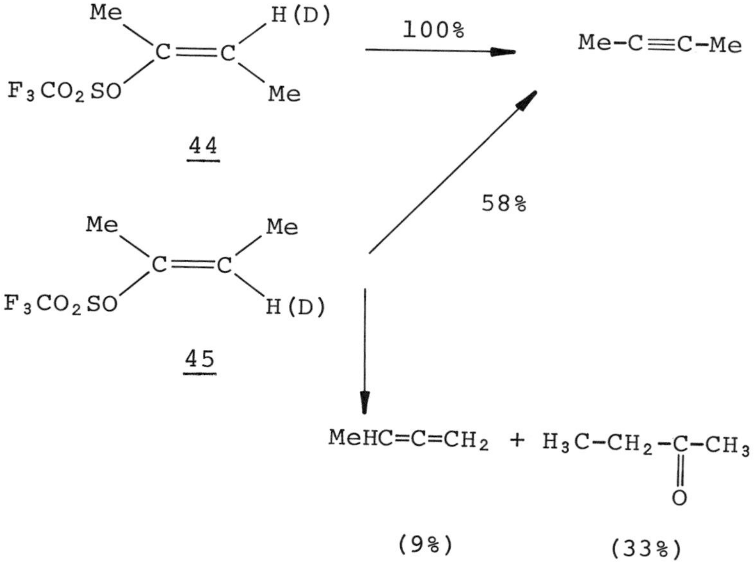

44

45

MeHC=C=CH$_2$ + H$_3$C-CH$_2$-C-CH$_3$
                                ‖
                                O

(9%)   (33%)

the possible intervention of an anti-E$_{C=C}$2 mechanism,

the cis-isomer being confined to rate-determining vinyl cation formation. Isotope studies confirm this interpretation. For the trans-isomer, k_H/k_D = 2.09 (76°C), this primary isotope effect indicating bond breaking in the transition state, the solvent acting as the base. For the cis-isomer, k_H/k_D is only 1.20 at 100°C, the magnitude expected for a β-deuterium secondary isotope effect in vinyl cation formation.

The triflate group is an extremely good leaving entity as shown by the k(triflate/k(tosylate) ratio of 41,700 in the acetolysis of the triarylvinyl derivatives.[139] As for the other potential vinyl-cation reactions discussed thus far, the major problem in a kinetic study is to design experiments to exclude alternative mechanisms. The acetolysis of triarylvinyl triflate gives the corresponding acetate. In addition to (1) vinyl cation formation followed by rapid solvent capture, the acetate could form via (2) a bimolecular substitution, (3) an acetate addition, followed by a sulfonate elimination, (4) a protonation of the double bond leading to an Ad-$E_{C=C}$ mechanism, and (5) a nucleophilic displacement at the S-O bond. Routes (2) and (3) are excluded as added acetate ion does not affect the rate. The inertness of phenyltriflate and p-methoxyphenyltriflate to acetic acid makes route (5) very unlikely.[140] Mechanism (4), the Ad-$E_{C=C}$ route can be excluded by solvent isotope studies. The solvolysis rate ratio, $k_{CH_3COOH}/k_{CH_3COOD}$ = 1.04 at 67° for the corresponding fluorosulfonic ester.[141,142] If proton addition to the double bond occurs in the slow step, a significant primary isotope effect is expected.[143] Alternatively, if a rapid preequilibrium protonation of the alkene followed by slow acetate addition occurs, then a large inverse isotope effect is predicted (see Section VI.B.2.b). The value observed is consistent with a medium effect on vinyl cation formation.

All the precursors for vinyl cations, under solvolytic conditions, have possessed good leaving groups, a prerequisite that always appears to be necessary. However, the presence of an aryl group, to afford cation delocalization is not essential. Grob and Spaar have studied the solvolyses of substituted butadienes (see Eq. 32).[144] Solvent effects, reaction order, independence of rate on added base, and substituent effects are consistent with the intermediacy of a vinyl cation, stabilized by mesomeric interaction with the adjacent olefinic bond (46). Maximum overlap of

$$
\begin{array}{c}
\underset{H}{\overset{H}{>}}C=C\underset{\underset{H}{>}C=C<\underset{CH_3}{\overset{CH_3}{}}}{\overset{Br}{\underset{CH_3}{}}}
\end{array}
\xrightarrow[\text{Et}_3\text{N(buffer)}]{\text{80\%ethanol-water}}
$$

$$
CH_2=C=CH-\underset{\underset{CH_3}{|}}{\overset{\overset{CH_3}{|}}{C}}-OC_2H_5 + HC\equiv C-CH=\underset{}{\overset{\overset{CH_3}{|}}{C}}-CH_3 \qquad (32)
$$

$$\underset{(55\%)}{} \qquad\qquad\qquad (29\%)$$

+ mesityl oxide

(16%)

$$
\underline{46} \quad \overset{}{>}C=\overset{+}{\underset{}{C}}-C\overset{}{\underset{}{<}}\underset{}{\overset{}{C}}<\quad \longleftrightarrow \quad >C=C=C<\overset{}{\underset{\underset{|}{C}-}{}} \quad \equiv \quad >C=C\cdots C<\overset{}{\underset{\underset{|}{C}-}{\overset{+}{}}}
$$

orbitals involved in the delocalization of positive charge should occur if the planes of the double bonds are perpendicular. The bromodienes should therefore react in a nonplanar conformation, and substituents that cause deformation of planarity of the double bonds enhance the solvolysis rate. The cyclopropyl substituent can also afford stabilization, and relatively stable vinyl cations are involved in the solvolysis of 1-cyclopropyl-1-X-ethylene (X = I,[145] X = Cl[146]). The solvolyses are catalyzed by Ag^+ ion, are independent of added bases, and are first-order reactions (see Scheme 8). Under comparable reaction conditions, the corresponding 1-isopropyl derivative is inert.[146] This observation emphasizes the ability of cyclopropyl groups to stabilize adjacent positive charge.

A wider range of solvolyses leading to vinyl cations have been compiled recently by Hanack.[142] Additional to vinyl halides and sulfonates as precursors, the deamination of vinylamines,[147] the photolysis of pyrazolenines,[148] and the decomposition of vinyl triazenes[149] have been interpreted in terms of vinyl cation intermediates.

There is no doubt that vinyl substrates ionize to vinyl cations much more reluctantly than the corresponding alkyl derivatives form saturated carbonium ions. Reactivity ratios of between 10^{-4} to 10^{-8} have

Scheme 8

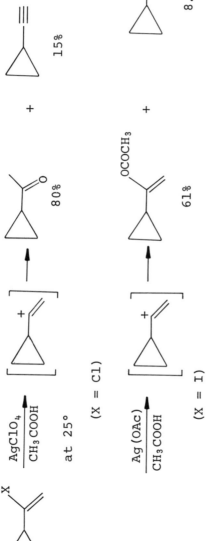

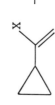

been estimated for a variety of systems.[142]   These differences in reactivity can be attributed either to the high energy of the vinyl cations or to the greater stability of vinyl than alkyl substrates.   However, the common ion effects and other kinetic phenomena discussed above suggest a similarity in behavior between vinyl and saturated carbonium ions, and a sufficient lifetime for vinyl cations to be trapped by added common ions.   Reactant stabilization offers the more likely explanation of the low reactivity.   Roberts and Chambers[150] have proposed partial π-character of the bond between the vinyl carbon and the leaving group (see 20, Section IX.B.2.a), whereas Moffit invokes a stronger σ-bond involving a vinyl carbon and the leaving group due to the hybridization state.[151]   Whichever explanation is accepted, physical measurements, such as bond lengths, bond dissociation energies, and ionization potentials confirm that C-X bonds in vinyl halides and esters are shorter and stronger than those in the corresponding alkyl structures.[81,142]

REFERENCES

1. E. Cordes and W. P. Jencks, J. Am. Chem. Soc., 85, 2843 (1963).
2. J. Hine, "Physical Organic Chemistry," 2nd ed., McGraw-Hill, New York, 1962, Chap. 8.
3. A. N. Bourns and P. J. Smith, Can. J. Chem., 44, 2553 (1966).
4. G. W. Cowell, A. Ledwith, and D. G. Morris, J. Chem. Soc., B, 697, (1967).
5. E. Ciuffarin and A. Fava, Prog. Phys. Org. Chem., 6, 81 (1968).
6. A. Schöberl, Ann., 507, 111 (1934).
7. A. J. Parker and N. Kharasch, Chem. Rev., 59, 583 (1959).
8. D. S. Tarbell and D. P. Harnish, Chem. Rev., 49, 11 (1951).
9. T. J. Wallace, J. E. Hofmann, and A. Schriesheim, J. Am. Chem. Soc., 85, 2739 (1963); 86, 1561 (1964).
10. N. A. Rosenthal and G. Oster, J. Am. Chem. Soc., 83, 4445 (1961).
11. U. Miotti, A. Sinico, and A. Ceccon, Chem. Comm., 724 (1968).
12. U. Miotti, U. Tonellato, and A. Ceccon, J. Chem. Soc., B, 325 (1970).
13. F. H. Westheimer, Chem. Rev., 61, 265 (1961).
14. W. H. Saunders, Jr., D. G. Bushmann, and A. F.

Cockerill, J. Am. Chem. Soc., 90, 1775 (1968).

15. R. Baker and M. J. Spillett, J. Chem. Soc., B, 482 (1969).

16. J. Banger, A. F. Cockerill, and G. L. O. Davies, J. Chem. Soc., B, 498 (1971).

17. H. H. Jaffé, Chem. Rev., 53, 191 (1953).

18. A. Ceccon, U. Miotti, U. Tonellato, and M. Padovan, J. Chem. Soc., B, 1084 (1969).

19. U. Miotti and A. Ceccon, Ric. Sci., 38, 824 (1968).

20. R. G. Hiskey, W. H. Bowers, and D. N. Harpp, J. Am. Chem. Soc., 86, 2010 (1964).

21. T. E. Stevens, J. Org. Chem., 32, 670 (1967).

22. F. A. Johnson, C. Haney, and T. E. Stevens, J. Org. Chem., 32, 466 (1967).

23. (a) W. H. Saunders, Jr., S. R. Fahrenholtz, E. A. Caress, J. P. Lowe, and M. R. Schreiber, J. Am. Chem. Soc., 87, 3401 (1965);
(b) C. H. DePuy and C. A. Bishop, J. Am. Chem. Soc., 82, 2535 (1960).

24. C. R. Hauser, J. W. LeMaistre, and A. E. Rainsford, J. Am. Chem. Soc. 57, 1056, (1935).

25. W. E. Jordan, H. E. Byas and D. G. Hill, J. Am. Chem. Soc., 63, 2383 (1941).

26. S. K. Brauman and M. E. Hill, J. Org. Chem., 34, 3381 (1969).

27. A. L. Logethetis and G. N. Sausen, J. Org. Chem., 31, 3689 (1966).

28. A. L. Logethetis and G. N. Sausen, J. Org. Chem., 32, 2261 (1967).

29. A. L. Logethetis J. Org. Chem., 31, 3686 (1966).

30. G. N. Sausen and A. L. Logethetis, J. Org. Chem., 33, 2330 (1968).

31. R. C. Petry and J. P. Freeman, J. Org. Chem., 32, 4034 (1967).

32. S. K. Brauman and M. E. Hill, J. Am. Chem. Soc., 89, 2131 (1967).

33. S. K. Brauman and M. E. Hill, J. Am. Chem. Soc., 89, 2127 (1967).

34. L. J. Lohr and R. W. Warren, J. Chromatogr., 8, 127 (1962).

35. G. Just and C. Pace-Asciak, Tetrahedron, 22, 1069 (1966).

36. M. Benger and O. L. Brady, J. Chem. Soc., 1221 (1950).

37. D. Ambrose and O. L. Brady, J. Chem. Soc., 1243 (1950).

38. O. L. Brady and R. F. Goldstein, J. Chem. Soc., 1918 (1926).

39. N. S. Kornblum and H. E. De la Mare, J. Am.

Chem. Soc., 73, 880 (1951).

40. J. W. Baker and D. M. Easty, Nat., 166, 156 (1950).

41. A. Ledwith, G. W. Cowell, and D. G. Morris, J. Chem. Soc., B, 697 (1967).

42. W. A. Waters, "Mechanisms of Oxidation of Organic Compounds," Methuen, London, 1964, Chap. 4.

43. P. D. Bartlett and J. D. McCollum, J. Am. Chem. Soc., 78, 1441 (1956).

44. H. B. Hass and M. L. Bender, J. Am. Chem. Soc., 71, 1967 (1949).

45. J. W. Baker and D. M. Easty, J. Chem. Soc., 1193 (1952).

46. J. W. Baker and D. M. Easty, J. Chem. Soc., 1208 (1952).

47. J. W. Baker and T. G. Heggs, J. Chem. Soc., 616 (1955).

48. J. W. Baker and A. J. Neale, J. Chem. Soc., 3225 (1954).

49. A. N. Bourns and E. Buncel, Can. J. Chem., 38, 2457 (1960).

50. A. N. Bourns and P. J. Smith, Can. J. Chem., 44, 2553 (1966).

51. F. H. Westheimer, Chem. Rev., 45, 419 (1949).

52. W. A. Waters, Q. Rev. (London), 12, 277 (1958).

53. N. Venkatasubramanian, J. Sci. Ind. Res. (India) 22, 397 (1963).

54. K. Wiberg "Oxidation in Organic Chemistry," Vol. 5A, Academic Press, 1965, pp. 140-172.

55. R. Stewart, "Oxidation Mechanisms," W. A. Benjamin and Co., New York, 1964, p. 37.

56. F. H. Westheimer and N. Nicolaides, J. Am. Chem. Soc., 71, 25 (1949).

57. J. Roček and J. Krupička, Coll. Czech. Chem. Commun., 23, 2068 (1958).

58. A. Leo and F. H. Westheimer, J. Am. Chem. Soc., 74, 4383 (1952).

59. J. Roček, Coll. Czech. Chem. Commun., 25, 1052 (1960).

60. H. Kwart and P. S. Francis, J. Am. Chem. Soc., 77, 4907 (1955).

61. J. Roček, F. H. Westheimer, A. Eschenmoser, L. Moldoványi and J. Schreiber, Helv. Chim. Acta, 45, 2554 (1962).

62. C. G. Swain, R. A. Wiles, and R. F. Bader, J. Am. Chem. Soc., 83, 1945, (1961).

63. N. C. Deno and N. H. Potter, J. Am. Chem. Soc., 89, 3555 (1967).

64. L. A. Kaplan, J. Am. Chem. Soc., 80, 2639 (1958).

65. I. R. L. Barker, W. G. Overend, and C. W. Rees,

J. Chem. Soc., 3263 (1964).

66. N. Venkatasubramanian and V. Thiagarajan,
    Tetrahedron Lett., 1711 (1968).

67. R. Filler, Chem. Rev., 63, 21 (1963).

68. P. F. Kruse, Jr., K. L. Grist, and T. A. McCoy,
    Anal. Chem., 26, 139 (1954).

69. J. Lecomte and H. Gault, Co. R., 238, 2538 (1954).

70. G. Langbein and B. Steinert, Ber., 95, 1873 (1962).

71. N. Venkatasubramanian and V. Thiagarajan,
    Tetrahedron Lett., 3349 (1967).

72. N. Venkatasubramanian and V. Thiagarajan, Can. J.
    Chem., 47, 694 (1969).

73. R. Natarajan and N. Venkatasubramanian,
    Tetrahedron Lett., 5021 (1969).

74. G. Köbrich, Angew. Chem. Int. Ed., 4, 49 (1965).

75. H. G. Vieke and E. Franchimant, Chem. Ber., 95,
    319 (1962).

76. H. Normant, Adv. Org. Chem., 2, 1 (1960).

77. C. A. Grob and R. Spaar, Tetrahedron Lett., 1439
    (1969).

78. C. A. Coulson, "Valence," Oxford University Press,
    Oxford, 1952, p. 206.

79. P. B. D. de la Mare, Qu. Rev. (London), 3, 126
    (1949).

80. E. D. Hughes, Trans. Faraday Soc., 34, 185 (1938);
    37, 603 (1941).

81. L. L. Miller and D. A. Kaufman, J. Am. Chem. Soc.,
    90, 7282 (1968).

82. Z. Rappoport, Adv. Phys. Org. Chem., 7, 1 (1969).

83. D. Y. Curtin, and E. E. Harris, J. Am. Chem. Soc.,
    73, 2716 (1951).

84. D. Y. Curtin, H. W. Johnson, Jr., and E. G.
    Steiner, J. Am. Chem. Soc., 77, 4566 (1955).

85. L. Maioli and G. Modena, Gazz. Chim. Ital., 89,
    854 (1959).

86. P. Beltrame, G. Favini, M. G. Cattania, and F.
    Guella, Gazz. Chim. Ital., 98, 380 (1968).

87. E. F. Silversmith and D. Smith, J. Org. Chem., 23,
    427 (1958).

88. P. Beltrame and G. Favini, Gazz. Chim. Ital., 93,
    757 (1963).

89. W. M. Jones and R. Damico, J. Am. Chem. Soc.,
    85, 2273 (1963).

90. E. Winterfeldt, Angew. Chem. Int. Ed. (in English)
    6, 423 (1967).

91. (a) Reference 82, p. 54-58;
    (b) R. Huisgen, B. Giese, and H. Huber,
    Tetrahedron Lett., 1883 (1967).

92. S. I. Miller and P. K. Yonan, J. Am. Chem. Soc.,
    79, 5931 (1957).

93. D. E. Jones, R. O. Morris, C. A. Vernon, and R. F. M. White, J. Chem. Soc., 2349 (1960).
94. Z. Rappoport, C. Degani, and S. Patai, J. Chem. Soc., 4513 (1963).
95. L. Maioli, G. Modena, and P. E. Todesco, Boll. Sci. Fac. Chim. Ind. Bologna, 18, 66 (1960).
96. A. Campagni, G. Modena, and P. E. Todesco, Gazz. Chim. Ital., 90, 694, (1960).
97. G. Modena, F. Taddei, and P. E. Todesco, Ric. Sci., 30, 894 (1960).
98. S. Ghersetti, G. Modena, P. E. Todesco, and P. Vivarelli, Gazz. Chim. Ital., 91, 620 (1961).
99. L. Di Nunno, G. Modena, and G. Scorrano, J. Chem. Soc., B, 1186 (1966).
100. S. I. Miller and R. M. Noyes, J. Am. Chem. Soc., 74, 629 (1952).
101. (a) S. J. Cristol, J. Am. Chem. Soc., 69, 338 (1947); 71, 1894 (1949);
     (b) S. J. Cristol and J. S. Meek, J. Am. Chem. Soc., 73, 674 (1951).
102. S. I. Miller and W. G. Lee, J. Am. Chem. Soc., 81, 6313 (1959).
103. G. S. Wheland and D. S. K. Chen, J. Phys. Chem., 24, 67 (1956).
104. L. C. Leitch and H. J. Bernstein, Can. J. Res., 28B, 35 (1950).
105. T. J. Houser, R. B. Bernstein, R. G. Miekka and J. G. Angus, J. Am. Chem. Soc., 77, 6201 (1955).
106. D. Landini, F. Montanari, G. Modena, and F. Naso, J. Chem. Soc., B, 243 (1969).
107. G. Marchase, G. Modena, and F. Naso, J. Chem. Soc., B, 958 (1968).
108. G. Marchese, F. Naso, N. Tangari and G. Modena, J. Chem. Soc., B, 1196 (1970).
109. G. Marchese, F. Naso, N. Tangari, and G. Modena, Boll. Sci. Fac. Chim. Ind. Bologna, 26, 209 (1968).
110. H. H. Jaffé, Chem. Rev., 53, 191 (1953).
111. G. Marchese, F. Naso and G. Modena, J. Chem. Soc., B, 290 (1969).
112. S. J. Cristol and R. S. Bly, J. Am. Chem. Soc., 83, 4027 (1961).
113. S. J. Cristol and C. A. Whittemore, J. Org. Chem., 34, 705 (1969).
114. J. G. Pritchard and A. A. Bothner-By., J. Phys. Chem., 64, 1271 (1960).
115. G. Köbrich, Angew. Chem. Int. Ed., (in English), 6, 41 (1967).
116. G. Köbrich and K. Flory, Chem. Ber., 99, 1773 (1966).

117. S. J. Cristol and R. F. Helmreich, J. Am. Chem. Soc., 77, 5034 (1955).
118. G. Wittig, W. Boll, and K.-H. Krück, Chem. Ber., 95, 2514 (1962).
119. M. Schlosser and V. Ladenberger, Tetrahedron Lett., 1945 (1964).
120. M. Schlosser and V. Ladenberger, Chem. Ber., 100, 3877, 3893, 3901 (1967).
121. R. P. Bell, "The Proton in Chemistry," Cornell University Press, Ithaca, 1959.
122. W. K. Kwok, W. G. Lee, and S. I. Miller, J. Am. Chem. Soc., 91, 468 (1969).
123. A. T. Blomquist, L. H. Liu, and J. C. Bohrer, J. Am. Chem. Soc., 74, 3643 (1952).
124. G. Wittig and R. Pohlke, Chem. Ber., 94, 3276 (1961).
125. A. Favorsky, M. F. Chestakovsky, and N. A. Domnine, Bull. Soc. Chim. Fr., 1727 (1936).
126. G. Wittig and U. Mayer, Chem. Ber., 96, 329 (1963).
127. A. T. Blomquist and L. H. Liu, J. Am. Chem. Soc., 75, 2153 (1953).
128. F. G. Willey, Angew. Chem. Int. Ed., 3, 138 (1964).
129. G. Wittig, Angew. Chem. Int. Ed., 1, 415 (1962).
130. (a) J. D. Roberts and F. Scardiglia, Tetrahedron, 1, 343 (1957);
     (b) L. K. Montgomery and J. D. Roberts, J. Am. Chem. Soc., 82, 4750 (1960).
131. (a) W. Tochtermann, Angew. Chem., 74, 432 (1962);
     (b) W. Tochtermann, K. Oppenländer, and U. Walter, Chem. Ber., 97, 1318 (1964).
132. D. R. Taylor, Chem. Rev., 67, 317 (1967).
133. P. E. Peterson and J. E. Duddey, J. Am. Chem. Soc., 88, 4990 (1966).
134. C. A. Grob and G. Cesh, Helv. Chim. Acta, 47, 194 (1964).
135. C. A. Grob, J. Csapilla and G. Cesh, Helv. Chim. Acta, 47, 1590 (1964).
136. Z. Rappoport and Y. Apeloig, J. Am. Chem. Soc., 91, 6734 (1969).
137. R. C. Fahey and D. J. Lee, J. Am. Chem. Soc., 88, 5555 (1966).
138. P. J. Stang and R. Summerville, J. Am. Chem. Soc., 91, 4600 (1969).
139. W. M. Jones and D. D. Maness, J. Am. Chem. Soc., 91, 4314 (1969).
140. T. M. Su, W. F. Sliwinski, and P. v. R. Schleyer, J. Am. Chem. Soc., 91, 5386 (1969).

141. W. M. Jones unpublished result quoted in ref. 142.

142. M. Hanack, Acc. Chem. Res., $\underline{3}$, 209 (1970).

143. D. S. Noyce and P. M. Pollack, J. Am. Chem. Soc., $\underline{91}$, 119 (1969).

144. C. A. Grob and R. Spaar, Tetrahedron Lett., 1439 (1969).

145. S. A. Sherrod and R. G. Bergmann, J. Am. Chem. Soc., $\underline{91}$, 2115 (1969).

146. M. Hanack and T. Bassler, J. Am. Chem. Soc., $\underline{91}$, 2117 (1969).

147. D. Y. Curtin, J. A. Kampmeier, and B. R. O'Connor, J. Am. Chem. Soc., $\underline{87}$, 863 (1965).

148. A. C. Day and M. C. Whiting, J. Chem. Soc., $\underline{B}$, 991 (1967).

149. W. M. Jones and F. W. Miller, J. Am. Chem. Soc., $\underline{89}$, 1960 (1967).

150. J. D. Roberts and V. C. Chambers, J. Am. Chem. Soc., $\underline{73}$, 5034 (1951).

151. W. Moffit, Proc. Roy. Soc. Ser. $\underline{A}$., $\underline{202}$, 548 (1950).

X   ELIMINATIONS REACTIONS OTHER THAN β ELIMINATIONS

A DEFINITIONS

The preceding chapters in this book are concerned with the mechanisms of β-elimination. Of course, the leaving groups in an elimination reaction do not necessarily have to be situated on adjacent atoms, and in the simplest case, when both leaving groups are removed from the same atom, the process is termed an α-elimination (α-E). Possibly the most well-known example is the reaction of chloroform with hydroxide ion, α-elimination giving rise to the electron deficient intermediate, dichlorocarbene, which rapidly reacts with the reaction medium (Eq. 1).[1] Because of their instability, the intermediacy of carbenes has to be inferred by a combination of kinetic studies and trapping experiments.

$$CHCl_3 + \bar{O}H \rightleftharpoons \bar{C}Cl_3 + H_2O \xrightarrow{\bar{C}l} \left[:CCl_2\right] \xrightarrow[H_2O]{\bar{O}H}$$

$$CO + HCO\bar{O} \qquad (1)$$

γ-Eliminations give rise to three-membered ring systems (Eq. 2),[2] which in some cases are unstable and react further under the conditions of the reaction (Eq. 3).[3] For such cases, the intermediacy of three-membered rings has to be inferred from a combination of kinetic and labeling techniques. Two types of δ-elimination are known, those that yield four-membered ring compounds, and others that involve a bond migration (Eq. 4).[4] Many of the features, which have been well defined in β-elimination, can be recognized in these less commonly encountered eliminations, and we use a similar terminology in this book; for example, the hydrolysis of chloroform, Eq. 1, is an example of an α-elimination occurring via a carbanion and is termed an α-ElcB reaction.

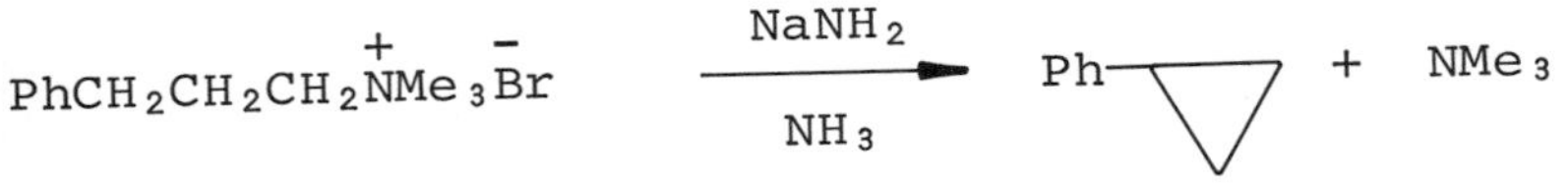

$$PhCH_2CH_2CH_2\overset{+}{N}Me_3\overset{-}{B}r \xrightarrow[NH_3]{NaNH_2} Ph{-}\triangledown + NMe_3 \qquad (2)$$

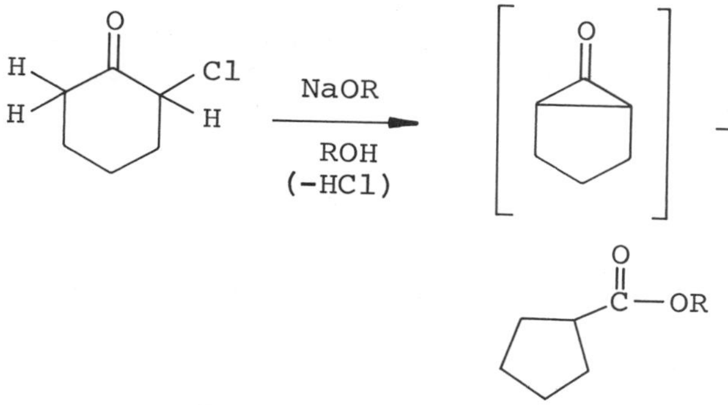

$$(3)$$

$$(4)$$

During the last decade, Grob[5] has recognized a new class of reactions called fragmentations (Eq. 5), which have many features in common with β-eliminations. Depending on the order in which the fragments are released, one- or two-step processes, analogous to the El, E2, and ElcB mechanisms, can be recognized.

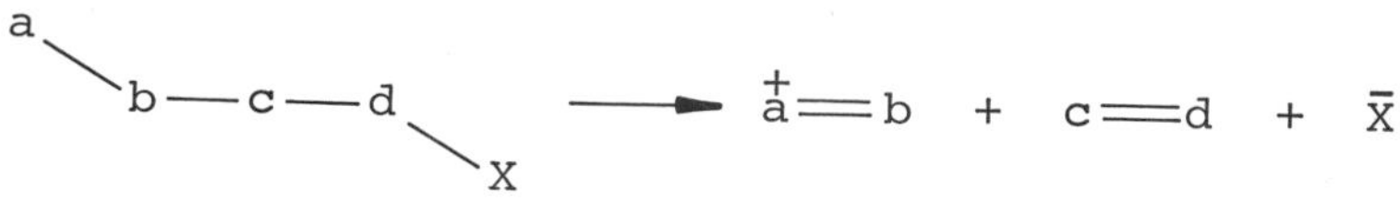

$$(5)$$

B   α-ELIMINATION

1   General Comments

In this section we discuss the mechanisms of a number of reactions which, from consideration of the structure of the reactants and the nature of the products, appear to be typical examples of α-elimination. However, as we see, appearances can be deceiving and a number of these reactions, which show qualitative features consistent with carbene intermediates, actually occur via concerted decompositions of anions or involve carbonium ion or organometallic intermediates. Reactions that exhibit the qualitative features of carbene reactions are now classified by the less committal name,

"carbenoid."[6]   The recognition of a genuine α-elimination involves an understanding of the properties of
intermediates in which a carbon atom possesses only a
sextet of electrons. Consequently, before we discuss
the mechanisms of α-elimination, we present a brief
outline of the structural features and typical reactions of carbenes.

2 Structure and Reactions of Carbenes

Our discussion of the structure and reactions of carbenes is very brief, and the reader is referred to the
multitude of references concerned with their chemistry
for more comprehensive coverage.[7-18] The term carbene
is a generic name used to describe organic molecules
possessing a divalent carbon atom. Such an electron-
deficient carbon has only six valence electrons, four
of which are involved in two covalent bonds. The remaining two electrons are paired in one orbital in
singlet carbenes (1) and are unpaired in degenerate
orbitals in triplet carbenes (2). Structure 3 is used
to describe a carbene of undefined multiplicity. Much

$$>C \!\uparrow\!\downarrow \qquad\qquad >C \!\uparrow\uparrow \qquad\qquad >C:$$

$$\underline{1} \qquad\qquad\qquad \underline{2} \qquad\qquad\qquad \underline{3}$$

of the interest in carbenes has been concerned with
proving the existence of these states of different
multiplicity and with determining differences in their
properties.
     The considerable improvement in instrumental
techniques in recent years has enabled the detection of
very unstable molecules.[17] Transient carbenes have
been observed by flash photolysis or by "matrix isolation," a technique in which the carbenes are generated
in dilute solution in rigid matrices, such as crystals
or glasses, at low temperatures, conditions under which
they are relatively stable. For the simplest carbenes, electronic and vibrational-rotational spectra
have been recorded, whereas electron-spin resonance
spectroscopy is useful in providing structural information on more complicated carbenes in the triplet state.
     The spectrum of singlet methylene has been ob-
served by Herzberg and Shoosmith in the flash photolysis of diazomethane in the gas phase.[19] A number of
authors have observed the electronic spectra of di-
fluoro- and dichloro-methylene.[17] The C-Cl frequencies
in the infrared spectrum of dichloromethylene are similar to those in dichloromethane, suggesting little

interaction between the lone pair electrons of the halogens and the vacant orbitals of the divalent carbon. However, the spectra of difluoromethylene suggest considerable π-interaction between the fluorine and vacant orbital on the carbon atom. It would seem reasonable to expect the difluoro carbene to be more stable, and mass spectrometer investigations of carbenes generated thermally support this deduction

(e.g., $X_3CSiX_3 \xrightarrow{\Delta} X_2C:$ + SiX_4). The heats of formation of the carbenes are as follows: CH_2 : CF_2 : CCl_2 = 86(kcal/mole) ; -39 : 57.[20] Clearly, the exothermic nature of the difluoromethylene formation suggests considerable stability relative to the other carbene-forming reactions.

The reactions of carbenes fall into three categories:

1. The insertion of a group CR^1R^2 into a single bond (Eq. 6), which often leads to the formation of a cyclopropane in intramolecular reactions.

2. The formation of cyclopropanes by addition to alkenes (Eq. 7).

3. The formation of dimeric olefins from CR^1R^2 groupings (Eq. 8), a reaction only encountered usually in the gas phase.[21] Dimeric olefins can arise by carbenoid reactions in solution, but are not usually formed by a carbene dimerization.

$$\diagup\!\!\!\!\diagdown C\!\!-\!\!X \; + \; :CR^1R^2 \longrightarrow \diagup\!\!\!\!\diagdown C\!\!-\!\!CR^1R^2X \tag{6}$$

(X = H, alkyl, halogen, etc.)

$$\diagup\!\!\!\!\diagdown C\!\!=\!\!C\diagup\!\!\!\!\diagdown \; + \; :CR^1R^2 \longrightarrow \underset{R^1 \quad R^2}{\diagup\!\!\!\!\diagdown C\!\!-\!\!C\diagup\!\!\!\!\diagdown} \tag{7}$$

$$2 \; :CR^1R^2 \longrightarrow R^1R^2C\!\!=\!\!CR^1R^2 \tag{8}$$

Of the three modes of reaction, the addition to alkenes offers the best prospect for distinguishing between singlet and triplet states. It is usually assumed that singlet carbenes add in a stereospecific manner, while triplet addition occurs in a nonstereospecific fashion.[22] The reasoning behind this view is that singlet carbene should add to the alkene

effectively in a one-step concerted process as the reaction can occur with conservation of spin (Eq. 9). Alternatively, if addition occurs in two steps, with the
carbene acting as an electrophile, attractive force between the opposite charges in the intermediate 4 prevents rotation about the C-C bond before bond formation
giving the cyclopropane. For the triplet reaction,
concerted addition is precluded as this would violate
the rule of conservation of spin. Addition occurs via
a diradical in which spin inversion followed by cyclization, and bond rotation occur at comparable rates,
leading to nonstereospecific reaction (Scheme 1).

The above early intuitive approaches have been
supported by recent theoretical treatments. It is
suggested that singlet carbene adds in a stereospecific
fashion to alkenes not because it is a singlet, but
because it can correlate with the lowest singlet configuration of trimethylene and, consequently, with the
ground state of a cyclopropane. On the other hand, the
complex between a triplet carbene and a ground state
ethylene must correlate with a triplet state of an
excited trimethylene in which there are no barriers to
free rotation around the terminal bonds. Provided that
the singlet carbene is in a linear, excited configuration, symmetrical addition is allowed.[23,24]

$$\text{>C} + \underset{R^2}{\overset{R^1}{>}}C = C\underset{R^4}{\overset{R^3}{<}} \longrightarrow \left[\underset{R^2}{\overset{R^1}{>}}C = C\underset{R^4}{\overset{R^3}{<}} \right] \longrightarrow$$

$$\underset{R^2}{\overset{R^1}{>}}C - C\underset{R^4}{\overset{R^3}{<}} \qquad (9)$$

$$\underset{R^2}{\overset{R^1}{>}}C - \overset{\oplus}{C}\underset{R^4}{\overset{R^3}{<}} \qquad \underline{4}$$

Scheme 1

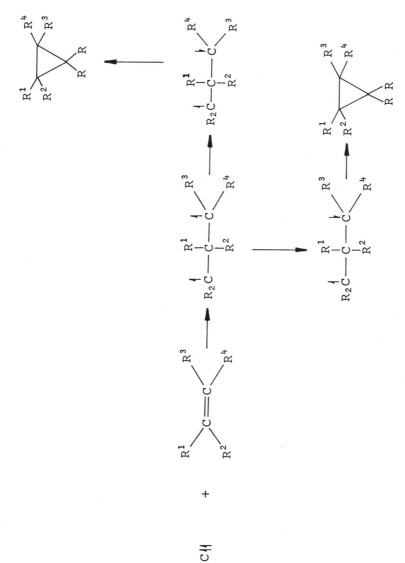

3   Reaction Mechanisms

(a)   <u>General Sources of Carbenoids</u>.   A wide range of reactants are potential precursors of carbenoids. Basic reagents appear to effect α-elimination from alkyl and vinyl halides lacking a β-hydrogen.   However, with metal alkyl bases, the intermediacy of an ion pair cannot be ignored,[15,25] and for reactions in protic solvents, decomposition may occur via the initially formed carbanion rather than involve a carbene.   Reducing agents, such as chromous ion,[26] and metal couples,[27] can be used to dehalogenate geminal dihalides, but organometallic intermediates seem most probable.   Photolysis and thermal decompositions offer the most likely sources of carbenes, and within this category are the decompositions of aliphatic diazo compounds,[28-30] diazirines,[31] ketenes,[32] epoxides,[33] and olefins.[34]   Reactions, which at first glance may constitute examples of α-E1 reactions, include acid catalyzed decompositions of aliphatic diazo compounds, pyrolyses of mercury and tin organometallics,[35] and the Bamford-Stevens alkali-catalyzed decomposition of tosylhydrazones.[36]   In the following sections, we consider the mechanisms of a few of these examples in more detail.

(b)   <u>Base-Induced α-Elimination</u>

(i)   <u>Haloform solvolyses</u>.   Geuther[37] first suggested that the alkaline hydrolysis of chloroform involved dichlorocarbene as an intermediate in 1862, but it was not until 1950 that Hine and his co-workers[1,38,39] provided the kinetic evidence to support the now-accepted mechanism of Scheme 2.

Scheme 2

$$CHCl_3 + \bar{O}H \rightleftharpoons \bar{C}Cl_3 + H_2O \qquad \text{fast}$$

$$\bar{C}Cl_3 \longrightarrow :CCl_2 + \bar{C}l \qquad \text{slow}$$

$$:CCl_2 + \bar{O}H/H_2O \longrightarrow CO + HCO\bar{O} \qquad \text{fast}$$

The hydrolysis is first order in both chloroform and hydroxide ion, an observation that excludes an S_N1 mechanism, but is consistent with two other mechanisms depicted in Schemes 3 and 4.   Chloroform undergoes base-catalyzed deuterium exchange very much faster than base-catalyzed hydrolysis,[40] as expected for both Schemes 2 and 3, but this does not exclude the S_N2

displacement to which exchange could be an irrelevant side reaction. For strongly basic reagents (e.g., $\bar{O}H$), the reactivity sequence is $CH_3Cl \gg CH_2Cl_2 \ll CHCl_3 \gg CCl_4$,[38] an order that does not accord with an S_N2 mechanism. More compelling evidence in favor of Scheme 2 is provided by the hydrolyses in the presence of added nucleophiles, which can divert the dichlorocarbene. For example, chloroform is almost inert to thiophenoxide ion, but is rapidly converted into triphenylorthothioformate in the presence of hydroxide ion. Chloride ion exerts a mass action effect, retarding the rate at

Scheme 3

$$CHCl_3 \quad + \quad \bar{O}H \rightleftharpoons \bar{C}Cl_3 \quad + \quad H_2O \qquad \text{fast}$$

$$H_2O + \bar{C}Cl_3 \longrightarrow H-\overset{+}{\underset{\underset{H}{|}}{O}}-\overset{}{\underset{\underset{Cl}{|}}{C}}-Cl \quad + \quad \bar{C}l$$

$$\text{slow}$$

$$H-\overset{+}{\underset{\underset{H}{|}}{O}}-\overset{-}{\underset{\underset{Cl}{|}}{C}}-Cl \quad \xrightarrow[H_2O]{\bar{O}H} \quad CO \quad + \quad HCO\bar{O} \qquad \text{fast}$$

Scheme 4

$$CHCl_3 \quad + \quad \bar{O}H \longrightarrow CHCl_2OH \quad + \quad \bar{C}l \quad \text{slow}$$

$$CHCl_2OH \quad \xrightarrow[H_2O]{\bar{O}H} \quad CO \quad + \quad HCO\bar{O} \qquad \text{fast}$$

concentrations under which weakly nucleophilic ions such as perchlorate have no effect.[1] Bromide and iodide ions behave similarly but are more effective than chloride, and convert the chloroform into bromodichloromethane and dichloroiodomethane, respectively, the latter actually being isolated in the early stages of reaction.[1]

In aqueous solution, dichlorocarbene is effectively trapped by the water molecules, but under aprotic conditions can be captured by added alkenes. Thus Doering and Hoffmann were able to prepare cyclopropanes from the reaction of alkenes when chloroform was treated with potassium t-butoxide in benzene.[41] Dihalocarbenes, generated in this manner, add to cis- and trans-2-butene in a stereospecific manner, consistent with the intermediacy of a singlet carbene.

Dihalomethylenes, formed by an α-ElcB mechanism,

are intermediates in the hydrolyses of a series of other haloforms, the reactivity sequence being as follows;[42] $CHBrClF >> CHBrCl_2 > CHBr_2Cl \simeq CHCl_2I > CHBr_3 > CHCl_3$. This sequence represents a combination of the acid-strengthening influence of the α-halogen and the ease of fragmentation of the C-halogen bond. However, mixed haloforms containing two fluorine atoms are subject to base-catalyzed hydrolysis at a rate faster than anticipated for initial carbanion formation, and the lack of deuterium exchange with the reaction medium during hydrolysis, is consistent with the operation of an α-E2 mechanism; that is, concerted elimination to difluorocarbene, without intermediate carbanion formation.[42] Compared with the other halogens, fluorine destabilizes the carbanion but stabilizes the carbene (Section X.B.2).

The reactions of dihalocarbenes with alkoxide ions are discussed under the mechanism of deoxidation (Section VII.b). Dichloromethylene has also been trapped in the Reimer-Tiemann reaction with phenols (Scheme 5).[43]

Scheme 5

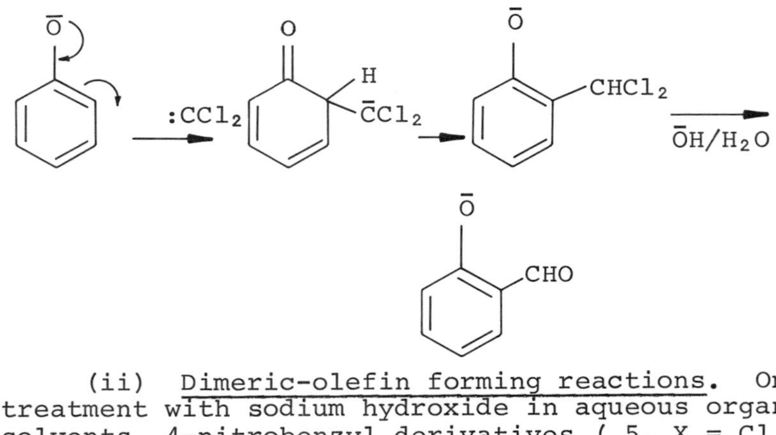

(ii) Dimeric-olefin forming reactions. On treatment with sodium hydroxide in aqueous organic solvents, 4-nitrobenzyl derivatives (5, X = Cl,[44,45] X = $\overset{+}{S}Me_2$,[46,47]) are converted in almost quantitative yield into 4,4'-dinitrostilbene. Similarly, 9-fluorenyl halides, 6, are converted into bifluorenylidene by a number of alkoxide bases,[48] and diphenylmethyl chloride, 7, gives tetraphenyl ethylene on treatment with potassium t-butoxide in dimethyl sulfoxide[49] or sodium amide in liquid ammonia.[50] Nominally, all of these conversions can be explained in terms of a carbene-forming reaction (Scheme 6, route c) in which,

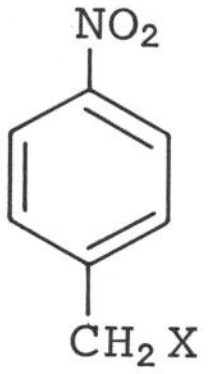

NO_2

CH_2X

5

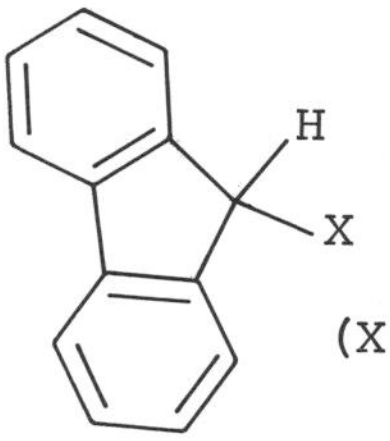

H

X

(X = Cl, Br, I)

6

CHX

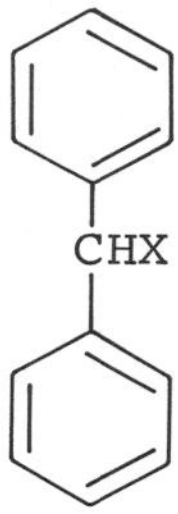

(X = Cl, Br)

7

since carbenes are highly energetic, the reaction should exhibit first-order kinetics in both the substrate and the base. In the alternative bimolecular displacement mechanism, the rate-determining step involves alkylation of the initially formed carbanion by a second molecule of the organic halide. In this case (Scheme 6, route b), the reaction is still first order in base concentration, but should now show a second-order dependence in the organic halide.

Scheme 6

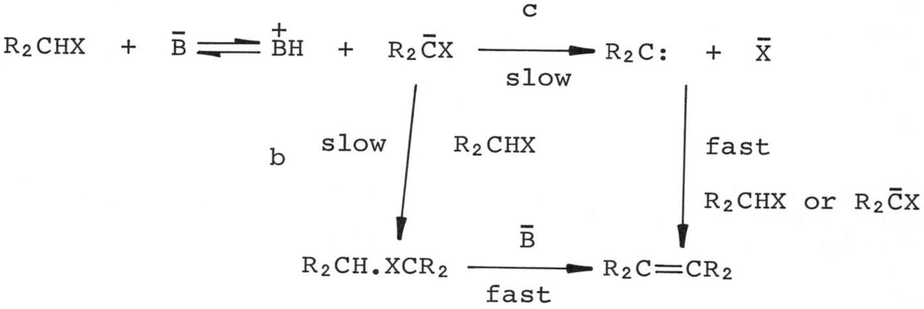

$$R_2CHX \;+\; \bar{B} \;\rightleftharpoons\; \overset{+}{B}H \;+\; R_2\bar{C}X \;\xrightarrow[\text{slow}]{c}\; R_2C: \;+\; \bar{X}$$

b   slow R_2CHX

fast

R_2CHX or $R_2\bar{C}X$

$$R_2CH.XCR_2 \;\xrightarrow[\text{fast}]{\bar{B}}\; R_2C{=}CR_2$$

In t-butyl alcohol containing its potassium or
sodium salt or a dilute solution of benzyltrimethylammonium hydroxide, the formation of bifluorenylidene
follows second-order kinetics in 9-bromofluorene and
first-order kinetics in the medium basicity, as measured by the ionization of aniline indicators.[48,51] Under
these reaction conditions, 9-deutero-9-bromofluorene
undergoes deuterium exchange with the solvent much more
rapidly than it is converted into bifluorenylidene,
and β-elimination of hydrogen bromide from the intermediate dimeric halide is a much faster process than
consumption of 9-bromofluorene.[52] These facts are
consistent with the operation of the bimolecular displacement mechanism (Scheme 6, route b).

The rate of elimination of 9-halofluorenes sometimes shows a first-order dependence in both the organic halide and the medium basicity when sodium or
potassium t-butoxides are used as bases in t-butyl
alcohol, even though exchange of 9-deuterium is still
a rapid process. Factors that promote this apparent
change in mechanism are the presence of powerful electron-withdrawing substituents in the fluorene nucleus
(e.g., 2-NO$_2$, 2-CN, 2-Cl, 2-Br, all for X = Br, in 6;
2-NO$_2$, 2-CN, for X = Cl, in 6), bromine rather than
chlorine as the leaving group, and the addition of dipolar aprotic solvents such as tetramethylene sulfone
and dimethyl sulfoxide.[51,53] These results all accord
with the carbene mechanism, but a number of additional
considerations deter one from this interpretation.

Whatever the reaction order, the yields of bifluorenylidene are quantitative, when the reaction is
carried out with the rigorous exclusion of oxygen, even
at spectroscopic concentrations. If a carbene is involved, it shows remarkable selectivity, combining only
with the carbanion (or organic halide) rather than with
the solvent molecules, which are present in considerable excess. For the first-order reactions, the leaving group effect, k_{Br}/k_{Cl}, is only 1.5 (2-cyano-9-
halofluorenes), a very small effect if C-X cleavage is
involved in the rate-determining step. For substrates
following the second-order law, k_{Br}/k_{Cl} = 40 (2-bromo-
9-halofluorenes). Finally under all reaction conditions, no cyclopropane adducts were formed when alkenes
were added to the reaction solutions. Consequently,[51]
rather than propose a carbene mechanism, the authors
modified the bimolecular displacement mechanism (see
Scheme 7) and suggested that the first-order dependence
in organic halide arises when ion-pair dissociation to
give more reactive solvent separated or free ions[54]

becomes rate determining (Eq. 10). In support of this idea, addition of small amounts of sodium tetraphenyl-boron or sodium perchlorate to the reaction of 2-bromo-9-bromofluorene with sodium t-butoxide in t-butyl alcohol, causes the reaction order to revert from unity, in the organic halide, to two, as $k_{-a}(\overset{+}{M})$ becomes greater than $k_c(Fl\overline{H}X)$.[51]

Scheme 7

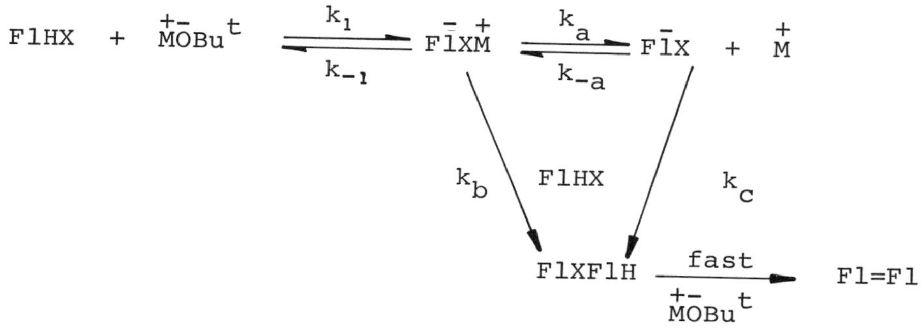

$$Fl HX \; + \; \overset{+-}{M}OBu^t \; \underset{k_{-1}}{\overset{k_1}{\rightleftharpoons}} \; Fl\overline{X}\overset{+}{M} \; \underset{k_{-a}}{\overset{k_a}{\rightleftharpoons}} \; Fl\overline{X} \; + \; \overset{+}{M}$$

$$k_b \searrow \quad FlHX \qquad \swarrow k_c$$

$$FlXFlH \; \xrightarrow[\overset{+-}{M}OBu^t]{fast} \; Fl{=}Fl$$

(FlH= 9-fluorenyl, X = Br or Cl, M = K or Na)

Contrary to an earlier preference for a carbene mechanism,[49] the recent observation of a second-order dependence of rate in diphenylmethyl chloride is consistent with the operation of the bimolecular displacement mechanism for the dimerization of this halide in strongly basic media.[54] On treatment with hydroxide ion in aqueous organic solvents, the 4-nitrobenzyl derivatives undergo α-hydrogen exchange much more rapidly than dimerization, the rate of the latter reaction following first-order kinetics in the substrate and base.[44,47] In the presence of 4-nitrobenzaldehyde, the carbanion intermediate can be diverted to give an epoxide. These results are consistent with a carbene mechanism (Scheme 6), but again carbene adducts have not been isolated when experiments are conducted with added alkenes. In these polar reaction media, rate-determining ion-pair dissociation seems unlikely, but first-order kinetic behavior in the substrate can be accommodated by a radical-anion mechanism. It is well known that C-alkylation products, formed in the reactions of o- and p-nitrobenzyl chloride with ambident nucleophiles (e.g., 2-nitropropane anion), arise via intermediate radical anions, which can be detected in solution by electron

$$\frac{d(Fl=Fl)}{dt} = \frac{k_1}{k_{-1}} \left\{ k_b + \frac{k_a k_c}{k_{-a}(\overset{+}{M}) + k_c(F1HX)} \right\} (F1HX)^2 \ (MOBu^t)^{+-} \qquad (10)$$

Order in F1HX is, two if $k_b \gg k_a k_c/[k_{-a}(\overset{+}{M}) + k_c(F1HX)]$.

or if $k_{-a}(\overset{+}{M}) \gg k_c(F1HX)$,

is one if $k_b \ll k_a k_c/[k_{-a}(\overset{+}{M}) + k_c(F1HX)]$.

and $k_{-a}(\overset{+}{M}) \ll k_c(F1HX)$.

spin resonance spectroscopy.[55]   Such reactions are
catalyzed by light, and precautions were not taken to
exclude this potential catalyst in the dimerizations of
the 4-nitrobenzyl derivatives. It is worth noting that
preliminary results show that o-dinitrobenzene, an
efficient trapping agent for radical anions, almost
totally prevents the formation of the stilbene when
4-nitrobenzyldimethylsulfonium ion is treated with
hydroxide ion; 4-nitrobenzyl alcohol is the major reaction product, but is formed much more slowly than is
the stilbene in the absence of the inhibitor.[56]   Scheme
8 depicts a possible radical-anion mechanism for stilbene formation, inhibition by o-dinitrobenzene arising
because it interrupts the chain reaction by capturing
the 4-nitrobenzyl chloride radical anion (step g).
However, in the absence of confirmation of the radical-
anion intermediate by spectroscopic means, possibly
because of too short a lifetime,[55a] this mechanism remains a matter of conjecture at the present time.

Scheme 8

(a) $Ar'CH_2Cl$ + $\bar{O}H$ $\xrightarrow{\text{fast}}$

$Ar'\bar{C}HCl$ + H_2O

(b) $Ar'\bar{C}HCl$ + $Ar'CH_2Cl$ $\longrightarrow$

$Ar'\dot{C}HCl$ + $(Ar'CH_2Cl)^{\cdot-}$

(c) $(Ar'CH_2Cl)^{\cdot-}$ $\longrightarrow$

$Ar'\dot{C}H_2$ + $\bar{C}l$

(d) $Ar'\dot{C}H_2$ + $Ar'\bar{C}HCl$ $\longrightarrow$

$(Ar'CH_2CHClAr')^{\cdot-}$

(e) $(Ar'CH_2CHClAr')^{\cdot-}$ + $Ar'CH_2Cl$ $\longrightarrow$

$(Ar'CH_2Cl)^{\cdot-}$ + $(Ar'CH_2CHClAr')$

(f) $Ar'CH_2CHClAr'$ $\xrightarrow[\bar{O}H]{\text{fast}}$

$Ar'HC=CHAr'$

(g) $(Ar'CH_2Cl)^{\cdot -}$ +

[structure: benzene ring with two NO_2 groups (ortho)] ⟶

[bracketed radical anion: benzene ring with two NO_2 groups (ortho)]$^{\cdot -}$

$Ar'CH_2Cl$ +

$(Ar' = p-NO_2-C_6H_4-)$

(iii) <u>The Fritsch-Buttenberg-Wiechell rearrange-</u>ment.[25] In 1894, Fritsch, Buttenberg and Wiechell,[57] found that 1,1-diaryl-2-chloroethylenes rearranged to diaryl acetylenes when treated with sodium ethoxide in ethanol at 180 to 200°C, and proposed a mechanism involving a vinylidene carbene (Eq. 11). The rearrangement can also be brought about by alkali metal amides in liquid ammonia,[58] and metal alkyls in nonpolar solvents.[59,67]

[Equation structures:]

$\underset{Ar}{\overset{Ar}{>}}C=C\underset{Cl}{\overset{H}{<}}$ ⟶ $\left[\underset{Ar}{\overset{Ar}{>}}C=C:\right]$ $\xrightarrow{EtOH}$

$Ar-C\equiv C-Ar$ + $\underset{Ar}{\overset{Ar}{>}}C=C\underset{H}{\overset{OEt}{<}}$

(11)

With alkoxide bases, α-hydrogen exchange with the medium is a rapid process compared to rearrangement, and occurs with retention of configuration.[60] Carbon-labeling experiments clearly demonstrate that the aryl group originally trans to the halogen migrates (Scheme 9).[61] This observation excludes a carbene mechanism as the symmetry of the carbene with respect to the olefinic axis should not give rise to such a stereoselective reaction. The order of reactivity with changing halogen (Br > I >> Cl) is explicable by ease of heterolysis of the carbon-halogen bond and the acidity of the vinylic hydrogen.[60] Electron-releasing substituents in the aryl group, situated trans to the halogen, increase the rate of rearrangement, suggesting that the migrating group develops some carbonium ion

Scheme 9

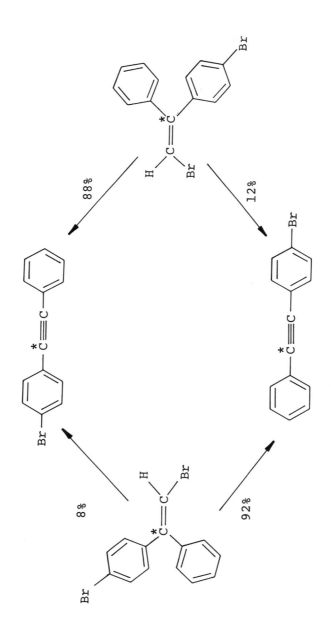

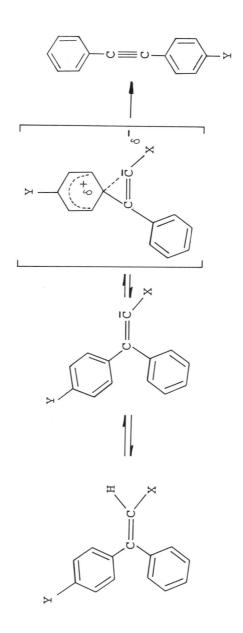

Scheme 10

character in accordance with the mechanism of Scheme 10.[62,63]

The α-deuterated material reacts more slowly than 1-chloro-2,2-diphenylethylene when treated with lithium alkyls in diethyl ether.[59] The lesser reactivity of an ion pair than a dissociated carbanion may account for the rate-determining proton abstraction as the isotope effect is not observed in the more solvating tetrahydrofuran, in which rearrangement is a slower process than proton removal. In fact, in this solvent, the halovinyllithium compounds can be trapped as their carboxylation derivatives when metallation is carried out at low temperatures, followed by treatment with carbon dioxide.[64] Metal-hydrogen exchange is also accompanied by metal-halogen exchange, but the latter reaction predominates only when bromine and iodine are the halogen groups (Scheme 11).[59] The choice of metal alkyl, salt effects, and substituents in the aryl groups are other factors that govern the partition between the two modes of metal exchange.[59,65]

Scheme 11

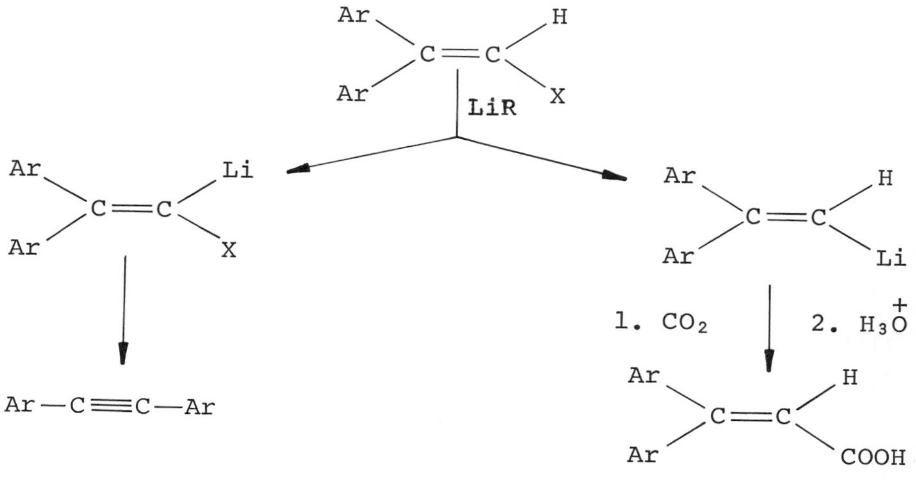

The Fritsch-Buttenberg-Wiechell rearrangement can be suppressed by linking the aryl groups. Thus 9-bromomethylenefluorene, 8, dimerizes to the cumulene 9, on treatment with potassium amide in liquid ammonia (presumably by a mechanism similar to that found for 9-bromofluorene, see previous section), steric considerations preventing the formation of the six-membered cyclic acetylene 10.[65,66] However, greater flexibility allows the intermediacy of the 8-membered cyclic acetylene 11, which undergoes an addition reaction with the base to give <u>12</u> (Scheme 12).[65]

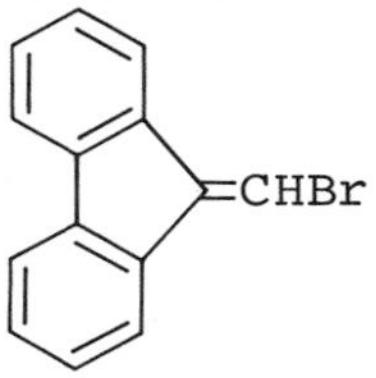

<u>8</u> 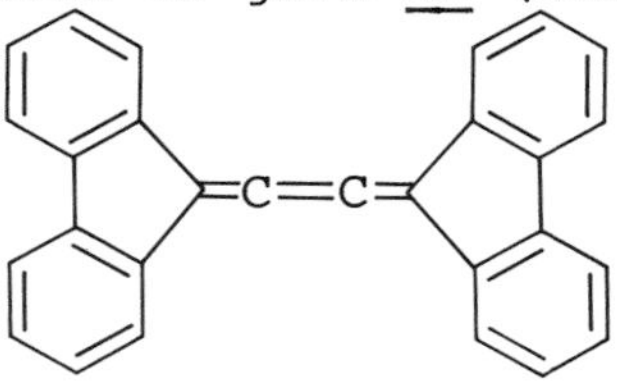   <u>9</u> 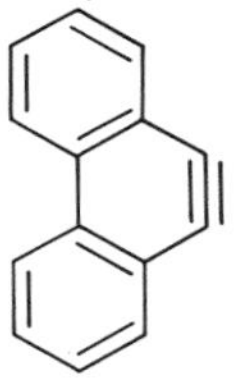   <u>10</u>

Scheme 12

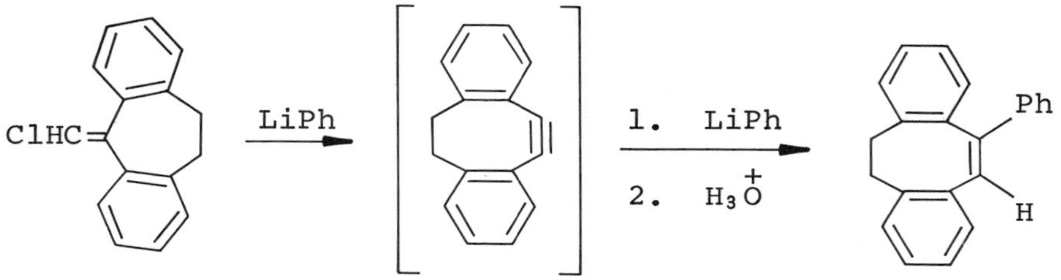

<u>11</u>   <u>12</u>

(c) <u>Decomposition of Diazoalkanes</u>. The chemistry of
diazoalkanes has been reviewed recently.[68] Under
appropriate conditions, diazomethane can behave as an
acid, a base, an electrophile, a nucleophile, a 1,3-
dipole, or as a carbene source. Spectroscopic techniques have been used to establish the intermediacy of
carbenes in photolytic and thermal decompositions of
aliphatic diazo compounds in the gas phase (Section
X.B.2). It is generally assumed that carbenes are
also formed in the thermal or photolytic decompositions
of more stable aryl-substituted diazo compounds in
solution. In aprotic solvents, the formation of dimeric products in significant yield, along with insertion products, can be rationalized in terms of carbene
intermediates. Thus azine arises from attack of carbene at the terminal nitrogen of a second molecule of
diazoalkane, whereas attack at the central carbon
gives the dimeric olefin (Scheme 13).[69] This view can
be an oversimplification as, in some cases, the consumption of diazoalkane follows both a first- and
second-order law in diazoalkane.[70,71] Thus dimeric
products can be formed directly from a dimerization of
the diazo compound without carbene formation,[71] and can

also arise via a catalytic process involving the azine
product[69b] or the dimeric olefin.[71] In one instance,
a surface-catalyzed decomposition of a diazoalkane has
been reported.[69c]

Scheme 13

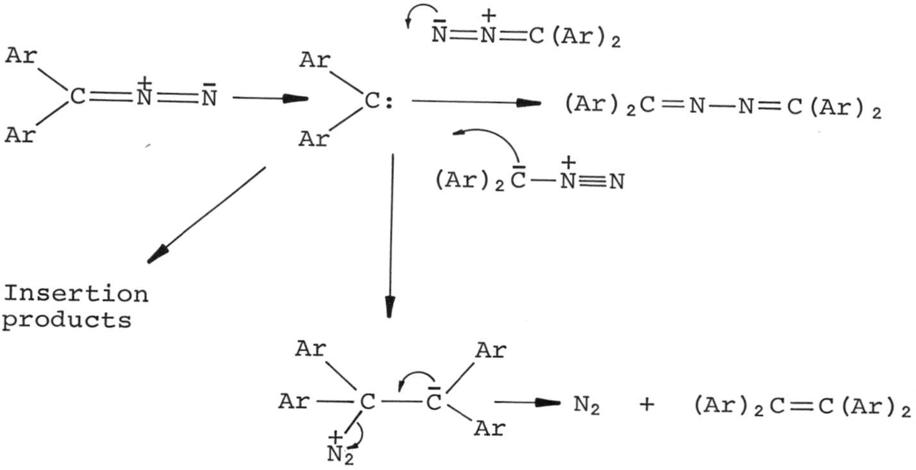

Acid catalysis of diazoalkanes can often lead to
apparent carbenoid products. In olefinic solvents containing trifluoroacetic acid, phenyldiazomethane gives
mainly phenylcyclopropanes, along with less abundant
products, which are characteristic of carbonium ion intermediates. The cyclopropanes are not carbenoid
products as they are not formed in the absence of the
added acid. Deuterium-labeling studies exclude a discrete benzyl cation as a precursor for the cyclopropanes, and the authors proposed that the role of the
acid was confined to a hydrogen-bonding interaction
with the substrate (13).[72]

13

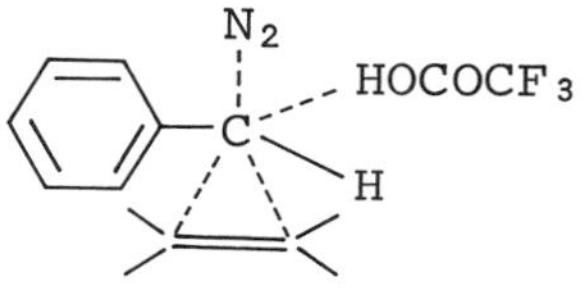

(d)   The Bamford-Stevens Reaction.[73] This reaction
concerns the base-induced decomposition of

tosylhydrazones (Eq. 12) in protic or aprotic solvents,
usually between 160 to 180°C. Kinetic studies reveal
a rate-determining cleavage of the tosylhydrazone
anion,[74] the resulting diazoalkane being isolable in a
few cases,[73,75] but more often decomposed rapidly under
the reaction conditions. Despite the medium basicity,
a carbonium-ion mechanism is usually favored for the
diazoalkane decomposition in protic solvents, whereas
a carbenoid route is preferred for reactions in aprotic
media.[74,76] Energetic cations and carbenes often behave similarly in skeletal rearrangements,[77] and the
necessary methodology to distinguish between these intermediates is amply illustrated by the elegant labeling studies of Nickon and Werstiuck (Scheme 14).[78]
For the carbonium-ion route, a nonclassical ion, 14,
is not the sole intermediate, as the label distribution
in the products differs for the two reactions.

$$R^1R^2C{=}N{-}NHSO_2C_7H_7$$

$$R\bar{O} \quad \Big|\Big| \quad ROH$$

$$R^1R^2C{=}N{-}\bar{N}SO_2C_7H_7$$

$$\Big| \quad slow \qquad\qquad (12)$$

$$R^1R^2C{=}N_2 \quad + \quad C_7H_7S\bar{O}_2$$

$$\Big| \quad fast$$

products

Scheme 14

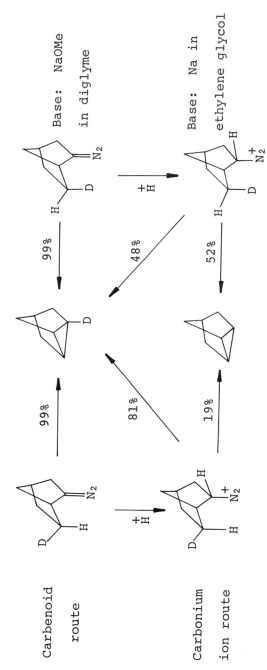

Carbenoid
route

Carbonium
ion route

Base: NaOMe
in diglyme

Base: Na in
ethylene glycol

99%

48%

52%

99%

81%

19%

$\underline{14}$

(The above diazoalkanes were generated from the corresponding tosylhydrazones <u>in</u>
<u>situ</u>. The yield of tricyclene was 99% in the aprotic solvent and 93% in the <u>protic</u>
solvent.)

561

A diazonium ion (protonation of diazoalkane) or carbonium ion has been proposed as an intermediate in the Bamford-Stevens reaction of 15, as the alcohol fractions in the product are almost identical to those found in the deamination of the corresponding amine.[79] In both these highly energetic reactions, a reactant-like transition state is expected in the reactions of the ion, and orbital overlap with the ethylenic double bond has little influence on the rearrangement to 17 and 18. However, a more productlike transition state is anticipated in the solvolysis of the tosylate 19, and the only rearranged alcohol formed is 17. Simple Hückel molecular orbital calculations indicate ion 20, the precursor of 17, is about 5 kcal/mole more stable than 21, in which overlap between the vacant carbonium-ion orbital and the olefinic bond is minimal.

Scheme 15

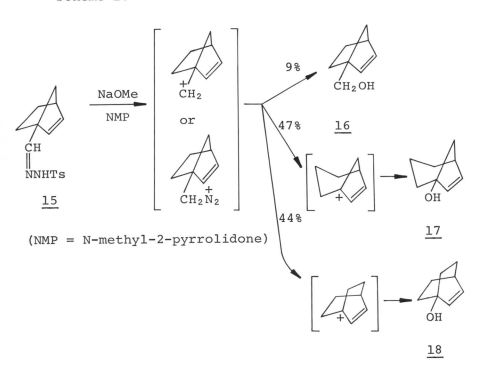

(NMP = N-methyl-2-pyrrolidone)

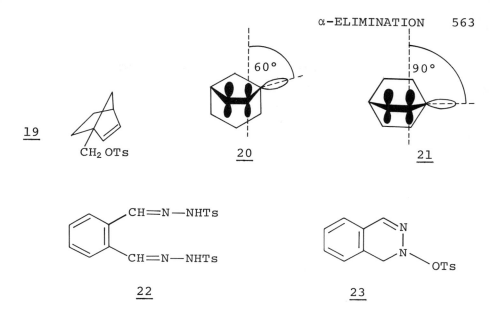

19   20   21

22   23

A number of interesting Bamford-Stevens reactions
on bistosylhydrazones have been reported, but dicar-
benoid intermediates do not appear to be involved.[80,81]
For example, 22 gives 23, rather than benzocyclobutene.

(e) Organometallic Reagents.[15,17,82] Methylene, generated in the photolysis or thermolysis of diazomethane
undergoes both addition to alkenes and insertion reactions, but that potentially formed by other routes
(CH_2N_2/Cu; CH_2I_2/Zn-Cu; CH_3Cl/NaPh) exhibits only the
addition-reaction mode. Similar results have been
found for carbethoxycarbene and chlorocarbene, both
reaction modes being observed in the simple diazoalkane
decompositions but only addition being observed for alternative methods of carbene generation
($CH(COOEt)N_2$/Cu; CH_2Cl_2/LiR).[83,84] The reactions giving only addition products appear to be more realistically regarded as carbenoid processes involving organometallic complexes. Certainly such a complex is involved in the Simmons-Smith reaction, as the reaction
is first order in both the olefin and the methylene
halide (Eq. 13), and the stereospecific syn-addition
is subject to a considerable steric effect.[85]

$$2CH_2I_2 + 2Zn \longrightarrow (ICH_2)_2ZnZnI_2 \xrightarrow[\text{slow}]{2 \searrow C=C \swarrow} 2 \triangle$$

$$+ 2ZnI_2 \qquad (13)$$

A carbene intermediate appears to be involved in the decomposition of $PhHgCCl_2Br$, as the rate of reaction is almost independent of the alkene concentration and is retarded by added PhHgBr (Eq. 14).[86]   In fact, it seems that independent of the method of generation, "dichlorocyclopropanation" involves free dichlorocarbene as Skell and Cholod have shown that the **selectivity** of this reagent toward alkene pairs is almost independent of the source of the divalent species.[87]

$$PhHgCCl_2Br \overset{slow}{\underset{fast}{\rightleftharpoons}} PhHgBr + :CCl_2 \overset{>C=C<}{\underset{fast}{\longrightarrow}}$$

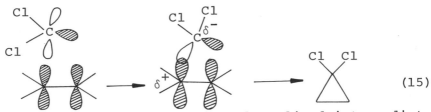

(14)

The reactivity of dichlorocarbene toward various alkyl-substituted olefins correlates with the Hammett-Taft equation; $\log k/k_O = -4.3 \sigma^* + 1.0E_S$, where E_S is the steric substituent constant. This numerical relationship bears a close resemblance to that for bromination of alkenes, another electrophilic addition reaction, for which $\log k/k_O = -5.43 \sigma^* + 0.96E_S$.[88] However, there is a lack of correlation between hydration and addition of dichlorocarbene to alkenes. Skell and Cholod attribute this difference to a significant variation in the reaction mode. Hydration involves a cation intermediate in which charge is localized on one of the olefinic carbon atoms, whereas the dichlorocarbene acts as a bridging electrophile, and only partial carbonium ion character develops on one of the olefinic carbon atoms (Eq. 15).[87]

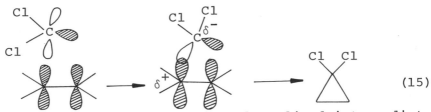

(15)

Di(phenylthio)carbene is a long-lived intermediate (presumably because of effective overlap involving the low-lying d-electrons of sulfur) in the decomposition of tri(phenylthio)methyl lithium (Scheme 16). Added lithium thiophenoxide reduces the rate of appearance of dimeric product, electrophiles capable of scavenging lithium thiophenoxide increase the rate, and added Me-C_6H_4-Li is incorporated into the product.[89]

Scheme 16

$$(PhS)_3CLi \xrightleftharpoons{\text{slow}} Ph\overset{\cdot\cdot}{S}CSPh \ + \ PhSLi \xrightarrow{(PhS)_3CLi}$$

$$(PhS)_2C=C(SPh)_2$$

4   α-Elimination at Nitrogen

Space restricts our discussion on this topic, and the reader is referred to a number of authorative reviews on the chemistry of nitrenes and rearrangements involving electron-deficient nitrogen atoms.[90-95] By using matrix isolation techniques, both singlet and triplet nitrenes have been characterized by spectroscopic methods. Azide decompositions constitute the most well-characterized nitrene-forming reactions. Triplet states usually arise in photolytic decomposition, and singlet states are formed thermally. A wide range of nitrenes has been prepared (e.g., RN: , ArN: , RSO_2N: , $R_2C=CHN$:), and these are much more useful in synthesis than the corresponding carbenes, as they show greater selectivity in reaction.[94] Acyl nitrenes have not been prepared, and reactions in which these species might be intermediates are regarded more realistically as concerted rearrangements. For example, electron-donating substituents markedly increase the rate of the Hofmann rearrangement, the reaction constant $\rho = -2.5$,[96] being similar to that observed in the solvolysis of 2-aryl-2-methyl-1-propyl p-bromobenzene sulfonates ($\rho = -3.0$) in which aryl participation is well documented.[97,98] The concerted rearrangement mechanism illustrated for the Hofmann reaction in Scheme 17 is typical of the mechanisms accepted for the related Schmidt, Curtius, Lossen, and Beckmann rearrangements.[98]

Scheme 17

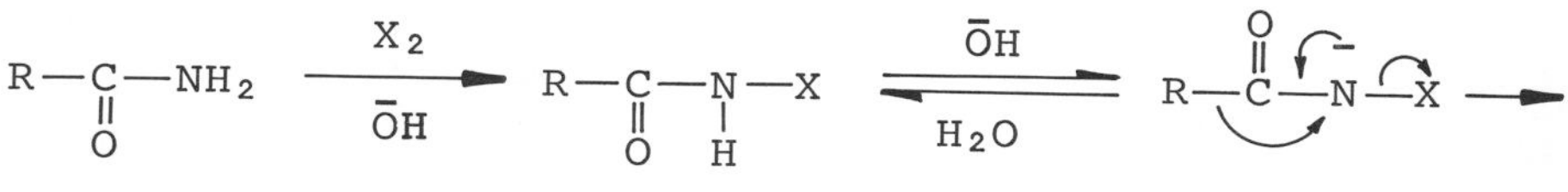

$$\left[\begin{array}{c} O \\ \parallel \\ C \\ R'\text{----}N\text{----}X \end{array} \right] \longrightarrow R\text{---}N\text{=}C\text{=}O \xrightarrow[\text{H}_2\text{O}]{\overline{\text{O}}\text{H}} RNH_2 \quad + \quad CO_2$$

C   γ-ELIMINATION

1   Nomenclature

Despite the increased interest in mechanisms of γ-elimination in recent years, the reviews available provide
specific[99-101] rather than general coverage.[102]   In
this book, we use a common nomenclature for all -elim-
inations. Within a five-center system, 24, X and Y
are the leaving groups and the atoms a, b, and c are
labeled with respect to the more electronegative group
Y, X in most instances being hydrogen.[102]   This approach differs from that used previously for reactions
such as the Favorskii,[103] for which the terms α' and
α (25) are equivalent to γ and α, respectively, in our
terminology.

24      $\overset{\alpha}{\text{Y}}\text{---}\overset{\beta}{\text{a}}\text{---}\overset{\gamma}{\text{b}}\text{---}\text{c}\text{---}\text{X}$ 25 $H\text{---}\overset{\alpha'}{C}\diagdown\quad\diagup\overset{\alpha}{C}\text{---}Y$

$$\begin{array}{c} C \\ \parallel \\ O \end{array}$$

     Free rotation about bonds a-b and b-c gives rise
to numerous conformations, and consequently there are
a greater number of stereochemical arrangements for
γ- than β-elimination. Considering only staggered conformations, Nickon and Werstiuck have suggested a con-
cise nomenclature to describe the five arrangements
for concerted γ-elimination (Table 1).[104] When the
reactant conformations are modified to transition
states, only four arrangements are possible. Of the
substrate conformations, only the W-form can go directly to a productlike transition state without conformational rotation.
     Two-step reactions are more likely for γ- than
for β-eliminations, because of the greater separation
of the leaving groups and there is considerable experimental evidence to support this view. Nickon and
Werstiucks' nomenclature to cater for γ-ElcB and γ-El
reactions is shown in Scheme 18, a rate-determining

TABLE 1 Stereochemical Nomenclature for Concerted γ-Eliminations[104]

Staggered Substrate	Productlike Transition State	Short Notation	Terminology	Stereochemical Consequence
1.			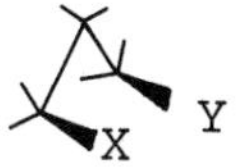 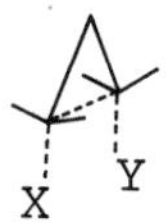 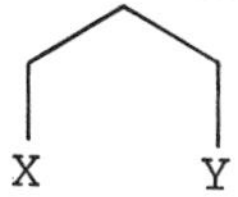 U	Retention
2.			   W	Inversion
3.			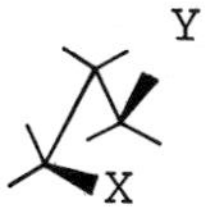  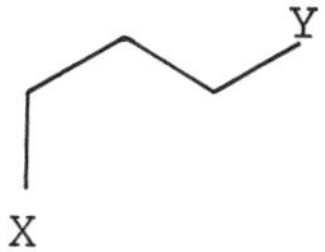 exo-Sickle	Retention at c[a] Inversion at a[a]
4.			 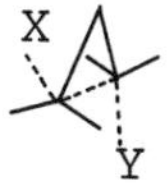 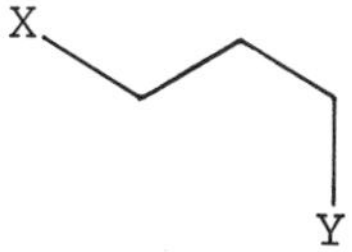 endo-Sickle	Retention at a[a] Inversion at c[a]

TABLE 1 Continued

Staggered Substrate	Productlike Transition State	Short Notation	Terminology	Stereochemical Consequence
5. 			apo-Sickle	Depends on which product-like transition state is adopted.

aFor definitions of a and c, see 24.

ring closure from the carbanion or carbonium ion being assumed, and detectable stereochemical changes now being confined to atom a and c, respectively. γ-Elimination involving a carbanion can also be regarded as an example of an internal nucleophilic displacement reaction.

Scheme 18   Nomenclature for the stereochemical Arrangements of γ-ElcB and γ-El reactions

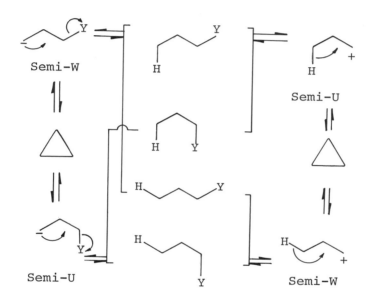

2   Reactions Giving Cyclopropanes

(a)   Base-Catalyzed Reactions.   3-Phenylpropyl derivatives undergo competitive β- and γ-elimination in strongly basic media (Eq. 16), β-elimination predominating when X = Br or Cl, and **cyc**lopropane formation being favored when X = F, tosylate, and $\overset{+}{N}Me_3$.[105], [106] Under Hofmann-elimination conditions (pyrolysis of the quaternary hydroxide), 3-phenylpropyltrimethylammonium ion decomposes only to the alkene, but 3,3-diphenylpropyltrimethylammonium ion gives 1,1-diphenylpropene (72%) and also 1,1-diphenylcyclopropane (28%). Thus cyclopropane formation is also promoted by a more acidic γ-hydrogen, a stronger base, and a poorer leaving group. 3-(o-Tolyl)propyltrimethylammonium ion eliminates to give the cyclopropane in 87% yield when sodium amide in ammonia is the base, but

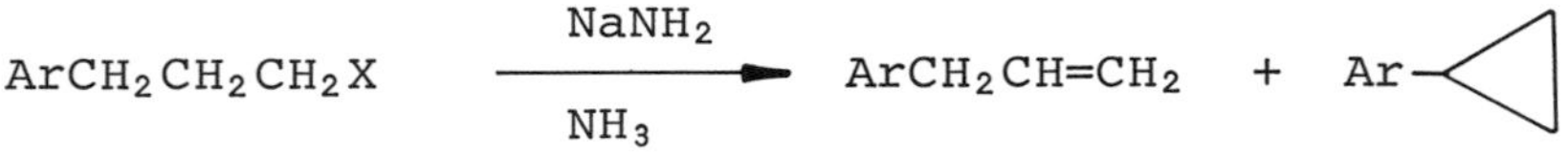

$$\text{ArCH}_2\text{CH}_2\text{CH}_2\text{X} \xrightarrow[\text{NH}_3]{\text{NaNH}_2} \text{ArCH}_2\text{CH=CH}_2 \;+\; \text{Ar}\!-\!\!\triangleleft \qquad (16)$$

the greater steric hindrance to proton abstraction (and possible subsequent displacement by the carbanion) precludes γ-elimination for the 3-mesityl derivative. This fact suggests that the γ-C-H bond is involved in a relatively selective step and rules against cyclopropane formation arising via an α-elimination followed by a carbene insertion. This mechanism is accepted for the reactions of phenyl sodium with deuterated n-butyl chloride (Eq. 17),[107] and triphenylmethyl sodium with i-butyl diphenylsulfonium fluoroborate.[108] However, more definitive information would be preferable, and labeling experiments combined with a study of the effect of aryl substituents could fulfill this requirement.

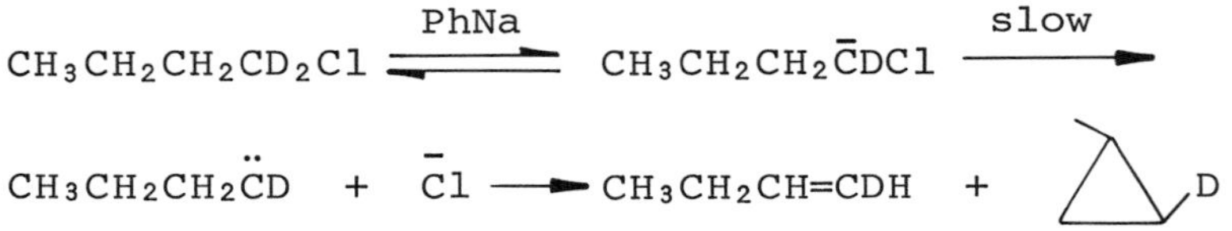

$$\text{CH}_3\text{CH}_2\text{CH}_2\text{CD}_2\text{Cl} \underset{\text{PhNa}}{\rightleftharpoons} \text{CH}_3\text{CH}_2\text{CH}_2\bar{\text{C}}\text{DCl} \xrightarrow{\text{slow}}$$

$$\text{CH}_3\text{CH}_2\text{CH}_2\overset{..}{\text{C}}\text{D} \;+\; \bar{\text{C}}\text{l} \longrightarrow \text{CH}_3\text{CH}_2\text{CH=CDH} \;+\; \triangle\!\!-\!\text{D}$$

$$(17)$$

Cristol and his co-workers have shown that γ-ElcB elimination occurs in some cases only if in the carbanion, the leaving group is situated in a semi-W arrangement (cf. Scheme 18). 2-Quadricyclylphenyl sulfone (26) is formed only from the exo-bromide (Eq. 18).[109] Similarly endo-chloride, 27, epimerizes

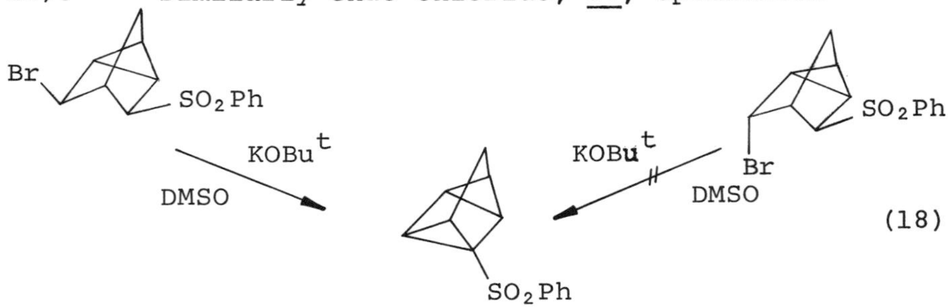

$$(18)$$

26

rapidly in ethanol containing sodium ethoxide at room temperature, the equilibrium slightly favoring epimer 28 (NMR). γ-Elimination occurs when the solution is heated under reflux.[110] On the other hand, the exo-chloride 29, epimerizes rapidly to give mainly its epimer 30, but γ-elimination is not observed even after heating under reflux is prolonged for several days. In more strongly basic media, the exo-chlorides do undergo γ-elimination, but this involves initial epimerization at the benzylic α-carbon to give the endo-isomers, 27 and 28, which then eliminate rapidly.[111]

Nickson and Werstiuck have studied the decomposition of norbornyl tosylates in t-butyl alcohol in the absence and presence of potassium t-butoxide.[112] Under solvolytic conditions (Eq. 19), the rate of decomposition of exo-norbornyl tosylate follows first-order kinetics and is not affected significantly by concentrations of added base of less than 0.04M. The exo- and endo-6-deuterated compounds, 31 and 32, both give the same sample mixture of exo-t-butyl ether (I.R. analysis) in which the deuterium is totally retained, and the same fraction of deuterium is lost in both samples of nortricyclene. These facts accord with the operation of a γ-El mechanism involving a bridged or rapidly equilibrating norbornyl cation. Greater concentrations of added base cause an increase in the elimination component and a change to second-order kinetics. However, the two samples of nortricyclene contain different amounts of deuterium (Eq. 20), and after allowance is made for the unimolecular reaction, it is clear that the exo-sickle arrangment is preferred to the W-arrangement for the base-catalyzed γ-elimination of exo-norbornyl tosylate.

70%   10%   20%

(19)

31   32   33

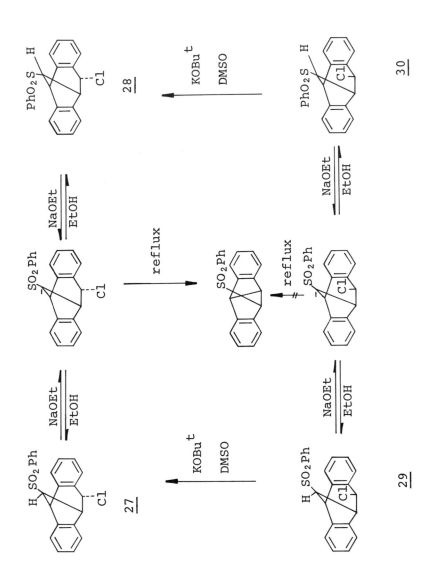

Scheme 19

(20)

(Base concentration = 0.6 - 0.9M.)

A similar procedure was used to show that for base-catalyzed γ-elimination from endo-norbornyl tosylate, 33, the U-arrangement is preferred to the endosickle arrangement.[113] The preference for the U-arrangement appears to contradict the findings of Cristol for 29 and 30, but it should be noted that these isotopic analyses refer to reaction products formed at only one temperature.

For hydrogen-deuterium exchange of the carbon-hydrogen bond adjacent to substituted aryl sulfones, a Hammett reaction constant of 2.8 is observed.[114] This value is greater than that of 2.32 found for the cyclization of aryl-3-chloropropyl sulfones by $KOBu^t$-t-BuOH, and originally a γ-E2 mechanism was proposed for this reaction.[115,116] However, a γ-hydrogen exchange is a faster process than elimination, and the isotope effect k_H/k_D = 0.5($ArSO_2CH_2CH_2CH_2Cl$ in t-BuOH/$ArSO_2CD_2CH_2CH_2Cl$ in t-BuOD).[116] This latter value could constitute a primary, secondary, and solvent isotope effect, but clearly reflects only the last term, the deuterated medium being more basic. Thus a stepwise mechanism (γ-ElcB) is more plausible, and the reaction constant of only 2.32 indicates the effect of aryl substituents on both the concentration and nucleophilicity of the carbanion.

Stirling and his co-workers[115-118] have compared the rates of cyclization of some alkyl halides bearing electron-withdrawing substituents on the γ- and ξ-carbons. The three-membered ring is formed more easily than the five-membered ring when a conjugative group is attached directly to the ring. Consequently, they have suggested that conjugative interaction between this substituent and the forming ring is greater for γ-eliminations, and aids stabilization of the transition state for cyclization. The partial double-bond character of cyclopropyl rings has been well characterized.[119] (For further discussion of the relative rates of closure of various sized rings, see Table 3 in Section X.d).

(b) Reactions Involving Cationic Intermediates.
Cyclopropanes are sometimes found among the reaction
products in the decomposition of certain carbonium
ions. The solvolysis of exo-norbornyl tosylate in t-
butyl alcohol, to yield nortricyclene, as discussed in
the previous section, is a typical example of a γ-E1
reaction. However, except for solvolyses of bridged-
type structures such as the norbornyl system, E1 reactions of alkyl halides and esters usually give only
olefinic and substitution products (see Section V).
On the other hand, the more reactive carbonium ions
formed in the acid-catalyzed decomposition of aliphatic diazo compounds, or in deamination or deoxidation reactions, are less selective and often give
significant yields of cyclopropanes. The mechanisms of
these reactions are discussed elsewhere (Section VII.
A,B; Section X.B.3.c). These reactions should not be
considered as good synthetic methods because of the
multitude of products often formed.

(c) Dehalogenations. The β-elimination of vicinal
dihalides on treatment with reducing agents or nucleophiles is well-known as a synthetic method for the
preparation of alkenes (Section VII.C) and alkynes (Section IX.B.1). In recent years, a number of γ-eliminations have been effected from 1,3-dihalides. An interesting example is the formation of bicyclo[1,1,0]-
butane from 3-chloro-1-bromocyclobutane (Eq. 21).[120]
Attempted cyclization with zinc in aqueous ethanol
gives only cyclobutene (23%), butadiene (69%), and 1-
butene (5%).

Br

Na
⟶
dioxane

+   (21)

93 to 96%   5 to 7%

(Cis or trans)

The stereoisomeric 2,4-dibromopentanes give low
yields of cyclopropanes on treatment with n-butyl
lithium (Table 2).[121] 2-Bromopentane is isolated as
one of the major reaction products, so it seems that
the elimination comprises a lithium-halogen exchange

to give a carbanion, which then cyclizes. Metal-halogen exchange usually occurs with retention[122] and, therefore, substitution must also occur with retention at the α-carbon to explain the stereochemistry of the prominent product in each case. Thus these reactions require a U-(or semi-U) conformation of the leaving groups, analogous to the syn-β-eliminations that predominate in the dehydrohalogenations induced by lithium-alkyl bases (Section IX.B.4).

Cristol, Dahl, and Lim,[123] have compared the relative yields of cyclopropane 34 to hydrocarbon 35 formed in the reactions of dibromides 36 and 37 with a

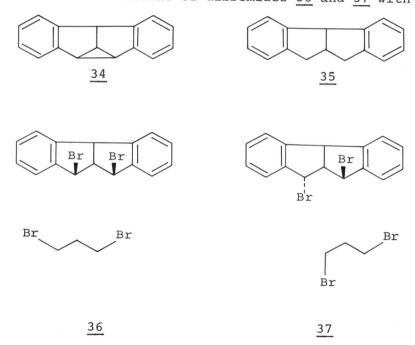

variety of reagents. Dibromide 36 is a model for W-stereochemistry, and 37 is a model for the exo-sickle alignment of the leaving groups. The radical reduction with triphenyltin hydride converts both dibromides into mainly hydrocarbon 35. With magnesium in ether or sodium in glyme, both dihalides give high yields of the cyclopropane 34. The yield of cyclopropane is greater from 37(80%) than from 36(42%) when elimination is effected with lithium aluminum hydride in tetrahydrofuran, showing here a preference for the exo-sickle geometry. However, in the absence of a knowledge of

TABLE 2 Products of γ-Elimination of 2,4-Dibromopentanes with n-Butyl Lithium[121]

Isomer	Temperature °C	% 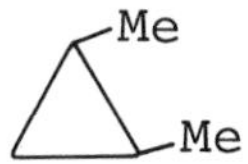	% 	% Yield of Cyclopropanes
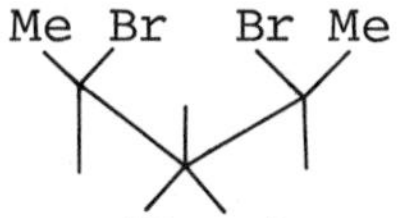 (Meso)	-30	82.2	17.8	2.3
	-78	90.4	9.6	1.7
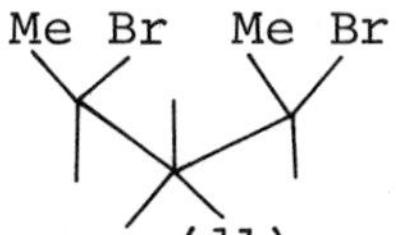 (dl)	-30	24.3	75.7	7.0
	-78	16.5	83.5	5.6

the relative rates of the two reductions, such a conclusion is tentative. Elimination from the exo-sickle conformation is also favored under protic conditions, isomer 37 giving only cyclopropane with zinc (or Zn-Cu couple) in ethanol, while 36 gives only hydrocarbon 35.
    Cyclopropane formation via dechlorination with metals[111] and deiodination with either lithium aluminum hydride[124] or peroxide catalysts[125] have also been reported.

(d) _Pyrolytic Eliminations_. The pyrolyses of endo-2-norbornyl acetate and endo-2-norbornyl methyl sulfone yield norbornene (7 and 16%) and nortricyclene (93 and 84%, respectively).[126] Presumably, by analogy with the accepted syn-stereospecificity for the corresponding β-eliminations, these reactions utilize the U-arrangement, thereby involving the removal of the 6-endo-hydrogen. The Cope and Hofmann eliminations of the endo-2-norbornyl derivatives give only the alkene.
    The mechanism of pyrolysis of bornyl chloride to tricyclene (55%), bornene (20%), and camphene (25%) has already been discussed (Section VIII.B.5.c).[127]

3 Reactions Giving Three-Membered Heterocyclic Rings

1,3-Eliminations of HY to give a three-membered heterocyclic ring occur with particular facility if the heteroatom carries the hydrogen and Y is a halogen or ester (Eq. 22).[99] For the more nucleophilic atoms (e.g., X = NH), the cyclizations often occur in the absence of added base, the rate of reaction being first order. Basic reagents are required to effect the elimination with the weaker nucleophilic heteroatoms, and either γ-E2 or γ-ElcB mechanisms operate. The carbanion mechanism has been demonstrated for the ethylene oxide formation from 2-chloroethanol and aqueous sodium hydroxide as the isotope effect, $k_H/k_D = 0.65$ at 25°C ($ClCH_2CH_2OH$ in $H_2O/\bar{O}H$ / $ClCH_2CH_2OD$ in $D_2O/\bar{O}D$).[128] The

$$\begin{array}{c} H \\ \diagup \\ X \end{array} \quad \overset{X}{\diagup\!\!\diagdown} \qquad (22)$$

$$>\!\!C\!\!-\!\!C\overset{\alpha}{\underset{Y}{\diagdown}} \longrightarrow >\!\!C\!\!-\!\!C\overset{\alpha}{-} \; + \; HY \qquad (22)$$

(X = NH, NR, O, S)

value of less than unity is clearly inconsistent with a primary isotope effect expected for a γ-E2 mechanism, but accords with a carbanion mechanism in which the carbanion concentration is greater in the more basic deuteroxide medium. For some time it has been known that cyclization occurs with inversion at C_α in the formation of epoxides,[129] aziridines,[130] and α-lactams.[131]

β-Iodocarbamates are inert to mild bases such as pyridine or sodium bicarbonate, but cyclize to aziridines in alcoholic hydroxide solution.[132] Cyclization occurs prior to carbamate hydrolysis as the substituted ariridine 38 has been isolated (Scheme 20). The hydrolysis of the aziridine carbamate proceeds much more rapidly than that of the β-iodocarbamate, in which the ease of formation of the anion opposes further attack by hydroxide ion which would produce a doubly-charged intermediate.

Scheme 20

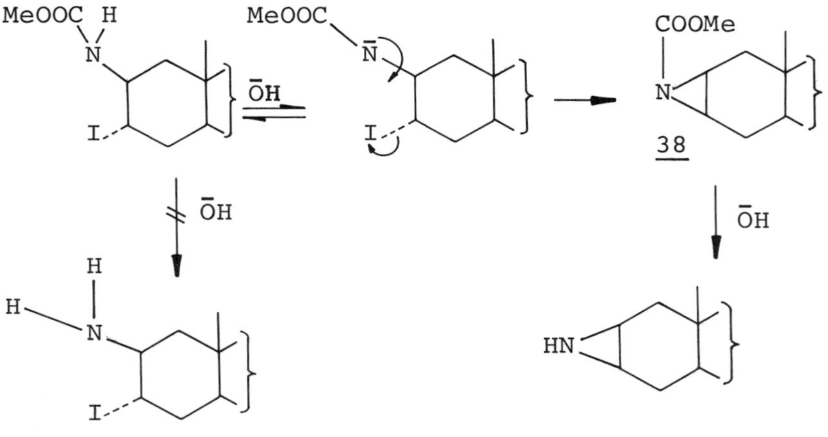

The solvolysis of α-bromocarboxylates occurs with retention of configuration, despite its being a first-order reaction (Eq. 23).[133] γ-Elimination gives an unstable α-lactone, which opens to the retention product. In methanolysis, 2-bromopropionate reacts 20 times faster than isopropyl bromide,[133] in which nucleophilic participation to ionization by an adjacent group is not possible. Reactions of this type have obvious common features with γ-elimination, but are more often classified under neighboring group

participation. To participate effectively, the neighboring group should be situated trans to the departing
group, a requirement amply illustrated by the 10^6-fold
greater rate of solvolysis of trans-2-chlorocyclohexyl-
phenyl sulfide than its cis-isomer in aqueous ethanol.[134]   Many groups possess lone pairs or π-electrons
that can aid displacements, and for more detailed
coverage readers are referred to the literature.[99,135]

$$R-\underset{\underset{O}{|}}{\overset{\overset{Br}{|}}{\underset{H}{C}}}-C=O \longrightarrow R-\underset{H}{\overset{O}{C}}-C=O \xrightarrow{H_2O}$$

$$R-\underset{H}{\overset{\overset{OH}{|}}{C}}-\underset{\underline{O}}{C}=O \qquad (23)$$

4 Reactions Involving Three-Membered Ring Inter-mediates[100]

In some cases the direct products of a γ-elimination
are unstable under the reaction conditions, and the intermediacy of a three-membered ring must be established
by indirect methods. Typical of this reaction category
are the well-known Favorskii, Ramberg-Bäcklund, and
Neber rearrangements.

(a) The Ramberg-Bäcklund Reaction.[101,136]   α-Halosulfones, possessing a γ-hydrogen, undergo 1,3-elimination
on treatment with base to give an episulfone, which
rapidly extrudes sulfur dioxide (Scheme 21). Halide-
ion release follows first-order kinetics in both base
concentration and the α-halosulfone,[137] and for reactions in deuterated media, the product alkenes are completely deuterated in the vinylic positions,[138] and exchange of both γ- and α-hydrogen is a faster process
than halide ion release.[138]   Carbenoid intermediates
can be excluded as the reaction works equally well for
both α,α-dihalo and α,α,α-trihalosulfones,[140] but it
fails for α-halosulfones that lack a γ-hydrogen. Thus
α-hydrogen exchange is a side reaction to the elimination. γ-Hydrogen exchange may also be a side reaction
to a concerted cyclization, but a γ-ElcB mechanism is
usually preferred. The leaving group effect, k_{Br}/k_{Cl}
= 620 (for $PhCH_2SO_2CHXPh$ at 0°C),[139] is much larger
than the ratio observed in β-eliminations in the

Scheme 21

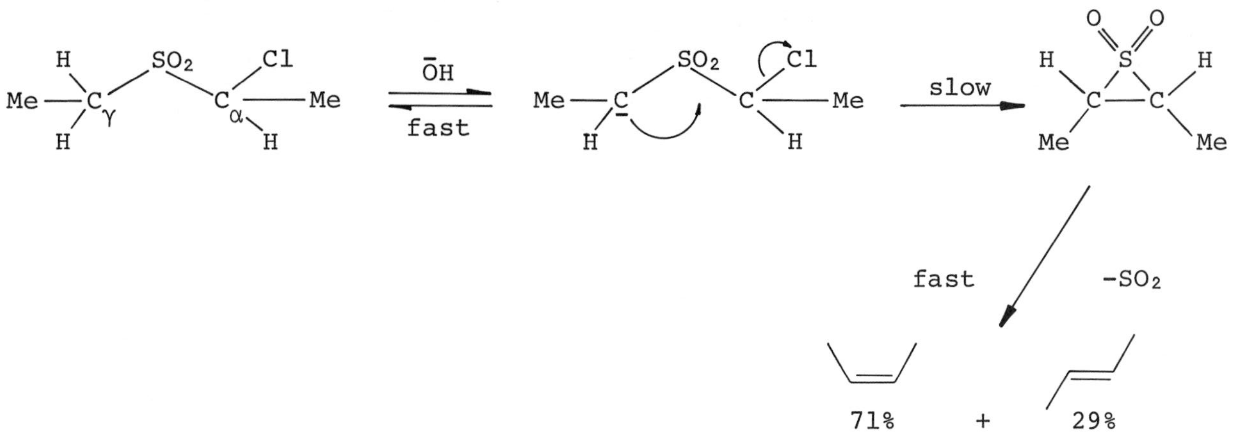

$-SO_2$

slow

fast

$\bar{O}H$

fast

71%   +   29%

2-phenylethyl series ($k_{Br}/k_{Cl} \simeq 40$). This implies considerable C-X bond breaking in the transition state for cyclization. However, $C_\gamma-C_\alpha$ bond formation appears to run ahead of C-X bond breaking, since a reaction constant of 0.8 is observed in the reaction of substituted α-halobenzylmethyl sulfones.[141]

The episulfones, prepared by an independent method, decompose more rapidly than their precursor α-halosulfones under the conditions of the Ramberg-Bäcklund reaction.[139,142] Unless very strongly basic conditions are used, or the episulfones possess particularly acidic hydrogens, the stereochemical arrangement of substituents in the episulfone is maintained in the alkene product. Thus the stereochemistry of the alkenes formed in the Ramberg-Backlund reaction is determined at the cyclization stage. The predominant formation of the thermodynamically less stable cis-olefin has been attributed to a number of factors. These include greater ease of solvation in the transition state leading to the cis-episulfone,[100] attractive London forces between alkyl groups becoming cis-related,[138c] and greater ease of rotation of the carbanion precursor in attaining the semi-W conformation for cyclization to the episulfone.[101]

One example [threo- PhCH(Me)SO$_2$CBr(Me)Ph with NaOMe/MeOH] of the Ramberg-Bäcklund reaction has been reported in which episulfone decomposition is rate determining.[143] Bromide ion release is first order in methoxide, but faster than trans-α,α'-dimethylstilbene formation, which is independent of methoxide concentration.

(b) The Favorskii Rearrangement.[101,103] A number of mechanisms have been considered to explain the Favorskii rearrangement of α-haloketones under basic conditions,[144-146] but we limit our discussion to the γ-elimination mechanism. This reaction sequence involves the formation of a cyclopropanone, which undergoes ring opening to give the more stable carbanion, and subsequently the carboxylic acid or ester (Scheme 22). This mechanism is usually accepted for α-halo-ketones which possess a γ-hydrogen.[147-149] Appearance of halide ion follows a first-order law in both base and α-haloketone. When halide ion release is rate determining, both γ- and α-hydrogens are rapidly exchanged.

The intermediacy of a cyclopropanone was first supported by the work of Loftfield on 2-chlorocyclo-hexanone-1,2-^{14}C with less than an equivalent of hydroxide ion (Scheme 22).[147] The label was initially

Scheme 22

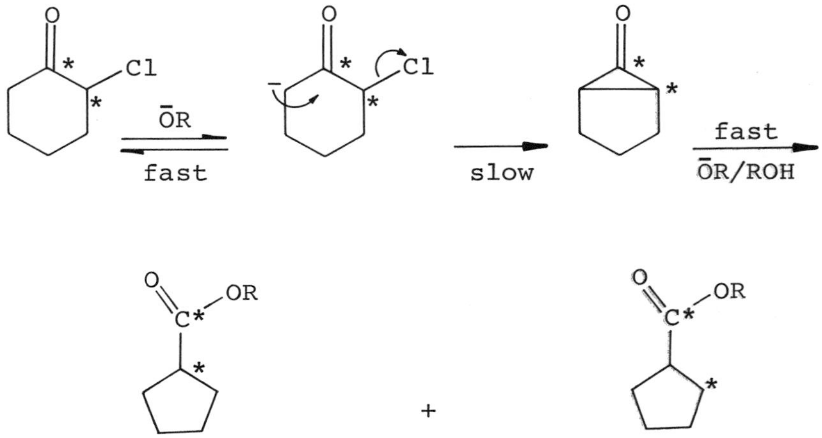

$$\ce{->[\bar{O}R][fast]} \qquad \ce{->[slow]} \qquad \ce{->[fast][\bar{O}R/ROH]}$$

+

distributed on the 1- and 2-carbons, and this isotope distribution was maintained in the recovered starting material. In the cyclopentane carboxylic acid product, the label was distributed 25% on both the α- and two β-carbons and 50% on the carbonyl carbon. The isomeric α-haloketones, 39 and 40, both give 3-phenylpropionic acid,[150] suggesting a common intermediate 41 is involved in both reactions. These observations exclude halogen migration before rearrangement and imply that the reaction product arises from an intermediate in which the α- and γ-carbons of the substrate have become equivalent.[147] Cyclopropenones have been isolated in the Favorskii reaction of dihaloketones (Eq. 24).[151]

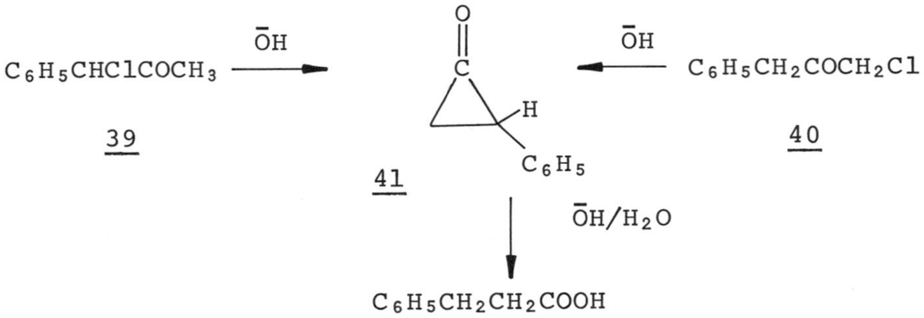

$$\ce{C6H5CHClCOCH3 ->[\bar{O}H]} \qquad \underset{41}{\qquad} \qquad \ce{<-[\bar{O}H] C6H5CH2COCH2Cl}$$

39   40

41

$$\ce{->[\bar{O}H/H2O]}$$

$$\ce{C6H5CH2CH2COOH}$$

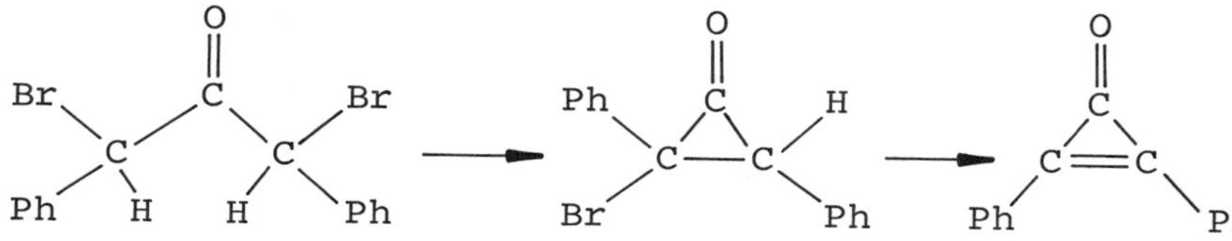

$$(24)$$

As shown in Scheme 22, displacement leading to
cyclopropanone should occur with inversion of configuration. The Favorskii reactions of 42 and its diaster-
eoisomer are consistent with this interpretation (Eq.
25).[152] Substitution with inversion of configuration
suggests an internal S_N2 reaction, but this seems unlikely for a tertiary halide for steric reasons. More-
over, a number of authors have argued that the geometry
of the enolate π-orbital is unsuitable for effective
S_N2-type overlap with the σ-orbital of the α-carbon
atom. They have proposed an intermediate zwitterion
(43),[153] stabilized by mesomerism, but as this possesses a plane of symmetry, unless it is asymmetrically

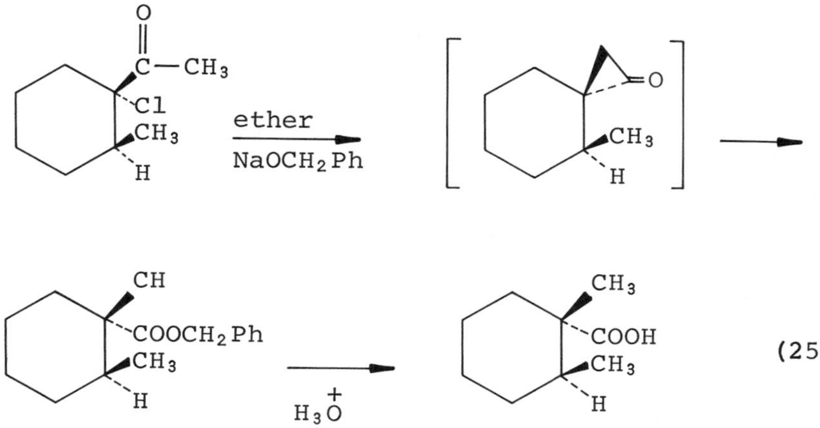

$$(25)$$

solvated, it cannot be involved in the necessary
stereospecific reaction.
    Recently, Bordwell and Scamehorn have demonstrated
the dipolar character of the transition state for
cyclization.[154] For the reaction of substituted α-
chlorobenzylmethyl ketones with methoxide ion in
methanol, γ-hydrogen exchange occurs faster than
halide ion formation, a large leaving group effect
(k_{Br}/k_{Cl}) is observed, and the Hammett reaction

constant $\rho^+ = -2.37$ at 0°C. These results are consistent with extensive carbonium ion character at C_α and, clearly, C_α-X bond breaking runs ahead of C_γ-C_α bond formation in the transition state.

The structural variation from $PhCH_2COCH_2Cl$ to $PhCH_2COCHClCH_3$ causes an increase in the rate of the Favorskii reaction with methoxide ion in methanol of 220 times.[155-156] γ-Deuterium exchange prior to rearrangement is no longer found, the leaving group effect, k_{Br}/k_{Cl}, decreases from 63 to 0.9, and the reaction constant changes from -5.0 to +1.4. These results accord with a change in the rate-determining step from halide ion loss to proton removal. Rate-determining proton removal in the Favorskii reactions of α-halocyclohexyl ketones have also been reported, but a change in the halogen from bromine to the more strongly bound chlorine causes the rate-determining step to revert to displacement.[157]

Strong evidence for a dipolar intermediate is provided by the formation of an indanone among the reaction products in the Favorskii reaction of 44 and 45 (Scheme 23),[158] and it has been suggested that the dipolar ion is in equilibrium in these cases with the cyclopropanone, the yield of indanone being greater at low methoxide ion concentration. For both of the above reactions, proton removal is rate determining.

Scheme 23

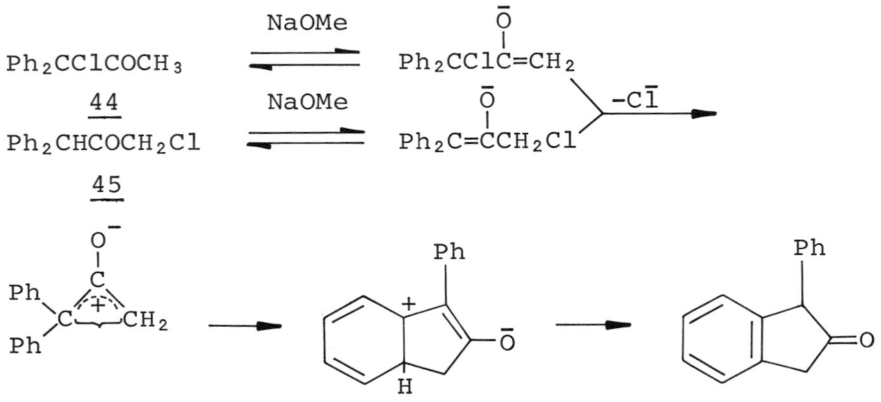

The dipolar nature of the transition state for the Favorskii reaction contrasts with the nucleophilic displacement for cyclization in the closely related Ramberg-Bäcklund reaction, for which a dipolar ion intermediate is less probable because of electrostatic repulsion between the adjacent electrophilic sulfur and the α-carbon (46).

46

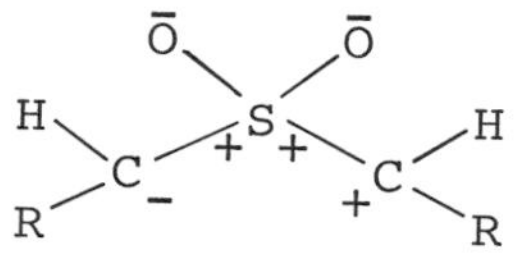

(c) The Neber Rearrangement.[159,160] Oxime tosylates rearrange to α-aminoketones when treated with basic reagents. A γ-hydrogen is essential and the amino-group is substituted for the more acidic γ-hydrogen, when more than one is present in the reactant, irrespective of the stereochemical arrangement.[159-162] The mechanism of Cram and Hatch[161] accomodates the few mechanistic details which are available most satisfactorily, 1,3-elimination giving a nitrene that then cyclizes (Scheme 24). House and Berkowitz's mechanism (Scheme 25) is less likely as it requires that the nitrene attack only the less nucleophilic γ-carbon.[162]

Scheme 24

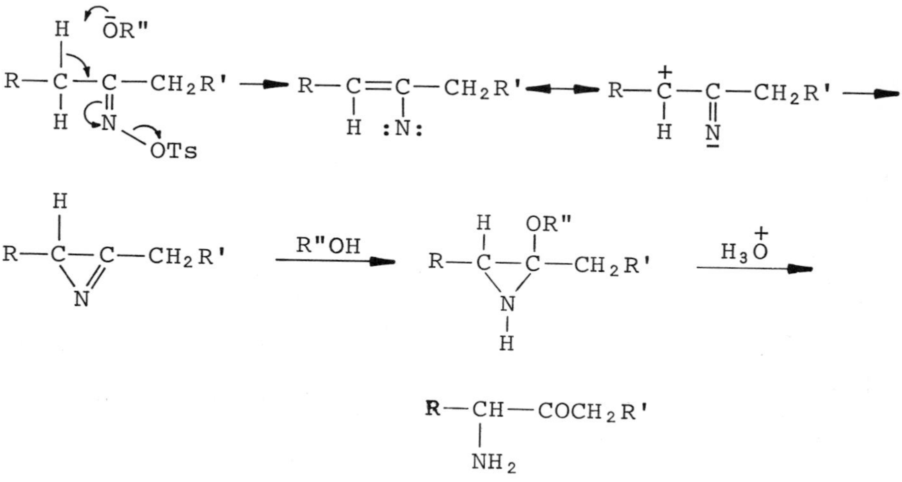

(R is more electron-withdrawing than R'.)

Scheme 25

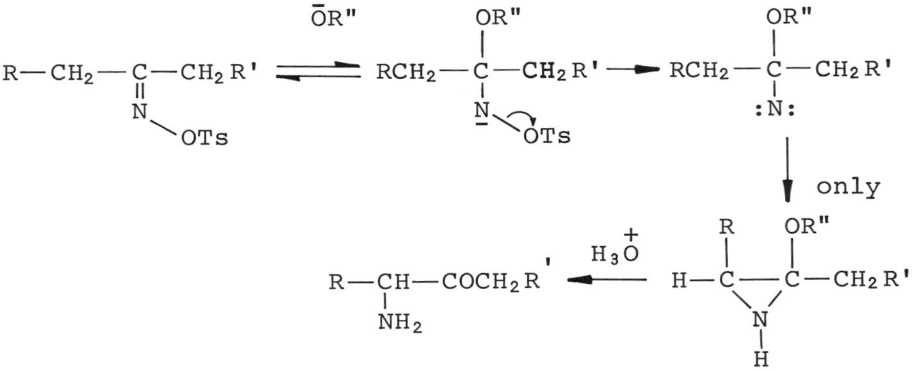

$$R-CH_2-\underset{\underset{OTs}{N}}{\overset{\overset{\bar{O}R''}{|}}{C}}-CH_2R' \;\rightleftharpoons\; RCH_2-\underset{\underset{OTs}{N}}{\overset{\overset{OR''}{|}}{C}}-CH_2R' \;\longrightarrow\; RCH_2-\underset{\overset{\cdot\cdot}{N}\cdot\cdot}{\overset{\overset{OR''}{|}}{C}}-CH_2R'$$

only

$$R-\underset{NH_2}{\overset{}{CH}}-COCH_2R' \;\xleftarrow{\;H_3\overset{+}{O}\;}\; H-\underset{\underset{H}{N}}{\overset{\overset{R}{|}}{C}}\!-\!\!\underset{}{\overset{\overset{OR''}{|}}{C}}-CH_2R'$$

5 Conclusions

ElcB mechanisms have been extensively characterized for
γ-eliminations under basic conditions. However, by
comparison to β-elimination, concerted γ-E2 reactions
are rarely encountered. There are few studies concerned with the mechanisms of pyrolytic γ-elimination,
and dehalogenations have also received scant attention.
As might be expected, greater separation of the eliminating groups in γ- than in β-elimination increases the
probability of multi-step reactions.

D HIGHER-ORDER ELIMINATIONS

Solvolytic and nucleophilic reactions giving alicyclic
rings can also be considered as higher-order eliminations. Ring formation causes a loss in rotational
freedom, the entropy loss increasing with ring size.
The angular strain factor on ring formation decreases
in going from three to six-membered and thereafter increases with ring size up to ten. In addition to these
effects, conjugative groups enhance three-membered ring
formation (the only ring in which partial double-bond
character is significant) as discussed previously
(Section X.C.2.a). The interplay of these factors can
be seen in the relative rates listed in Table 3.
    β-Haloketones, lacking a hydrogen alpha to the
carbonyl, undergo a homo-Favorskii reaction with bases.
Unlike the cyclopropanone intermediates of the
Favorskii reaction, the cyclobutanone is often stable
under the reaction conditions (Eq. 26).[164]

TABLE 3   Relative Rates of Ring Closure

Substrate	Conditions	Product	Temperature °C	Relative Rate (Ring Size)				Reference
				3	4	5	6	
$H_2N(CH_2)_nBr$	H_2O	 $HN\langle(CH_2)_n\rangle$	10	1	0.017	850	15	163
$(EtO-CO)_2CH(CH_2)_nCl$	$KOBu^t$/t-BuOH	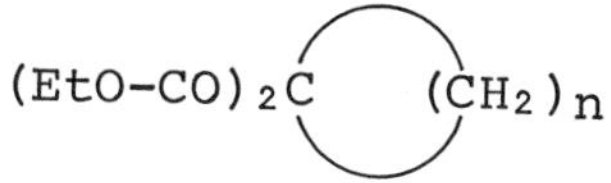 $(EtO-CO)_2C\langle(CH_2)_n\rangle$	25	1	1.5×10^{-6}	0.01	8×10^{-6}	117
$HO(CH_2)_nCl$	H_2O	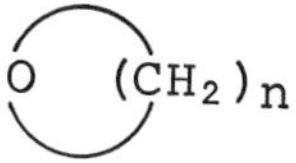 $O\langle(CH_2)_n\rangle$	70.5	1	4.5	9.3×10^2	41	99
$p-Me-C_6H_4SO_2(CH_2)_nBr$	$KOBu^t$/t-BuOH	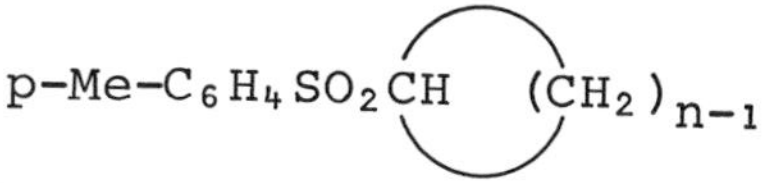 $p-Me-C_6H_4SO_2CH\langle(CH_2)_{n-1}\rangle$	55	1	$\sim 10^{-4}$	0.01	$\sim 10^{-4}$	115

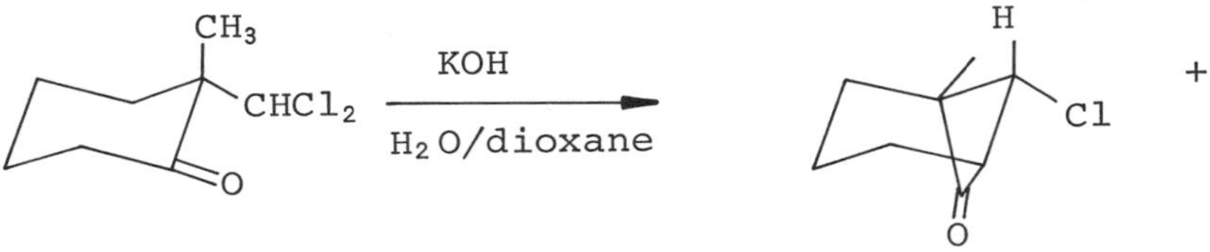

other products   (26)

Another category of δ-eliminations involves double-bond migration rather than ring closure (Eq. 27).[165] Base-catalyzed δ-elimination occurs up to 1200 times faster from trans-9,10-di-X-dihydroanthra-cenes (47) than from the cis-compounds.[166] Syn-δ-elim-ination is preferred as it comprises the coupling of two anti-β-eliminations, whereas anti-δ-elimination in-volves one syn-β-elimination and one anti-β-elimina-tion.

47

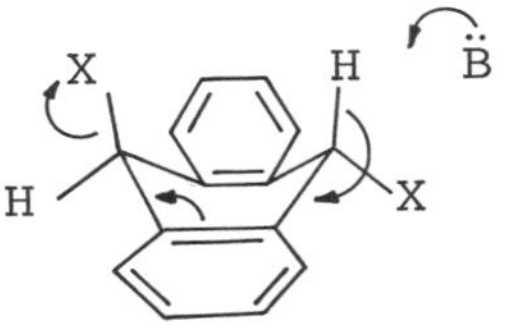

X = halogen, OH, OAc, OCOPh.

The products of dehydrochlorination of the naph-thalene tetrachlorides have been interpreted as in-dicative of the following reactivity order; anti-1,2 > syn-1,2 > syn-1,4. Anti-1,4-elimination was regarded as the least feasible. The product analysis for the ξ-isomer is shown in Scheme 26.[167]

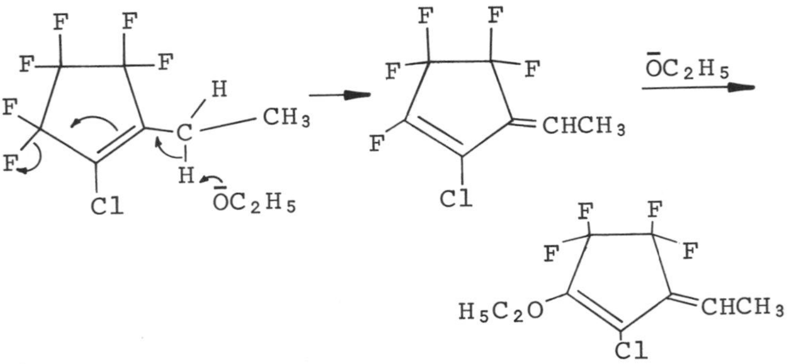

(27)

Scheme 26

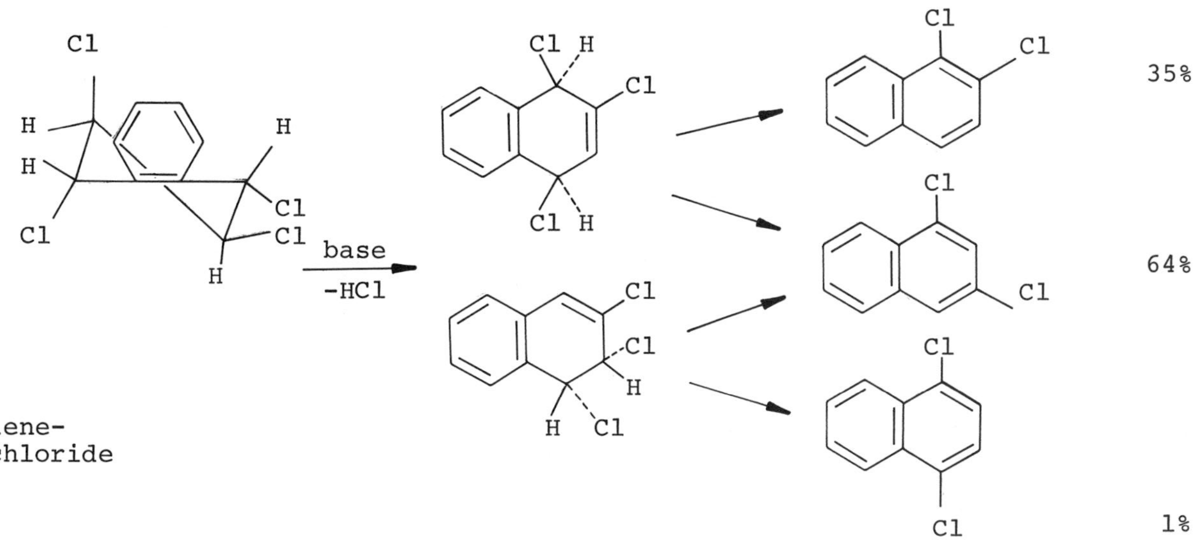

Naphthalene-ξ-tetrachloride

base
−HCl

35%

64%

1%

δ-Elimination from 48 occurs much faster than β-elimination from 49 in the polar reaction medium of aqueous acetone (base; OH).[168] On the other hand, dehydrochlorination (OEt/EtOH) of 1,2-dichloroethane occurs slightly faster than δ-elimination from trans-1,4-dichloro-2-butene, and a similar rate ratio is found for the iodide-promoted dehalogenation of 1,2-di-iodoethane relative to trans-1,4-di-iodo-2-butene. These results were interpreted by transition states with considerable carbonium-ion character for the dehydromethanesulfonations, but of a concerted nature for the other two reaction pairs. The rationale behind this interpretation was as follows: (1) a large ratio, k_δ/k_β is expected for transition states with charge development as the allylic anion or cation should be more stable than the localized ions, (2) for synchronous eliminations, k_δ/k_β should be somewhat greater than unity as butadiene has a greater resonance energy than an isolated double bond, (3) rotational entropy loss is greater in attaining the transition state for δ-elimination, thereby partly canceling the other factors. It will be interesting to see if this reactivity ratio stands up to more rigorous testing. The effect of the leaving group(s) on the reactivity of the

$$CH_2OSO_2Me$$

$$MeO_2SOH_2C$$

48

$$MeO_2SOCH_2CH_2OSO_2Me$$

49

β-hydrogen is likely to be greater than on the δ-hydrogen, and this feature could add further complication to the anticipated reactivity profile.

The Woodward-Hoffmann rules on the conservation of orbital symmetry represent the outstanding contribution to our understanding of mechanisms of concerted reactions.[169] It was predicted that concerted noncatalytic dehydrogenation should occur 1,4 rather than 1,2. Thermal fragmentation of 50 gives monodeuterobenzene by a symmetry-allowed loss of HD, but 51 gives a mixture of deuterobenzenes by a radical mechanism.[170] An analogous example concerns 2,5-dihydrofuran which loses hydrogen in a unimolecular process (E_A = 48.5 kcal/mole), whereas 2,3-dihydrofuran reacts in a nonconcerted manner.[171] Hydrogen elimination from labeled cyclopentenes has also been shown to occur preferably by 1,4-rather than 1,2-elimination.[172]

50

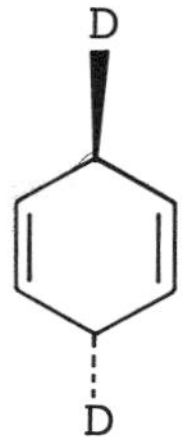

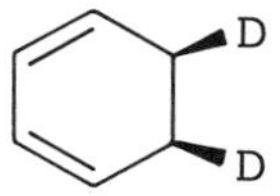

51

E   TRANSANNULAR REACTIONS[173,174]

Derivatives of medium-sized (8-11 membered) alicyclic rings exhibit quite different chemical properties from those of smaller and larger rings (>12). For example, bridged products are isolated in the hydrocarbon fraction in addition to the expected dehydration products, cis- and trans-cyclodecene, in the reaction of cyclo-decanol and hydrobromic acid (Eq. 28).[175] Chemical transformations of this kind, in which a bond is formed

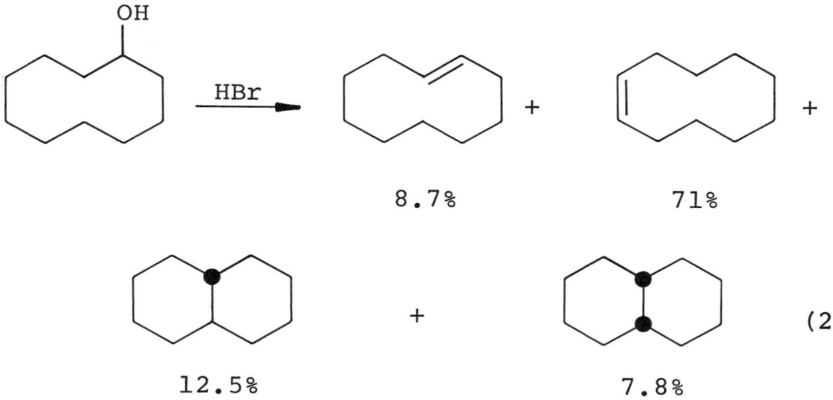

OH

HBr

8.7%   71%

12.5%   7.8%

(28)

between atoms on opposite sides of the ring, are called transannular reactions, and are of steric origin in the main. X-ray data demonstrate that in medium-sized rings the valence angles are larger than tetrahedral, and groups situated 1,2 are partially eclipsed. Steric interference between nonadjacent atoms is apparent, and this I-strain, as termed by Brown,[176] is relieved on substitution of an sp^2-carbon for an sp^3-carbon. Thus reactions that involve this change in hybridization are enhanced by the medium-ring framework.[174] In addition, in certain conformations, the opposite sides of the

ring come into very close proximity, giving rise to
electronic interactions and, consequently, to greater
probability of chemical reaction.

Cis-cyclodecene oxide gives rise to intramolecular
alkylation products when treated with the strong base,
lithium diethylamide (Eq. 29).[177]   The carbene (Scheme
27) and carbanion mechanisms (Scheme 28) can easily be
distinguished by deuterium-labeling studies.   The
transannular products, 52 and 53, contain only one
deuterium per molecule when prepared from 54, but two
deuterium atoms per molecule in the reaction of 55.[178]
Both these analyses accord with the carbene mechanism,
but transannular interaction may aid formation of the
carbene as the reactions are stereospecific.

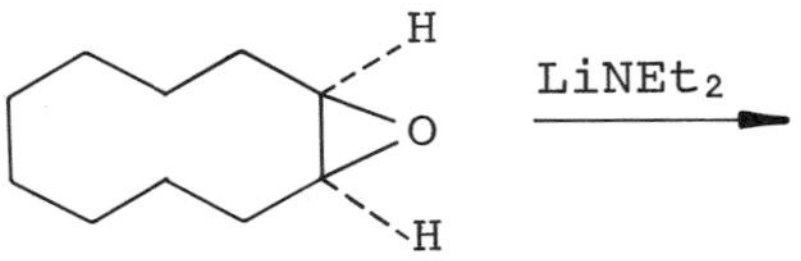

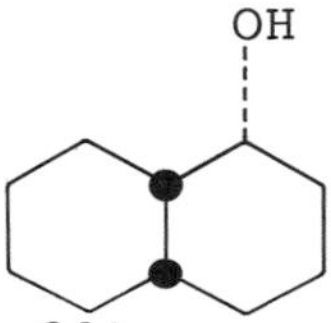

83%

52

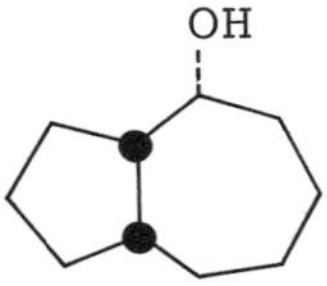

9%

53

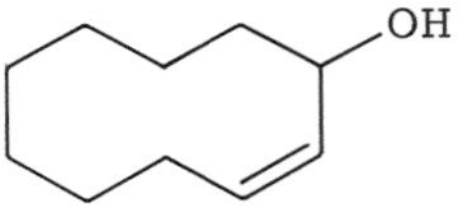

8%

(29)

Scheme 27

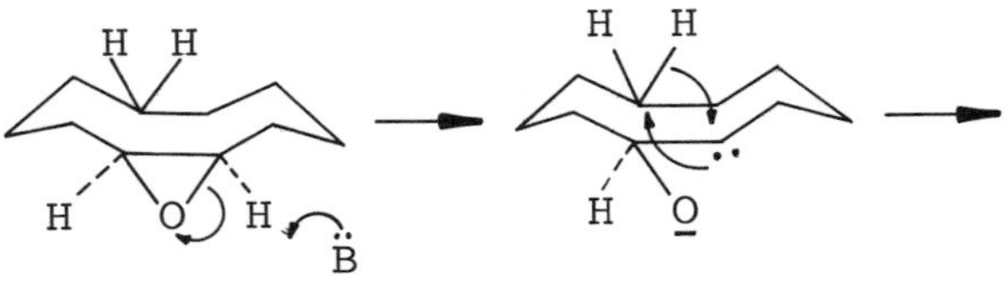

52   +   53

Scheme 28

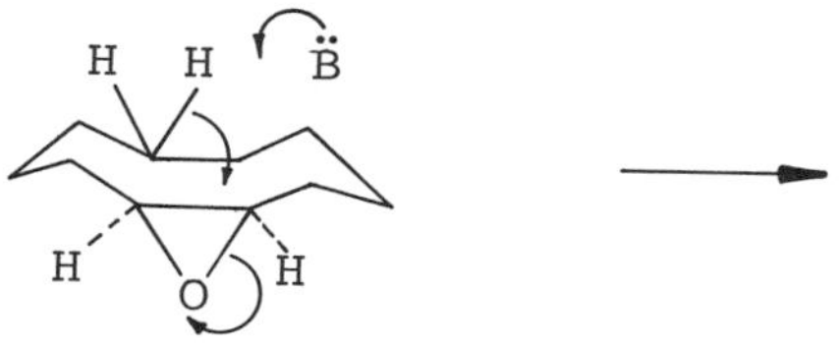

52 + 53

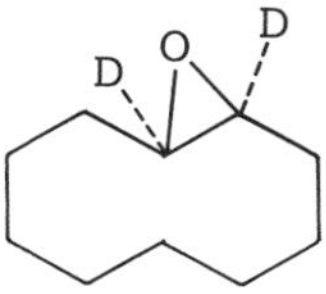

54

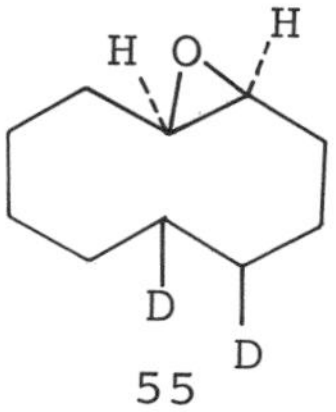

55

Transannular reactions involving radical intermediates are exemplified by the photolytic decomposition of N-chloro-N-methylcyclo-octylamine in acidic
solution (Scheme 29).[179] However, most transannular
reactions, like other rearrangements, have been induced
under solvolytic or nucleophilic displacement conditions and involve cationic intermediates. A nitrenium
ion is believed to be involved in the cyclization of 56
(Eq. 30).[180] The greater assistance of transannular
hydride interaction than of π-participation in cyclooctyl brosylate solvolysis is illustrated by the 40-
fold greater reactivity of 57 than of 58.[181] However,
rate-determining hydride abstraction is not involved
in solvolyses of this type as cyclodecyl tosylate,
deuterated at C-6 and C-5, reacts only 1.08 times more
slowly than the unlabeled tosylate.[182] Prelog has
attributed this small effect to a decrease in ground
state strain arising from the smaller spatial factor
of the C-D bond. An isotope effect of 3.0 is observed
during the decomposition of the destabilized dibenzo-
cycloheptyl cation (Eq. 31).[183] Transannular reactions
of other benzo derivatives have been reviewed.[184,185]
    In acidic solution, cis-cyclo-octene oxide de-
composes to give the epoxide-ring opened derivative and
a range of transannular products (Eq. 32).[186] Labeling
studies with deuterium reveal that the major product,
cis-1,4-cyclo-octane diol arises by a combination of
1,5-(61%) and 1,3-hydride shifts.[187] The extent of
transannular reaction is markedly influenced by the

Scheme 29

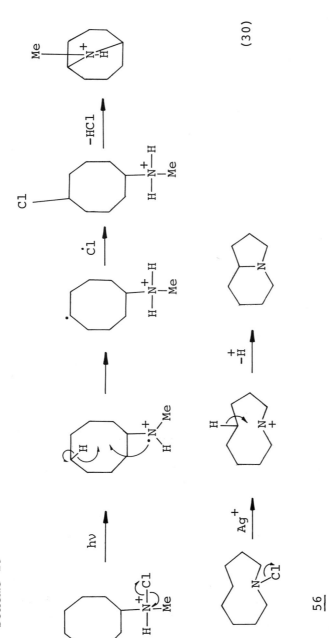

(30)

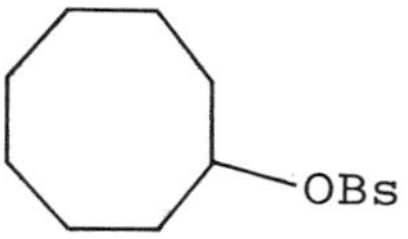

57

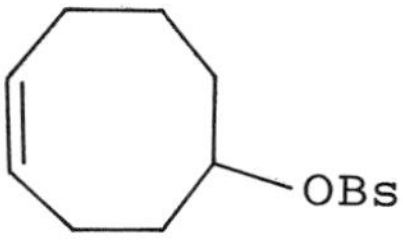

58

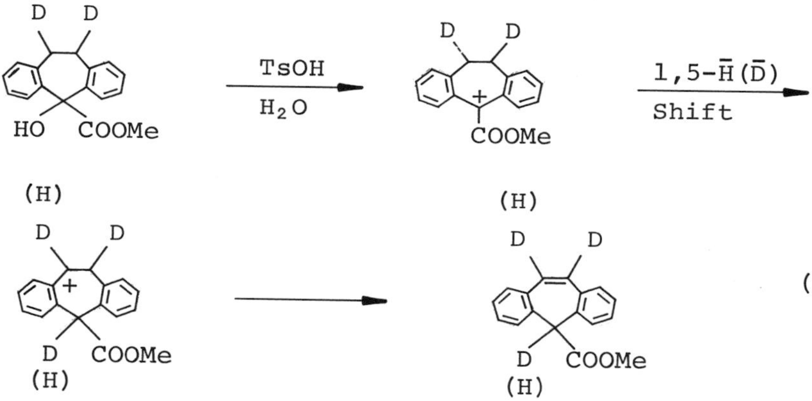

(31)

acidic catalyst.[188]   Thus, in trifluoroacetic acid,
only transannular products are formed, while in acetic
acid containing sodium acetate, 76% of the reaction
products are formed via the normal opening reaction of
epoxides, namely 1,2-trans-addition.   Clearly, as the
solvent becomes less ionizing and less nucleophilic,
the contribution of the migrating transannular hydride
ion gains in importance.   This trend accords with the
medium effects observed in neighboring group participation.[99,189]

Prelog and Urech have elegantly demonstrated the
involvement of 1,5- and 1,6-hydride shifts in the solvolysis of ^{14}C-labeled cyclodecyl tosylate.[190]   The
product (Scheme 30), cyclodecene was converted into
sebacic acid by hydroxylation with osmium tetroxide
followed by treatment with lead tetra-acetate.   Stepwise degradation of the acid enabled determination of
the label distribution in the cyclodecene.   The absence
of label on the γ-carbon cannot be explained in terms
of 1,2- 1,3- and 1,4-shifts only.

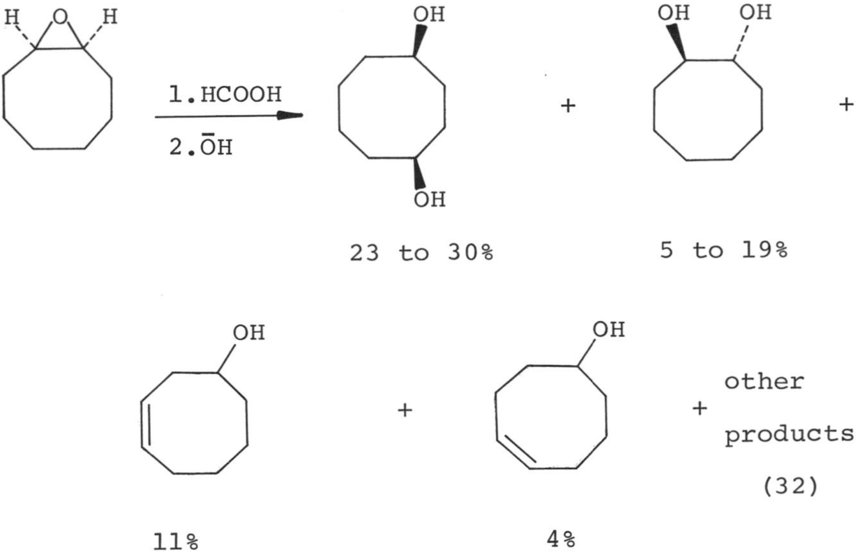

23 to 30%   5 to 19%

11%   4%

Scheme 30

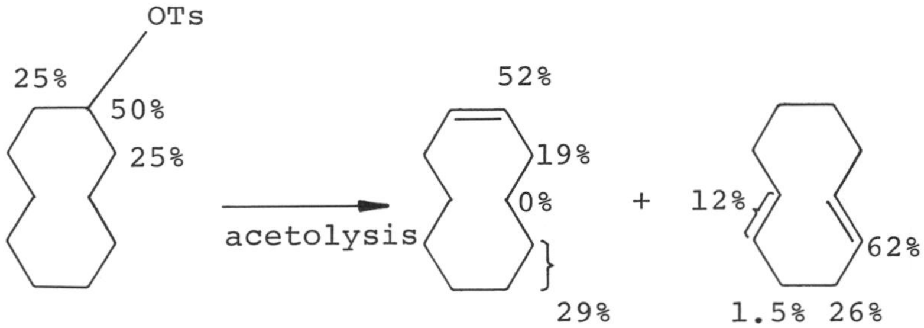

(Percentages refer to the distribution of ^{14}C.)

Transannular hydride shifts appear to play a lesser role in deamination than in tosylate solvolysis.[191] This result accords with the exothermic nature of the decomposition of the aliphatic diazonium ion precursors, and the low selectivity of reaction of the resultant carbonium ions (cf. Section VII.A). Interestingly cyclodecyl tosylate solvolyzes to give mainly trans-cyclodecene, while cis-cyclodecene is the major component in the alkene fraction arising from deamination of cyclodecylamine.[191]

F FRAGMENTATION REACTIONS

1 Reaction types and Definitions

Many organic chemists associate the term "fragmentation" with the molecular disintegrations that occur
when an organic compound is irradiated in a mass spectrometer. Grob has used the same term to describe the
series of cleavages that occur in the general formula,
a-b-c-d-X (Eq. 33). The terminology adopted here is
that outlined in previous reviews.[192-197] In heterolytic fragmentation, the electrofugal group, a-b,
leaves without the bonding electron pair, being equivalent to a hydrogen atom in most β-eliminations. The

$$a \underline{\qquad} b \underline{\qquad} c \underline{\qquad} d \underline{\qquad} X \longrightarrow \overset{+}{a} =\!=\!= b \quad +$$

| Electrofuge | Middle group | Nucleofuge | Electrofugal fragment |

$$c =\!=\!= d \quad + \quad \bar{X} \qquad\qquad (33)$$

| Unsaturated fragment | Nucleofugal fragment |

middle group, c-d, provides the unsaturated fragment,
and the nucleofugal group X, which leaves with the
bonding electrons, is often typical of the leaving
groups encountered in β-elimination. Grob and Schiess
have compiled the most comprehensive collection of
fragmentation reactions,[192] and rather than repeat
their compilation in this book, we concentrate on a few
of the reactions for which detailed mechanisms have
been reported. Some of the general synthetic aspects
of fragmentation are shown in Table 4, and their uses
in the synthesis of acetylenes[198] (Eq. 34) and cyclodecyl derivatives[199] (Eq. 35) have been reviewed. The
categories listed in Table 4 usually require acidic or
alkaline catalysis.

TABLE 4 Heterolytic Fragmentations

$$a\!-\!b\!-\!c\!-\!d\!-\!X \longrightarrow \overset{+}{a}\!=\!b + c\!=\!d + \bar{X}$$

a—b	c—d	X
Alkene-forming fragmentations		
HOCR$_2$—	—CR$_2$CR$_2$—	Cl, OH, $\overset{+}{N_2}$
HOOC—	—CR$_2$CR$_2$—	Br, Cl, I
O=CR—	—CR$_2$CR$_2$—	Br, Cl, I, $\overset{+}{NMe_3}$
—N=CR—	—CR$_2$CR$_2$—	Cl, OH
R$_2$NCR$_2$—	—CR$_2$CR$_2$—	Cl, OH, OTs
R$_3$C—	—CR$_2$CR$_2$—	OTs, Cl, Br, OH
HRNNR—	—CR$_2$CR$_2$—	Cl, OTs, OMes
HOP—	—CR$_2$CR$_2$—	Br, Cl
R$_3$Si—	—CR$_2$CR$_2$—	Cl, Br
—B=	—CR$_2$CR$_2$—	Cl, Br
Alkyne-forming fragmentations		
HOOC—	—CR=CR—	Cl, OTs, Br
O=CR—	—CR=CR—	Cl, Br
HOP—	—CR=CR—	Cl, Br

Imine-forming fragmentations

$HOCR_2-$	$-CR_2NR-$	$Cl,\ Br,\ NMe_3^+,\ NHMe_2^+$
$HOOC-$	$-CR_2NR-$	$Cl,\ Br,\ SO_2Ar$

Nitrile-forming fragmentations

R_3C-	$-N=CR-$	$Cl,\ Br$
$R-N=N-$	$-N=CR-$	$Cl,\ Br$

Carbonyl-forming fragmentations

R_3C-	$-R_2C-O-$	$H,\ CrO_3H$
$HOCR_2-$	$-R_2C-O-$	$OH,\ Pb(OAc)_3$
$O=CR-$	$-R_2C-O-$	OH
$RN=CR-$	$-R_2C-O-$	OH

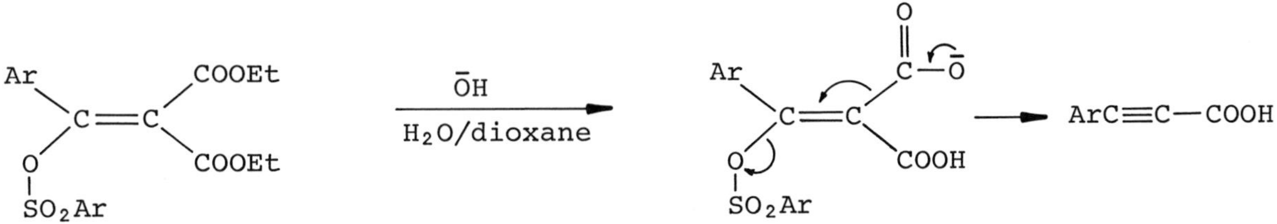
Ar
COOEt
COOEt
O
SO2Ar
ŌH
H2O/dioxane
Ar
O
C—Ō
COOH
O
SO2Ar
ArC≡C—COOH

(34)

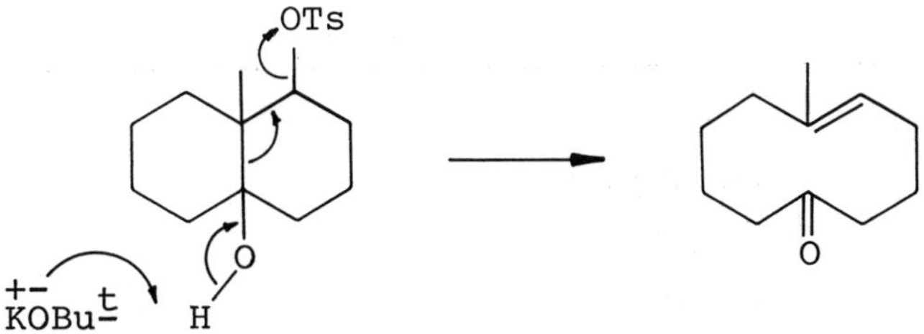
OTs
O
H
KOBu-t
O

(35)

Depending on the order in which the fragments are lost, mechanisms analogous to the E1, E2, and E1cB schemes for β-elimination can be recognized (Section I.A.1; A.2; B.1). We term these fragmentations F1, F2, and F1cB, to simplify the discussion when these reactions compete alongside the more well-known substitution, elimination, and ring-closing reactions. As we noted previously, only a few detailed kinetic studies of fragmentations have been made, and the majority of these refer to γ-amino-derivatives.[195]

2   Unimolecular Fragmentations

Unimolecular fragmentation, substitution, and elimination share a common rate-determining step, namely formation of a carbonium ion (Scheme 31). These mechanisms are favored by a good nucleofuge and stable carbonium ion, facets satisfied by the solvolysis of 3-chloro-N,N-3-trimethylbutylamine in 80% ethanol-water.[197,200] The low selectivity of the carbonium ion results in its decomposition by F1, E1, S_N1, and ring-closure pathways. The product proportions are virtually unchanged when the nucleofuge is changed to bromide, but the rate increases as the nucleofuge goes from Cl → Br → I. Thus, in this solvent, the effect of the nucleofuge is confined to the rate-determining step. It appears that the role of the nitrogen atom is confined to its electron-withdrawing inductive effect, neighboring-group assistance to ionization being minimal as the solvolysis rate ratio k(59)/k(60) = 0.52 at 25°C. Grob and his co-workers[195] have often made use of this type of rate comparison, involving the so-called homomorphous compound (e.g., 60), in mechanistic assignment of fragmentation reactions.

$$Me_2CHCH_2CH_2CMe_2Cl \qquad \underline{60}$$

Scheme 31

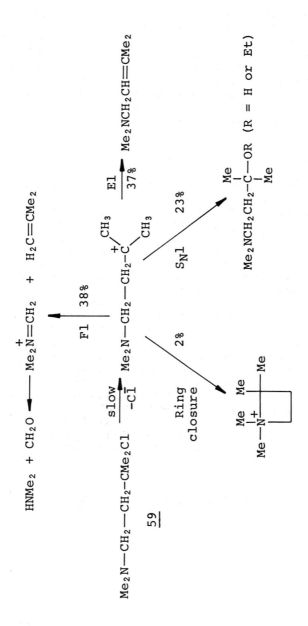

3   FlcB-Fragmentations

This mechanism requires the electrofugal group can
separate to leave a stabilized carbanion, possessing a
reluctant nucleofuge. The decompositions of the bis-
p-nitrophenyl derivatives (Scheme 32) are consistent
with an FlcB mechanism, in which the ratio, k_2/k_{-1}
varies slightly with the nucleofuge. The influence of
substituents in the benzoate group on the rate is very
small, ruling against Fl and F2 mechanisms[201] and,
in the absence of the nitro-substituents, only ester
hydrolysis occurs.

Scheme 32

$$Me_2N—CH_2—\overset{\displaystyle Ar}{\underset{\displaystyle Ar}{C}}—CH_2X \underset{k_{-1}}{\overset{k_1}{\rightleftharpoons}} Me_2\overset{+}{N}=CH_2 \; + \; Ar_2\bar{C}—CH_2X \overset{k_2}{\longrightarrow}$$

$$Ar_2C=CH_2 \; + \; \bar{X}$$

(Ar = p-nitrophenyl; X = OH, p-Y-C_6H_4COO in which Y =
H, p-Me, p-MeO, $\underline{p}$-NO$_2$.)

4   Concerted Fragmentations

Of the three most common heterolytic fragmentations,
the stereochemical consequences are only significant
for the concerted reactions. By analogy with the preferred anti-stereospecificity for concerted E2 reactions, concerted fragmentations are expected only where
the C_α-X bond and the lone pair of electrons arising
from the electrofuge are aligned anti and parallel
(anti-periplanar) to the C_β-C_γ bond. These stereochemical requirements have been demonstrated for
γ-amino-derivatives by Grob.[195] Thus the stereochemical arrangement is satisfactory for concerted
fragmentation in staggered conformation 61 as well as
all its rotamers about the C_β-C_γ bond (e.g. 62, 63).
Rotations about the C_α-C_β and C_γ-N bonds allow only
reduced overlap between the forming unsaturated frag-
ments, and the synchronous process becomes less favorable than stepwise mechanisms.

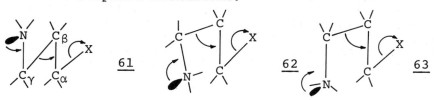

To establish a concerted fragmentation, it is necessary to demonstrate that the reaction follows a single pathway and not competing processes, and that changes in bonding to both the electrofugal and nucleofugal groups are occurring in the transition state. These requisites are satisfied by the fragmentation of 4-substituted quinuclidines (Eq. 36).[202] 4-Bromo-quinuclidine reacts 50,000 times more rapidly than its homomorph 64, and the nucleofuge effect, k_{OTs}/k_{Br} runs at about 10^3 (cf. Section II.C.2.b), factors indicating considerable involvement of the nitrogen lone pair and extensive nucleofuge ionization in the transition state. This example is of interest as the stereochemistry is equivalent to the eclipsed form 63.

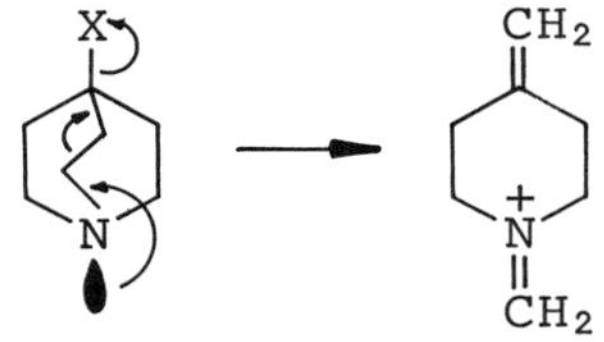

$$+ \quad \bar{X} \qquad\qquad (36)$$

(X = Br, OTs)

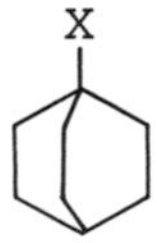

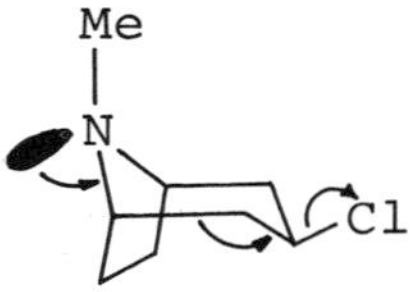

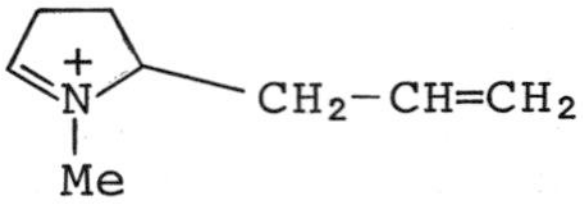

64   65   66   67

The importance of the anti-parallelism principle is amply illustrated by the different reaction products formed from the stereoisomeric 3-chlorotropanes 65 and 66.[203] Compound 65, has skew conformation equivalent to 62 and undergoes synchronous fragmentation to 67. The rate of reaction is also 1.3×10^4 times faster than that of the corresponding homomorphous chloride. However, compound 66 reacts only slightly faster than its homomorph and gives only elimination and substitution products. In this case, the C-Cl bond is not anti-parallel to the $C_\beta-C_\gamma$ bond, and dissociation involves rate-determining carbonium-ion formation. In acyclic systems F1 mechanisms are more likely than F2 mechanisms for good nucleofuges as attainment of the required stereochemistry leads to loss of rotational freedom.

G   GENERAL CONCLUSIONS

From the numerous examples discussed in this chapter, it is obvious that the mechanisms that have been systematically established for olefin-forming β-eliminations are also found in more complex eliminations. The extensive investigations of β-elimination thus provide a framework on which we can extend our understanding of these more complex reactions. The major contrast between these higher order eliminations (γ,δ,ε, transannular, fragmentations) and β-elimination is the tendency for reactions to occur in a stepwise rather than in a concerted manner, because of the greater separation of the eliminating fragments, which causes reduced proximity effects and a decrease in the probability of attaining the required stereochemistry for reaction.

REFERENCES

1. J. Hine and A. M. Dowell, Jr., J. Am. Chem. Soc., 76, 2688 (1954).
2. D. J. Cram, "Fundamentals of Carbanion Chemistry," Academic Press, New York, 1965, pp. 243-256.
3. A. S. Kende, Org. React., 11, 261 (1960).
4. R. F. Hudson, J. Arendt, and A. Mancuso, J. Chem. Soc., B, 1069 (1967).
5. C. A. Grob and P. W. Schiess, Angew. Chem. Int. Ed., 6, 1 (1967).
6. L. Friedman and H. Shechter, J. Am. Chem. Soc., 81, 5512 (1959).
7. P. Miginiac, Bull. Soc. Chim. Fr., 2000 (1962).
8. E. Chinoporos, Chem. Rev., 63, 235 (1963).
9. J. Hine, "Divalent Carbon," Ronald Press Co., New York, 1964.
10. J. A. Bell, Prog. Phys. Org. Chem., 2, 1 (1964).
11. H. M. Frey, Prog. React. Kinet., 2, 131 (1964).
12. W. B. DeMore and S. W. Benson, Adv. Photochem., 2, 219 (1964).
13. C. W. Rees and C. E. Smithen, Adv. Heterocyc. Chem., 3, 57 (1964).
14. W. Kirmse, Angew. Chem. Int. Ed., 4, 1 (1965).
15. G. Köbrich et al., Angew. Chem. Int. Ed., 6, 41 (1967).
16. P. J. Wagner and G. S. Hammond, Adv. Photochem, 5, 21 (1968).
17. D. Bethell, Adv. Phys. Org. Chem., 7, 153 (1969).
18. A. Ledwith, RIC Lect. Ser., No 5, (1964).
19. G. Herzberg and J. Shoosmith, Nat., 183, 1801 (1959).

20. (a) J. A. Kerr, Chem. Rev., 66, 465 (1966);
    (b) I. P. Fisher, J. B. Homer, and F. P. Lossing,
    J. Am. Chem. Soc., 87, 957 (1965);
    (c) R. F. Pottie, J. Chem. Phys., 42, 2607 (1965);
    (d) J. S. Shapiro and F. P. Lossing, J. Phys.
    Chem., 72, 1552 (1968).
21. (a) I. A. D'yakonov, Chem. Abstr., 45, 7023i
    (1951);
    (b) I. A. D'yakonov, and M. I. Komendantov, Chem.
    Abstr., 52, 2762i (1958).
22. P. S. Skell and R. C. Woodworth, J. Am. Chem. Soc.,
    78, 4496 (1956).
23. R. Hoffmann, J. Am. Chem. Soc., 90, 1475 (1968).
24. A. G. Anastassiou, Chem. Comm., 991 (1968).
25. G. Kobrich, Angew. Chem. Int. Ed., 4, 49 (1965).
26. C. E. Castro and W. C. Kray, J. Am. Chem. Soc.,
    88, 4447 (1966).
27. H. E. Simmons, E. P. Blanchard, and R. D. Smith,
    J. Am. Chem. Soc., 86, 1347 (1964).
28. R. Huisgen, Angew. Chem., 67, 439 (1955).
29. H. Zollinger, "Azo and Diazo Chemistry,"
    Interscience, London, 1961.
30. R. A. More O'Ferrall, Adv. Phys. Org. Chem., 5,
    331 (1967).
31. E. Schmitz, "Dreiring mit Zwei Heteroatomen,"
    Springer Verlag, Verlin, Heidelberg, New York,
    1967, pp. 114-167.
32. S. Y. Ho and W. A. Noyes, J. Am. Chem. Soc., 89,
    5091 (1967).
33. (a) A. M. Trozzolo, W. A. Yager, G. W. Griffen,
    H. Kristinsson, and I. Sarkar, J. Am. Chem. Soc.,
    89, 3357 (1967);
    (b) H. Kristinsson and G. W. Griffen, Angew.
    Chem. Int. Ed., 4, 868 (1965). J. Am. Chem. Soc.,
    88, 1579 (1966).
34. D. Saunders and J. Heicklen, J. Am. Chem. Soc.,
    87, 2088 (1965).
35. D. Seyferth, J. Y. P. Mui and J. M. Burlitch, J.
    Am. Chem. Soc., 89, 4953 (1967).
36. W. R. Bamford and T. S. Stevens, J. Chem. Soc.,
    4735 (1952).
37. A. Geuther, Ann., 123, 121 (1862).
38. J. Hine, J. Am. Chem. Soc., 72, 2438 (1950).
39. J. Hine, "Physical Organic Chemistry," McGraw-
    Hill, New York, 2nd ed., 1962, Chap. 24.
40. (a) J. Hine, R. C. Peek, Jr., and B. D. Oakes,
    J. Am. Chem. Soc., 76, 827 (1954);
    (b) Y. Sakamoto, J. Chem. Soc. Jap., 57, 1169
    (1936);
    (c) J. Horiuti and Y. Sakamoto, Bull. Chem. Soc.
    Jap., 11, 627 (1936).

41. W. von E. Doering and A. K. Hoffmann, J. Am. Chem. Soc., 76, 6162 (1954).

42. J. Hine and P. B. Langford, J. Am. Chem. Soc., 79, 5497 (1957).

43. J. Hine and J. M. van der Veen, J. Am. Chem. Soc., 81, 6446 (1959); J. Org. Chem., 26, 1406 (1961).

44. S. B. Hanna, Y. Iskander, and Y. Riad, J. Chem. Soc., 217 (1961).

45. D. M. Doleib and Y. Iskander, J. Chem. Soc., B, 1154 (1967).

46. I. Rothberg and E. R. Thornton, J. Am. Chem. Soc., 86, 3296, 3302 (1964).

47. C. G. Swain and E. R. Thornton, J. Am. Chem. Soc., 83, 4033 (1961).

48. D. Bethell, J. Chem. Soc., 666 (1963).

49. A. Ledwith and Y. Shih-Lin, Chem. Ind. (London), 1868 (1964).

50. C. R. Hauser, S. W. Kantor, P. S. Skell, and W. R. Brasen, J. Am. Chem. Soc., 78, 1653 (1956).

51. D. Bethell, A. F. Cockerill, and D. B. Frankcum, J. Chem. Soc., B, 1287 (1967).

52. D. Bethell and A. F. Cockerill, J. Chem. Soc., B, 917 (1966).

53. D. Bethell and A. F. Cockerill, Proc. Chem. Soc., 283 (1964).

54. S. B. Hanna and L. G. Wideman, Chem. Ind. (London), 486 (1968).

55. (a) G. A. Russell and W. C. Danen, J. Am. Chem. Soc., 90, 347 (1968);
(b) N. Kornblum et al., J. Am. Chem. Soc., 87, 4520 (1965); 88, 5660 (1966); 88, 5662 (1966); 89, 725 (1967).

56. E. R. Thornton, personal communication of unpublished results.

57. (a) P. Fritsch, Ann., 279, 319 (1894);
(b) W. P. Buttenberg, Ann., 279, 324 (1894);
(c) H. Wiechell, Ann., 279, 337 (1894).

58. G. H. Coleman, W. H. Holst, and R. D. Maxwell, J. Am. Chem. Soc., 58, 2310 (1936).

59. D. Y. Curtin and E. W. Flynn, J. Am. Chem. Soc., 81, 4714 (1959).

60. J. G. Pritchard and A. A. Bothner-By, J. Phys. Chem., 64, 1271 (1960).

61. A. A. Bothner-By, J. Am. Chem. Soc., 77, 3293 (1955).

62. W. M. Jones and R. Damico, J. Am. Chem. Soc., 85, 2273 (1963).

63. (a) P. Beltrame and S. Carra, Gazz. Chim. Ital., 91, 889 (1961);

(b) P. Beltrame and G. Favini, Gazz. Chim. Ital.,
93, 757 (1963).

64. G. Köbrich and H. Trapp, Z. Naturforsch., 18b,
1125 (1963).

65. D. Y. Curtin and W. H. Richardson, J. Am. Chem.
Soc., 81, 4719 (1959).

66. C. R. Hauser and D. Lednicer, J. Org. Chem., 22,
1248 (1957).

67. G. Köbrich and F. Ansari, Chem. Ber., 100, 2011
(1967).

68. (a) G. W. Cowell and A. Ledwith, Q. Rev.
(London), 24, 119 (1970).
(b) R. A. More O'Ferrall, Adv. Phys. Org. Chem.,
5, 331 (1967).

69. (a) H. Reimlinger, Angew. Chem., 74, 153 (1962);
Ber., 97, 339, 3503 (1964);
(b) H. E. Zimmermann and D. H. Paskovich, J. Am.
Chem. Soc., 86, 2149 (1964);
(c) D. Bethell, J. D. Callister, and D. Whitaker,
J. Chem. Soc., B, 2466 (1965).

70. C. G. Overberger and J. -P. Anselme, J. Org.
Chem., 29, 1188 (1964).

71. D. Bethell and D. Whittaker, J. Chem. Soc., B,
778 (1966).

72. G. L. Closs, R. A. More, and S. H. Goh, J. Am.
Chem. Soc., 88, 364 (1966).

73. W. R. Bamford and T. S. Stevens, J. Chem. Soc.,
4735 (1952).

74. J. W. Powell and M. C. Whiting, Tetrahedron, 7,
305 (1959).

75. (a) G. L. Closs, L. E. Closs, and W. A. Böll,
J. Am. Chem. Soc., 85, 3796 (1963);
(b) G. M. Kaufman, J. A. Smith, G. G. Van der
Stouw, and H. Shechter, J. Am. Chem. Soc., 87,
935 (1965).

76. (a) L. Friedman and H. Shechter, J. Am. Chem.
Soc., 81, 5512 (1959); 83, 3159 (1961);
(b) C. H. DePuy and D. H. Froemsdorf, J. Am.
Chem. Soc., 82, 634 (1960).

77. (a) J. W. Wilt, J. M. Kasturik, and R. C.
Orlowski, J. Org. Chem., 30, 1052 (1965);
(b) J. W. Wilt et al., J. Org. Chem., 31, 1543
(1966);
(c) J. A. Smith, H. Shechter, J. Bayless, and
L. Friedman, J. Am. Chem. Soc., 87, 659 (1965).

78. A. Nickon and N. H. Werstiuck, J. Am. Chem. Soc.,
88, 4543 (1966).

79. J. W. Wilt et al., J. Org. Chem., 33, 694 (1968).

80. C. C. Leznoff, Can. J. Chem., 46, 1152 (1968).

81. W. Kirmse and H. W. Bücking, Ann., 711, 31 (1968).

82.   "Organic Reaction Mechanisms," B. Capon and C. W. Rees, Eds., Wiley, London, 1969, p. 395.

83.   P. S. Skell and M. S. Cholod, J. Am. Chem. Soc., 91, 6035 (1969).

84.   (a) P. S. Skell and R. M. Etter, Proc. Chem. Soc., 443 (1961);
      (b) W. von E. Doering and L. H. Knox, J. Am. Chem. Soc., 78, 4947 (1956);
      (c) W. R. Moser, J. Am. Chem. Soc., 91, 1135, 1141 (1969);
      (d) P. Yates, J. Am. Chem. Soc., 74, 4376 (1952).

85.   (a) E. P. Blanchard and H. E. Simmons, J. Am. Chem. Soc., 86, 1337 (1964);
      (b) H. E. Simmons, E. P. Blanchard, and R. D. Smith, J. Am. Chem. Soc., 86, 1347 (1964);
      (c) H. E. Simmons and R. D. Smith, J. Am. Chem. Soc., 81, 4256 (1959).

86.   D. Seyferth, J. Y. P. Mui, and J. M. Burlitch, J. Am. Chem. Soc., 89, 4953 (1967).

87.   P. S. Skell and M. S. Cholod, J. Am. Chem. Soc., 91, 7131 (1969).

88.   G. Mouvier and J. E. DuBois, Bull. Soc. Chim. Fr., 1441 (1968).

89.   (a) D. Seebach, Angew. Chem. Int. Ed., 6, 442 (1967);
      (b) D. Seebach and A. K. Beck, J. Am. Chem. Soc., 91, 1540 (1969).

90.   "Nitrenes," W. Lwowski, Ed., Wiley, New York, 1968.

91.   J. H. Boyer, "Mechanisms of Molecular Migrations," Vol. 2, B. S. Thyagarajan, Ed., Wiley, New York, 1969, p. 267.

92.   R. A. Abramovitch and B. A. Davis, Chem. Rev., 64, 149 (1964).

93.   L. Horner and A. Christmann, Angew. Chem. Int. Ed., 2, 599 (1963).

94.   R. K. Smalley and H. Suschitzky, Chem. Ind. (London), 1338 (1970).

95.   T. L. Gilchrist and C. W. Rees, "Carbenes, Nitrenes and Arynes," Nelson, Bath, England, 1969.

96.   C. R. Hauser and W. B. Renfrow, Jr., J. Am. Chem. Soc., 59, 121 (1937).

97.   R. Heck and S. Winstein, J. Am. Chem. Soc., 79, 3432 (1957).

98.   J. Hine, "Physical Organic Chemistry," McGraw-Hill, New York, 2nd Ed., 1962, Chap. 15.

99.   B. Capon, Q. Rev. (London), 18, 45 (1964).

100. D. J. Cram, "Fundamentals of Carbanion Chemistry," Academic Press, New York, 1965, pp. 243-256.

101. L. A. Paquette, Acc. Chem. Res. (American Chemical

Society Publication), 1, 209 (1968).

102. A. F. Cockerill, "Elimination Reactions," in
     "Comprehensive Chemical Kinetics," Vol. 9, Ch. 3
     1973 C. H. Bamford and C. F. H. Tipper,
     Eds., Elsevier, Amsterdam.
103. A. S. Kende, Org. React., 11, 261 (1960).
104. A. Nickon and N. H. Werstiuck, J. Am. Chem.
     Soc., 89, 3914 (1967).
105. C. L. Bumgardner, Chem. Comm., 374 (1965).
106. C. L. Bumgardner, J. Am. Chem. Soc., 83, 4420,
     4423 (1961).
107. L. Friedman and J. G. Berger, J. Am. Chem. Soc.,
     83, 492 (1961).
108. V. Franzen, H. J. Schmidt, and C. Mertz, Chem.
     Ber., 94, 2942 (1961).
109. S. J. Cristol, J. K. Harrington, and M. S.
     Singer, J. Am. Chem. Soc., 88, 1529 (1966).
110. S. J. Cristol and B. B. Jarvis, J. Am. Chem.
     Soc., 88, 3095 (1966).
111. S. J. Cristol and B. B. Jarvis, J. Am. Chem.
     Soc., 89, 401 (1967).
112. A. Nickon and N. H. Werstiuck, J. Am. Chem.
     Soc., 89, 3915 (1967).
113. A. Nickon and N. H. Werstiuck, J. Am. Chem.
     Soc., 89, 3917 (1967).
114. H. Hogeveen, G. Maccagnani, F. Montanari, and
     F. Taddei, J. Chem. Soc., 4101 (1964).
115. A. C. Knipe and C. J. M. Stirling, J. Chem.
     Soc., B, 808 (1967).
116. R. Bird and C. J. M. Stirling, J. Chem. Soc.,
     B, 111 (1968).
117. A. C. Knipe and C. J. M. Stirling, J. Chem.
     Soc., B, 67 (1968).
118. A. C. Knipe and C. J. M. Stirling, J. Chem.
     Soc., B, 1218 (1968).
119. For a list of the relevant references, see
     Refs. 115 and 117.
120. K. Wiberg and G. M. Lampmann, Tetrahedron Lett.,
     2173 (1963).
121. B. M. Trost, W. L. Schinski, and I. B. Mantz,
     J. Am. Chem. Soc., 91, 4320 (1969).
122. (a) B. M. Trost and S. Ziman, Chem. Comm., 181
     (1969);
     (b) E. L. Eliel, "Stereochemistry of Carbon
     Compounds," McGraw-Hill, New York, 1962, Chap.
     13.
123. S. J. Cristol, A. R. Dahl, and W. Y. Lim, J.
     Am. Chem. Soc., 92, 5670 (1970).
124. M. S. Newman, J. R. LeBlanc, H. A. Karnes, and
     G. Axelrad, J. Am. Chem. Soc., 86, 868 (1964).

125. L. A. Kaplan, J. Am. Chem. Soc., 89, 1753 (1967).
126. F. M. Sonnenberg and J. K. Stille, J. Org. Chem.,
      31, 3441 (1966).
127. R. C. Bicknell and A. Maccoll, Chem. Ind.
      (London), 1912 (1961).
128. C. G. Swain, A. D. Ketley, and R. F. W. Bader,
      J. Am. Chem. Soc., 81, 2353 (1959).
129. (a) P. D. Bartlett, J. Am. Chem. Soc., 57,
      224 (1935);
      (b) S. Winstein and H. J. Lucas, J. Am. Chem.
      Soc., 61, 1576 (1939).
130. (a) A. Weissberger and H. Bach, Ber., 64, 1095
      (1931);
      (b) F. H. Dickey, W. Fickett, and H. J. Lucas,
      J. Am. Chem. Soc., 74, 944 (1952);
      (c) O. E. Paris and P. E. Fanta, J. Am. Chem.
      Soc., 74, 3007 (1952).
131. H. E. Baumgarten, R. L. Zey, and U. Krolls, J.
      Am. Chem. Soc., 83, 4469 (1961).
132. A. Hassner and C. Heathcock, Tetrahedron, 20,
      1037 (1964); J. Org. Chem., 29, 3646 (1964);
      30, 1748 (1965).
133. (a) W. A. Cowdrey, E. D. Hughes, and C. K.
      Ingold, J. Chem. Soc., 1208, 1243 (1937);
      (b) K. C. Kemp and D. Metzger, J. Org. Chem.,
      33, 4165 (1968).
134. H. L. Goering and K. L. Howe, J. Am. Chem. Soc.,
      79, 6542 (1957).
135. D. Bethell and V. Gold, "Carbonium Ions,"
      Academic Press, London, 1967, pp. 225-229.
136. L. Ramberg and B. Bäcklund, Chem. Abstr., 34,
      4725 (1940).
137. F. G. Bordwell and G. D. Cooper, J. Am. Chem.
      Soc., 73, 5187 (1951).
138. (a) F. G. Bordwell and N. P. Neureiter, J. Am.
      Chem. Soc., 85, 1209 (1963);
      (b) N. P. Neureiter, J. Am. Chem. Soc., 88,
      558 (1966);
      (c) L. A. Paquette, J. Am. Chem. Soc., 86,
      4085 (1964).
139. F. G. Bordwell and J. M. Williams, Jr., J. Am.
      Chem. Soc., 90, 435 (1968).
140. L. A. Paquette, L. S. Wittenbrook, and V. W.
      Kane, J. Am. Chem. Soc., 89, 4487 (1967).
141. F. G. Bordwell and R. G. Scamehorn, J. Am.
      Chem. Soc., 90, 6751 (1968).
142. F. G. Bordwell, J. M. Williams, Jr., E. B. Hoyt,
      Jr., and B. B. Jarvis, J. Am. Chem. Soc., 90,
      429 (1968).
143. F. G. Bordwell, E. Doomes, and P. W. R. Corfield,
      J. Am. Chem. Soc., 92, 2581 (1970).

144. J. M. Conia and J. Salann, Bull. Soc. Chim. Fr., 1957 (1964).

145. E. M. Warnhoff, C. M. Wong, and W. T. Tai, J. Am. Chem. Soc., 90, 514 (1968).

146. C. R. Engels, S. K. Roy, J. Capitaine, J. Bilodeau, C. McPherson-Foucar, and P. Lachance, Can. J. Chem., 48, 361 (1970).

147. R. B. Loftfield, J. Am. Chem. Soc., 73, 4707 (1951).

148. W. B. Hammond and N. J. Turro, J. Am. Chem. Soc., 88, 2880 (1966).

149. C. Rappe and L. Knutsson, Acta Chem. Scand., 21, 2205 (1967).

150. W. McPhee and E. Klingsberg, J. Am. Chem. Soc., 66, 1132 (1944).

151. R. Breslow, J. Posner and A. Krebs, J. Am. Chem. Soc., 85, 243 (1963).

152. G. Stork and I. Borowitz, J. Am. Chem. Soc., 82, 4307 (1960).

153. (a) A. A. Sacks and J. G. Aston, J. Am. Chem. Soc., 73, 3902 (1951);
      (b) J. G. Burr and M. J. S. Dewar, J. Chem. Soc., 1201 (1954);
      (c) A. W. Fort, J. Am. Chem. Soc., 84, 2620, 2625 (1962).

154. F. G. Bordwell and R. G. Scamehorn, J. Am. Chem. Soc., 90, 6751 (1968).

155. F. G. Bordwell and M. W. Carlson, J. Am. Chem. Soc., 92, 3370 (1970).

156. F. G. Bordwell, M. W. Carlson and A. C. Knipe, J. Am. Chem. Soc., 91, 3949 (1969).

157. (a) H. R. Nace, and B. A. Olsen, J. Org. Chem., 32, 3438 (1967);
      (b) F. G. Bordwell, R. R. Frame, R. G. Scamehorn, J. G. Strong, and S. Meyerson, J. Am. Chem. Soc., 89, 6706 (1967).

158. (a) F. G. Bordwell, R. G. Scamehorn and A. C. Knipe, J. Am. Chem. Soc., 92, 2172 (1970);
      (b) F. G. Bordwell and R. G. Scamehorn, J. Am. Chem. Soc., 93, 3410 (1971).

159. P. W. Neber, A. Burgard, and W. Thier, Ann., 526, 277 (1926).

160. C. O'Brien, Chem. Rev., 64, 81 (1964).

161. M. J. Hatch and D. J. Cram, J. Am. Chem. Soc., 75, 33, 38 (1953).

162. H. O. House and W. E. Berkowitz, J. Org. Chem., 28, 307, 2271 (1963).

163. H. Freundlich and H. Kroepelin, Z. Phys. Chem., 122, 39 (1926).

164. E. Wenkert, P. Bakuzis, R. J. Baumgarten, C. L.

Leicht, and H. P. Schenk, J. Am. Chem. Soc., <u>93</u>, 3208 (1971).

165. J. D. Park and R. J. McMurtry, J. Org. Chem., <u>32</u>, 2397 (1967).

166. (a) S. J. Cristol, N. L. Hause, and J. S. Meek, J. Am. Chem. Soc., <u>73</u>, 674 (1953);
      (b) S. J. Cristol and D. D. Fix, J. Am. Chem. Soc., <u>75</u>, 2647 (1953).

167. P. B. D. de la Mare, R. Koenigsberger, and J. S. Lomas, J. Chem. Soc., <u>B</u>, 834 (1967).

168. (a) R. F. Hudson and R. J. Withey, J. Chem. Soc., <u>B</u>, 237 (1966);
      (b) R. F. Hudson, J. Arendt, and A. Mancusco, J. Chem. Soc., <u>B</u>, 1069 (1967).

169. R. B. Woodward and R. Hoffmann, Angew. Chem. Int. Ed., <u>8</u>, 781 (1969).

170. I. Fleming and E. Wildsmith, Chem. Comm., 223 (1970).

171. C. A. Wellington and W. D. Walters, J. Am. Chem. Soc., <u>83</u>, 4888 (1961).

172. J. E. Baldwin, Tetrahedron Lett., 2953 (1966).

173. A. C. Cope, M. M. Martin, and M. A. McKervey, Q. Rev. (London), <u>20</u>, 119 (1966).

174. J. Sicher, Prog. Stereochem., <u>3</u>, 202 (1962).

175. A. C. Cope, M. Brown, and G. L. Woo, J. Am. Chem. Soc., <u>87</u>, 3107 (1965).

176. H. C. Brown, J. Chem. Soc., 1248 (1956).

177. A. C. Cope, M. Brown, and H. H. Lee, J. Am. Chem. Soc., <u>80</u>, 2855 (1958).

178. A. C. Cope, G. A. Berchtold, P. E. Peterson and S. H. Sherman, J. Am. Chem. Soc., <u>82</u>, 6370 (1960).

179. (a) S. Wawzonek and P. J. Thelen, J. Am. Chem. Soc., <u>72</u>, 2118 (1950);
      (b) E. J. Corey and W. R. Hertler, J. Am. Chem. Soc., <u>82</u>, 1657 (1960).

180. D. E. Edwards, D. Vocelle, J. W. Apsimon and F. Haque, J. Am. Chem. Soc., <u>87</u>, 678 (1965).

181. W. D. Closson, J. L. Jernow, and D. Grey, Tetrahedron Lett., 1141 (1970).

182. Unpublished results of V. Prelog and S. Borcić, quoted in Ref. 173.

183. D. E. Horning and J. M. Muchowski, Can. J. Chem., <u>46</u>, 3665 (1968).

184. P. T. Lansbury, Acc. Chem. Res., <u>2</u>, 210 (1969).

185. M. A. Davis, T. A. Dobson, and J. M. Jordan, Can. J. Chem., <u>47</u>, 2827 (1969).

186. (a) A. C. Cope, S. W. Fenton, and C. F. Spencer, J. Am. Chem. Soc., <u>74</u>, 5884 (1952);
      (b) A. C. Cope and B. C. Anderson, J. Am. Chem.

Soc., 79, 3892 (1957);
(c)   A. C. Cope and A. Fournier, J. Am. Chem.
Soc., 79, 3896 (1957);
(d)   A. C. Cope and R. W. Gleason, J. Am. Chem.
Soc., 84, 1928 (1962).

187. A. C. Cope, G. A. Berchtold, P. E. Peterson,
and S. H. Sharman, J. Am. Chem. Soc., 82, 6366
(1960).

188. A. C. Cope, J. M. Grisar, and P. E. Peterson,
J. Am. Chem. Soc., 81, 1640 (1959).

189. A. Streitwieser, Jr., Chem. Rev., 56, 571 (1956).

190. (a)   H. J. Urech and V. Prelog, Helv. Chim. Acta,
40, 477 (1957);
(b)   V. Prelog, W. Kisng, and T. Tomljenović,
Helv. Chim. Acta, 45, 1352 (1962).

191. V. Prelog, H. J. Urech, A. A. Bothner-By, and
J. Wursch, Helv. Chim. Acta, 38, 1095 (1955).

192. C. A. Grob and P. W. Schiess, Angew. Chem. Int.
Ed., 6, 1 (1967).

193. C. A. Grob and W. Baumann, Helv. Chim. Acta,
38, 594 (1955).

194. C. A. Grob, Exper., 13, 126 (1957).

195. C. A. Grob, Angew. Chem. Int. Ed., 8, 535
(1969).

196. C. A. Grob, "Theoretical Organic Chemistry,"
report on the Kekulé Symposium, Butterworths,
London, 1958, p. 114.

197. C. A. Grob and F. Ostermayer, Helv. Chim. Acta,
45, 1119 (1962).

198. J. E. Craig, M. D. Bergenthal, I. Fleming, and
J. Harley-Mason, Angew. Chem Int. Ed., 8, 429
(1969).

199. (a)   J. A. Marshall, Rec. Chem. Prog., 30, 3
(1969);
(b)   J. A. Marshall and J. H. Babler, J. Org.
Chem., 34, 4186 (1969).

200. C. A. Grob, F. Ostermayer and W. Randenbusch,
Helv. Chim. Acta, 45, 1672 (1962).

201. C. A. Grob, F. M. Unger, E. D. Weiler, and A.
Weiss, unpublished results quoted in Ref. 195.

202. C. A. Grob, H. P. Fischer, W. Randenbusch, and
J. Zergenyi, Helv. Chim. Acta, 47, 1003 (1964).

203. A. Bottini, C. A. Grob, E. Schumacher, and J.
Zergenyi, Helv. Chim. Acta, 49, 2516 (1966).

XI PHOTOCHEMICAL AND HOMOLYTIC ELIMINATIONS

A SCOPE OF PHOTOELIMINATIONS

While many photochemical reactions can be regarded as
eliminations in the sense that fragmentation occurs,
nearly all of the examples of olefin-forming photoeliminations involve carbonyl compounds with a γ-car-
bon atom bearing at least one hydrogen, and can be
represented schematically by Eq. 1, where X is either
a carbon atom or an oxygen atom. By far the most

$$R\overset{\overset{\displaystyle O}{\|}}{C}-X-\overset{\overset{\displaystyle H}{|}}{\underset{|}{C}}-\overset{|}{\underset{|}{C}}- \quad \xrightarrow{h\nu} \quad R\overset{\overset{\displaystyle O}{\|}}{C}-X-H \quad + \quad {>}C{=}C{<} \tag{1}$$

thoroughly investigated reaction of this type is that
involving ketones. It was first reported by Norrish
and Appleyard[1] in 1934, and is often referred to as the
"Norrish Type II Split" because it was the second of
a number of primary photochemical processes that he
established for simple ketones (the Type I process was
a free-radical split). Esters also undergo the reaction[2-4] and have been the subject of a moderate number
of recent investigations that are recounted below

(Section X.I.C).   Information on the Type II reactions
of other carbonyl compounds is sparse. They have been
reported to occur with acid anhydrides,[5] amides,[6]
acids,[7] and the acid portion of an ester.[8] More recent
workers have been unable to repeat the claimed[6]
Type II reactions of amides.[7]

Other examples of olefin-producing photoeliminations are very seldom encountered, and we cite only a
few from the recent literature. Isocyanates undergo
eliminations that are probably mechanistically related
to the Type II process.[9] Somewhat more distantly related are the photolyses or α-alkyl quinolines to
produce α-methyl quinolines and an olefin.[10] The
photolysis of ethyllithium produces ethylene.[11]

B   PHOTOELIMINATIONS FROM KETONES

1   Basic Mechanism

The main features of the presently accepted mechanism
for the Type II photoelimination were first suggested
by Davis and Noyes[12] in 1947, but considerable time
elapsed before definitive evidence was presented.
The mechanism (Eq. 2) requires the production of an
enol whose oxygen atom bears a hydrogen atom originat-
ing from the γ-position of the original ketone. The
photolysis of 2-hexanone-5,5-d_2 (Eq. 3) yields

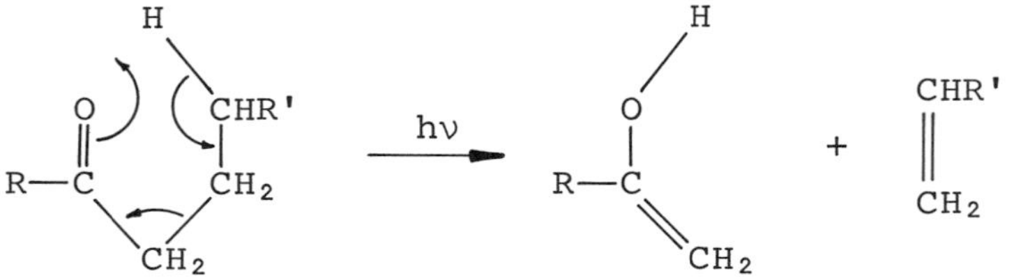

$$CH_3\overset{O}{\overset{\|}{C}}CH_2CH_2CD_2CH_3 \xrightarrow{\;3130\text{Å}\;} CH_2{=}CDCH_3 \;+\; CH_3\overset{O}{\overset{\|}{C}}CH_2D$$

45%

$$CH_3\overset{O}{\overset{\|}{C}}CH_3$$

55%

(3)

propylene deuterated in the expected position, but the
acetone was not completely deuterated as required by
the mechanism.[13] This was shown to have been caused by
exchange of the initially formed enol with water adsorbed on the walls of the reaction vessel, since
photolysis of undeuterated 2-hexanone in a vessel which
had been exposed to deuterium oxide vapor gave partially deuterated acetone under conditions where acetone itself does not exchange. Later the presence of
acetone enol in the reaction mixture from 2-pentanone
photolysis was demonstrated directly by observation
of its infrared spectrum.[14]
    Although the mechanism of Eq. 2 shows the reac-
tion as a concerted removal of γ-hydrogen by the carbonyl oxygen and a split to yield olefin, it is now
believed that most, if not all, examples of the reaction involve a diradical intermediate (1), which can
either split into olefin and enol or can undergo a
back donation of hydrogen from oxygen to the γ-carbon

$$
\begin{array}{c}
\text{H} \\
\diagup \\
\text{O} \qquad \cdot\text{CHR}' \\
| \qquad\quad | \\
\text{R—C} \cdot \qquad \text{CH}_2 \\
\diagdown \quad\diagup \\
\text{CH}_2
\end{array}
$$

1

so as to regenerate the original ketone. The diradical
can also cyclize to yield a cyclobutanol. Although the
cyclization is not of interest in a discussion on elimination, it can sometimes be an important competitor
to the elimination process. While various pieces of
evidence, to be discussed subsequently (Section XI.B.
2,3), point to a diradical intermediate, its presence
was demonstrated only recently by trapping with
mercaptan.[15]

2 Electronic States of Reactive Species

We begin by summarizing briefly, and without evidence,
some basic concepts of photochemistry which are necessary to an understanding of the subsequent material.
Those who wish further information on this background
topic are referred to the book by Turro.[16]
    A molecule may be promoted to an electronically

excited state either by absorption of light or by interaction with another excited species (energy transfer). Essentially all organic molecules of photochemical interest are in the singlet state; that is,
they possess no unpaired electrons (Section X.B.2).
Absorption of light by a singlet ground state molecule
always produces a singlet excited state species. If
the species produced by absorption of light is not the
lowest excited singlet, it will generally decay to the
lowest excited singlet before it can do anything else.
The excited singlet may then undergo reaction, lose
energy with emission of radiation (fluorescence), lose
energy without emission of radiation (radiationless
decay), or undergo intersystem crossing to the triplet
state, which possesses two unpaired electrons. The
triplet in turn can undergo reaction, lose energy with
emission of radiation (phosphorescence), or lose energy
without emission of radiation. The triplet state of a
molecule can also be produced directly by energy transfer from the excited triplet state of another molecule.
The donor is quenched (loses its energy) in the process. A singlet excited state can similarly be produced by energy transfer from the excited singlet of another molecule. In summary, then, a photoreaction can
occur from either an excited singlet or excited triplet
state of a molecule, and deciding whether one or both
states react is one of the first questions to be answered in elucidating a photochemical mechanism.

The early literature on alkyl ketone photolysis
contained considerable controversy over whether reaction occurred via singlet or triplet states.[17-19]
Ausloos and Rebbert[20] suggested that the Type II process might occur from the singlet in some cases and
from the triplet in others, a suggestion that was soon
confirmed by the discovery that 2-pentanone and 2-
hexanone gave olefin via both states.[21,22] The evidence was based on energy-transfer studies in which
varying concentrations of piperylene (1,3-pentadiene)
were added to the photolysis mixture. Piperylene does
not absorb appreciably in the region where the ketones
absorb, but it has a lower triplet energy than simple
ketones, and should accept energy from ketone triplets
at a diffusion-controlled rate. Thus piperylene should
lower the quantum yield of photoproducts from the ketone by quenching the ketone triplet. In fact, it can
be shown that a plot of Φ_0/Φ (the ratio of quantum
yield in the absence to that in the presence of piperylene) versus the concentration of piperylene should be
linear for a mechanism involving competition between
quenching and unimolecular reaction (and/or other

unimolecular decay processes) of the ketone triplet.
Such plots are linear for 2-pentanone and 2-hexanone at
low piperylene concentrations, but curve toward limiting values at high piperylene concentrations.   In other
words, it is impossible to quench the reaction completely even at very high piperylene concentrations.
The unquenchable part of the reaction is most reasonably interpreted as involving the singlet state, and
amounts to about 40% of the total reaction for 2-hexa-
none and 10% of the total reaction for 2-pentanone.
Aryl ketones seem to react entirely via the triplet
state; plots of the type described above (Stern-Volmer
plots) are entirely linear for butyrophenone and
valerophenone.[21]
       The end product of a photochemical reaction is a
molecule or molecules in the ground (unexcited) electronic state, and it is of interest to ask when the
electronic excitation is lost.   In principle, it is
possible either for it to be lost during reaction of
the excited state, or for an excited product to be
formed and then lose energy.   The former seems to be
true in the photolysis of 2, since it forms stilbene
(3) which is, at least, 98% trans.[23,24]   Triplet

$$C_6H_5\overset{\overset{\text{O}}{\|}}{C}CH_2CHCH_2C_6H_5$$
$$|$$
$$C_6H_5$$

2

3

stilbene is known to decay to a 60:40 mixture of cis
and trans.[25]

3   Relative Reactivities and Detailed Mechanism

Substantial effects of structure on reactivity show up
in both intramolecular and intermolecular comparisons.
For example, 4 gives about 9 times as much 2-butene as
1-butene on photolysis at 3130Å, and 5 gives 12 to 18
times as much 2-pentanone plus propylene as 2-hexanone

$$CH_3\overset{\overset{\text{O}}{\|}}{C}CH_2CHCH_2CH_3$$
$$|$$
$$CH_3$$

4

$$CH_3CH_2CH_2\overset{\overset{\text{O}}{\|}}{C}CH_2CH_2CH_2CH_3$$

5

plus ethylene.[26,27] Of the 2-butene from 4, the trans
isomer is preferred over the cis by about 3:1. The
photoeliminations evidently follow the Saytzev rule,
which predicts that the more stable (usually the more
substituted) olefin will form in larger yield (Section
IV.A.5). At first sight, these observations might seem
to favor a concerted decomposition of the electronic-
ally excited reactant because this transition state
for such a decomposition would presumably possess
double-bond character. On the other hand, it is known
that alkyl substitution on a carbon atom increases the
rate of abstraction of hydrogen from that carbon atom
by free radicals.[28] The relative olefin yields may
also depend on the relative efficiencies with which the
respective biradicals cleave to the olefins, and here
a product-stability effect could come into play. Thus,
these relative reactivities are consistent with either
a biradical intermediate (1), or a concerted mechanism.
     The selectivities of the triplet and singlet of
5 were disentangled by performing the photolysis in the
presence of quenchers. The resulting ratios of 2-
pentanone to 2-hexanone were around 12, compared to
15 to 18 in the absence of quencher.[27] Similarly,
photolyses of compounds of the type of 6 in the presence and absence of quenchers so as to separate the
singlet and triplet processes revealed that the singlet

$$CH_3\overset{\displaystyle O}{\overset{\displaystyle \|}{C}}-CH_2CH_2CHR_1R_2$$

$R_1 = R_2 = H$

$R_1 = CH_3, \; R_2 = H$

$R_1 = R_2 = CH_3$

6

is somewhat more reactive and less selective than the
triplet.[29,30] Although these differences could indicate different mechanisms for the singlet and triplet
ketones, increased reactivity is expected to be
accompanied by decreased selectivity even when there is
no change of mechanism.
     Deuterium isotope effects offer a special case of
relative reactivities that can give very useful mechanistic information (Section II.C.3). Photolysis of
deuterated 2-hexanones in the presence and absence of
quenchers gives k_H/k_D of 2.7 for the singlet and 6.7
for the triplet. This would be consistent with a
hydrogen transfer that was near the half-way point in
the transition state for the triplet reaction, but was
closer to reactant with the more reactive singlet

(Section II.B.1, C.3.a). An intriguing observation was that the quantum yield for photoelimination from the triplet was actually increased by deuteration. A number of explanations can be adduced for this phenomenon, but the most likely seems to be that deuteration slows elimination and nonradiative decay from the singlet, and thereby increases the efficiency of intersystem crossing to the triplet.[29]

Additional evidence for a biradical intermediate was afforded by a study of solvent effects on photoelimination from valerophenone.[32] The quantum yield for photoelimination (from the triplet, Section XI.B. 2) runs markedly higher in alcohols than in benzene or hexane. Addition of a little t-butyl alcohol to a hexane solution of the ketone raises the quantum yield sharply at first, followed by a slow rise to the value characteristic of pure t-butyl alcohol as the concentration of the alcohol is increased. It was suggested that return of the biradical intermediate to reactant was hindered by solvents capable of hydrogen bonding to the hydroxyl group of the intermediate.

Photolysis of the optically active ketones 7 and 8 is accompanied by racemization, but for the reaction of 7, addition of enough pentadiene to the reaction mixture to quench all reaction via the triplet stops

$$CH_3\overset{\overset{\textstyle O}{\|}}{C}CH_2CH_2CH\diagup^{CH_3}_{\diagdown C_2H_5} \qquad C_6H_5\overset{\overset{\textstyle O}{\|}}{C}CH_2CH_2CH\diagup^{CH_3}_{\diagdown C_2H_5}$$

$$\underline{7} \qquad\qquad\qquad \underline{8}$$

the racemization completely.[30,33] The observation of racemization in the triplet-state reactions is excellent evidence for the biradical intermediate, since rotation in a biradical followed by back-donation of hydrogen is by far the most reasonable mechanism for racemization. In the photolysis of 8 in benzene, racemization is faster than photoelimination, but racemization is almost entirely suppressed in t-butyl alcohol, thus supporting the explanation of the solvent effect described in the preceding paragraph.[30] The absence of racemization in the singlet-state photoelimination from 7 is consistent with a concerted elimination from the singlet, but it would also be consistent with a singlet biradical that either did not undergo return to reactant or collapsed so rapidly, either to reactant or product, that there was no time

for rotation to occur. In support of a discrete biradical intermediate is the fact that the singlet
states of the erythro and threo isomers of methyl 3,4-
dimethyl-6-ketoheptanoate undergo photoelimination in
a not entirely stereospecific manner (cis:trans ratios
of 90:1 and 1:90, respectively).[56]

One further interesting aspect of the reactivity
problem is the effect of ring substituents on reactivities of butyrophenones[34] and valerophenones.[35] Most
electron-withdrawing substituents have little effect
on the quantum yields, but alkoxy and acetoxy groups
tend to lower it, sometimes drastically. For example,
m-methoxy and m-acetoxy groups lower the quantum yield
to a few hundredths of its value for the unsubstituted
compound. The explanation of this unusual behavior
seems to be that two different triplet states are involved. The triplet of a simple alkyl ketone is called
an n,π* triplet, because it results from the promotion
of an unshared (nonbonding) electron on oxygen to an
antibonding π-orbital between oxygen and carbonyl carbon. The resulting species has an electron-deficient
oxygen that is well suited for abstraction of a γ-hy-
drogen. Groups capable of resonance interaction with
the carbonyl group, such as phenyl or substituted
phenyl, raise the energy of this transition by increasing electron density on carbon and thus making it more
difficult to move the nonbonding electron closer to
carbon. If its energy is raised enough, it may become
of higher energy than the π,π* triplet, which results
from promotion of a bonding π-electron to the π*-orbi-
tal. The π,π* state does not have an electron-deficient oxygen, and should be ineffective at abstracting a γ-hydrogen. The unreactive ketones also have
much longer phosphorescence lifetimes than the reactive
ones, another characteristic of π,π* states. More is
said on this subject when we discuss photoeliminations
from esters (Section XI.C.1). It is unclear whether
the small amount of reaction that is observed comes
from the higher n,π* state, or from the π,π* state,
possibly as the result of some admixture of n,π*
character.[30] Nitrogen-containing ketones with reactive
π,π* triplets have been reported, but the mechanism
of photoelimination appears to involve an electron
transfer to oxygen from nitrogen followed by a proton
transfer to oxygen, rather than a simple hydrogen atom
transfer.[36,37]

C   PHOTOELIMINATION FROM ESTERS

1   Mechanism and Nature of Excited States

Most esters possessing a β-hydrogen in the alcohol portion of the ester undergo Type II photoelimination to
produce olefin and the corresponding carboxylic acid
(Eq. 4). In the gas phase and, under some conditions,

$$
\underset{\substack{| \ | \\ | \ |}}{R-C-O-C-C-H} \quad \xrightarrow{h\nu} \quad R-\overset{O}{\overset{\|}{C}}-OH \quad + \quad \overset{>}{\underset{/}{C}}=\overset{<}{\underset{\backslash}{C}} \tag{4}
$$

in the liquid phase, the photoelimination can be accompanied by radical processes leading to carbon monoxide,
carbon dioxide, alcohols, and saturated hydrocarbons.[3],[38] The basic mechanism is doubtless analogous to that
of the Type II process in ketones (Eq. 2), and the
same questions of mechanistic detail must be considered.
        Much less is known about the excited states of
esters than of ketones. The triplet state of simple
aliphatic esters has been estimated to lie somewhat
over 100 kcal/mole.[39] Recent investigations of ester
photolysis have concentrated on phenylacetates and
benzoates, where the situation can be quite complex,
since singlet and triplet states with excitation cen-
tered on the carbonyl group, the aromatic ring, or both
must be taken into consideration. Photoelimination
from the phenylacetates is quenched by piperylene, a
fact that at first was interpreted to indicate reaction from a triplet state.[40] Later investigations
showed, however, that the quenching of elimination was
accompanied by quenching of fluorescence, indicating
that the state being quenched was a singlet rather
than a triplet.[41],[42] Simple olefins quench photoelimination from 2-pentyl phenylacetate in alcohol solvents.[43] In methanol there is a break in the plot of
Φ_0/Φ versus cyclopentene concentration, suggesting that
two different excited states may be involved in the
photoelimination (Section XI.B.2).[21],[22] Quantum yields
for olefin formation from the phenylacetates are fairly
good (0.1 to 0.3), decreasing with increasing concentration of ester.[43] This latter observation suggests that some deactivation occurs via interaction of
excited with unexcited ester molecules (self-quenching).
        The quantum yield of olefin in the photolysis of
aromatic esters runs about one tenth that for the
phenylacetates (0.01 to 0.02).[43-45] In most cases a
Π,Π^* singlet state appears to be the precursor of

olefin, although a few aromatic esters were found to
react via the triplet state.[44]   It was suggested that
the reaction involves a diradical intermediate (9), and
that the inefficiency arises from a greater rate of

$$
\begin{array}{c}
\text{H} \\
\diagdown \\
\text{O} \qquad \text{·CHR'} \\
| \qquad\qquad | \\
\text{R——C·} \qquad \text{CH}_2 \\
\diagdown \qquad \diagup \\
\text{O}
\end{array}
$$

9

return to reactants relative to decomposition to products for the aromatic esters than for the phenylacetates.[44]

2   Relative Reactivities and Detailed Mechanism

When elimination can occur in either of two directions,
elimination into the branch bearing the greater number
of alkyl groups seems to be preferred. Photolysis of
2-butyl acetate gives about twice as much 2-butene as
1-butene in the vapor phase and 6 to 8 times as much in
the liquid phase.[46]   There is also a preference for
trans- over cis-2-butene, about 2:1 in the gas and
3:1 in the liquid phase. At very low temperatures a
curious increase in the proportion of 1-butene was
found, the 1-butene constituting about 40% of the olefin fraction at -150° and 60% at -195°. The photolysis
of 2-pentyl phenylacetate and benzoate in solvents
ranging from methanol to t-butyl alcohol yields 80 to
85% 2-pentene of which about 70 to 75% is the trans
isomer.[43]   While there are slight variations with solvent and with the nature of concentration of reactant
or quencher, there are few, if any, clear trends.
     The preference for the more substituted olefin in
nearly all cases is, as noted above (Section XI.B.3),
consistent with a diradical intermediate, but it is
also consistent with a concerted process. The concerted process should be stereospecific, while the
stepwise process may not be if the diradical survives
long enough to rotate before reaction. The threo and
erythro diastereomers of 1,2-dimethylbutyl acetate
(10) should yield trans- and cis-3-methyl-2-pentene,
respectively, in a concerted stereospecific syn elimination. In fact, both diastereomers yield mixtures
(though not identical mixtures) of the cis and trans

$$\underset{\substack{| \quad | \\ \text{Me Me}}}{\overset{\overset{\displaystyle O}{\overset{\displaystyle \parallel}{}}}{\text{Me-C-O-CH-CH-Et}}}$$

10

olefins, suggesting partial, but not complete, equilibration of a diradical intermediate through rotation.[47]
A subsequent investigation of 1,2-dimethylbutyl phenylacetate, however, showed that most, but not all, of
the lack of stereospecificity was due to isomerization
of the first-formed olefin by the ester triplet,[48]
leaving less than 10% nonstereospecific elimination.
In both cases, there was no isomerization of the reactant, indicating no return of rotationally equilibrated
diradical to reactant. Evidence on whether part of
the nonstereospecificity in the photolysis of the acetate (10) arises from subsequent isomerization of the
product has not been presented. cis-2-Methylcyclohexyl
phenylacetate photoeliminates mainly to 3-methylcyclo-
hexene, the expected product of syn elimination, but
there is about 10% of 1-methylcyclohexene in the olefin mixture.[42] As there was no isomerization of reactant, a mainly, but not entirely, stereospecific elimination process via a diradical is again suggested.
The photolysis of optically active 2-methylbutyl benzoate is accompanied by racemization, in contrast to
results with the phenylacetates.[49] It is not unreasonable to expect return from a diradical to be more important with the benzoates than the phenylacetates, in
view of the much lower quantum yields of photoelimination from the former (Section XI.C.1).

D   HOMOLYTIC ELIMINATIONS

The photoeliminations discussed thus far in this chapter might be classified as homolytic eliminations,
since most or all of them seem to involve diradical intermediates. One might also inquire whether a process
analogous to the E2 reaction, but with a radical instead of a base attacking the β-hydrogen, can be observed (Eq. 5). There is no doubt that processes
corresponding to Eq. 5 in overall stoichiometry are
possible. The abstraction of hydrogen atoms from
organic molecules, including those β to functional
groups, is one of the best-known reactions of free
radicals. The other half of the process, loss of a
functional group β to a radical center, is also well-
known. For example, halogen atoms and thiyl radicals

$$R\cdot \; + \; H\text{-}\overset{\displaystyle |}{\underset{\displaystyle |}{C}}\text{-}\overset{\displaystyle |}{\underset{\displaystyle |}{C}}\text{-}X \longrightarrow RH \; + \; >\!C=\!C\!< \; + \; X\cdot \qquad (5)$$

are known to add reversibly to olefins to yield β-halo or β-thio radicals.[50]   A carbon-carbon bond β to a radical center may also cleave to yield an olefin and another radical, especially if the molecule is strained.[51]   Since both steps are clearly feasible, the question of mechanistic interest is whether they occur separately (Eqs. 6 and 7), or can combine into a single process (Eq. 5) that might be called an E_H2 reaction (bimolecular elimination, homolytic).

$$R\cdot \; + \; H\text{-}\overset{\displaystyle |}{\underset{\displaystyle |}{C}}\text{-}\overset{\displaystyle |}{\underset{\displaystyle |}{C}}\text{-}X \longrightarrow RH \; + \; \cdot\overset{\displaystyle |}{\underset{\displaystyle |}{C}}\text{-}\overset{\displaystyle |}{\underset{\displaystyle |}{C}}\text{-}X \qquad (6)$$

$$\cdot\overset{\displaystyle |}{\underset{\displaystyle |}{C}}\text{-}\overset{\displaystyle |}{\underset{\displaystyle |}{C}}\text{-}X \longrightarrow >\!C=\!C\!< \; + \; X\cdot \qquad (7)$$

There is, to our knowledge, only one research group that has made a careful study of this problem. The reaction of phenylazotriphenylmethane, a source of phenyl radicals, with t-butyl and phenyl t-butyl sulfides proceeds cleanly according to Eq. 8 to yield equivalent amounts of benzene and isobutylene.[52]   On

$$C_6H_5\cdot \; + \; (CH_3)_3C\text{-}S\text{-}R \longrightarrow C_6H_6 \; + \; (CH_3)_2C=CH_2 \; +$$
$$RS\cdot \qquad (8)$$

the basis of other work, the authors concluded that the β-hydrogens of t-butyl sulfide were markedly more reactive than hydrogens in other simple aliphatic systems, and that this enhanced reactivity might result from concerted cleavage of the carbon-hydrogen and carbon-sulfur bonds. The same suggestion had been made earlier to explain the high chain-transfer constant for t-butyl sulfide in styrene polymerization.[53] Curiously, phenyl t-butyl sulfide did not show this enhanced reactivity.

This difference between t-butyl and phenyl t-butyl sulfides prompted further investigation of the elimination reactions of phenyl alkyl sulfides.[54,57]   The relative reactivities of primary, secondary, and tertiary β-hydrogens were determined by examining products from 11 and 12, and from mixtures of 13 and 14. The result was a prim.:sec.:tert. ratio of 1:8.5:43, almost the same as that observed for hydrocarbon

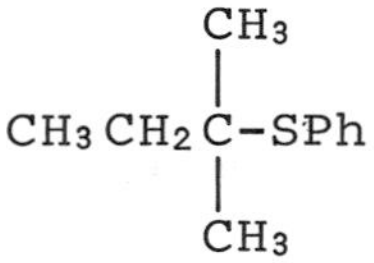

$$CH_3CH_2\overset{\overset{\displaystyle CH_3}{|}}{\underset{\underset{\displaystyle CH_3}{|}}{C}}-SPh \qquad\qquad (CH_3)_2\overset{\overset{\displaystyle CH_3}{|}}{CH}CH-SPh$$

11 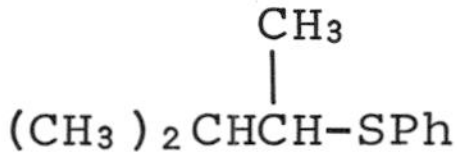   **12**

$$CH_3CH_2\overset{\underset{\underset{\displaystyle CH_3}{|}}{}}{CH}CH_2-SPh \qquad\qquad (CH_3)_2CHCH_2CH_2-SPh$$

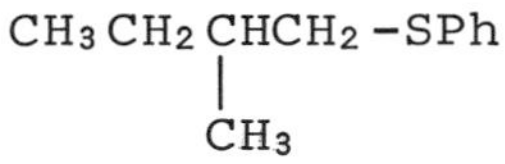

13 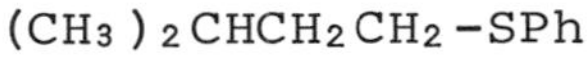   **14**

hydrogens $(1:9.2:44)$.[55] Furthermore, a competition experiment showed that the methyl hydrogens of t-butyl sulfide and of 11 had the same reactivity. In addition, the work that had suggested a high reactivity for t-butyl sulfide toward phenyl radicals was reinvestigated and found to be incorrect, probably because of the presence of some thiol in the sulfide.

These results show conclusively that the β-hydrogens of sulfides show no enhancement of either intramolecular or intermolecular reactivity relative to the hydrogens of related compounds where no driving force resulting from elimination is possible. In principle it is possible that such driving force might be masked by compensating factors in a concerted process (Eq. 5), but the most reasonable conclusion is that the two-step process (Eqs. 6 and 7) operates in these reactions. Thus far no demonstrably concerted radical elimination has been discovered, and it seems unlikely that radical eliminations will offer so fertile a field of investigation as have ionic and cyclic eliminations.

REFERENCES

1. R. G. W. Norrish and M. E. S. Appleyard, J. Chem. Soc., 874 (1934).
2. P. Ausloos and R. E. Rebbert, J. Phys. Chem., 67, 163 (1963).
3. P. Ausloos, J. Am. Chem. Soc., 80, 1310 (1958).
4. P. Ausloos, Can. J. Chem., 36, 383 (1958).
5. P. Ausloos, Can. J. Chem., 34, 1709 (1956)..
6. G. H. Booth and R. G. W. Norrish, J. Chem. Soc., 188 (1952).
7. C. H. Nicholls and P. A. Leermakers, J. Org. Chem., 35, 2754 (1970).

8. A. A. Scala and G. E. Hussey, J. Org. Chem., 36, 598 (1971).

9. N. J. Friswell and R. A. Black, Can. J. Chem., 47, 4169 (1969).

10. F. R. Stermitz and C. C. Wei, J. Am. Chem. Soc., 91, 3103 (1969).

11. W. H. Glaze and T. L. Brewer, J. Am. Chem. Soc., 91, 4490 (1969).

12. W. Davis, Jr. and W. A. Noyes, Jr., J. Am. Chem. Soc., 69, 2153 (1947).

13. R. Srinivasan, J. Am. Chem. Soc., 81, 5061 (1959).

14. G. R. McMillan, J. G. Calvert, and J. N. Pitts, Jr., J. Am. Chem. Soc., 86, 3602 (1964).

15. P. J. Wagner and R. G. Zepp, J. Am. Chem. Soc., 94, 287 (1972).

16. N. J. Turro, "Molecular Photochemistry," Benjamin, New York, 1965.

17. V. Brunet and W. A. Noyes, Jr., Bull. Soc. Chim. Fr., 121 (1958).

18. J. L. Michael and W. A. Noyes, Jr., J. Am. Chem. Soc., 85, 1027 (1963).

19. P. Borrell, J. Am. Chem. Soc., 86, 3156 (1964).

20. P. Ausloos and R. E. Rebbert, J. Am. Chem. Soc., 86, 4512 (1964).

21. P. J. Wagner and G. S. Hammond, J. Am. Chem. Soc., 87, 4009 (1965); 88, 1245 (1966).

22. T. J. Dougherty, J. Am. Chem. Soc., 87, 4111 (1965).

23. R. A. Caldwell and P. Fink, Tetrahedron Lett., 2987 (1969).

24. P. J. Wagner and P. A. Kelso, Tetrahedron Lett., 4151 (1969).

25. G. S. Hammond et al., J. Am. Chem. Soc., 86, 3197 (1964).

26. P. Ausloos, J. Phys. Chem., 65, 1616 (1961).

27. P. J. Wagner, Tetrahedron Lett., 5385 (1969).

28. C. Walling and M. J. Mintz, J. Am. Chem. Soc., 89, 1515 (1967).

29. N. C. Yang, S. P. Elliott and B. Kim, J. Am. Chem. Soc., 91, 7551 (1969).

30. P. J. Wagner, Acc. Chem. Res., 4, 168 (1971).

31. D. R. Coulson and N. C. Yang, J. Am. Chem. Soc., 88, 4511 (1966).

32. P. J. Wagner, J. Am. Chem. Soc., 89, 5898 (1967).

33. N. C. Yang and S. P. Elliott, J. Am. Chem. Soc., 91, 7550 (1969).

34. J. N. Pitts, Jr., D. R. Burley, J. C. Mani and A. D. Broadbent, J. Am. Chem. Soc., 90, 5902 (1968).

35. P. J. Wagner and A. E. Kemppainene, J. Am. Chem.

Soc., 90, 5898 (1968).

36. A. Padwa, W. Eisenhardt, R. Gruber, and D. Pashayan, J. Am. Chem. Soc., 91, 1857 (1969).

37. A. Padwa, W. Eisenhardt, T. Gruber, and D. Pashayan, J. Am. Chem. Soc., 93, 6998 (1971).

38. M. H. J. Wijnen, J. Am. Chem. Soc., 80, 2394 (1958).

39. R. Simonaitis and J. N. Pitts, Jr., J. Am. Chem. Soc., 90, 1389 (1968).

40. H. Morrison, R. Brainard, and D. Richardson, Chem. Commun., 1653 (1968).

41. R. Brainard and H. Morrison, J. Am. Chem. Soc., 93, 2685 (1971).

42. M. Yarchak, J. C. Dalton, and W. H. Saunders, Jr.. unpublished results.

43. K. H. Brown and W. H. Saunders, Jr., unpublished results.

44. J. A. Barltrop and J. D. Coyle, J. Chem. Soc., B, 251 (1971).

45. J. G. Pacifici and J. A. Hyatt, Mol. Photochem., 3, 267 (1971).

46. R. Borkowski and P. Ausloos, J. Am. Chem. Soc., 83, 1053 (1961).

47. J. E. Gano, Tetrahedron Lett., 2549 (1969).

48. J. E. Gano, Mol. Photochem., 3, 79 (1971).

49. J. G. Pacifici and J. A. Hyatt, Mol. Photochem., 3, 271 (1971).

50. C. Walling, "Free Radicals in Solution," Wiley, New York, 1957, pp. 302, 322.

51. C. Walling in P. de Mayo, Ed., "Molecular Rearrangements," Wiley, New York, 1963, Chap. 7.

52. J. A. Kampmeier, R. P. Greer, A. J. Meskin, and R. W. D'Silva, J. Am. Chem. Soc., 88, 1257 (1966).

53. W. A. Pryor and T. L. Pickering, J. Am. Chem. Soc., 84, 2705 (1962).

54. J. T. Hepinstall, Jr., Ph.D. thesis, University of Rochester, 1971.

55. R. F. Bridger and G. A. Russell, J. Am. Chem. Soc., 85, 3754 (1963).

56. L. M. Stephenson, P. R. Cavigli, and J. L. Parlett, J. Am. Chem. Soc., 93, 1984 (1971).

57. J. T. Hepinstall Jr. and J. A. Kampmeier, J. Am. Chem. Soc., 95, 1904 (1973).

INDEX